Bridge Evaluation, Repair and Rehabilitation

NATO ASI Series

Advanced Science Institutes Series

A Series presenting the results of activities sponsored by the NATO Science Committee, which aims at the dissemination of advanced scientific and technological knowledge, with a view to strengthening links between scientific communities.

The Series is published by an international board of publishers in conjunction with the NATO Scientific Affairs Division

A Life Sciences **B Physics**	Plenum Publishing Corporation London and New York
C Mathematical and Physical Sciences **D Behavioural and Social Sciences** **E Applied Sciences**	Kluwer Academic Publishers Dordrecht, Boston and London
F Computer and Systems Sciences **G Ecological Sciences** **H Cell Biology**	Springer-Verlag Berlin, Heidelberg, New York, London, Paris and Tokyo

Series E: Applied Sciences - Vol. 187

Bridge Evaluation, Repair and Rehabilitation

edited by

Andrzej S. Nowak
Department of Civil Engineering,
University of Michigan,
Ann Arbor, MI, U.S.A.

Kluwer Academic Publishers

Dordrecht / Boston / London

Published in cooperation with NATO Scientific Affairs Division

Proceedings of the NATO Advanced Research Workshop on
Bridge Evaluation, Repair and Rehabilitation
Baltimore, Maryland, U.S.A.
April 30–May 2, 1990

Library of Congress Cataloging-in-Publication Data

NATO Advanced Research Workshop on Bridge Evaluation, Repair, and
Rehabilitation (1990 : Baltimore, Md.)
Bridge evaluation, repair, and rehabilitation : proceedings of the
NATO Advanced Research Workshop on Bridge Evaluation, Repair, and
Rehabilitation, Baltimore, Maryland, USA, April 30-May 2, 1990 /
edited by Andrzej S. Nowak.
p. cm. -- (NATO ASI series. Series E, Applied sciences, no.
187)
"Published in cooperation with NATO Scientific Affairs Division."
Includes index.
ISBN 0-7923-0999-5
1. Bridges--Maintenance and repair--Congresses. 2. Bridges-
-Evaluation--Congresses. I. Nowak, Andrzej S. II. Title.
III. Series.
TG315.N33 1990
624'.2--dc20 90-48022

ISBN 0–7923–0999–5

Published by Kluwer Academic Publishers,
P.O. Box 17, 3300 AA Dordrecht, The Netherlands.

Kluwer Academic Publishers incorporates the publishing programmes of
D. Reidel, Martinus Nijhoff, Dr W. Junk and MTP Press.

Sold and distributed in the U.S.A. and Canada
by Kluwer Academic Publishers,
101 Philip Drive, Norwell, MA 02061, U.S.A.

In all other countries, sold and distributed
by Kluwer Academic Publishers Group,
P.O. Box 322, 3300 AH Dordrecht, The Netherlands.

Printed on acid-free paper

Printed in the Netherlands

TABLE OF CONTENTS

V - REPAIR AND REHABILITATION

PREFACE

Evaluation, repair and rehabilitation of bridges are increasingly important topics in the effort to deal with the deteriorating infrastructure. For example, in the United States about 40 percent of the nation's 570,000 bridges are classified, according to the Federal Highway Administration's (FHWA) criteria, as deficient and in need of rehabilitation and replacement. In other countries the situation is similar. FHWA estimates the cost of a bridge replacement and rehabilitation program at 50 billion dollars. The major factors that have contributed to the present situation are: the age, inadequate maintenance, increasing load spectra and environmental contamination. The deficient bridges are posted, repaired or replaced. The disposition of bridges involves clear economical and safety implications. To avoid high costs of replacement or repair, the evaluation must accurately reveal the present load carrying capacity of the structure and predict loads and any further changes in the capacity (deterioration) in the applicable time span.

Accuracy of bridge evaluation can be improved by using the recent developments in bridge diagnostics, structural tests, material tests, structural analysis and probabilistic methods. There is a need for an international exchange of advanced experience to increase the research efficiency. The Workshop is organized on the premise that the exchange of existing American and European experience in the area of bridge evaluation, repair and rehabilitation is beneficial for both parties involved.

The objective of the Workshop was to review recent advances in the areas of materials, member behavior, loads, diagnostic procedures, structural analysis and probabilistic methods, in order to serve as a basis for the development of guidelines for bridge evaluation, repair and rehabilitation.

- Bridge management systems (optimization models, decision-making strategy, BMS data set, bridge expert systems, strategic long range planning),
- Analytical evaluation of bridges (loads, system identification methods, damage accumulation models, time history models),
- Diagnostic procedures (bridge tests, weigh-in-motion, performance monitors, nondestructive techniques, signal processing, integrated sensor systems, microwave, acoustic and optical techniques),
- Materials and techniques for repair and rehabilitation of bridges (steel, reinforced and prestressed concrete, masonry arches, timber, new materials).

Accordingly the Workshop provided a review of the state of the art of the main topics as viewed on one hand in North America and, on the other hand, in Europe. The result is a more unified picture of the problems of bridge evaluation, repair and rehabilitation.

The approach to bridge maintenance was presented by the representatives of bridge authorities in the North America and Europe. A growing number of deficient structures in national inventories is combined with limited and insufficient funds. The repairs and replacements must be prioritized. This work requires an efficient decision making process. A considerable effort is made to develop computer-based bridge management systems. A need was identified for deterioration models and practical tools for field assessment of the degree of deterioration.

Diagnostic procedures were reviewed, including optical techniques (fiber-optics), radar, microwave, acoustic, strain gages and others. Applications of some of these high-technology approaches were presented in technical papers. There is a need to focus attention on development of efficient diagnostic procedures for evaluation of deterioration due to corrosion (steel, reinforced concrete and prestressed concrete), fatigue (steel and concrete), monitoring live loads and long-term deformations.

In the area of bridge evaluation there are some differences in North America and Europe. The controversial issues include quantification of bridges strength (Rating Factor in the USA, Reduction Factor in the UK) and redistribution of loads. The new generation of bridge evaluation criteria is based on probabilistic methods and corresponds to the load and resistance factor design (LRFD) in the United States of limit states design (LSD) in Europe and Canada. Computer aided design (CAD) technology requires a new approach to structural analysis. Traditional safety reserves are often reduced and there is a need for selection of a rational risk level. Bridge tests (static and dynamic) can be used as an efficient tool in the evaluation process. The test results can verify calculated parameters. Research needs were identified including inspection methods for the "inside" of a bridge, nonlinear methods of analysis, system reliability models, strength of cables with localized corrosion, evaluation of foundations.

Bridge repair and rehabilitation methods were discussed. There is a need for extensive use of nondestructive test (NDT) methods. Repair methods depend on structural type, traffic volume, bridge location and other parameters. Case studies may serve as a good basis for the development of efficient procedures. Important new developments are in the area of new materials and technologies. The issues discussed included external prestressing, treatment of concrete decks and repair of substructure.

The papers are grouped in five major chapters on bridge management, diagnostics and monitoring, loads and analysis, evaluation and tests, repair and rehabilitation. Also included are the Working Group reports with lists of topics selected for further US-European cooperation.

The papers were reviewed by members of the International Scientific Committee and other bridge engineering experts.

Andrzej S. Nowak

ACKNOWLEDGEMENTS

The US-European Workshop on Bridge Evaluation, Repair and Rehabilitation was held at the Inner Harbor Omni Hotel in Baltimore, Maryland, USA on April 30 - May 2, 1990.

The Workshop was sponsored by the NATO Scientific Affairs Division with ARW Program Directors Craig Sinclair and G. A. Venturi. The support of the NATO Science Committee is gratefully acknowledged.

The Workshop was Co-sponsored by the American Society of Civil Engineers with ASCE Structures Congress Co-Chairs Ross B. Corotis and Bruce Ellingwood, and The University of Michigan.

Thanks are due to Assistant Editors Sandra Pagnette-Kayser, Maciek A. Nowak, Dorota M. Nowak and Wojtek Nowak. Most of the papers were diligently retyped by Kathryn Helm which is gratefully acknowledged.

INTERNATIONAL SCIENTIFIC COMMITTEE

Dr. Elie Absi (CEBTP, Paris, France)
Dr. Baidar Bakht (Ministry of Transportation, Downsview, Ontario, Canada)
Professor Robert N. Bruce (Tulane University, New Orleans, LA, USA)
Professor Renaud Favre (Ecole Polytechnique Federale, Lausanne, Switzerland)
Professor John W. Fisher (Lehigh University, Bethlehem, PA, USA)
Mr. Stanley Gordon (Federal Highway Administration, Washington, DC, USA)
Professor Gert Konig (Technische Hochschule, Darmstadt, W. Germany)
Dr. John M. Kulicki (Modjeski & Masters, Harrisburg, PA, USA)
Professor Franco Levi (Politecnico di Torino, Torino, Italy)
Professor Andrzej S. Nowak (University of Michigan, Ann Arbor, MI, USA), Chair
Dr. John B. Scalzi (National Science Foundation, Washington, DC, USA)
Professor Gerhart I. Schueller (Universitat Innsbruck, Innsbruck, Austria)
Mr. Graham P. Tilly (Transport and Research Labs., Crowthorne, Berks, United Kingdom)

ORGANIZING COMMITTEE

Dr. Baidar Bakht (Ministry of Transportation, Downsview, Ontario, Canada)
Professor Gert Konig (Technische Hochschule, Darmstadt, W. Germany)
Professor Andrzej S. Nowak (University of Michigan, Ann Arbor, MI, USA), Chair
Mr. Graham P. Tilly (Transport and Research Labs., Crowthorne, Berks, United Kingdom)

LIST OF PARTICIPANTS

Mr. John J. Ahlskog
Federal Highway Administration
400 Seventh St. S.W. Room 3107
Washington, DC 20590 USA

Ms. Kalliopi Aligizaki
Philaretou 62
176-72 Kallithea
Athens, GREECE

Mr. Niels H. Andersen
Danish Ministry of Transport,
Vejdirektoratet, Lyngbyvej 17
DK-2100 Copenhagen, DENMARK

Professor S. H. Ahmad
Department of Civil Engineering
North Carolina State University
Raleigh, NC 27695 USA

Dr. Baidar Bakht
Research and Development Branch
Ministry of Transportation
1201 Wilson Avenue
Downsview, ON M3M 1J8, CANADA

Professor F. W. Barton
Department of Civil Engineering
University of Virginia
Charlottesville, VA 22903 USA

Dr. Rolf Bez
Swiss Federal Institute of Tech.
CH-1015 Lausanne
SWITZERLAND

Professor M. Biswas
Transport. and Infrastrure Res. Center
Duke University
Durham, NC 27706 USA

Professor Mark Bowman
Deparment of Civil Engineering
Purdue University
W. Lafayette, IN 47907 USA

Mr. Nicholas J. Carino
Bldg. 226, Room B 168
NIST
Gaithersburg, MD 20899 USA

Professor Habib J. Dagher
Department of Civil Engineering
University of Maine
Orono, ME 04401 USA

Dr. Allen G. Davis
Testconsult CEBTP LTD
11 Trinity Ct.
Warrington WA3 6QT, ENGLAND

Professor Mohammad R. Ehsani
Department of Civil Engineering
University of Arizona
Tucson, AZ 85721 USA

Mr. Robert Eymard
Laboratoire Central des Ponts
et Chaussees
58 Boulevard Lefebvre
75732 Paris, FRANCE

Mr. Robert E. Falconer
Structural Survey Partnership
Brunel House, The Sidings
Meopham, Kent DA 13 OND
ENGLAND

Mr. Bert Farago
Ministry of Transport.-Ontario
7th Floor Atrium Tower
1201 Wilson Ave., M3M 1J8
CANADA

Professor Renaud Favre
Swiss Federal Institute of Tech.
CH-1015 Lausanne
SWITZERLAND

Professor John W. Fisher
Lehigh University
117 ATLSS Drive, H Building
Bethlehem, PA 18015 USA

Dr. R.G. Flesch
Bundesversuchs und Forschung.
Arsenal, Wien 3
Faradaygasse 3, AUSTRIA

Professor Dan Frangopol
Department of Civil Engineering
University of Colorado at Boulder
Campus Box 428
Boulder, CO 80309 USA

Mr. Jean Philippe Fuzier
Freyssinet International
52-54, rue de la Belle Feuille
92100 Boulogne, FRANCE

Professor H.V.S. GangaRao
Civil Engineering Department
West Virginia University
Morgantown, WV 26506 USA

Mr. Bruno Godart
Laboratoire Central des Ponts
et Chaussees
58 Boulevard Lefebvre
75732 Paris, FRANCE

Mr. Kamar Golabi
Optima Inc.,
517 Washington St., Suite 205
San Francisco, CA 94111 USA

Professor J. Leroy Hulsey
Department of Civil Engineering
University of Alaska
Fairbanks, AK 99775 USA

Professor F. Wayne Klaiber
Iowa State University
416 Town Engineering Building
Ames, IA 50011 USA

Professor Gert Konig
Technische Hochschule Darmstadt
61 Darmstadt, Alexanderstrasse 5
Institue fur Massivbau
WEST GERMANY

Dr. John M. Kulicki
Modjeski and Masters
P.O. Box 2345
Harrisburg, PA 17105 USA

Professor Roberto Leon
University of Minnesota
Dept. of Civil and Mineral Eng.
122 C&ME Bld. 500 Pillsbury Dr.
Minneapolis, MN 55455 USA

Mr. Clellon Loveall
Structures, Tennessee DOT
1200 James K. Polk Building
Nashville, TN 37219 USA

Mr. James G. Lutz
Greiner, Inc.
2219 York Road, Suite 200
Timonium, MD 21093 USA

Professor Giuseppe Mancini
University of Cagliari
Via Castelgomberto, 75
10137 Torino, ITALY

Professor Sami F. Masri
University of Southern California
Civil Engineering Department
Los Angeles, CA 90089 USA

Dr. Dennis R. Mertz
Modjeski and Masters
P.O. Box 2345
Harrisburg, PA 17105 USA

Dipl.-Ing. Hans-Joachim Miesseler
Strabag Bau-AG
Siegburger Str. 241
D-5000, Koln 21 WEST GERMANY

Professor Fred Moses
Department of Civil Engineering
Case Western University
Cleveland, OH 44106 USA

Professor Antoine E. Naaman
Department of Civil Engineering
University of Michigan
Ann Arbor, MI 48109-2125 USA

Professor Andrzej S. Nowak
Department of Civil Engineering
University of Michigan
Ann Arbor, MI 48109-2125 USA

Mr. Christian Y. Pichot
Public Works Research Lab.-BP 8
54510 Tomblaine
FRANCE

Dr. Tomasz T. Pietraszek
CN Railways, 3950 Hickmore Ave.
St. Laurent, Quebec H4T 1K2
CANADA

Mr. Daniel Poineau
Service d'Etuess Techniques
des Routes et Autoroutes
46, Avenue Aristide Briand
Boite Postale 100
92223 Bagneux Cedex, FRANCE

Mr. David Prine
Chamberlain GARD
7449 North Natches Ave.
Niles, IL 60648 USA

Dr. Joan R. Casas Rius
Civil Engineering Department
Technical University of Catalunya
08034 Barcelona, SPAIN

Professor M. "Saiid" Saiidi
Civil Engineering Department
University of Nevada
Reno, NA 89557 USA

Professor Gerhart I. Schueller
Universitat Innsbruck
Technikerstrasse 13
A-6020 Innsbruck, AUSTRIA

Dr. Long Jacques Trinh
CEBTP B.P. 37
78470 Saint Remy Les Chevreuse
FRANCE

Professor Theodore Tassios
National Tech. Univ. of Athens
42, Patission St.
GR-10682 Athenes, GREECE

Professor Erdogan Uzgider
Istanbul Technical University
Civil Engineering Faculty
Maslak 80626 - Istanbul
TURKEY

Mr. Daniel J. Thiele
5406 S. 107th St.
Omaha, NE 68127 USA

Dr. Reinhard Wolff
Strabag BAU-AG
Postfach 21 11 20
5000 Koln 21, WEST GERMANY

Mr. Graham P. Tilly
Head Bridge Division
Highways and Road Research Lab.
Old Wokingham Rd.
Crowthorne Berks RG11 6AU, ENGLAND

PART-TIME PARTICIPANTS AND OBSERVERS

Professor Pedro Albrecht
Department of Civil Engineering
University of Maryland
College Park, MD 20742 USA

Professor T. T. Baber
Department of Civil Engineering
University of Virgina
Charlotteville, VA 22903 USA

Mr. Per Clausen
Ministry of Transport
The Road Directorate
Lyngbyvej 17
DK-2100 Copenhagen, DENMARK

Mr. Ian M. Friedland
Senior Program Officer
Transportation Research Board
2101 Constitution Ave., NW
Washington, DC 20418 USA

Professor Michel Ghosn
Civil Engineering
City College of New York
138th St. and Convent
New York, NY 10031 USA

Mr. Stanley Gordon
Federal Highway Administration
400 Seventh St. S.W. Room 3113
Washington, DC 20590 USA

Professor Roger Green
Department of Civil Engineering
University of Waterloo
Waterloo, Ontario N2L 3G1
CANADA

Dr. Ian Markey
Swiss Federal Institute of Tech.
CH-1015 Lausanne
SWITZERLAND

Mr. W. T. McKeel
VA Transportation Research Council
P.O. Box 3817
University Station
Charlottesville, VA 22903 USA

Mr. Daniel O'Connor
Federal Highway Administration
Washington, DC 20590 USA

Dr. John B. Scalzi
National Science Foundation
1800 G Street, NW
Washington, DC 20550 USA

Professor Janice J. Trautner
Department of Civil Engineering
University of Utah
Salt Lake City, UT 84112 USA

In front of Inner Harbor Omni Hotel (left: Wolff, Miessler, Pietraszek, Lutz, Nowak, Schueller, Biswas, Scalzi, Fisher, Loveall, Clausen, Anderson, Farago, Trautner, Mancini, Trinh, Tilly, Ehsani, Aligizaki, Rius, Favre, Konig, Klaiber, Fuzier, Ghosn, Naaman, Saiidi, Leon, Eymard, Carino, Hulsey, Markey, Godart)

Workshop Reception (left: Konig, Tilly, Baker, Nowak, Fisher, Favre)

I - Management

BRIDGE MANAGEMENT - THE ANSWER TO THE CHALLENGE

J. J. AHLSKOG
Chief, Review and Design Branch
Federal Highway Administration
400 Seventh Street, S.W.
Washington, D.C. 20590 USA

ABSTRACT. About 42 percent of the 578,000 highway bridges carrying public roads in the United States are classified as deficient by the Federal Highway Administration (FHWA). Recent FHWA studies show that bridges between 15 and 35 years old have a total deck area of more than one billion square feet, or 40 percent of the total area of existing bridges. This huge cohort will require the best, most innovative, engineering and bridge management practices to be applied to meet the challenge of continued safety, mobility and economic opportunity that the United States' highway system currently provides.

The distribution of bridge needs and probable future accruing needs are discussed. The role of the various engineering and management science disciplines is suggested and a brief discussion of the past, current and future status of comprehensive bridge management systems is presented.

Possible Federal legislative or regulatory requirements for bridge management systems are discussed.

1. Introduction

Many new and innovative techniques for the replacement or rehabilitation of existing bridges have been presented during the 1990 American Society of Civil Engineers Structures Congress. The ingenuity, sophistication and practicality of these advanced techniques will improve significantly the safety, economy and useful life span of our existing bridges. Collectively, these new design, analysis and construction innovations will provide a multitude of possibilities for improving individual bridges. The engineers responsible for sustaining the safe useful life of large groups of bridges will welcome all possible bridge improvement techniques.

However, a basic dilemma will still confront the engineer. That is -- which bridges should be replaced, which should be rehabilitated, which should be lightly maintained and which should be benignly neglected?

It is certain that the budget available to any engineer or administrator for sustaining the usefulness of a large number of bridges will not be adequate to improve all bridges that are deficient in load capacity, width or vertical clearance.

A. S. Nowak (ed.), Bridge Evaluation, Repair and Rehabilization, 3–10.

The national statistics vividly illustrate the fact that there are too many substandard bridges and not enough money to improve them all. Of the Nation's 578,000 highway bridges on public roads, more than 40 percent of them or 238,000 are deficient. In a good year, sufficient resources ($5 billion annually) are available to improve about 10,000 bridges. Unfortunately about 10,000 other bridges become newly deficient during the same period. For the past few years, the Nation has just been holding its own - improving just enough bridges to stay even. The FHWA has recently estimated that $92 billion will be needed to correct all existing and accruing highway bridge needs during the period between 1987 and 2005. [1]

2. The Future

A quick look at the future leads to the unhappy conclusion that the situation may get worse during the next few years. Recent studies by the FHWA indicate that the average life span of highway bridges is about 70 years and rehabilitation is needed when the average bridge reaches midlife. Using these averages and looking at a histogram which presents the age distribution versus the number of bridges (Figure 1) or the square footage of existing bridges (Figure 2) some important conclusions can be made. [2]

1. The majority of existing bridges were constructed after 1945.
2. A full 40 percent of the total square footage of existing bridges is represented by bridges between 15 and 35 years old.
3. If historic bridge life spans continue, 40 percent of our existing bridge square footage will require replacement or rehabilitation during the next decade.

These conclusions lead to several other observations:

1. More future resources must be directed towards bridge improvements.
2. New and more economical bridge improvement techniques must be used as they become available.
3. Because of resource limitations, rational and sophisticated procedures are needed to select which bridges to replace and which to be rehabilitate or maintain.

This last conclusion has profound implications. Basic to the need for improved decision making procedures is the requirement that agreement be reached as to what basis is used to define bridge needs and what basis is used to define benefits resulting from bridge improvement projects.

In periods of resource shortage and large bridge needs, bridge improvement decisions must be made on a network basis rather than on an individual bridge basis. A simple example might be the case where it is more cost-effective to rehabilitate the decks of five bridges than spend all of the available funds on one bridge replacement project.

While all experienced bridge engineers can evaluate individual bridges to make the most economical rehabilitation or replacement decisions, it is difficult, if not impossible for most to make the best group of improvement decisions when confronted with hundreds or thousands of bridges which are all substandard.

3. Systematic Network Wide Approaches Are Needed

A systematic bridge management system which has the support of the top level decision makers in a political entity and is based on sound, simple to understand definitions of needs and on sound economic principles is the best answer to the bridge challenge before us in the 1990's and beyond. Without such systems, governmental entities and engineers responsible for maintaining the safety and integrity of substantial groups of bridges will be subjected to intolerable pressures to improve bridges from misguided political coalitions, erroneous perceptions by the public and errors in engineering judgment.

4. Comprehensive Bridge Management Programs Provide The Answer

Comprehensive bridge management systems which incorporate the most current engineering, managerial and system technology will provide the basis for meeting the challenge. A comprehensive bridge management system must include the following elements:

- Data collection and management
- Data base
- Analysis
- Needs, predictions, options and costs
- Program formulation and planning

Fortunately, most States can use the National bridge Inventory data as a beginning for their bridge management systems. Methods for manipulating and evaluating the data base are available. Several States have defined minimum tolerable levels of service of bridges and have developed formulas, based upon engineering judgment and empirical studies, to measure relative deficiencies of their highway bridges. figure 3 illustrates some minimum and desirable levels of service used by the State of North Carolina. [3]

It cannot be overstressed that the validity of any system depends largely upon the uniformity, accuracy and currency of the data base. If the inventory data is flawed, the entire system will suffer. The integrity of any management system for a large array of structures depends principally upon the validity of the inventory and appraisal data.

5. Needs Definition

Needs cannot be determined without a universally accepted definition. One governmental unit's definition of needs will be anothers luxury. One State's inadequate bridge may be entirely adequate in another State where traffic and industrial or commodity shipping requirements are less demanding. The best definition of needs that most have been able to agree on is one which defines needs on the basis of benefits to the user of the facilities. This definition of needs is universal to any array of structures. Two forms of this definition of needs are:

- Needs are the least cost actions to make up the gap between existing conditions and standards which are socially optimal by virtue of maximizing net benefits to society.
- Needs are the actions that maximize the net benefits for each bridge, and thus represent the socially optimal choices from a broad range of alternatives.

Applying either of these definitions to a number of improvement options for an individual bridge should result in a unique choice for improvement which is optimal for the bridge users.

A comprehensive bridge management systems will provide a systematic procedure for making bridge programming decisions which is markedly different from applying engineering expertise on a bridge-by-bridge basis.

6. Incremental Benefit/Cost Ratio

One method of applying bridge management techniques to bridge improvement decisions is to determine the alternative improvement options for each substandard bridge, estimate the improvement cost for each alternative and estimate the user costs incurred by the public for each of the alternatives.

The incremental benefit/cost ratio is determined by taking each increment of benefit and dividing it by each increment of cost. At some point there will be an increment of benefit which equals the increment of cost. This the optimal point for improvement. Figure 4 illustrates this procedure.

If this process is repeated for all substandard bridges and the projects are listed in the order which lists the highest incremental benefit/cost ratio projects first, the resulting list will be the optimal list of projects in priority order. [4]

The process can be readily computerized. Recently, researchers from North Carolina State University applied the process to a group of 25 bridges and made some interesting observations. [5] Some of the conclusions are:

- The process is sound and superior to use of empirical priority ranking methods.
- The process provides near optimal sets of alternatives under budget constraints and optimal project sets under conditions of no budget restraints.

7. Policy Implications

The use of the incremental benefit/cost techniques to select the best bridge improvement options on a system wide basis has some interesting policy implications. Some of these are: [1 and 3]

- For the normal range of discount rates, it is almost always better to rehabilitate or perform heavy maintenance on a bridge than replace it.
- It is in the public interest to spend about 6 percent of the replacement cost of a bridge each year to keep it in service. Put another way, if the discount rate is 6 percent, it is in the pub-

lic interest to spend up to 6 percent of the replacement cost of bridges each year to keep them in service for an additional year.

- The long term trend in the United States should be to increase bridge maintenance budgets dramatically. Bridge rehabilitation budgets should be significantly increased and bridge replacement budget needs should be drop correspondingly.
- Better bridge rehabilitation and maintenance techniques coupled with comprehensive bridge management systems are required to meet the challenge of bridge operations in the 1990's and beyond.

8. Future Legislation

In Public Law 100-17, the 1987 Surface Transportation and Uniform Relocation Assistance Act, the U.S. Congress required the U.S. Secretary of Transportation to study and report to the Congress on the need for bridge management systems at the State level. The Secretary of Transportation reported in 1988 that bridge management systems were desirable but sufficient tools had not then been developed to mandate their use.

Most recently, the U.S. Secretary of Transportation has developed and endorsed a National Transportation Policy which recommends the use of bridge and pavement management systems by the States. [6] Most certainly the Congress will soon revisit this issue while developing the new legislation required to continue the Federal-aid highway program. (Current Federal authorizations run out in 1991.) It is probable that legislation will be enacted or regulations issued to require States to formulate and enact comprehensive bridge management systems as a condition for continued use of Federal-aid highway funds.

As the States develop and implement bridge management systems, it is likely that future Federal legislation or regulations will permit more flexibility in the use of available Federal funds for bridge improvement.

It is also likely that future Federal-aid highway funds for bridge improvements will be distributed to the States based upon needs as measured by uniform level of service definitions. This will be a significant change from the current method where FHWA definitions of deficient bridges are used and are subject to individual interpretation by each State. The level of service approach will evaluate all bridges based on three primary service requirements; (1) load capacity, (2) roadway width and (3) vertical clearance. [7]

The minimum acceptable levels of service will differ for each functional classification of highways but will be uniformly applied to all bridges in all States for purposes of Federal-aid bridge fund apportionment.

9. Conclusions

Network wide bridge management systems are needed to sustain the safety, economic opportunity and mobility provided by our Nation's bridge system. The prospect of scarce funds and increasing bridge needs requires that all bridge owners manage their systems as efficiently as possible.

Past practices for bridge improvement decisions have not always provided optional results.

10. Current Status

Two States, Pennsylvania and North Carolina, have comprehensive bridge management systems in place. The Pennsylvania system does not have a fully integrated economic optimization model and the North Carolina system, while containing an economic optimization element, is not yet operational on a routine basis. Approximately 20 other States are developing elements of bridge management systems.

The California Department of Transportation with FHWA support and assisted by a panel of experienced bridge engineers from five other States is developing a comprehensive bridge management system which should be amenable to rapid adoption and implementation by other interested States. It will be available in late 1991.

11. References

1. Federal Highway Administration, U.S. Department of Transportation (1989). The Status of the nation's Highways and Bridges: Condition and Performance and Highway Bridge Replacement and Rehabilitation 1989, Program, Washington, D.C.
2. Ahlskog, J.J. (1988). "Challenge of Highway Bridge Evaluation, Operation and Maintenance." Post Congress Report, 13th Congress, International Association of Bridge and Structural Engineers, Zurich.
3. Federal Highway Administration, U.S. Department of Transportation (1987): Bridge Management Systems, Demonstration Project Number 71, Washington, D.C.
4. McFarland, W., Rollings. J. and Dheri, R., Documentation for Incremental Benefit-Cost Technique, Technical Report prepared for Federal Highway Administration, U.S. Department of Transportation, by Texas Transportation Institute, Texas A&M University System, College Station, Texas, December 1983.
5. Farid, F. Johnston, D. Chen, C. Laverde, M., Rihani, B., Feasibility of Incremental Benefit-Cost Analysis for Optimal Allocation of Limited Budgets to Maintenance, Rehabilitation and Replacement of Bridges, North Carolina State University, Raleigh, North Carolina, 1988.
6. U.S. Department of Transportation, "Moving America - New Directions, New Opportunities," 1990.
7. Federal Highway Administration, U.S. Department of Transportation (1988): Recording and Coding Guide for the Structure Inventory and Appraisal of the Nation's Bridges.

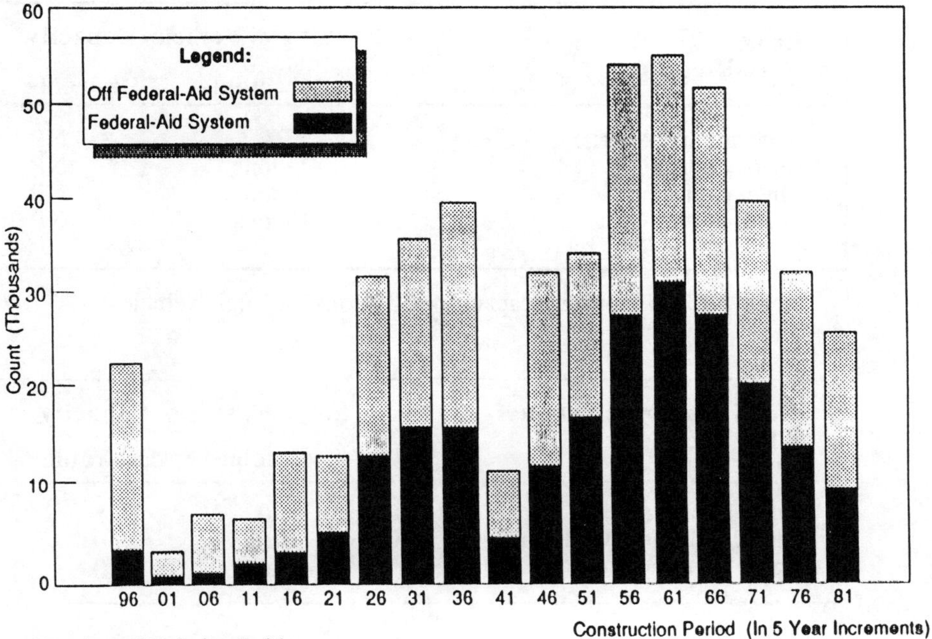

Fig. 1 Number of Existing Bridges.

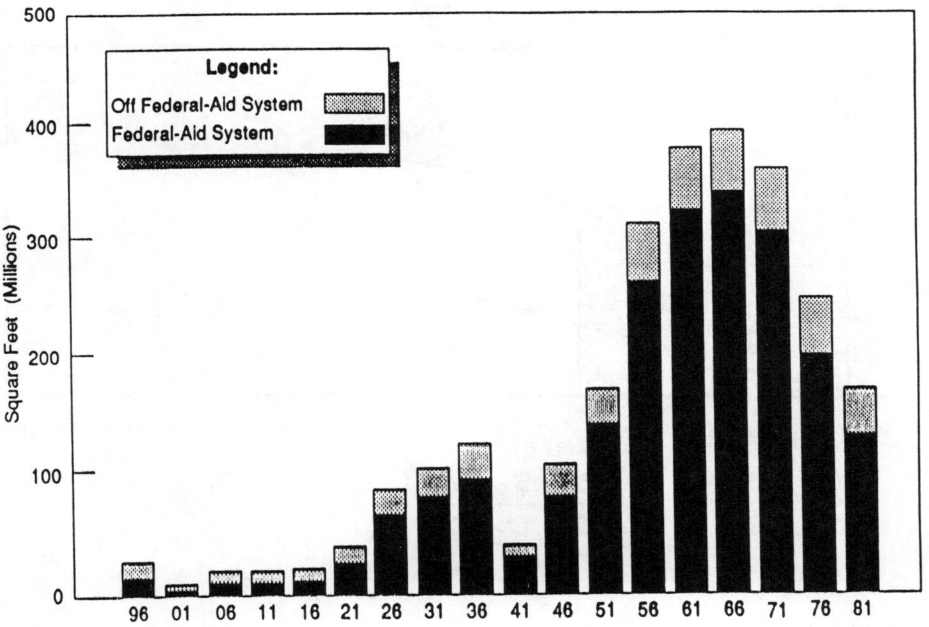

Fig. 2 Existing Deck Area - All Bridges.

a)

Road Over Functional Classification	Single Vehicle Capacity	
	Acceptable	Desirable
Interstate and Arterial	NP	NP
Major Collector	25 tons	NP
Minor Collector	16 tons	NP
Local	16 tons	NP

Note: NP = Not posted (capacity = 33.6 tons for single vehicles)

b)

Road Under Functional Classification	Single Vehicle Capacity	
	Acceptable	Desirable
Interstate and Arterial	14.0	16.5
Major and Minor Collectors	14.0	15.0
Local	14.0	15.0

Fig. 3 NCDOT Bridge Capacity and Vertical Clearance Goals: (a) Capacity, (b) Vertical Clearance (Overclearance and Underclearance).

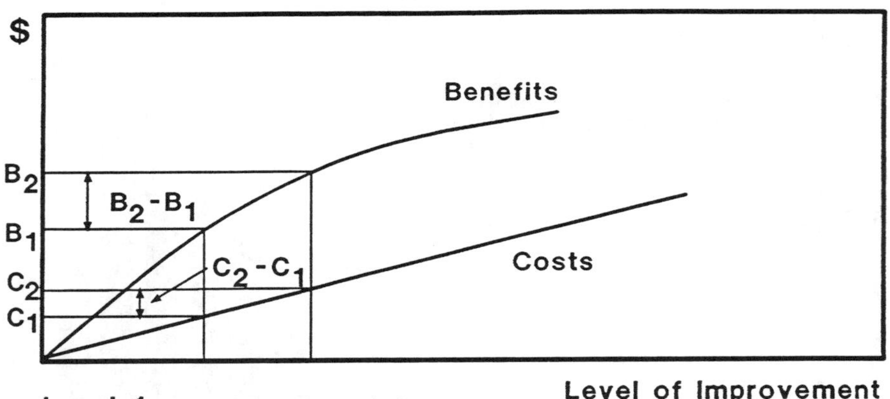

Level 1

$$\frac{\Delta B}{\Delta C} = \frac{B_1}{C_1}$$

Level 2

$$\frac{\Delta B}{\Delta C} = \frac{B_2 - B_1}{C_2 - C_1}$$

Fig. 4 Comparing Alternatives.

DANBRO - A BRIDGE MANAGEMENT SYSTEM FOR MANY LEVELS

NIELS H. ANDERSEN
The Danish Road Directorate
Bridge Department
Lyngbyvej 17
DK-2100 Copenhagen 0
DENMARK

ABSTRACT. Danbro is a computer-supported BM-system that is in use in Denmark today. This paper takes a look at the system from the user side with emphasis on the various levels at which the system can be utilized to support bridge management. The concept of the system - how its components work together - is explained. So is the organization which the system is designed to support. The activities which are covered by the system are mentioned, and the build-up of the EDP part is described - what data the data bases contain and what computer programs are developed to process the data. The setup of the EDP system for both central and decentral use is explained together with the way information is exchanged between the various locations where the system is used. Finally, the possibility of using the system in a modified version in other countries is discussed.

1. Introduction

Whenever the concept of BM-systems is discussed, we tend to focus on the computer programs and the deterioration models which are part of the system. These are, or course, of great importance as they enable us to analyse and make estimates of the future costs of our bridge stock. However, we must not forget that other parts of the system are also vital.

In NCHPR 300 Bridge Management Systems the following definition is given "Specifically, a bridge management system (BMS) is a rational approach to organizing and carrying out the activities related to planning, designing, constructing, maintaining, rehabilitating and replacing bridges vital to the transportation infrastructure". I have taken the liberty of quoting this definition as it covers well the way we look upon BMS in my country.

The BM-system I am going to describe is the Danbro system which has recently been developed in Denmark. The system is a second generation BM-system which is based on the experience gained from an older system run by the Road Directorate.

Danbro exists in two versions. One was tailored to suit the requirements of the Danish State Railways, the other those of the Danish Road Directorate. These two public agencies created the system in collaboration with the consulting firm COWI-consult.

A. S. Nowak (ed.), Bridge Evaluation, Repair and Rehabilization, 11–21.

The maintenance organizations of the two agencies are set up differently; this is one of the main reasons for the need for two versions.

As an employee of the Danish Road Directorate, who has participated in the development of the version now in use at this agency, I will in the following explain the functions of this version.

One of the unique features of our system is that it can be used for maintenance work at several levels. This is of great importance as our routine maintenance is decentralized and both the central part of our organization and the decentralized parts need the system as a tool in decision making.

2. The Levels

Before going into detail about the build-up of the system, let us look at who the users of the system are at the various levels, and at what we can get out of the system at these levels.

We can define the following four levels:

A. The executive level
B. The planning level
C. The administrative level
D. The maintenance level

2.1 THE EXECUTIVE LEVEL

This is the level at which the decision-makers decide on future policy for bridge maintenance and thereby about future budgets. The decision-makers are primarily the management group at the Road Directorate which is responsible for overall policy, but also the politicians who approve the annual State Budget for road administration may be considered as belonging to this group.

In Denmark we work with 5-year budgets; when proposals for the annual budget are sent to the parliament, preliminary budgets for the following four years are attached. As part of these budgets, the funds ear-marked for bridge maintenance are given.

Funds for maintenance works are limited, as in most countries of the world. Therefore the budgets we get do not cover the actual needs.

The information we can feed back to the executive with the help of the system includes:

1. The current condition of the bridge stock.
2. The need for funds, if we choose to use the optimal repair strategies on all bridges.
3. The backlog we build up, if we do not exceed the given funds over the next few years, together with the additional cost incurred.
4. The consequences for the road standard and traffic flow if the budgets are kept unchanged or are suddenly changed.

The latter has happened frequently in past years; our government has cut the funds for road administration several times during one year.

2.2 THE PLANNING LEVEL

The chief of the Bridge Maintenance Division is in charge of the planning of bridge maintenance. This has to be done according to the policy determined at the executive level.

At the planning level the following decisions are taken:

1. How much to use on routine maintenance out of the total budget.
2. An optimization of the repair and rehabilitation works, to suit the allocation set aside for this type of work.
3. A decision on what new types of repair methods and material are to be introduced and what older types are to be abandoned.

2.3 THE ADMINISTRATIVE LEVEL

This is the level at which data about the bridges are handled. Inspectors, office technicians and budget supervisor collect and use these data for their work.

The data can include:

1. Administrative data on the bridges.
2. Technical data on construction of the bridges.
3. Inspection data on the condition of the bridges.
4. Repair strategies for bridges in need of remedial works.
5. Construction costs and repair costs of bridges and bridge components.
6. Test data about deterioration i.e. location, type, cause and development.

2.4 THE MAINTENANCE LEVEL

The practical routine maintenance is dealt with at this level. The local bridge engineer plans and monitors the work.

He uses the system for the follow purposes:

1. Work orders for routine maintenance.
2. Work orders for periodic maintenance
3. Control of the quantities for the above-mentioned works.
4. Control of budgets and costs for routine maintenance.
5. Monitoring the quality, costs and service life of materials used for maintenance.

3. The Components of the System

In the previous sections we have looked at the use we can make of the system; we will now take a look at the components of the system. These are the following:

1. A set of interrelated activities for bridge handling.
2. A set of codes and rules for the activities.
3. A set of data bases where data resulting from the activities are stored.

4. A set of EDP-tools to process the collected data.

The system is tailored to the maintenance organization set up to handle all the highway bridges in Denmark, but it has also proven suitable for regional organizations, as some of the countries have introduced it on their own road networks.

3.1 INTERRELATED ACTIVITIES

These are activities concerning bridge handling e.g. design, construction, inspection, maintenance, repair and budgetary work, including the storing of data resulting from the activities.

3.2 CODES AND RULES

These are the basic documents that initiate the activities.

Firstly we have the road legislation, that is the basis for all activities on the roads, as it states who is responsible for the construction and maintenance of the roads.

Next we have design codes and standards for bridge construction and repairs together with standards and guidelines for inspections and special investigations.

Finally we have manuals which have been specifically written for the system, that give guidelines for activities not covered by the other documents mentioned.

3.3 THE DATA BASE

To store information e.g. data resulting from the activities, a set of data bases are used as part of the system.

3.4 THE EDP-TOOLS

To process the data sorted in the data bases, a set of EDP programs have been developed. These programs give information and recommendations about further activities to be started as part of the bridge management.

The EDP part of the system will be explained later.

4. The Maintenance Organization

4.1 PARTIES RESPONSIBLE FOR THE ROAD NETWORK

The administrative and economic responsibility for the road network in Denmark is divided so that the Road Directorate is responsible for the main roads, the counties for the regional roads, and the municipalities for the local roads.

4.2. THE BRIDGE DEPARTMENT'S RESPONSIBILITIES

In the Road Directorate the Bridge Department is responsible for the maintenance of all highway bridges in the country. The 14 counties in the country assist the Department by monitoring running maintenance and repair works on ordinary structures, whereas the Department itself monitors all activities concerning operation, maintenance and repair works on ordinary structures, whereas the Department itself monitors all activities concerning operation, maintenance and repairs on the special structures (all bridges longer than 200 m, tunnels and ferry berths are defined as special structures).

All activities concerning bridge maintenance are the responsibility of the Bridge Maintenance Division.

4.3 THE SET-UP OF THE MAINTENANCE ORGANIZATION

The head of the Bridge Maintenance Division runs the maintenance organization. He has a staff to assist him with the planning and the budgetary work carried out at the planning level.

Furthermore he has three inspectors who take care of the ordinary structures and five bridge maintenance experts who take care of the special structures. Each of these is responsible for their part of the daily work at the administrative level.

The set-up of the organization is shown in Figure 1.

5. Activities Covered by the System

The following activities are covered by the system:

- Inventories
- Principal inspections
- Special inspection
- Routine maintenance and inspection
- Optimization and budgeting
- Design of remedial works
- Tendering
- Execution of remedial works
- Budget and cost control
- Administration of special transports
- Feed-back

A set of manuals describes how these activities are to be carried out and how data about them are to be registered in the data base.

Some of the activities which are specially related to BMS will be further explained.

5.1 INVENTORIES

Inventories are carried out whenever a bridge has been completed or when remedial works have been carried out.

At the inventories all documents such as drawings, calculations, soil reports, tender documents and construction reports are collected, an index for all archive materials is set up and the most important data are extracted. These data can be of administrative or technical nature.

The administrative data which are of interest include: bridge name and number, owner, designer, construction year, number and name of roads and their relation to the structure.

The technical data can be bridge type, construction materials, geometry including clearances, and load-carrying capacity.

5.2 PRINCIPAL INSPECTIONS

Principal inspections are visual inspections which are carried out at set intervals to register the condition of the structure. Important damage is registered and all components are rated on a scale from 0 to 5. A rough estimate of the costs of remedial works to be carried out over the next ten years are made and the times when they should be carried out are registered. All registration is made on tape recorders.

5.3 SPECIAL INSPECTIONS

Special inspections are thorough investigations where the size and type of deficiencies are established together with their causes. The present and future influence of the defects on the stability of the structure are evaluated.

Two or three repair strategies are considered for a 25 year period and their economic consequences evaluated. The consequences of a 5 year postponement are also evaluated. The corresponding road user costs are taken into account, for each strategy.

5.4 ROUTINE MAINTENANCE AND INSPECTION

Many minor cleaning and maintenance works e.g. cleaning of manholes, maintenance of bitumen joints, repair of pot holes and of concrete spallings have to be carried out at suitable intervals. These types of jobs are carried out routinely together with superficial inspections, where only major abnormal conditions are considered.

5.5 OPTIMIZATION

For bridges in need of remedial works an optimization is made. This involves setting up a priority list that states which bridges are to be repaired when, and according to what repair strategy. The optimization is based on the 5 year budgets together with repair strategies from special inspections and the forecasts from principal inspections.

5.6 ADMINISTRATION OF SPECIAL TRANSPORTS

Special permission has to be given for a road transport if it is heavier, wider or higher than the limits given in the traffic regulations. Permission is given by the police authorities, if all the bridges on the road stretch to be used permit the transport to pass without problems.

The police use lists retrieved from the BM-system for the checking.

All bridges have been rated and given load bearing classes which are given in the list the police use.

The heavy vehicles have to be classified so the load of the vehicle can be compared to the bridge classes. The classification is done routinely at the Bridge Department in a couple of hours after a request from a truck owner is received.

6. The EDP Part of the System

The EDP part of the system is developed in the DOS version of a third generation language called Dataflex. It runs on IBM-compatible personal computers. The system is built up of the following modules:

6.1 THE BASIC MODULE

The basic module contains the administrative and structural data bases and the principal inspection data bases.

Further, a set of programs is included which gives the user of the system access to screen pictures composed of information from the various data bases. One can choose a screen picture that gives a chronological review, showing the most important "events" for a bridge since construction. It is also possible to "run through" a road in this module, so that one can obtain information on clearances and load-carrying capacities for the structures on a given road stretch. This can be used for the administration of high and heavy transports.

Also programs for print-out of inspection reports and other reports are part of the module.

6.2 THE MAINTENANCE MODULE

The maintenance module contains data bases with information about components of the bridges which have to be cleaned or maintained at regular intervals. Also a data base containing possible remedial works on the components is included.

The programs incorporated in this module can automatically print out work orders at set intervals for the local bridge engineer when he has decided upon which bridge components are to be maintained, which maintenance works are to be carried out on these components, the starting date of a maintenance job and the time interval between repetitions.

This module has been developed in collaboration with the bridge engineers of the counties. Data have been entered in all counties during the winter of 1989-1990, so this part of the system starts to run in the spring of 1990.

6.3 THE PRICE CATALOGUE

This module is set up to support the estimation of the costs of repairs in connection with general inspection, special inspection, and design.

The data base in this module contains unit prices for various works, based on tender prices.

A built-in program secures that all prices in the module are adjusted to present-day prices before they are used for estimation.

Unit prices for specified items classified by geographical area or by the size of the work can be calculated and printed. Also combined prices for major works such as deck rehabilitation can be obtained in this module.

Finally, bills of quantities can be printed for tender documents.

6.4 THE OPTIMIZATION MODULE

The data bases in this module contain the economic data for the various repair alternatives set up at the special inspections. These data, together with the economic data from the principal inspection, are processed by the optimization program that is part of this module.

The program operates at the network level but incorporates information from the project level.

By means of an iterative process, the structures which it is most economic to repair, and the repairs which will cost least to postpone when budgetary limits do not permit carrying out all repairs at the optimum time, can be found.

The net present value method is used, and the optimum use of budgetary allocations for a 5 year period is found by iteration.

A program can calculate the consequences of having budgetary allocations lower than the actual demand. The consequences are additional expenditure because of insufficient resources to carry out repairs at the optimum time, and also accumulation of postponed repairs.

6.5 THE BUDGET AND COST MODULE

The data bases in this module contain information about budgets for maintenance and repairs. Furthermore, account figures on the running works are registered.

Changes in works and budgets are registered continuously over the financial year. Budgets are typically changed after a tendering or after a change in the work decided on during its execution. The module can give information at any time on the expected total expenditure for the year.

The module has been used for several years. A new version which can interact directly with the other modules is at present under development. It is the goal to have this version completed in 1990.

6.6 THE EXPERIENCE MODULE

Another module is also expected to supplement the system this year. This is the experience module. It will include data on damage as well as economic data.

Data on damage consist of information on the position, type, extent, and cause of the damage, as well as the materials involved. Also the results of measurements made in the field or on specimens in laboratories will be stored. All the data are collected in connection with special investigations.

The economic data include registration of the expenditures on each structure over a number of years.

The stored data will be used for analysis to obtain information about service life and service life costs of bridge and material types.

7. The Set-Up of the System

The EDP programs are installed on personal computers at the Bridge Department's two offices. At these places all modules except the maintenance module are continuously updated.

The programs are also installed at the offices of the local engineers in the counties. Only the maintenance module can be updated at these places, but access is available to the other modules for information and analysis of the data.

A back-up of the system is run on the main-frame used by the Computer Department at the Directorate.

The back-up is connected to the Road Data Bank. All personal computers on which the system runs are connected via modem to the main-frame computer. Data can thus be transferred between the various users.

8. The Running of the System

The following is an attempt to explain how the interrelated activities interact with each other and how the data for these activities are stored in the data base, and used in the computer program.

A flow chart of the system is shown in Figure 2. Superficial inspections and routine maintenance are carried out by roadmen according to lists prepared by local bridge engineers with the aid of the maintenance module.

Inventories are carried out at the head office of the Bridge Department. The information collected is stored in the basic module.

The three bridge inspectors from the Bridge Department carry out principal inspections of all bridges at regular interval. The interval between inspections of a bridge depends on the condition of the bridge; it can vary between two and six years. The data from the inspections are stored in the basic module. Estimates of costs are obtained with the help of the price catalogue.

Special inspections are carried out by experienced materials engineers from selected consulting companies. The inspections or investigations as they are also called, are carried out whenever the need is recognized during principal inspections or if an unusual event has happened, which may have caused a change in a structural part of the bridge.

Proposals for remedial works and the costs of these are stored in the optimization module. The costs are based on the unit prices from the price catalogue. Information about the damage and measurements are stored in the experience module.

Optimization is carried out once a year when funds are made available, or whenever changes in the allocated funds are made during the year. Budgets for remedial works to be carried out are stored in the budget and cost module, and the local engineers are informed about funds made available for bridges to be repaired in their district.

Design of remedial works are carried out by consulting engineers according to the repair method selected at the optimization, and instructions from the local engineers. The design is approved by the inspection engineer before tendering.

Remedial works are carried out by contractors after tendering. The economic data from the tendering are stored in the budget and cost module. The data from the two lowest bids are stored in the price catalogue.

Long-term budgets are planned on the basis of experience and the forecast of the optimization module.

Special transports are handled by the police authorities, with the help of information from the basic module.

9. The Flexibility of the System

As the system is built up in modules it is possible to adapt it to conditions in other countries by omitting modules not suitable for local conditions and incorporating new modules e.g. another type of long-term prognostic module could be included if desired.

10. References

1. Hudson, S.W., Carmichael, III, R.f., Moser, L.O. and Hudson, W.R. (1987) NCHPR 300 Bridge Management Systems, Transportation Research Board.

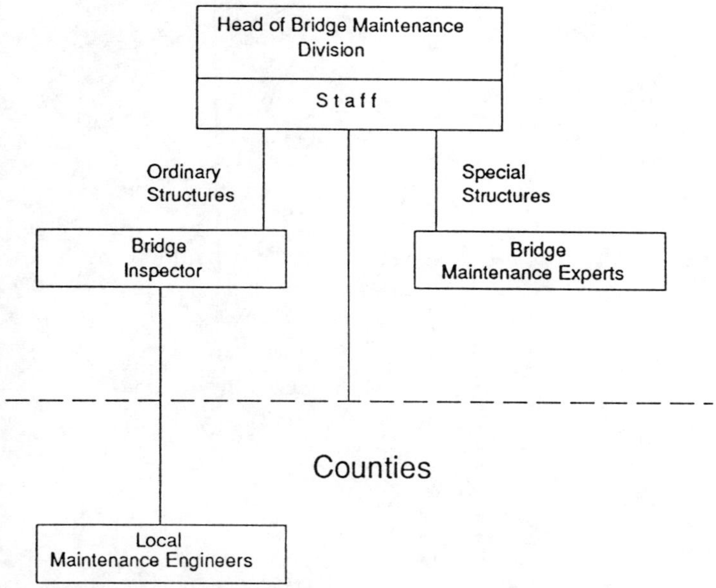

Fig. 1 Flow Chart of Bridge Maintenance System.

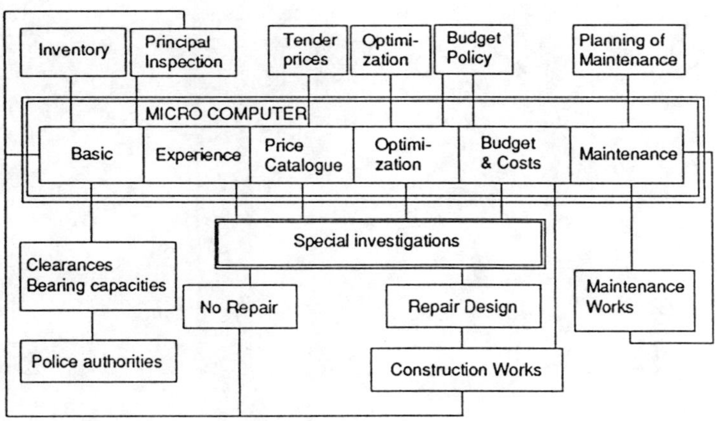

Fig. 2 Flow Chart of Bridge Management System.

EMERGING METHODOLOGIES FOR BRIDGE MANAGEMENT SYSTEMS

K.C. SINHA, M.D. BOWMAN, Y. JIANG, S. MURTHY, M. SAITO
and A.-B. TEE
School of Civil Engineering
Purdue University
West Lafayette, IN 47907 USA

ABSTRACT. Managing a bridge system is a complex decision-making process. There is a host of emerging tools and techniques that can be effectively applied in bridge management systems. This paper presents one particular application in which some of these techniques were used in the development of a statewide bridge management system in Indiana. Theories of fuzzy sets were used in bridge structural condition assessment and bridge traffic safety evaluation, while the analytic hierarchy process was adapted for priority ranking. Techniques of dynamic and integer programming were applied to develop an optimization module where bridge structural conditions were dynamically updated by using Markov chain principles. Examples are presented to illustrate some of the applications.

1. Introduction

The purpose of a comprehensive bridge management system is to provide bridge engineers with a tool for managing bridge systems under the condition of imbalance between bridge repair and replacement needs and available fiscal resources. Such a system was developed for the Indiana Department of Transportation, in which the theories or techniques of fuzzy sets, Markov chain, analytic hierarchy process (AHP), integer linear programming and dynamic programming were employed. Many of these techniques are new in the area of bridge management.

2. Overview of the Indiana Bridge Management (IBMS)

A graphical presentation of the primary features of the Indiana Bridge Management System (IBMS) is shown in Figure 1. The system has eight modules that are used to evaluate various bridge system factors, and incorporate their influence on the overall bridge performance and integrity.

Fuzzy set mathematics was used in the condition rating assistance and bridge traffic safety evaluation modules. A detailed statistical analysis was conducted to develop appropriate information for the life-cycle cost analysis submodule. A modified analytic hierarchy process (AHP)

A. S. Nowak (ed.), Bridge Evaluation, Repair and Rehabilization, 23–39.

was developed incorporating utility curves in the bridge project ranking submodule. The use of dynamic programming, in combination with integer linear programming and Markov chain, provided an efficient tool to select bridge projects with the system effectiveness or benefit being maximized subject to budget constraints over a given program period. The remaining modules primarily involve input and output data of the bridge management system. The application of these techniques made the input data consistent and reliable, and the IBMS effective and flexible. The core of the system consists of condition rating assessment, bridge traffic safety evaluation, ranking and optimization. These four modules are described in the following sections.

3. Application of Fuzzy Sets in IBMS

3.1 FUZZINESS AND RANDOMNESS

Information available to bridge inspectors as to the structural condition or traffic safety is generally imprecise and often can be separated into objective and subjective components. The objective component concerns measurable, countable or quantitative information such as the diameter of a reinforcing bar or the width of a concrete crack. The subjective components, on the other hand, include intangible or quantitative information such as the wisdom, judgment and experience of a bridge inspector. Fuzzy set theory [Zadeh 1965], a systematic technique to quantify imprecise information in the subjective component, can be effectively used to improve the bridge condition and traffic safety evaluation process.

Traditionally, to deal quantitatively with imprecision, the concepts and techniques of probability theory are often employed. By using probability theory it is assumed that the imprecision is caused by randomness of variables. However, the validity of employing probability theory to deal with imprecision in decision making is questionable because imprecision should often be attributed to fuzziness rather than randomness [Zadeh 1965].

Randomness has to do with uncertainty concerning membership or nonmembership of an object in a nonfuzzy set. Fuzziness, on the other hand, has to do with classes in which there may be grades of membership intermediate between full membership and nonmembership. For instance, to answer the question of "is the shape a circle?", in the concept of randomness of nonfuzzy set, both an ellipse and a square are considered as "not a circle". However, in the concept of fuzziness, the value of membership belonging to a circle for a square could be 0.0 because a square has no similarity to a circle, and the value of the membership for an ellipse could be close to the full membership of 1.0, such as 0.90, because an ellipse could be very similar to a circle in shape.

In the bridge inspection process, an inspector must often judge the significance of corrosion of a bridge component as "severe" or "not severe". In a nonfuzzy set approach, the inspector's response can be either "severe" or not severe". Using fuzzy set approach, however, it is possible for the inspector to quantify a significance level somewhere between "severe" and "not severe" using a numerical rating.

3.2 BRIDGE CONDITION ASSESSMENT MODULE

Bridge condition rating is one of the key parameters that is used in determining the types of repair necessary for a bridge. The usefulness of a bridge management system depends upon the reliability and accuracy of the bridge inspection information. The current bridge inspection practices, however, suffer from three inherent shortcomings [Tee et al. 1989]:

1. The parameters in bridge inspections are not completely defined or cannot be precisely measured;
2. Personal judgment bias and subjectiveness are often included but not systematically accounted for in the evaluation process;
3. There is a lack of guidelines establishing the relationsHip between the extent of deterioration and the assignment of values of condition rating.

The purpose of the condition assessment module was to filter the field inspection data of any inconsistencies before entering the bridge management system. The fuzzy set theory [Zadeh 1975] was utilized to minimize the bias of human judgment and to enhance the overall accuracy of the bridge condition evaluation. One of the existing methods of combining fuzzy knowledge, the fuzzy weighted average [Juang and Elton 1986], was used in the module. The fuzzy weighted average has a simple mathematical form:

$$R = \frac{1}{\Sigma W_i} * \Sigma (W_i * r_i) \tag{1}$$

where, R is a fuzzy number denoting a bridge component's resultant rating, r_i denotes the fuzzy rating of the ith subcomponent and W_i denotes the fuzzy importance factor for the ith subcomponent.

The importance factors are imprecise quantities and are represented as fuzzy sets. In this module, three bridge components (deck, superstructure and substructure) were further divided into subcomponents, as shown in Table 1. Equation 1 represents the fuzzy relationship between the condition of a bridge component and those of its subcomponents. W_i represents the degree of importance of a subcomponent relative to other subcomponents. A detailed discussion of this module and the method of determining fuzzy importance factors can be found elsewhere [Tee et al. 1989].

A microcomputer based software package for the condition assessment module was developed. Using this package, typical run time on an IBM XT machine with a math co-processor ranges from a few seconds to a minute per bridge, depending on the complexity of the problem. To investigate the accuracy of the module with available bridge assessment histories, inspection records from different bridges were selected to run the software package. An example of the results of this investigation is presented in Table 2. It was found that the results were in good agreement with the assessment of experienced bridge inspectors. Consequently, the package can also be used as an expert system for bridge inspection that can provide considerable assistance to new and inexperienced bridge inspectors.

3.3 BRIDGE TRAFFIC SAFETY EVALUATION MODULE

Bridge traffic safety information is often ignored in decisions regarding bridge improvement projects. Similar to bridge condition evaluation, bridge traffic safety evaluation can also be affected by many factors and subjective judgments may be necessary to assess it. Kochen and Badre [1974] found that in subjective evaluations, responses are more consistent when verbal descriptions are used than when precise numeric evaluations are required. Therefore, the bridge safety evaluation module [Murthy et al. 1989] was developed based on fuzzy set theory and the concept of linguistic variables.

In the module, all subjective ratings are made as fuzzy sets on the basis of a linguistic scale. These ratings are used as fuzzy numbers in the computations. The module takes into consideration the cumulative effect of the three groups of contributing factors: bridge, approach roadway, and environmental conditions. The weighted average for the fuzzy weights and fuzzy ratings are calculated for each group of factors. These weighted averages are then logically combined to obtain the Bridge Safety Index that represents the degree of bridge traffic safety. The module allows bridge inspectors to translate their subjective judgments on bridge traffic safety related factors into a quantifiable value. In order to assist inspectors making consistent judgments about bridge traffic safety, an interactive computer program was developed. In the program, the bridge inspector is asked to input work ratings for the bridge components associated with traffic safety and the program computes an appropriate index based on fuzzy set principles [Murthy and Sinha 1989]. Table 3 presents an example application of this module. The indices are expressed as a numerical value on a scale 0-10, with 10 being very safe.

4. Project Selection Module

To select projects primarily on the basis of costs, the life-cycle cost analysis sub-module is used. However, when other factors, such as structural condition, traffic safety, community impact and so on, are considered, the ranking and optimization sub-modules are employed. The ranking sub-module can be used to compare bridge projects on the basis of multiple objectives. Even though the ranking program may not give a systemwide optimal solution, the list of ranked bridges can be a guideline for the bridge manager to identify which bridge projects should be placed in the following programming period. The optimization sub-module, however, yields a systemwide optimal solution for any given budget level and program period.

4.1 RANKING SUB-MODULE

Decisions about bridge rehabilitation and replacement depend on a set of criteria, such as bridge structural condition rating, traffic safety, and impact on the surrounding community. This situation is a typical multi-attribute decision-making problem, requiring simultaneous evaluation of several related factors. The technique of the analytic hierarchy process (AHP), developed by Saaty [Saaty 1980], was adopted for the ranking module. The AHP method is a useful tool to rank projects when subjective judgments are involved. However, a direct application of the method may not be practical when the number of projects is large. In the present study, the problem was resolved by using appropriate utility functions. In the context of bridge management, utility is the level of overall effectiveness that can be achieved by undertaking a project. If an appropriate utility is assigned to projects with respect to each of the given evaluation criteria, the expected aggregate utility of each alternative project can be evaluated [Saito and Sinha 1989].

Figure 2 shows the hierarchy system of the ranking sub-module for IBMS. This four-strata hierarchy consists of the overall goal of the ranking exercise, objectives that bridge managers may like to achieve, evaluation criteria and their respective utility curves, and individual bridges. The criteria weights can be obtained by applying the eigenvector approach proposed by Saaty [1980].

The results of the ranking model run with fifty state highway bridges in Indiana that needed rehabilitation or replacement are shown in Table 4. Assuming a given budget of $11,128,000, only the first six bridges can be selected from the top of the list. Thus, with this selection the total cost is $10,081,180, and the total gain of the utility is 272.6.

4.2 OPTIMIZATION SUB-MODULE

The optimization sub-module applies dynamic programming and integer linear programming to select projects while the effectiveness or benefit of a bridge system is maximized subject to the constraints of available budgets over a given program period [Jiang and Sinha 1989]. Markov chain transition probabilities of bridge structural conditions are used in the model to predict or update bridge structural conditions at each stage of the dynamic programming. The dynamic programming divides the federal and state budgets of each year into several possible spending portions, and the integer linear programming selects projects by maximizing yearly system effectiveness subject to different budget expenditures. The dynamic programming chooses the optimal spending policy, which maximizes the system effectiveness over a program period, by comparing the values of effectiveness of these spendings resulted by the integer linear programming.

At each stage of the dynamic programming, a decision must be made as to the optimal solution from stage 1 to the current stage. When a decision is made, a gain in effectiveness is obtained and the system undergoes a transformation to the next stage. The bridge conditions are updated for the next stage by Markov transition probabilities [Jiang et al. 1988].

In order to be consistent with the ranking module the dynamic programming formulation considers the effectiveness of a project i as the reduction in disutility (the difference in disutility values before and after a project is implemented), as shown below:

$$E_i = U_{ib} - U_{ia} \tag{2}$$

Considering that budgets can be carried over from year to year, the mathematical model for maximizing the overall effectiveness of various activities over a program period T was formulated:

$$\max \sum_{t=1}^{T} \left[\sum_{i} \sum_{a} X_{i,t}(a) * E_i * d_i (t) \right] \tag{3}$$

Subject to the following constraints:

a. available federal budget,

$$\sum_{t=1}^{T} \left[\sum_{i} \sum_{a} X_{i,t}(a) * c_i * F_i \right] \leq C_{BF} \tag{4}$$

b. available state budget,

$$\sum_{t=1}^{T} \left[\sum_{i} \sum_{a} X_{i,t}(a) * c_i * (1-F_i) \right] \leq C_{BS} \tag{5}$$

c. one activity can not be undertaken more than once on one bridge in T years,

$$\sum_{t=1}^{T} X_{i,t}(a) \leq 1 \tag{6}$$

Constraints (d) to (h) correspond to an integer linear programming:

d. maximize system effectiveness of year t,

$$\max \sum_{i} \sum_{a} [X_{i,t}(a) * E_i * d_i(t)] \tag{7}$$

e. spending constraint of year t for federal funds available,

$$\sum_{i} \sum_{a} [X_{i,t}(a) * c_i(a) * F_i] \leq \lambda_{tF} \tag{8}$$

f. spending constraint of year t for state funds available,

$$\sum_{i} \sum_{a} [X_{i,t}(a) * c_i(a) * (1-F_i)] \leq \lambda_{tS} \tag{9}$$

g. no more than one activity can be chosen on one bridge in year t,

$$\sum_{a=1}^{3} X_{i,t}(a) \leq 1 \tag{10}$$

h. decision variable,

$$X_{i,t} = 0 \text{ or } 1 \tag{11}$$

i. update bridge conditions by Markov chain transition probabilities if bridge i is not selected in year t,

$$R_{i,t+1} = R_{i,t} * p_i(R,t) + (R_{i,t}-1) * (1-p_i(R,t)) \tag{12}$$

j. improvement of bridge condition if bridge i is selected in year t for activity a,

$$R_{i,t+1} = R_{i,t} + R_i(a) \qquad (13)$$

where:

$X_{i,t}(a)1$ = if bridge is chosen for activity a;
$X_{i,t}(a)0$ = otherwise.
$d_i(t)$ = the absolute tangent value on performance curve of bridge i at time t;
C_{BF} = total available federal fund for the program period;
C_{BS} = total available state fund for the program period;
F_i = federal fund share of bridge i;
$1-F_i$ = state fund share of bridge i;
$c_i(a)$ = estimated cost of bridge i of activity a;
λ_{tF} = spending limit of federal fund in year t;
λ_{tS} = spending limit of state fund in year t;
$R_{i,t}$ = condition rating of bridge i in year t;
$p_i(R,t)$ = Markov condition transition probability of bridge i with condition rating R in year t;
$R_i(a)$ = condition rating gained by bridge i for activity a.

Equations 2 through 12 constitute a dynamic programming which includes an integer linear programming (Equation 6 to 10) as a part of the constraints. The objective of the model is to obtain optimal budget allocations and corresponding project selections over T years so that the system effectiveness can be maximized.

The results of the optimization module with the same fifty bridges as used in the ranking module example are shown in Table 5. The available budget was $11,128,000 or about 25% of the total budget needed for repairing and replacing all of the 50 bridges. The total gain in systemwide effectiveness was 900. By dividing the total gain of utility by its corresponding total cost, the gain of utility per million dollar for the proposed approach is:

$$\frac{900.0 \text{ (utility)}}{11128000(\$)} = 81 \text{ units}$$

and that for the ranking method is:

$$\frac{272.6 \text{ (utility)}}{10081180 \text{ (\$)}} = 27 \text{ units.}$$

Therefore, the optimization approach provides a much better allocation of resources than the ranking approach.

Figure 3 is a comparison of the results from the two approaches in terms of system effectiveness and available budget. It can be seen that the optimization approach always gives better solution than the ranking approach when the available funds are less than 100% of the need.

5. Conclusions

The paper presented a brief review of the application of some of the emerging tools and techniques being used in developing bridge management systems. To provide a reliable bridge management system, the theories of fuzzy sets, Markov chain, ranking and a combination of integer programming and dynamic programming have been applied. The use of fuzzy sets minimizes the bias of human judgment and enhances the accuracy of the input data of the system. With reliable input data, the project selection module can be used to program bridge activities based on either priority order or optimization results with respect to system benefit. The optimization approach, however, provides an optimal allocation of available resources over a period of time.

6. References

Jiang, Y., Saito, M. and Sinha, K.C. (1988), "Bridge Performance Prediction Model Using the Markov Chain,"
Transportation Research Board 1180, National Research Council, Washington, D.C., January.

Jiang, Y. and Sinha, K.C. (1989), "Dynamic Optimization Model for Bridge Management Systems," Paper presented at the 68th Transportation Research Board Annual Meeting, Washington, D.C., January.

Juang, C.H. and Elton, D.J. (1986), "Fuzzy Logic for Estimation of Earthquake Intensity Based on Building Damage Records,"
Civil Engineering Systems, Vol. 3, December, pp. 187-191.

Kochen, M. and Badre, A.N. (1974), "On the Precision of Adjectives Which Denote Fuzzy Sets,"
Journal of Cybernetics, Vol. 4, No. 1, pp. 49-59.

Murthy, S. and Sinha, K.C. (1990), "A Fuzzy Set Approach for Bridge Traffic," Paper presented at the 1990 Annual Meeting of Transportation Research Board, Washington, D.C.

Saaty, T. (1980), The Analytic Hierarchy Process: Planning, Priority Setting, and Resource Allocation, McGraw-Hill, Inc., New York.

Saito, M. and Sinha, K.C. (1989), "The Development of Optimal Strategies for Maintenance, Rehabilitation and Replacement of Highway Bridges Final Report Vol. 5: Priority Ranking Method," Report No. FHWA/IN/JHRP-89/12, Joint Highway Research Project, School of Civil Engineering, Purdue University, West Lafayette, Indiana.

Tee, A.B., Bowman, M.D. and Sinha, K.C. (1988), "A Fuzzy Mathematical Approach for Bridge Condition Evaluation,"
Civil Engineering Systems, Vol. 5, pp. 17-24.

Tee, A.B., Bowman, M.D. and Sinha, K.C. (1989), "A System for Bridge Structural Condition Assessment," Report No. FHWA/IN/JHRP-89-9, Joint Highway Research Project, School of Civil Engineering, Purdue University, West Lafayette, Indiana.

Zadeh, L.A. (1965), "Fuzzy Sets," Information and Control, Vol. 8, Academic Press, New York, pp. 338-353.

Zadeh, L.A. (1975), "The Concept of a Linguistic Variable and its Application in Approximate Reasoning," Parts 1 and 2, Information Science 8, pp. 199-249 and 301-357.

Table 1 Bridge Subcomponents

BRIDGE

DECK	SUPERSTRUCTURE	SUBSTRUCTURE
Wearing Surface	Bearing Devices	Bridge Seats
Deck-Structure	Stringers	Wings
Curbs	Girders	Backwall
Median	Floor Beams	Footings
Sidewalks	Trusses	Piles
Parapets	Paint	Erosion
Railings	Machinery	Settlements
Paint	Rivets-Bolts	Pier-cap
Drains	Welds	Pier-Column
Lighting Standards	Rust	Pier-Footing
Utilities	Timber Decay	Pier-Piles
Joint-Leakage	Concrete Cracks	Pier-Settlement
Expansion Joints	Collision Damage	Pile-Bents
	Deflection	Concrete Cracks
	Alignment of Members	Steel Corrosion
	Vibrations	Timber Decay
		Debris-seats
		Paint
		Collision Damage

Table 2 An Example of Results from the Condition Assessment Module.

Deck Elements	Condition Rating: Type of Structure[a] 1.	2.	3.	4.	5.
1. Wearing Surface	3	7	5	4	4
2. Deck	–	6	6	5	5
3. Curbs	7	7	6	6	6
4. Median	–	7	–	–	6
5. Sidewalks	7	7	–	6	7
6. Parapets	7	7	6	–	7
7. Railings	7	6	6	6	7
8. Paint	7	6	–	7	–
9. Drains	7	7	–	6	6
10. Lighting Standards	–	–	–	–	–
11. Utilities	–	–	–	–	–
12. Joint Leakage	6	7	5	5	6
13. Expansion Joints	6	7	5	5	7
Rating From Inspectors	5	7	5	5	5
Rating From Model	6	7	5	5	6

[a] Type of structure included the following:

1. Continuous Reinforced Concrete Girder
2. Steel Beam with Reinforced Concrete Approach
3. Reinforced Concrete Girder
4. Continuous Steel Beam
5. Continuous Reinforced Concrete Girder

Table 3 An Example of Results from the Traffic Safety Evaluation Module.

STRUCTURE NO: I65-178-5509B (NBL)	
CONTRIBUTING FACTORS	SAFETY RATING
BRIDGE RELATED FACTORS	
Roadway width	*safe*
Relative roadway width	*safe*
Shoulder width	*safe*
Shoulder width reduction	*safe*
Vertical clearance	*not applicable*
Approach guardrails and bridge rails	*moderately unsafe*
APPROACH ROADWAY FACTORS	
Approach sight distance	*very safe*
Approach roadway curvature	*very safe*
Approach gradient	*safe*
ENVIRONMENTAL FACTORS	
Volume/capacity ratio	*safe*
Percentage of trucks	*safe*
Lighting, signing and delineation	*safe*
Presence of nearby ramps, merges or intersections	*not present*
Presence of nearby lane drops or pavement transitions	*not present*

Estimated Bridge Safety Index estimated = 7.64

Table 4 Output of the Ranking Model.

Bridge No.	Priority	U_i	E_i	ΣE_i	C_i	ΣC_i	Activity
31	1	72.9	52	52	2159	2159	BRP
30	2	72.5	50	102	1210	3369	BRP
47	3	72.3	46	148	1549	4918	BRP
27	4	70.4	53	201	5000	9918	BRP
49	5	69.9	51	252	65	9983	DRC
24	6	69.0	21	273	98	10081	DRC
25	7	68.9	18	291	1993	12074	DRC
26	8	68.0	52	343	500	12574	BRP
46	9	67.6	50	393	965	13539	BRP
33	10	65.0	50	443	545	14084	BRP
50	11	65.0	50	483	280	14364	BRP
28	12	63.2	48	541	6228	20593	BRP
37	13	61.2	50	592	840	21433	BRP
48	14	60.5	50	642	420	21853	BRP
32	15	60.1	51	693	3409	25262	BRP
42	16	60.1	50	743	1571	26833	BRP
40	17	59.4	50	793	193	27026	BRP
17	18	59.0	46	839	1090	28116	DRC
43	19	59.0	50	889	1029	29145	BRP
44	20	59.0	50	939	388	29533	BRP
45	21	59.0	50	989	288	29821	BRP
23	22	55.7	40	1029	296	30117	DRC
35	23	52.7	42	1071	635	30759	BRP
34	24	51.7	42	1113	1297	32049	BRP
10	25	51.7	29	1142	330	32379	DRC
38	26	50.0	41	1183	3153	35532	BRP
36	27	49.2	38	1221	840	36372	BRP
39	28	46.0	37	1258	295	36667	BRP
11	29	42.0	15	1273	269	36936	DRC
29	30	42.0	33	1306	2192	39128	BRP
13	31	36.0	23	1329	201	39329	DRC
22	32	36.0	13	1342	164	39493	DRC
41	33	35.9	29	1372	385	39878	BRP
9	34	32.6	15	1387	257	40135	DRC
12	35	32.0	19	1406	201	40336	DRC
21	36	31.9	9	1415	247	40583	DRC
8	37	30.3	15	1430	1300	41883	DRC
18	38	28.6	8	1438	66	41949	DRC
7	39	28.4	12	1450	210	42159	DRC
3	40	28.2	12	1462	387	42546	DRC
6	41	27.0	10	1472	121	42668	DRC
20	42	26.9	8	1480	119	42787	DRC
14	43	26.8	4	1484	476	43262	DRC
16	44	26.6	10	1494	154	43416	DRC
15	45	23.0	4	1498	74	43491	DRC
4	46	22.3	8	1506	281	43771	DRC
5	47	21.9	8	1513	107	43878	DRC
1	48	20.4	8	1521	235	44113	DRC
2	49	20.0	8	1529	276	44389	DRC
19	50	19.9	4	1533	124	44512	DRC

Note: U_i = Utility Value of Bridge i.
E_i = Effectiveness, or Change of Utility Value, of Bridge i.
C_i = Cost of the Activity of Bridge i, in $1000.
BRP = Bridge Replacement.
DRC = Deck Reconstruction.

Table 5 Output off the Optimization Model.

Available Budget = $11,128,000
Total Objective Value (Benefit) = 900.0

Bridge No.	E_i	C_i	Activity
1	8	235	DRC
4	8	281	DRC
6	10	121	DRC
7	12	210	DRC
9	15	257	DRC
10	29	330	DRC
11	15	269	DRC
12	19	201	DRC
13	23	201	DRC
15	4	74	DRC
16	10	154	DRC
17	46	1090	DRC
18	8	66	DRC
19	4	124	DRC
20	8	119	DRC
21	9	247	DRC
22	13	164	DRC
23	40	296	DRC
24	21	98	DRC
26	52	500	BRP
33	50	545	BRP
35	42	635	BRP
36	38	840	BRP
37	50	840	BRP
39	37	295	BRP
40	50	193	BRP
41	29	385	BRP
44	50	388	BRP
45	50	288	BRP
46	50	964	BRP
48	50	420	BRP
50	50	280	BRP

Note: E_i = Effectiveness, or Change of Utility Value, of Bridge i.
C_i = Cost of the Activity of Bridge i, in $1000.
BRP = Bridge Replacement.
DRC = Deck Reconstruction.

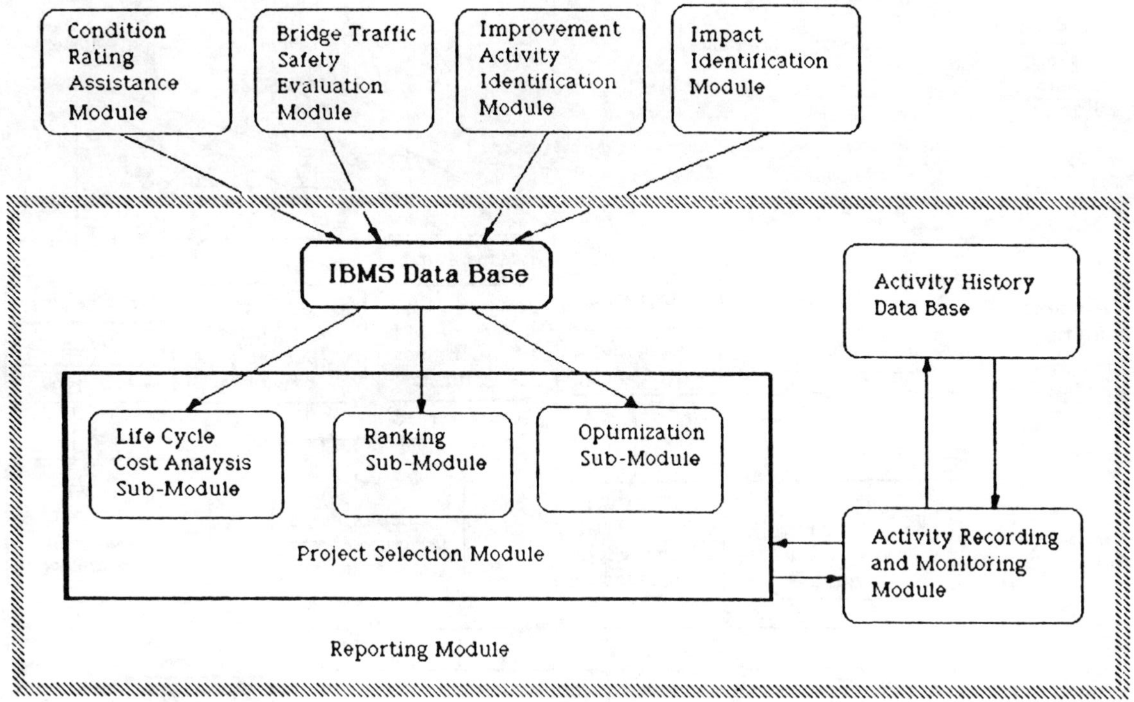

Fig. 1 Graphical Representation of the Indiana Bridge Management System (IBMS).

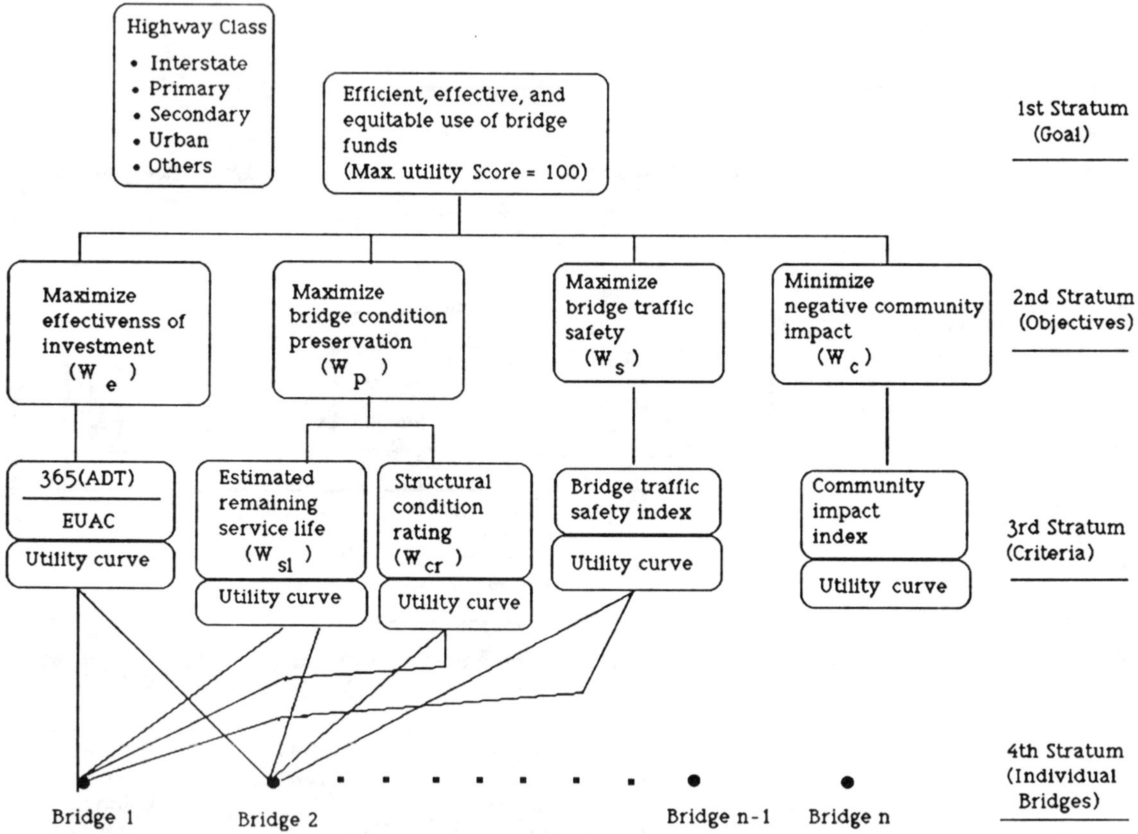

Fig. 2 Hierarchy Structure of the Ranking Module.

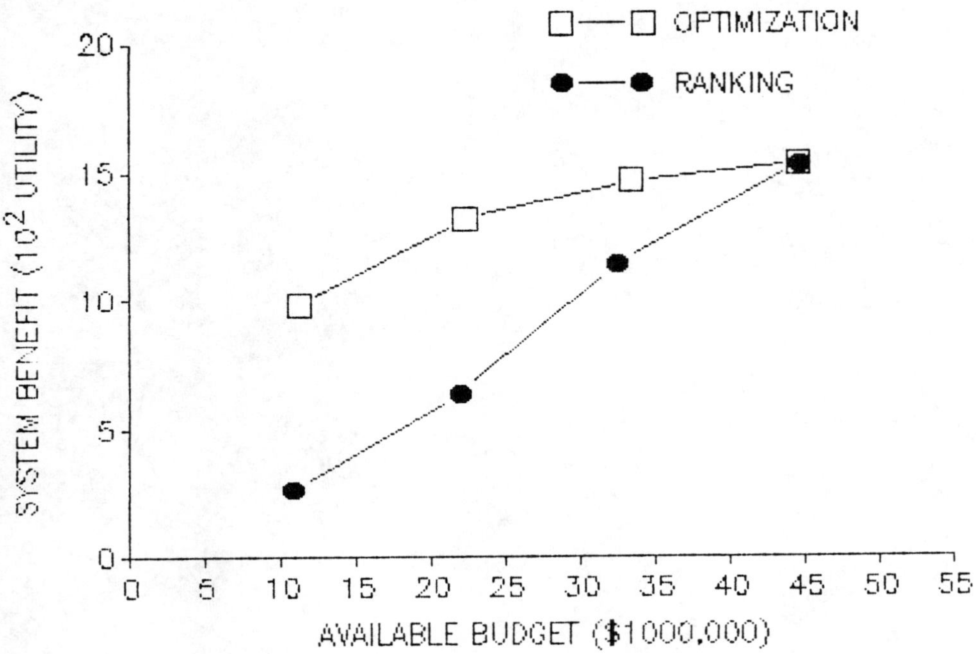

Fig. 3 Comparison of Rnaking and Optimization Approaches.

A NETWORK OPTIMIZATION SYSTEM FOR MAINTENANCE AND IMPROVEMENT OF CALIFORNIA'S BRIDGES

KAMAL GOLABI
Optima, Inc.
517 Washington Street
Suite 205
San Francisco, CA 94111 USA

PAUL D. THOMPSON
Cambridge Systematics, Inc.
222 Third Street
Cambridge, MA 02142 USA

ABSTRACT. In August 1989, Optima, Inc. and Cambridge Systematics, Inc. were awarded a contract to develop a network optimization system to address the improvement and maintenance of California's bridges. The goal of the FHWA-funded project is to develop a bridge management system with sufficient flexibility for implementation in various states over the next few years, replacing current procedures which are largely based on subjective sufficiency ratings. When completed, the system would not only provide each state with a decision tool to help in planning for maintenance, rehabilitation, improvement and replacement of its bridges, but would also provide a common basis for allocation of resources to states by federal agencies, according to improvement needs and bridge conditions of each state.

The Optima/CSI system, named Pontis, addresses the dynamic, multi-objective bridge network maintenance/improvement problem through a set of interrelated predictive, optimization and economic models. The system integrates the objectives of public safety and risk reduction, user convenience, and preservation of investment, with budgetary and program policies, and effectively provides a vehicle for comparing preventive versus corrective maintenance policies. It gives a systematic procedure for allocation of resources to the improvement and maintenance of the various bridges in the system, and considers the costs and benefits of maintenance policies versus investments in improvement or replacement. The model will recommend maintenance strategies within budgetary (and technical manpower and equipment, if necessary) constraints, and will specify the minimum funds required to carry out the improvement and maintenance program. It will also allow considerable flexibility for engineers to choose the appropriate rehabilitation actions.

This paper presents an overview of Pontis, the modeling approach, the distinguishing features of bridge maintenance management and the relevant issues, and will discuss how the system addresses those issues and concerns.

A. S. Nowak (ed.), Bridge Evaluation, Repair and Rehabilization, 41–55.

1. Background

1.1. INTRODUCTION

The United States road network consists of 3.8 million miles of roadways and 565 thousand bridges, as well as thousands of smaller highway structures for drainage, pedestrian crossing and other special uses.

The intensity of road construction during the late 1800's and early 1900's resulted in construction of a great number of bridges, so that more than 400,000, or 70 percent of the nation's existing bridges, were built prior to 1935. Some of them, still carrying large volumes of traffic, date back to the 1890's. Although none of the interstate bridges fall into this category, a great majority of county and local bridges are more than 50 years old. The age of the bridges, the fact that most were designed for less traffic, slower speeds and lighter loads, combined with the gradual effects of weather and de-icing chemicals, have caused substantial problems for the nation's bridge system. Unavailability of adequate funds, which has been responsible for many of the decisions to defer maintenance, has added to the problem. Today, approximately 40 percent of the nation's bridges are considered either structurally deficient or functionally obsolete. In spite of the allocation of large sums of money through the 4R Funds (for restoration, rehabilitation, resurfacing and reconstruction), and the Special Bridge Replacement and Rehabilitation Program (in addition to the recently enacted Strategic Highway Research Program) the gaps between bridge replacement improvements and maintenance needs on one hand and available funds on the other has increased. Even in states such as California where the bridges are relatively new and well maintained, adequate funds to address all needs do not exist.

Given the magnitude of the problem and the enormous funds needed to correct all present deficiencies in bridges in the United States (50 billion dollars according to a 1983 Federal Highway Administration (FHWA) estimate), it is obvious that available funds will not permit the replacement/improvement or rehabilitation of all deficient bridges. What is needed is a careful and systematic allocation of funds and the formulation of an efficient maintenance and improvement policy that would derive the maximum benefit from the use of the limited funds. Formulation of such a maintenance/improvement policy requires a networkwide analysis that would evaluate the needs of each bridge, and the options available to meet those needs, in the context of overall network benefits, budgets and restrictions.

This paper describes, in general terms, the structure of a system being developed to achieve these objectives. The system, Pontis (from Latin pons: bridge), being developed by Optima, Inc. and Cambridge Systematics, Inc. is a FHWA funded project for the California Department of Transportation with plans for implementation in several other states.

1.2. NEED FOR A COMPREHENSIVE SYSTEM

The development of bridge management systems has lagged behind that of pavement management. While pavement management systems have undergone several evolutionary and revolutionary mutations leading to comprehensive, rigorous network-wide systems [7,8], for bridges many states still rely on an approach developed by the FHWA [4] many years ago for subjectively rating the sufficiency of the various bridge components, and for prioritizing their repair needs.

Recently several federal and state agencies have either initiated studies or taken steps towards development of bridge management systems. The Transportation Research Board, through its National Cooperative Highway Research Program has funded a study [9], and the FHWA has completed a demonstration project aimed towards assessing the various methodologies and educating state agencies on the applicability of such methodologies [19].

The State of Pennsylvania has recently completed the development of a comprehensive database management system for bridges and has embarked on an ambitious data collection effort [2]. The FHWA and the North Carolina Department of Transportation have also initiated several studies at North Carolina State University [12,18,3,1] and the State of Indiana has funded a study by Purdue University [11]. Internationally, a comprehensive effort is now being undertaken in Finland by Cambridge Systematics, Inc. [14,16].

The first North Carolina study [12] proposes a level of service approach to evaluate the adequacy of North Carolina bridges in serving public needs by selecting three easy-to-quantify bridge characteristics as the most direct of these needs. These are (i) load capacity, (ii) clear deck width, and (iii) vertical roadway underclearance and overclearance. Since then, several other states have taken steps to adopt service level goals for evaluating their bridge deficiencies. In [18] they use an optimization algorithm developed by Golabi and his collaborators for maintenance of nonpavement elements [6,13] to establish maintenance activities. In [3] they assess the feasibility of using an incremental benefit-cost method [17] to bridge improvement decisions. In [1] they build on some of their previous work to quantify the costs (including user costs) and propose a simple year-by-year decision rule for improvement actions for each bridge. The Purdue study is an empirical study aimed at developing Markov chains for bridge deterioration, and performance curves as a function of age. The states of the Markov chain are the 0-9 condition ratings of the bridges. The Wisconsin Department of Transportation (WisDOT) analyzed statewide bridge repair and replacement needs through a computer simulation model [10]. The model uses life cycle cost analysis, and life cycle activity profiles, to find the expected number of bridges that need to be repaired and/or replaced in every period (up to 20 years).

Each of the above studies addresses one, or at best, a limited number of dimensions of the problem, and therefore are limited in their capabilities for addressing the bridge management problem. What is needed is the development of a comprehensive system consisting of interrelated subsystems which would collectively address the real concerns of the maintenance and improvement decision making process and realistic constraints. The models underlying the system should be flexible to address the many bridge management issues faced by each state, and be based on sound mathematical and economic concepts; the computer systems should be designed with efficiency and flexibility in mind; and the data requirements should be realistic. The criteria should not be that a model or solution method is available and can be used (and therefore fitting the problem into the model) but to develop the needed models and methodologies tailored to solving the multifaceted bridge maintenance and optimization problem.

1.3. DISTINGUISHING FEATURES OF BRIDGE MAINTENANCE MANAGEMENT

An aspect of any maintenance problem, and in particular bridge maintenance, is the dynamic nature of the decision process. Maintenance (in a general sense of the word) is a continuous process with the action taken in any time period influencing conditions, decisions and costs in subsequent periods. The maintenance planning tool must address the dynamic nature of the maintenance problem and the planning process. In addition, it must be able to operate within a program of gradual replacement of functionally obsolete bridges and within budgetary constraints.

The bridge maintenance problem has many similarities with pavement maintenance [7,8] and finding optimal repair/replacement strategies for pipelines [5]. However, bridge maintenance is a more complex problem. The complexity arises not only from the more critical nature of the bridges, the greater replacement costs and the risk to public safety in case of failure, but also because of several special characteristics. These include the following:

- During the last decade, a considerable amount of data and empirical equations have been developed to address the relationship between various pavement maintenance actions and deterioration rates. An extensive body of knowledge and data does not exist for bridges.
- There are more types and designs of bridges than pavements and pipelines.
- The various components deteriorate at differing rates.
- The economies-of-scale involved in repairing the components of a bridge are more pronounced than those involved in repairing, say, adjacent segments of a highway or pipelines.
- The funding situation is more complex for bridges than for highways and pipelines.

A major distinguishing factor between the bridge management problem and most other maintenance problems is the fact that "improvement" implies two sets of distinct activities. The first set constitutes maintenance and corrective actions. Such actions, as correcting scour conditions or replacing wearing surfaces, improve the condition of the bridge, which may deteriorate again with time. The second set essentially deals with functional aspects, and once performed do not change with time. This set includes actions such as deck widening or bridge raising to gain vertical clearance. In our presentation we distinguish between these two sets by calling the first set "maintenance, corrective and rehabilitation". The second set of actions we denote as "improvements". Recommended improvement actions are driven by state-specified level-of-service goals, or by the optimal levels of those standards recommended by a consideration of budget availabilities, the respective costs and benefits of various standards, and the backlog generated by any set of level-of-service standards. The set of recommended actions dealing with maintenance, and their budget requirements, need to be generated through a dynamic model, as there the condition is time-dependent and at least for some of the actions, there is a tradeoff between their cost and the time until the next action is necessary. To be specific, if two actions achieve a desirable condition at this time, and one is less expensive than the other but has a higher probability of deterioration in the future, the model must be able to address that tradeoff and incorporate it in its calculations. Based on this understanding, a bridge management system should have two distinct, but interrelated components to address maintenance and improvement.

Although these optimization models would be distinct, there is a substantial interrelationship among the two sets of activities. The interrelationships are not only through budget requirements, but also through coordination issues which may affect the prioritization and sequencing of improvements. For example, the improvement of a bridge needing both improvement and major maintenance may be accelerated to save cost and satisfy immediate maintenance needs even if ordinarily, the bridge would not have been scheduled for improvement for some time.

2. Objectives of Pontis

Budgeting for expected expenditures to address current and future rehabilitation, maintenance and corrective measures (collectively denoted as "maintenance" in this paper), as well as improvements and possible replacements and additions, are intimately related to the desire of decision makers to meet several objectives. An effective planning system must systematically address these objectives and prescribe the actions necessary to achieve and maintain desirable standards with respect to these objectives. In addition to cost and budgetary issues, the main objectives that planners and administrators are concerned with are as follows:

- Meeting and maintaining the highest standards of safety for the traveling public,
- Improving riding comfort and convenience of the public,
- Preserving the considerable investment in structures,
- Providing efficient routes for emergency services,
- Minimizing disruptions and delays, and costs to users,
- Providing economical routes for transport of industrial goods and agricultural products,
- Correcting deficiencies within reasonable time,
- Equitable allocation of resources to the various geographical areas and bridge activities,
- Avoidance of costly repairs through appropriate preventive maintenance,
- Efficient utilization of the engineering and maintenance personnel,
- Efficient utilization of funding sources, and
- Minimization of total expected costs over the long run.

Obviously, budget requirements are a function of envisioned future needs, and these needs are dictated by the present conditions of the structures, future expansion needs (increase in traffic) and the uncertainties associated with deterioration, as well as the complex interrelationship between present preventive and future corrective actions. In addition, a budgeting and maintenance planning system must be an iterative tool allowing adjustments dictated by actual conditions over time (instead of an assumed series of actions envisioned now for a long future period). It must also have sufficient flexibility to allow an analysis of the effect of various budgetary policies on future conditions which in turn would dictate future costs. The interrelationships among conditions, maintenance actions, and budget needs necessitates the development of a comprehensive, dynamic planning model that would effectively address the above objectives, as well as providing defensible near-term and long-term budget requirements. Furthermore, it should provide the decision makers with the flexibility for predicting and analyzing the effects of budget cuts on future conditions of the structures. The formulation of Pontis as a multi-objective, dynamic, stochastic decision model allows us to effectively address the above objectives. More specifically, Pontis is being designed to achieve the following:

1. Provide a systematic procedure for setting repair and improvement priorities.
2. Provide a systematic procedure for helping decision makers set realistic level-of-service goals for each of 12 predefined functional highway classifications.
3. Incorporate level-of-service goals in assessing bridge improvement needs and budget requirements.
4. Provide a capability to consider the entire bridge network simultaneously in arriving at recommendations, while retaining the flexibility to address any subset of bridges.
5. Provide priority orders and sequencing for bridges in need of improvement.
6. Coordinate maintenance planning decisions with future improvement decisions.
7. Be based on a methodology that is rigorous, general and flexible. The two latter characteristics are essential for transferability of the system from one state to another when state-specific subsystems are modified.
8. Require a reasonable and realistic amount of data for model input, provide procedures for data collection and checking the reliability of the collected data, assess the relative importance of data items, establish reliability standards commensurate with relative importance, and allow for incorporation of engineering judgment with objective data whenever necessary.
9. Consider the differing inspection and repair needs of the major structural components for bridges as well as the differing needs of the various types of bridges.
10. Address the probabilistic nature of bridge deterioration and the differences in the deterioration patterns of various components. The model would be able to consider the effect of design characteristics, geometry, environmental factors and maintenance history on future deterioration of bridge components.
11. Allow for updating of predictive probabilities as the necessary data would become available over time.
12. Consider the relative importance of the various bridges in terms of safety, risk exposure, public convenience and regional economics.
13. Consider the immediate and future costs and benefits of the various courses of action and their effect on future conditions. In particular, the model would weigh the benefits of preventive maintenance versus costlier (but less frequent) corrective actions.
14. Allow sensitivity analysis of the recommended policies in terms of future conditions of the bridge network and maintenance and replacement cost requirements. This would be flexible to accommodate different state-specified maintenance and fiscal policy issues.
15. Provide least-cost estimates for correcting deficiencies if budget constraints are removed. Compute costs over a planning horizon and prepare annual budget requirements to address those needs.
16. Provide a basis for short-term and long-term maintenance and improvement budget planning and resource allocation.
17. Consider the possible availability of Federal and special funds, the restriction on those funds and the availability of matching funds.

18. Consider economies-of-scale involved in repairing/replacing either adjacent bridges or several components of a single bridge.
19. Be easy to use by maintenance planners and engineers. Using the system would not require knowledge of operations research, statistical methodologies or computer science.
20. Provide flexibility for bridge engineers to exercise their professional judgment in making day-to-day decisions.
21. Provide a framework for reconciling divergent interests (e.g. increasing the ability of the system to transport more loads versus preserving historical bridges).
22. Provide a basis for equitable funding to States' bridge programs by the Federal government, and assist in allocating funds to different districts by each state.
23. Centralize the maintenance decision process and reduce gaps between the condition of similar bridges within the network.
24. Provide rigorous procedures and an analytical framework for incorporation of expert engineering judgement in the model.

Pontis consists of several interrelated modules that, taken together, effectively address the above issues and requirements.

3. System Organization

Figure 1 shows schematically the major components of Pontis and the manner in which they connect to one another in order to meet all of these objectives. As a result of the interconnectedness of the system, the organization of the optimization submodels has a determining effect on the data and organization of the other submodels in the system. Two divisions within the problem which are most fundamental are:

- Near-term vs. long-term programming. Long-term programming concerns itself with the agency's vision of where it would like to be, eventually. It begins with a set of general goals and, for an optimization problem, defines these goals in a manner which can be succinctly stated and quantified (e.g. "minimize social costs"). The outputs of long-term programming are maintenance and improvement policies in a general sense; funding allocations as they apply to long-term strategies to ensure reliable funding levels; and near-term objectives for the condition of the bridge infrastructure. Near-term programming starts with these objectives and finds a way to "get there from here." It is concerned with the identification and resolution of specific identified or predicted bridge needs; the listing of specific construction and maintenance projects; the development of a program that can be shown to maximize movement of the infrastructure toward the long-term goals (or minimize backsliding), within known budget constraints; and short-term efforts to ease those constraints.
- Maintenance and rehabilitation vs. improvement and replacement. Maintenance management is the pursuit of a most efficient way to keep the existing bridges in operation at their current level of service (except when the posted load has been reduced because of deterioration of bridge components). It begs the question of what the required level of service should be, or even whether the bridge should remain open; instead, it assumes that operations must

continue and that deterioration must be detected and remedied before operations are affected, at minimal cost. Improvement management, on the other hand, analyzes what the level-of-service standards should be, identifies instances where the standards are not met, develops strategies to meet them, and prioritizes and sequences such improvements.

The main components of the system are as follows:

3.1. INPUT DATA BASE

The input data base allows for storing all information needed to run the model as well as its updating as more information becomes available over time. The data base retrieves and stores the inventory, the physical characteristics of bridges, condition survey data and other relevant information from the agency's data files. It will also store traffic information, load characteristics, and cost information; as well as the main elements of each bridge, the condition states defining the possible conditions of each element, and the set of feasible corrective, maintenance and improvement actions associated with those conditions.

A key feature of Pontis will be a module which augments the existing agency's database with engineering judgement, in the form of user input and built-in decision rules, to generate the needed information. The system database to be designed for this project will therefore be a model for the kind of database which is both desirable and achievable for bridge program optimization. The project's recommendations will include the steps which a State must take to begin an improved data collection program to directly collect the needed data.

3.2. SELECTION OF FEASIBLE ACTIONS

The optimization procedures address the selection of feasible actions in a consistent, integrated way. Algorithms determining feasible actions represent two basic relationships involving maintenance and improvement activities.

1. They establish a correct mapping between the current condition of a bridge and the set of activities that are technologically appropriate to correct the type, severity, and extent of damage that is present, or to prevent damage that would otherwise occur in the future.
2. They embody criteria or preferences that help resolve overlaps among activities, and that establish a clear hierarchy among the activities to be applied.

These algorithms are needed because there is not a simple, one-to-one relationship between bridge damage and the activity available to correct or prevent that damage. Furthermore, major activities may subsume the effects of one or more subsidiary activities. Consider the example of bridge decks to illustrate these points. At the level of minor or routine maintenance, damage to bridge decks (involving, say, cracks or spalls) can be corrected by either crack sealing or patching. More extensive damage might be addressed either by extensive cutting and patching or a deck overlay. Not only do these activities again illustrate the first relationship above, but they also invoke the second relationship. A deck overlay, for example, not only corrects cracking and spalls, but addresses other types of surface damage, and therefore subsumes other routine maintenance activities besides crack filling and patching. At a still higher level of bridge rehabilitation, one could consider deck replacement, an activity that would correct or eliminate not only cracking and spalling, but also other categories of deck structural deficiencies. The combination of these two relationships -- the technological coupling between repair techniques and bridge deficiencies, and the hierarchical structure of the preventive and corrective technologies available -- are not clearcut, and a number of external or judgmental factors affect how they apply [15].

3.3. COST SUBMODELS

The Cost Submodels consist of the following components:

1. Improvement Costs: Calculates the costs associated with improvement of each bridge to achieve the level of service goals based on material quantities for each feasible improvement and the State's unit cost data. These costs are inputs to the Improvement and Replacement Needs Submodel.
2. Replacement Costs: Calculates the replacement cost for each bridge when replacement is a feasible option. These costs are input to the Improvement and Replacement Needs Submodel and possibly to the Long-Term Maintenance Optimization Submodel.
3. User Cost Submodel: Calculates the user costs associated with the present and improved condition of each bridge. The User Cost Submodel provides inputs to the Long-Term and Near-Term Maintenance Optimization Submodels as well as to the Improvement and Replacement Needs Submodel. Some of the measures (attributes) that are considered in the development of the user costs are:
   - Travel time and fuel consumption due to detours
   - Travel time and fuel consumption due to traffic congestion
   - Travel time, fuel consumption, and vehicle depreciation due to deck roughness
   - Travel time due to sharp turns or other geometric problems
   - Accident costs
4. Maintenance, Rehabilitation & Corrective Action Costs: This submodel calculates the costs of the feasible actions associated with preventive, rehabilitative and corrective measures by computing the necessary material quantities and using the State's unit cost data. The model consists of a set of detailed equations that would attempt to estimate the various costs as accurately as possible, associated with each element, condition, and feasible action. Either bridge-specific or aggregate data may be used.

A desirable feature of the system is its ability to quantify the cost of deferred maintenance and improvement. Generally, the costs associated with deferring actions that need to be taken now, but because of budgetary restrictions need to be postponed, are quantified and taken into consideration by the optimization models. However, the results can also be directly provided as outputs. If the bridge is in need of improvement the deferred costs consist mainly of user costs (including expected accident costs) until the deficiency is removed. These costs are quantified for every bridge and provided for bridges which are not chosen, as well as those chosen, for improvement. In its determination of the optimal maintenance policy for each bridge, Pontis considers the possibility of future needs for expensive corrective actions (including replacement) if appropriate corrective actions are not taken at the present time. It would also consider the risks associated with different conditions, and the conditions to which they are likely to deteriorate if proper maintenance actions are not taken. These costs and risks are also quantified for those bridges whose maintenance action has been deferred for budgetary reasons, and in particular for those bridges where no tradeoffs between the various maintenance actions exist.

3.4. DETERIORATION PREDICTION SUBMODEL

This model estimates the deterioration rates associated with each major bridge element as a function of its present condition, and the feasible maintenance and corrective actions (including "routine maintenance only"). The deterioration of bridges is affected by many factors. These factors can be classified as those related to the initial design of the bridge, quality of construction methods and materials, and the external environment of the bridge. To these one may add age and previous major maintenance and corrective actions.

Bridges are complex entities with interconnections among many elements. An understanding of the causes of deterioration is very important, much more so than for pavement management. The reasons for this are twofold. The first is that the existing data base consists of aggregate subjective ratings (between 0 and 9, usually according to the FHWA condition rating) for the three main components of bridges, namely, deck, substructure and superstructure. Since a component may be comprised of many elements, the reasons for such ratings can be defects in any number of elements or the severity of the defects, and different actions may be prescribed for bridges with the same rating. Therefore, past ratings, while useful for presentational purposes, are not very useful predictive tools for bridge management. The second reason is that past ratings are connected to past maintenance practices and budgets which have neither stayed constant, nor are they indicators of future practices.

Current models of deterioration tend to focus on the time-dependent change in some index of bridge condition, and reflect very little of the cause and effect relationship that results in damage and deterioration; therefore, they are not only subject to variability in their predictions, but they also require further subjective assessments to relate levels of deterioration to the activities that may be required. Realistically speaking, cause-and-effect models of bridge damage mechanisms cannot be developed for all mechanisms on all bridge components. The problem is too complex, and sufficient field data are not available. Nevertheless, such theoretical or empirical relationships do exist (or can be derived from published data) for certain components and for certain damage mechanisms; these are employed in conjunction with expert opinion and statistical analyses of historical data to develop a more informed judgment and calibration of the deterioration prediction model.

3.5. OPTIMIZATION OF MAINTENANCE, REHABILITATION, AND CORRECTIVE ACTIONS

This optimization submodel has two parts, long-term and near-term. In the long-term, the model combines the deterioration models with agency cost (and to a less important extent, user cost) models to analyze the long-term performance of each individual structure and the network as a whole. If replacement becomes necessary in this model, it is in-kind replacement with a facility meeting the same level-of-service criteria. If a facility is to be replaced by one with different capabilities, the question is left to the improvement and replacement optimization submodels, described in the next section; this line of separation is important conceptually because it allows the two separate analyses to be more focused. The products of the long-term maintenance optimization are expected long-term conditions, expected maintenance budget needs, and optimal policies for corrective and maintenance actions.

In the near-term, the focus is on finding optimal maintenance policies. The near-term is much more specific about individual projects, and considers existing condition and economies of scale. Although the processes, and therefore the models, are stochastic, deterioration and environmental effects are much more predictable in this model. Outputs in the near-term are concerned with binding commitments which the agency must make: scheduled maintenance actions and budgets. Both of these models emphasize the role of agency expenditures in relieving deterioration, thereby extending the life of the structure and reducing future agency costs. The models consist of smaller interrelated optimization models and are capable of addressing a variety of policy and engineering issues, including the lag time between planning and construction.

An issue which frequently is raised in the context of capital budgeting questions is the comparison of competing projects on an equal basis. For maintenance projects this is frequently accomplished by the choosing of a common time horizon for all analyses, and the assignment of salvage values for facilities which live beyond the time horizon. It is important to understand at this point why the structuring of the problem as a dynamic model obviates the need for a salvage value analysis. The key point is that the bridge infrastructure is being modelled as a "going concern". What this means in the long-term is that the scope of the analysis is limited to bridges which continue operating at the same level of service indefinitely, with maintenance and replacement actions taken when needed to ensure that this is so. While policy changes and cataclysmic events come and go, the assumption is made that the long-term is far enough away into the future that random occurrences average out: the network is said to be in a steady state, because year-to-year variations are not modelled but sustainability of the policy is. What the going-concern assumption means in the near-term is that the emphasis is on reaching a level of overall network condition that makes the long-term optimal policy possible. Success is measured not in terms of life-cycle costs, but in terms of progress made in the condition of the infrastructure. Thus, salvage values are not necessary in either model.

3.6. OPTIMIZATION OF IMPROVEMENTS AND REPLACEMENT

In this model level-of-service goals are considered explicitly. By definition, projects in this category are the only ones which change the level of service offered by a structure. The two submodels within this component of the system consider the possibility of allowing a bridge to be posted or closed, and they consider the raising of posted limits or the improvement of geometrics or other functional attributes, including the replacement of a bridge with one having better level-of-service characteristics. Obviously, there is an interaction between this component and the maintenance optimizations, since in the overall scheme of things the decisions made here set the levels of service to be maintained; but the factors considered as inputs, and the optimization methods, differ greatly. The ability of the improvement submodels to also prioritize down-grading of structures is an important part of the simplification achievable by the conceptual division between maintenance and improvements.

The two outputs of this model are a list of improvement projects, and their prioritization. In the maintenance models, there is a long-term vs. near-term dichotomy, because the perceived needs are always far greater than what can be done in the near future. The model proceeds in the generation of improvement needs by first settling on a set of "necessary" and "desired" level of service goals, which may vary according to various definitions of sub-networks (e.g., by functional class). Next, it aggregates the bridges in the network into units of analysis for the application of these goals.

The model is capable of functioning in two modes: (i) which bridges should be upgraded to the "desired" level and which should be postponed, and (ii) what is optimal level-of-service in the context of overall network benefits.

3.7. CAPITAL PROGRAM INTEGRATION

Since the above-mentioned models have as outputs two separate programs, one for maintenance and one for improvements, what remains is to integrate the two into combined long-term and short-term capital programs. The improvements optimization will have identified bridges for upgrading and downgrading, and the maintenance optimization will define a maintenance program for all of the remaining structures. Overall, the integration process is a matter of coordinating the optimization results with each other and with prior decisions, emergency work, new construction, and other over-rides. This is accomplished by a separate model that considers the scheduling and Federal constraints on improvement and major maintenance, as well as engineering dependencies of structural components on each other.

4. Acknowledgments

This project is funded by the Federal Highway Administration and California Department of Transportation. The authors wish to thank Dan O'Conner of FHWA, Eldon Klein and Richard Shepard of Caltrans for directing the effort.

5. References

1. Chen, C. and Johnston, D.W., "Bridge Management under a Level of Service Concept Providing Optimum Improvement Action, Time, and Budgetary Prediction," Center for Transportation Studies, North Carolina State University, Raleigh, North Carolina, September 1987.
2. Commonwealth of Pennsylvania, "The Pennsylvania Bridge Management System," Research Project 84-28A, Bureau of Bridge and Roadway Technology, Pennsylvania Department of Transportation, Harrisburg, Pennsylvania, February 1987.
3. Farid, F., Johnston, D.W., Chen, C., Lavarde, M.A. and Rihani,B.S., "Feasibility of Incremental Benefit-Cost Analysis for Optimal Allocation of Limited Budgets to Rehabilitation and Replacement of Bridges," Center for Transportation Eng. Studies, North Carolina State University, Raleigh, North Carolina, November 1987.
4. Federal Highway Administration, "Recording and Coding Guides for the Structure Inventory and Appraisal of the Nation's Bridges," U.S. Department of Transportation, Washington, D.C., 1979.
5. Golabi, K., "Steel Main Planning Model: A System for Maintenance, Protection and Replacement of Gas Pipelines," Optima Report to New York Gas Group, 1990.
6. Golabi, K., "A Branch-and-Bound Algorithm for the Solution of Nonlinear Integer Programs", Working Paper-WCC, June 1978.
7. Golabi, K., Kulkarni, R.B. and Way, G., "A Statewide Pavement Management System," Interfaces 12 (6), (Dec. 1982), 5-21.
8. Golabi, K., "A Markov Decision Modeling Approach to a Multi-Objective Maintenance Problem," Essays and Surveys on Multiple Criteria Decision Making, P. Hansen (Ed.), Springer-Verlag, New York, (1983) 115-125.
9. Hudson, S.W., Carmichael, R.F., Moser, L.O., Hudson, W.R. and Wilkes, W.J., "Bridge Management Systems," Transportation Research Board, National Research Council, Washington, D.C., December 1987.
10. Hyman, W.A. and Hughes, D.J. "A Computer Model for Life Cycle Cost Analysis of Statewide Bridge Repair and Replacement Needs," Division of Planning and Budget, Wisconsin DOT, December 1982.
11. Jiang, Y., Saito, M. and Sinha, K.C., "Bridge Performance Prediction Model Using the Markov Chain," School of Civil Engineering, Purdue University, West Lafayette, Indiana, August 1987.
12. Johnston, D.W. and Zia, P., "A Level of Service System for Bridge Evaluation," Transportation Research Record, 962, Transportation Research Board, Washington, D.C., (1984), 1-8.
13. Kulkarni, R.B., Golabi, K., Finn, F. N. and Johnson, R., "A Systematic Procedure for the Development of Maintenance Levels of Service," Transportation Research Record, 781, Transportation Research Board, Washington, D.C., (1980) 6-14.

14. Kähkönen, Ari, and Marshall, A.R., "Optimization of Bridge Maintenance Appropriations with the Help of a Management System - Development of a Bridge Management System in Finland," International Conference on Bridge Management, Surrey, England, March 1990.
15. Markow, M.J. and Brademeyer, B.D., EAROMAR Version 2: Final Technical Report, Federal Highway Administration, FHWA-RD-82/085, April 1984.
16. Marshall, A.R., and Söderqvist, M., "Local Agency Experience with the Utilization of Bridge Management Systems in Finland and the United States," International Conference on Bridge Management, Surrey, England, March 1990.
17. McFarland, W., Rollings, J. and Dheri, R., "Documentation for Incremental Benefit-Cost Technique, Program INCBEN", Technical Report prepared for the Federal Highway Administration, U.S. Department of Transportation, by Texas Transportation Institute, Texas A&M University System, College Station, Texas, December 1983.
18. Nash, S.C. and Johnson, D.W., Level of Service Analysis for Bridge Maintenance Activities in N. Carolina, Dept. of Civil Engineering, North Carolina State Univ. Report, November 1985.
19. O'Conner, D.S. and Hyman, W.A., "Bridge Management Systems", Report No. FHWA-DP-71-01, U.S. Department of Transportation, Forthcoming.

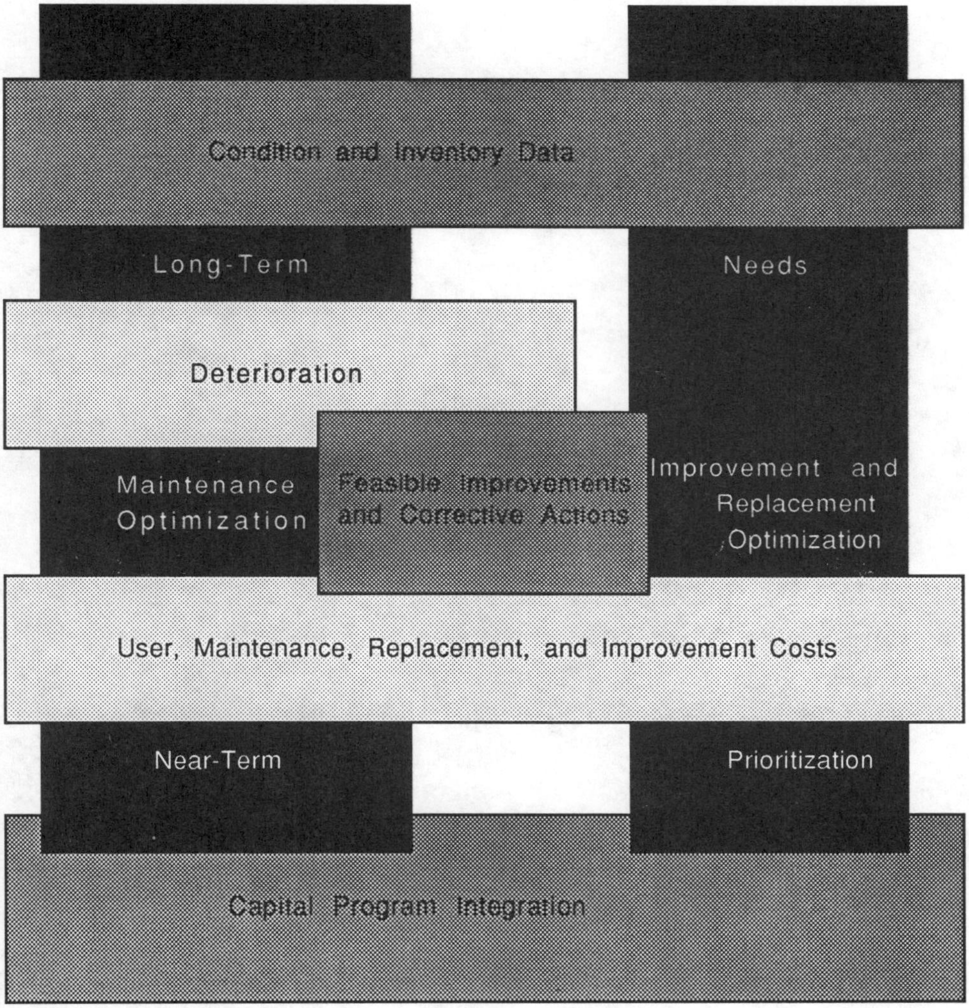

Fig. 1 Major Elements of PONTIS.

STATE OF THE ART FOR CONTROL OF BRIDGE APPROACH SETTLEMENT

DANIEL J. THIELE
Geotechnical Services
5730 South 86th Circle
Omaha, NE 68127 USA

MAHER K. TADROS and JOSEPH V. BENAK
Dept. of Civil Engrg.
Univ. of Nebraska-Lincoln
Omaha, NE 68182 USA

ABSTRACT: Settlement of bridge approaches has been a persistent problem for many state highway departments. In order to gain a better understanding of the problem, a research project was undertaken to identify the contributing factors and the state of the art practice for design and maintenance of bridge approaches. This research included a review of the available literature relating to approach embankments and pavements. A survey was conducted of state highway departments and other agencies involved in bridge design, construction, and maintenance. Results of the survey as well as suggestions for alleviation of the problem are presented.

1. Introduction

Settlement of bridge approaches has been recognized as a problem for many years, and considerable effort has been directed toward solving this problem with varying degrees of success. The bridge approach is a transition area between a slab on grade and the bridge deck. Approach settlement essentially is a problem of differential movement between these dissimilar structural systems. Settlement and subsequent deterioration of the approach results in a rough transition onto and off of the bridge.

Faulting in the approach area is primarily a nuisance to the public, but in extreme cases can become a significant safety hazard. In addition, it can result in gradual structural damage to bridge elements, including the backwall, expansion joints, and bridge deck. Periodic maintenance of deteriorating bridge approaches can also be a significant expense.

It is generally recognized that this problem is not attributable to any one factor. Rather, approach settlement is a generic term describing the many conditions that manifest themselves as a rough transition onto the bridge. These factors may occur alone or in combinations that magnify the approach fault.

This paper summarizes research that was undertaken to identify the current state of the art for control of approach settlements. The scope of this project included a review of available literature relating to approach embankments and pavements, a survey of organizations involved in the construction and maintenance of bridges in the U.S. and Canada, an inspection of bridges in

A. S. Nowak (ed.), Bridge Evaluation, Repair and Rehabilization, 57–70.

eastern Nebraska, and a review of available methods for possible implementation. The available methods included several innovative concepts developed by the researchers.

2. Literature Review

Considerable research has been done on approach settlement. Many of these studies consisted of comparisons between different structural systems, such as structural approach slabs versus flexible approach pavements, without regard to the underlying causes. Several of the studies, however, included case histories with long-term measurement of approach settlement and detailed investigation of the contributing factors. Performance studies of various methods intended to eliminate approach settlement have also been conducted or are currently under way. Some of the more significant research is summarized herein.

2.1. KENTUCKY INVESTIGATIONS

The Kentucky Department of Highways has conducted several studies into the causes of bridge approach settlement (Hopkins and Deen, 1969; Hopkins, 1969; Hopkins and Scott, 1970; Hopkins, 1973; and Hopkins, 1985). These studies included field surveys of the condition of bridge approaches in 1964 and 1968, and detailed investigations at six sites including subsequent observations over a period of up to 19 years.

A field survey of existing bridges in Kentucky was conducted during the summer of 1964. This survey included 782 approaches, of which 51 percent had required some form of maintenance. A similar survey over many of the same structures was conducted in 1968, and found that 78 percent of the approaches had required maintenance. Since many of the structures were less than two years old at the time of the 1964 survey, it was apparent that the then current practice was not satisfactorily ameliorating the problem.

Detailed site investigations were performed for five bridges built between 1966 and 1978. The purpose was to determine if approach settlement was primarily a result of volume changes in the embankment or foundation soils. Based on the investigations at these five sites, the Kentucky researchers made the following conclusions:

1) Consolidation of the approach embankment foundations can contribute significantly to overall approach settlement if adequate delay is not provided between embankment construction and paving. Secondary consolidation may also be a significant factor.
2) Improper compaction of approach embankments may lead to settlement of the approach pavement.
3) Lateral movements (creep) due to shear strain may lead to approach settlement when the factor of safety of the slope is low.
4) Erosion of materials from behind and around the abutment can contribute to the settlement of the pavement.
5) Bridge approaches that exhibit small post-construction settlements and cause minimal approach pavement distress can be constructed if the above factors are recognized.

6) Reinforced approach slabs do not eliminate the differential settlement between the approach pavement and the bridge deck, but they can provide a smoother transition.

2.2. CALIFORNIA RESEARCH

The California Department of Transportation (Caltrans) conducted research to determine the contributing factors to the problem of rough riding bridge approaches (Stewart, 1985). This research included field surveys of existing bridges and the performance monitoring of 60 experimental bridge approaches.

In 1973, a field survey was performed that included every fifth structure on an 1,800 mile loop of major roadways through nearly every geographical area of the state. A total of 820 approaches were inspected. The survey found that 74 percent of the asphalt approaches had been or were in need of repair, compared with 43 percent of the reinforced concrete approach slabs.

Between 1972 and 1975, 60 experimental reinforced concrete approach slabs were constructed and monitored. The slabs included various lengths and thicknesses of reinforced slabs, 30 and 50 foot long T-beams, stepped slabs (increased thickness closer to the structure), and reinforced slabs doweled into the abutment. All of the reinforced approach slabs rested on a paving notch, or a corbel on the back of the abutment, and many of the slabs also rested on a sleeper slab at the pavement end. Performance data for these approaches were collected through 1982.

Of the 60 experimental approaches, all of the approach fills showed measurable settlements, in the range of 0.2 to 2.6 inches. In nearly every case, the amount of settlement was fairly uniform from about 15 to 20 feet from the abutment to as far as the measurements were made. Since the slabs rested on a paving notch on the back of the abutment, the surface profile exhibited the greatest rate of change within 20 feet of the abutment. There was measurable separation of the slab from the subgrade in about 92 percent of the experimental slabs, in the range of 0.1 to 2.3 inches. Over 80 percent showed separations of at least 1/2 inch. Based on these measurements, it was concluded that original ground subsidence and fill settlement are the primary causes of approach slab settlement.

2.3. WYOMING RESEARCH

Since 1983, the Wyoming Highway Department has experimented with Geosynthetic Reinforced Soil (GRS) walls behind bridge abutments to attempt to eliminate recurring maintenance problems with settlement of bridge approaches and rotational movements of abutments caused by high lateral soil pressures. GRS walls consist of compacted granular fill with embedded layers of geotextiles. This results in a fill with greater stiffness, and helps distribute traffic loads over a greater area. The geotextile provides horizontal tensile reinforcement, and reduces the lateral soil pressure on the wall.

To evaluate actual performance, the University of Wyoming instrumented a bridge retrofitted with GRS walls (Edgar et al., 1988). In general, the GRS walls reduced the lateral forces on the abutment. However, embankment settlements of up to 1.5 inches were recorded for the reinforced embankments. This indicates that while GRS walls can be effective at reducing the lateral pressure on abutments, this method does not completely eliminate settlement of the approach.

3. Survey of Highway Departments

In an attempt to determine the magnitude of the problem and to identify the current state of practice for design and maintenance of bridge approaches, a survey of highway departments was conducted. The survey was sent to both design personnel and maintenance personnel to obtain different points of view regarding the problem. Fifty-two responses were received representing 36 states, the District of Columbia, and three Canadian provinces.

The survey responses generally indicated that approach settlement is a widespread problem. The problem occurs in most every terrain, but is most prevalent where high embankments are required for the approach. Several respondents indicated that settlement is a greater problem on high volume roads.

Structural reinforced approach slabs are used by nearly all of the agencies that responded to the survey. The length of the slabs varied from 10 to 100 feet, but were mostly in the range of 20 to 40 feet. The use of a sleeper slab in conjunction with the reinforced approach slab is also common. The majority of the respondents specify select backfill material behind the abutment. The use of embankment surcharges or wick drains to speed consolidation of foundation soils and the removal of unsuitable foundation soils in the approach area are also common. Several states are also incorporating flowable or lightweight backfill, geotextile/geogrid reinforced backfill, or pile supported approaches, but use of these methods is not widespread. The respondents generally indicated only moderate improvement due to the practices mentioned.

Once settlement occurs, asphalt overlays are routinely used to smooth the approach fault. Slab-jacking is also utilized by many of the respondents. These methods are generally a temporary solution, and most of the respondents indicated that repeat maintenance is necessary within three to five years. Thus, it is apparent that maintenance of approaches can be a significant annual expenditure.

4. Bridge Inspections

Several bridges in the Omaha metropolitan area were inspected. The purpose of these inspections was to quantify the severity of the problem in this area and begin to develop a criteria for tolerable settlement. This survey involved a visual inspection of 53 bridges and measurement of settlement at 7 of these sites.

Settlement ratings were based on the perception of the investigators of the transition roughness, and were classified as good, moderate, or severe. Approximately 40 percent of the approaches were rated as moderate or severe. It is important to note that this reflects the current condition of the approach, and 80 percent of these approaches had already undergone some type of maintenance to repair an approach fault.

The factors that were observed to affect tolerable settlement were angular variance, speed of vehicle, direction of traffic, type of vehicle, and human perception. Angular variance is defined as follows (Fig. 1):

$\alpha 1 = S_1/L_1$ $\alpha_3 = (S_1 - S_2)/L_2$

$\alpha_2 = \alpha_1 + \alpha_3 = S_1/L_1 + (S_1 - S_2)/L_2$

The limited number of measurements made in this study were not sufficient to develop a correlation between angular variance and traffic speed to identify ranges of tolerable and intolerable settlement. However, it is believed that this type of correlation could be useful in evaluating the need to perform maintenance and in determining the tolerable amount of approach camber (discussed in section 5).

An alternative means of evaluating the need to perform maintenance on an approach would be to incorporate a test vehicle such as one used in the California research (Stewart, 1985). This method involves a test vehicle that measures displacement between the axle and the body. A standard test vehicle equipped with an accelerometer may also be effective in conducting maintenance surveys. These types of measurements would provide a more objective evaluation of approach roughness. Further research will be necessary to determine tolerable ranges of settlement and angular variance, as well as developing a convenient and reliable means of measuring roughness.

5. Causes and Remedies of Approach Settlement

As stated previously, the problem of approach settlement is really one of differential settlement between dissimilar structural systems. The bridge deck, through the abutment, is usually supported on a deep foundation, such as piling driven to a firm bearing stratum. Settlement of abutments founded on piling is generally not significant. Conversely, the approach pavement is supported directly on the approach embankment and is free to move with the embankment. A typical embankment may settle several inches as the underlying foundation soil consolidates under the added embankment weight.

5.1 PLANNING, DESIGN, AND CONSTRUCTION PRACTICES TO MINIMIZE SETTLEMENT

Through the literature search and survey responses, several contributing factors were identified where application of current technology could appreciably reduce the number of structures that experience intolerable approach settlements. Although many highway departments have successfully incorporated available technologies, these methods have not been universally implemented.

5.1.1 *Geotechnical Investigations.* In order to properly identify the conditions which may be conducive to approach settlement at any bridge site, it is critically important to perform a detailed subsurface investigation in the approach area. Soil investigations are routinely performed for design of bridge foundation elements. However, to address the unique aspects of approach embankments, these investigations must include test borings in the embankment area. Approach borings should be included in the geotechnical investigation for any structure to be designed with an embankment fill depth of 5 feet or more or with proposed finish slopes steeper than 4 horizontal to 1 vertical.

5.1.2 *Foundation Consolidation.* Foundation consolidation is a major contributor to bridge approach settlement throughout the U.S. (Fig. 2) A reinforced concrete approach slab that rests on a paving notch is often used to provide a smooth transition from the settled embankment onto the bridge deck, as opposed to an abrupt bump. However, the relatively short length of the approach slab limits the range of settlement where this transition remains tolerable. Based on an approach slab length of 20 to 30 feet, post-construction embankment settlements of approximately 1 inch will generally result in a noticeable bump, while settlements of 3 inches or more typically result in an intolerable transition onto the bridge.

When a problem is identified with compressible foundation soils, there are a wide variety of corrective measures available. A few methods to reduce post-construction settlements are described here:

1. Time Delays. Based on the estimated time rate of settlement, final construction of the approach pavement can be delayed until primary consolidation is complete. This is often feasible on small or non-critical projects or when most of the consolidation is expected to occur in a relatively short time. This method may not be an option on large primary projects where the schedule does not allow a delay of 2 to 3 months, or when the time of primary consolidation is longer than 3 to 6 months.
2. Surcharging. The placement of additional surcharge weight to speed consolidation of the foundation soils is a common practice. Once settlement under the combined fill and surcharge weight reaches the level of estimated settlement the fill alone, the surcharge can be removed and construction of the approach may resume. Surcharging involves a waiting period while the surcharge is in place, but will significantly reduce time to effectively complete primary consolidation. This method has been successful in reducing the amount of post-construction settlement on numerous projects.
3. Wick Drains. The use of wick drains can further reduce the time surcharging alone requires for the consolidation of foundation soils, perhaps by 50 to 75. Wick drains are a prefabricated drainage ribbon that is driven vertically into the soil using a mandrel. These drains are typically installed through the compressible stratum on a grid pattern at spacings of 5 to 10 feet. By reducing the length of the drainage path, the wick increases the rate of consolidation. Wick drains are frequently used in conjunction with surcharging to greatly reduce the length of the delay between embankment construction and placement of the approach pavement.

4. Replacement of Unsuitable Surface Deposits. When embankment settlements will be significantly affected by compressible soils near the surface, these deposits can be excavated and replaced with controlled fill. Excavation and replacement is limited by economic factors to fairly shallow depths, perhaps 10 feet deep or less. The presence of a high groundwater table may also preclude the use of this method.
5. Lightweight Fill. To reduce the weight of the embankment, and thus, the amount of embankment settlement, lightweight fill material can be used for all or part of the approach embankment. Lightweight fill materials are commercially available in most parts of the country. These materials often consist of expanded shales. However, the strength and long term durability of some lightweight materials may preclude their use in highway embankments. Other lightweight products, such as cementatious foams, offer low unit weights and relatively high strengths. Two examples of cementatious foams are Elastizell manufactured by the Elastizell Corporation of America in Ann Arbor, Michigan, and Celcore manufactured by Celcore, Inc. in Fort Lauderdale, Florida. These proprietary products can be manufactured to various weight and strength requirements.
6. Pre-Camber of the Approach Pavement. By constructing the approach pavement to a somewhat higher profile than the bridge, the approach can experience fairly small settlements without a deterioration of riding quality. For instance, if a settlement of 2 inches from normal profile is deemed to be tolerable, the approach could be cambered 2 inches high resulting in a tolerable total settlement of 4 inches (Fig. 3). Pre-camber could be implemented on a widespread scale at a minimal cost. Grades are simply adjusted over a few hundred feet to a profile that is artificially high at the end of the approach slab. This camber is then eliminated over the distance of the approach slab.
7. Temporary Paving. If embankment settlements cannot be satisfactorily accounted for, either due to extremely long settlement periods, demanding project schedule, or prohibitive costs, a temporary pavement can be used in the approach area. Some maintenance will most likely be necessary as settlement occurs during the life of this temporary pavement. Once settlements have ceased, the temporary pavement can be removed and the permanent pavement can be constructed. This method does result in disruption of traffic at the time of final paving, but will result in a smoother riding surface over the remaining life of the structure.

5.1.3 *Embankment Stability*. Progressive slope movements frequently occur where high embankments are constructed over soft soil foundation soils (Fig. 4). These progressive failures may result in large settlements over the course of several years. It has been reported that creep type movements will occur in cohesive deposits when the shear stress exceeds approximately half of the residual shear strength. Therefore, the factor of safety for slope stability, as determined by a total stress analysis, should be 2 as a minimum with a value of 2.5 or higher desirable. It is recommended that slope stability be investigated for any endslope or sideslope steeper than 4 horizontal to 1 vertical.

In cases where slopes are marginally stable, it may be necessary to flatten the slope or to provide stability beams. The use of geogrid reinforced slopes has been shown to increase the factor of safety of steep slopes, but long term research into the creep behavior of geogrid reinforced soils is needed. Replacement of soft surface deposits will also improve slope stability, as will the use of lightweight fill for the embankment.

Another important factor in the stability of embankments is the slope of the existing ground surface upon which the embankment fill is placed. Sliding failures will often occur at this interface between original ground and fill. This is a particular problem if seepage occurs along this contact surface. Therefore, when constructing fills on slopes, it is important to bench the existing ground surface to eliminate this failure plane.

5.1.4 *Erosion.* Erosion of soil from behind and below abutments has been a problem in many areas. Loss of soil from these critical areas has resulted in settlement of the abutment and approach. It is important that surface runoff be carried on hardened surfaces. Careful attention must be directed toward sealing cracks and joints in the approach area to reduce infiltration.

5.1.5 *Abutment Backfill.* The use of select granular backfill by most states has undoubtedly led to improved compaction of the abutment backfill wedge. Continued awareness and attention by contractors and inspectors is important in continuing this positive trend. The use of reinforced concrete approach slabs that are capable of withstanding a loss of support in this area is also important in preventing faults due to improperly compacted backfill.

To further facilitate compaction of backfill in the area behind the abutment, it is important that all protrusions be removed from the rear of the abutment. The backwall should have a straight vertical face to avoid obstructing the use of mechanical compaction equipment.

Where tall abutments are used, the excavated area is much larger, and thus the backfilling process becomes more critical. The use of a cellular abutment reduces this problem by eliminating most of the backfill.

5.1.6 *Lateral Forces on Abutments.* Excessive lateral soil loads on abutments have resulted in displacement of the abutment and closure of expansion joints on seat type abutments. Some subsidence behind the abutment is believed to occur in conjunction with this lateral displacement. In extreme cases, this movement may be sufficient to allow the approach slab to drop off the paving notch. Expansion of the approach pavement may also result in excessive lateral forces on the abutment if an adequate expansion joint is not provided.

To reduce the occurrence of displaced abutments, it is important to properly consider the lateral soil forces on the abutment. Since deflection is to be held to a minimum, at-rest earth pressures of the compacted backfill should be considered as a minimum. Where high degrees of compaction are required, passive earth pressures may be appropriate to estimate the lateral forces. It is also critical that an adequate expansion joint be provided, preferably at the pavement end of the approach slab, to reduce the forces associated with pavement growth.

The widespread use of integral abutments has eliminated the problem of expansion joint closure, since these structures contain no expansion joint at the abutment. However, very little is known about the lateral forces on this type of abutment. Research into the behavior of integral abutment bridges and the soil pressures on the backwall is needed.

The use of geotextiles and geogrids should also be considered as an economical method of reducing lateral soil loads on abutments. Geosynthetic reinforced walls have been effectively used in Wyoming to eliminate problems with abutment displacement.

5.2 METHODS FOR CORRECTION OF EXCESSIVE SETTLEMENT

Through diligence in planning, design, and construction processes, the frequency and severity of approach settlement problems can be greatly reduced. However, it is impractical if not impossible to completely eliminate the source of trouble. It is inevitable that some problems will occur in this transition between two dissimilar structural systems.

Since some settlement must be expected in the bridge approach, it is desirable to incorporate an adjustment system into construction of the approach pavement. Thus, if settlements create a rough approach, there will be a convenient but effective method of adjusting grades in the approach area to restore a smooth transition onto the bridge.

Several concepts were considered that could be incorporated into new construction to facilitate maintenance in the approach area. These include some procedures that have been utilized on a limited basis as well as the so-called "wild ideas" of the research team.

5.2.1 *Preformed Grout Holes*. During construction of the approach pavement, grout holes can be preformed in the reinforced slabs. Grouting (mud-jacking) operations can then be done through these preformed holes. This method speeds the slab raising process and eliminates the problem of coring through reinforcing steel. In addition, since the locations of the holes are known, additional reinforcing can be placed around the holes to reduce the chance of cross cracking between holes. This method is practiced in a handful of states, including Missouri. The holes are typically filled with sand and topped with joint sealing material to prevent moisture infiltration. If the reinforced approach slabs are cast in single lane widths, mud-jacking could be performed in one lane while permitting traffic in the remaining lane. Thus, maintenance is performed quickly, neatly, and with a minimum of traffic disruption.

5.2.2 *Polyurethane Jacking*. A proprietary method is available to raise slabs in a procedure similar to mud-jacking. This method injects polyurethane foam instead of a cement-based grout. The process was developed to raise floor slabs and, to date, has not been used on pavements. A demonstration project may be necessary to evaluate its effectiveness at raising pavement slabs. This is a proprietary process developed in Denmark, and is distributed by Uretek, USA of Fairfield, Iowa.

5.2.3 *Physical Jacking*. Since mud-jacking is a difficult procedure, best described as an art rather than a science, it may be desirable to make provisions for physical jacking of the approach. Physical jacking would most likely consist of hydraulic rams or mechanical jacks that could be used to temporarily lift a sleeper slab between two consecutive reinforced slabs. Permanent support for the slabs could be provided by undergrouting through precast holes.

It is expected that it would be somewhat easier to train maintenance crews for this type of repair. The major drawback is that the entire roadway width would have to be jacked at once, making it necessary to reroute traffic during maintenance. This assumes that lifting must be accomplished from the exposed edges of the sleeper at the shoulders.

5.2.4 *Pneumatically Adjustable Sleeper.* This method is a variation of physical jacking involving a two-piece precast sleeper that contains two inflatable tubes (Fig. 5). The approach would be adjusted by inflating the tubes to raise the top half of the sleeper and lift the slabs that rest on it. The void space between the top and bottom sleeper sections could then be grouted with a quick setting grout. The reinforced approach slabs should be capable of spanning their entire length without intermediate subgrade support. If further settlement occurs, the approach could be adjusted again by reinflating the tubes.

The advantage of this system is that the approach slabs and sleeper could be built in single lane width, allowing repairs in one lane while permitting traffic in the adjacent lane. Adjustments could be completed quickly and traffic flow restored to normal by using quick setting grouts.

6. Recommendations

Several steps can be taken that will lessen the severity of approach settlement. It is recommended that guidelines for the conduct of geotechnical investigations be developed or modified to incorporate unique aspects of approach embankments. Approach investigations should be conducted for any embankment with over 5 feet of fill or slopes steeper than 4H:1V. These investigations must include estimates of total settlement and time rate of settlement, and evaluation of slope stability.

When a foundation consolidation problem is anticipated, measures should be incorporated to control the amount of post-construction settlement. These measures may include one or more of the following: time delays, surcharging, wick drains, or replacement of unsuitable surface deposits. The use of lightweight fill or temporary paving may also reduce the amount or effect of embankment settlement.

A relatively high degree of embankment stability should be maintained to prevent the occurrence of progressive slope failure. The minimum factor of safety against slope failure, as determined by total stress analysis, should be taken as 2, with 2.5 or greater being preferred. In addition, it is important that existing slopes be benched before placement of embankment fills to reduce the likelihood of sliding failures.

Select granular backfill should continue to be used behind abutments. In addition, it is recommended that standard details be changed to eliminate protrusions and sloped faces from abutment backwalls in order to facilitate compaction of the backfill.

Along with these practices, it is recommended that pre-camber of the approach pavement be incorporated into new construction. The amount of tolerable pre-camber is believed to be approximately 1 to 2 inches for an approach slab length not less than 20 feet. The approach pavement should be built to this higher profile, with the difference taken over the length of the approach slab.

Several possible approach adjustment methods have been identified or developed. These include mud-jacking, polyurethane jacking, and physical and pneumatic jacking of the sleeper supporting the approach slabs. Further research into the technical and economic merits of these methods is recommended through full scale testing.

7. Acknowledgements

Sponsors of this project were the Nebraska Department of Roads (NDOR) and the University of Nebraska-Lincoln (UNL). Many individuals contributed to the project. Special thanks are due to Ken Cheney, Lyman Freemon, Jim Howerton, Moe Jamshidi, Leona Kolbet, Dalyce Ronnau, and George Schmid, all of NDOR, for their support and guidance. The support of Dr. Martha Gilliland, Director of the Center for Infrastructure Research, UNL, and graduate student Mark Flott is also acknowledged.

8. References

Edgar, Thomas V.; Puckett, Jay A.; and D'Spain, Rodney B. (1988) "Effects of Geotextiles on Lateral Pressure and Deformation in Highway Embankments", University of Wyoming.

Hopkins, Tommy C. (1969) "Settlement of Highway Bridge Approaches and Embankment Foundations", Kentucky Department of Highways.

Hopkins, T.C. and Deen, R.C. (1969) "The Bump at the End of the Bridge", Kentucky Department of Highways.

Hopkins, T.C. and Scott, G.D. (1970) "Estimated and Observed Settlements of Bridge Approaches", Highway Research Record, No. 302, 76-86.

Hopkins, Tommy C. (1973) "Settlement of Highway Bridge Approaches and Embankment Foundations", Kentucky Department of Highways.

Hopkins, Tommy C. (1985) "Long-Term Movements of Highway Bridge Approach Embankments and Pavements", University of Kentucky Transportation Research Program.

Stewart, Carl F. (1985) "Highway Structure Approaches", California Department of Transportation.

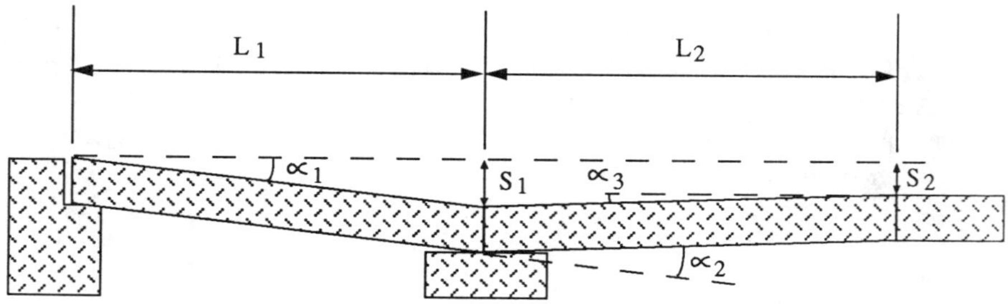

Fig. 1 Angular Variance of Approach Fault.

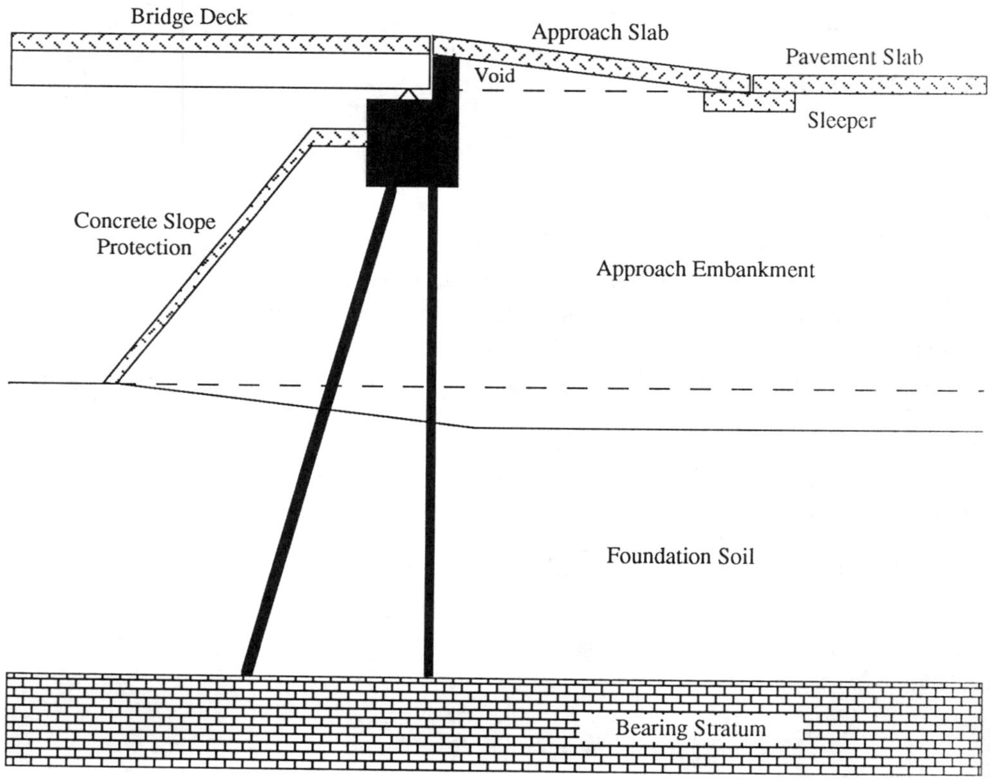

Fig. 2 Settlement Due to Foundation Consolidation.

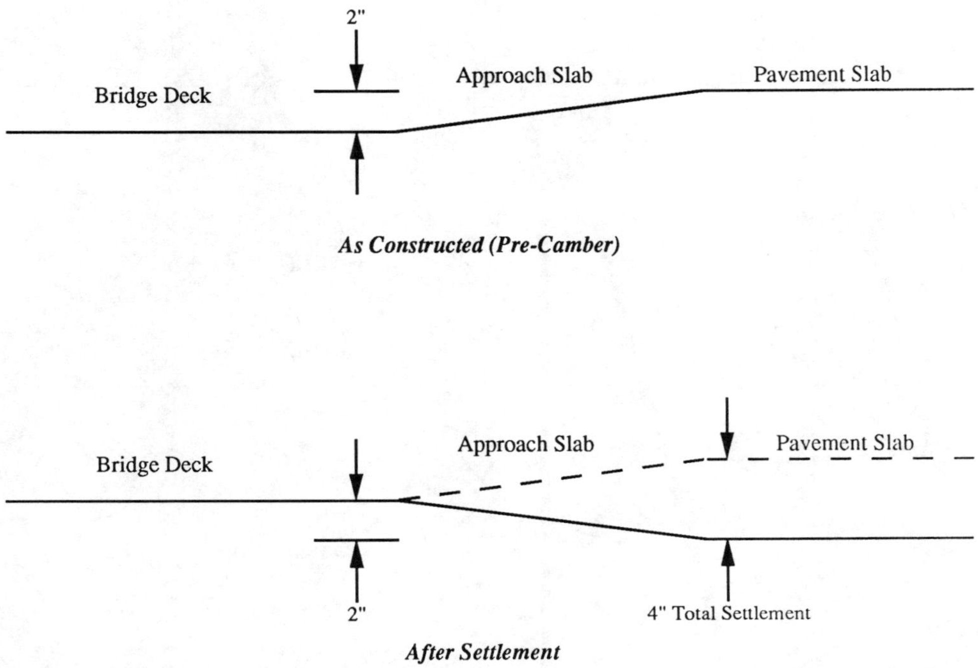

Fig. 3 Pre-Cambered Approach Profile.

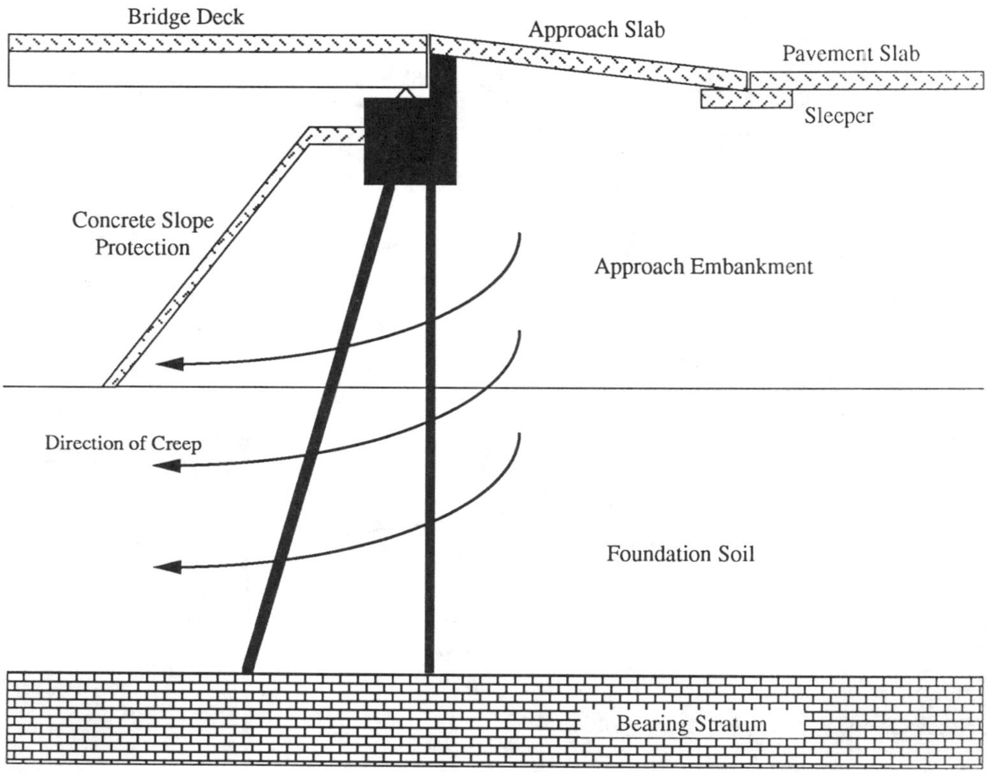

Fig. 4 Progressive Slope Failure (Creep).

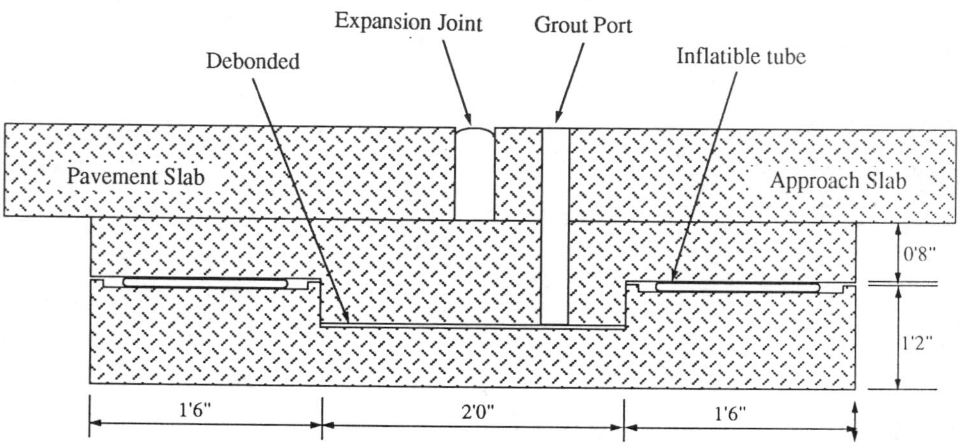

Fig. 5 Pneumotically Adjustable Sleeper.

II - Diagnostics and Monitoring

DIAGNOSTIC PROCEDURES FOR BRIDGES

ANDRZEJ S. NOWAK
Department of Civil Engineering
University of Michigan
Ann Arbor, Michigan 48109 USA

ABSTRACT. There is a need for efficient diagnostic procedures for evaluation of the current capacity of a structure, monitoring of load and resistance history, and evaluation of the accumulated damage. The available diagnostic techniques are reviewed. They include the methods based on various interrogating media: electromagnetic wave, stress wave, electricity, magnetism, and charged particles. Also considered are the approaches to signal processing. Failure detection and isolation procedures developed in other branches of engineering are summarized. The current trends in the development of new techniques are discussed. The areas considered include microwave, acoustic and optical techniques, integrated sensor systems and motion measurements. Currently used bridge load monitoring procedures are also summarized.

1. Introduction

Bridges constitute an important part of the national infrastructure. These assets should be protected. Yet, the degree of deterioration is dangerously high. Major decisions must be made to allocate the limited funds available for maintenance, repair, rehabilitation and replacement. In this process it is important to consider the life-cycle cost of the structure, rather than just the initial cost of design and construction.

The decision process involves prioritization of needs, in particular the selection of structures to be repaired, rehabilitated or replaced, and selection of the repair, rehabilitation or replacement methods. The basis for these decisions should be an evaluation of the actual strength (load carrying capacity) and remaining life of the structure. The major tools are structural analysis, load models and resistance models. However, load and structural resistance are random variables dependent upon the evaluation process which involves a considerable degree of uncertainty. Special techniques are required to limit the uncertainty to an acceptable level.

Diagnostic procedures play a very important role in evaluation of structural performance. They include techniques to identify the critical parts or elements of the structure, identify the cause of distress, monitor structural performance, warn against failure, and provide statistical data for the development of design and evaluation criteria. Traditionally, structural engineers

A. S. Nowak (ed.), Bridge Evaluation, Repair and Rehabilization, 73–84.

used the available methods to monitor performance of tested materials and members in the laboratory.

The need for efficient (accurate, inexpensive, long-lasting, easy-to-install, non-obstructive to occupants or users) monitoring devices and techniques is obvious. They would be used in various types of constructed facilities, including bridges. Additional information about structural resistance or load would allow for a reduction of uncertainty and thus a more rational decision. There is a need for procedures to evaluate the following parameters:

- actual state of the structure (stresses, strains, deflections, cracks, their presence, location, width, permanent deformations),
- current load carrying capacity of the structure; various techniques are available, however, it may be difficult to determine the capacity without damaging the structure, therefore, indirect methods should also be considered,
- accumulated damage (state of corrosion, section loss, changes in material properties, fatigue damage accumulation, deformations),
- remaining life of the structure,
- load history of the members, connections or structures (loads can be monitored on an ongoing basis, incrementally periodic or just when problems arise),
- extreme load events for the statistical data base.

There is also a need to provide a warning against excessive load occurrence (e.g. an extremely heavy truck on the bridge) or against collapse to allow prevention or rapid evacuation.

In the present practice there is a growing separation between various branches of engineering. Narrow specialization often prevents a communication and exchange of experience. Recent dynamic advancements in electrical engineering and in other branches of engineering can be used as a basis to develop a new generation of efficient diagnostic procedures. In particular, the new areas include electronic sensors, (measurements using laser beam, radar, microwave and acoustic techniques) and also new approaches to signal processing, in-field instrumentation, for example. A review of the available methodologies from the point of view of application to bridges is needed. Some of these techniques are readily available, others need modifications or improvements. Still other procedures may result from combining approaches and ideas.

The purpose of diagnostic procedures is to evaluate the existing state of the structure, establish causes of problems and select the remedy. The procedures include analytical methods, nondestructive tests, other tests, and combinations of analytical methods with experimental approaches. Structural analysis provides efficient tools for calculation of forces and deformations. However, there is usually a considerable degree of uncertainty associated with load and resistance parameters. Particularly important is the knowledge of defects or weak links in the structure. The accuracy of information about the actual material properties, defects, load sharing and bridge performance in general can be improved by special field or laboratory observations and tests.

The diagnostic procedures were discussed at several recent workshops and conferences, including NSF/CEBTP Workshop on Bridges in Saint-Remy-Les-Chevreuse (Nowak and Absi 1987), NSF Workshop on Nondestructive Evaluation for Performance of Civil Structures"

(Agbabian and Masri 1988), and NSF Workshop on Sensors and Measurement Techniques for Assessing Structural Performance (Marshall 1989).

2. Present Nondestructive Techniques

The available nondestructive techniques are well summarized in the report by Clifton et al. (1982), Collacott (1985) and in the NCHRP report (NCHRP Project 20-5, Topic 15-03, 1985). In general, the defects are detected by observation of changes in the response of the interrogating medium. Interrogating media in nondestructive tests include electromagnetic wave, stress wave, electricity, magnetism, and charged particles. The methods use different wave lengths or frequencies of vibration.

In nondestructive testing the practical range of wavelengths is from about 300 mm (microwaves) to less than 10^{-14} mm (gamma rays). Microwaves are mostly used to determine the moisture content of porous materials. Lasers, with wavelengths in the visible and infrared ranges, in structural engineering are mostly used in precision measuring devices. X-rays and gamma rays make up the shortest wavelength and have the greatest penetrating power. Thousands of X-ray radiographs taken from all sides of an object can be used to examine various properties of engineering materials. This new technique is called a computed tomography.

The stress wave spectrum covers a frequency range from hypersonic (10^{12} Hz) to sub-audible (0.01 Hz). Stress waves propagate through solid materials. Flaws or discontinuities in a material disturb the passage of the waves and therefore can be detected. There are two groups of sonic testing techniques: resonant and ultrasonic pulse measurement methods. The first one involves the measurement of the natural frequencies of vibration of a component. Ultrasonic testing involves the application of high-frequency sound waves (above the hearing range, 16 kHz).

Other interrogating media include electricity, magnetism and charged particles. Electrical resistance measurements are most often associated with evaluation of moisture content (in concrete and timber). Magnetic methods are used primarily for examination of steel structures.

Selection of a testing method depends on the property, structural material and cost. The most common forms of defects and deterioration that occur in concrete components include cracking, scaling of surface mortar, corrosion of reinforcement (ordinary rebars or prestressing strands), honeycombing and air pockets, popouts (usually frost-susceptible aggregates), surface deposits, chemical attack (acids), wear and erosion. On the other hand, in steel structures the most common forms of deterioration include corrosion (uniform, pitting, galvanic, stress or fatigue), cracking and bent or twisted members. The presence of defects and extent of deterioration can be determined using some of the available testing procedures, either in the field or in the laboratory.

Visual inspection often provides a good overview of the condition of the structure. In particular, it allows identification of the presence of cracks, corrosion or surface deposits. Special techniques can be used for inspection of inaccessible areas (borescope or periscope). However, visual inspection is inefficient in case of many serious defects such as corrosion of reinforcing and prestressing steel, pitting corrosion, fatigue damage accumulation, structural cracking and others.

Hardness of concrete and its strength can be measured by rebound and penetration methods (Swiss hammer and Windsor probe). Other ways to determine the modulus of elasticity (and strength) include sonic and ultrasonic pulse-velocity methods. These methods are based on the measurement of travel time of waves; in sonic methods the waves are generated by a single hammer blow or by repetitive blows, and in ultrasonic methods they are generated by electronically induced mechanical pulses.

Magnetic methods are mostly used to determine the position of reinforcement (cover meters or pachometers). Electrical methods are used to determine moisture content and corrosion rate (through resistance and potential measurements). Chemical methods are applied to evaluate the depth of carbonation and chloride ion content. Concrete delamination can be checked using infrared thermography, by measuring differences in surface temperatures. Radar can be used to detect deterioration in concrete bridge decks. There are also two new techniques available, radiography and radiometry. In radiography the radiation is detected by a photographic emulsion (film), and in radiometry, variations in the gamma intensity are detected by radiation detectors such as Geiger or scintillation counters.

Some defects on the surface of the structure can be detected by dye-penetrant method, magnetic particles (particles align themselves in the magnetic field following the outline of the crack) or eddy currents (defect is detected by perturbation in an electrical field).

There is a need to improve the efficiency of the available methods by increasing their accuracy, further simplification of use and reduction of cost.

3. Motion Measurement Techniques

There exist today many transducers which are appropriate for monitoring structural motion. These transducers usually measure particle velocity or acceleration and come in frequency ranges covering the entire span necessary to appropriately monitor structures, including foundations and free-field ground motion. The selection of specific transducers for monitoring structures depends on the environment into which the transducer will be installed. For example, locations where high degrees of moisture are present will call for hermetically sealed units and will require special treatment of cabling to ensure continued operation.

In some locations vandalism may be a problem for cables and for the transducers themselves. It is feasible with current technology to make telemetric transducers which would not require cabling thereby reducing vandalism and also allowing better concealment of transducers.

Tracking of deterioration of structural components is another goal of diagnostic procedures. Non-destructive techniques are essential if comparisons are to be made between competent structures and deteriorating structures over significant durations. One technique which has been developed significantly over the past few years is that of Spectral-Analysis-of-Surface-Waves (SASW). With two transducers, a dual channel FFT wave form analyzer, and a small hammer, the modulus and continuity (cracking) of structural members can be evaluated throughout the member by taps at the surface (Nazarian 1984, Hiltunen and Woods 1988). This technique is particularly valuable for checking concrete slabs, beams and columns, and for evaluation of concrete foundations (piles and piers).

Other seismic wave propagation tests are also available for non-destructive testing of long members like columns, piles and piers. These methods should be evaluated and tested to determine their applications in bridge evaluation (Olson and Thompson 1985).

An important application of the motion measurement techniques is monitoring of truck loads. Actual truck load is needed for evaluation of the structural performance and prediction of the remaining life. The measurements can be done in motion (WIM, weigh-in-motion) or using stationary scales. An example of the WIM system developed by the Bridge Weighing System is shown in Fig. 1. Tape switches are used to trigger the process. Measured parameters include the speed, axle spacing, truck configuration (type) and axle loads. The equipment is calibrated using a vehicle with known axle weights. Strains are measured (strain gages). Readings are taken in all the girders simultaneously. The results can be used to develop the live load model or to verify the load distribution properties. Other devices can also be used, including accelerometers or deflection gages.

4. Microwave, Acoustic and Optical Techniques

Recent developments in microwave technology and signal processing chips have led to a new generation of vector network analyzers capable of operating at high microwave frequencies. This has, in turn, opened the door to the use of wide-bandwidth radar techniques to obtain high range-resolution capabilities in probing through low-loss media, such as concrete. One of the revolutionary attributes of the new vector network analyzers is that it is now possible to "program" the desired radar design configuration with software, in contrast to having to do it with hardware, which was the case only several years ago. This allows the user to determine the optimum design configuration for a given problem prior to constructing a low-cost hardware design that achieves the desired goal and performance.

Three basic approaches are considered: (a) Microwave techniques, (b) Acoustic techniques, and (c) Optical techniques.

Since microwave energy can penetrate concrete with relative ease, high resolution radar systems, using available vector network analyzers, can be used to detect cracks and metal reinforcements in the concrete. The detection of the oxidation processes is possible because they change the dielectric properties of a metal target.

With microwaves, the following systems approaches can be considered:

1. Monostatic radar mode (transmitter and receiver co-located and using the same antenna).
2. Bistatic radar mode (transmitter and receiver placed at different locations). This includes transmission through as well as reflection at an azimuth angle.
3. Holographic techniques, wherein a transmitter and several receivers are used in order to construct a three-dimensional image of the target.

The bistatic and monostratic acoustic soundings can be used alone and in conjunction with the microwave systems. Hydrophone-type experiments (similar to geophysical testing) can be evaluated to determine how well abnormal discontinuities in the medium can be detected. Interferometric (optical) techniques have proven quite useful in detecting stress points and fractures in tires and materials of known geometry and shape. By Fourier transforming an interference pattern and then applying appropriate filtering in the spatial frequency domain, it may be possible to detect weak strains on concrete and metal structures.

5. Integrated Sensor Systems

Integrated sensor systems have demonstrated the ability to provide an interface between many different types of signals (acoustical, mechanical, thermal, magnetic, etc.) and the electronic instrumentation usually used in analysis and control. In addition, the compatibility of many of the transducer fabrication processes with conventional VLSI (very large scale integration) technology provides the means for direct communication between analysis and control systems, often with local signal conditioning (amplification, temperature compensation, digital/analog conversion, etc.) (Wise 1982). These factors often combine to make such integrated sensor systems a viable alternative for many applications.

Several types of sensors may be applicable to the problems of monitoring the condition of large structures such as bridges. Sensors that measure mechanical inputs such as force, position, velocity, and acceleration have been demonstrated. In addition, integrated circuits that respond to the different types of radiant energy currently used for diagnostic purposes (electromagnetic image sensors, ultrasound receivers, etc.) may be able to extend the sensitivity and accuracy of these techniques.

The ability to directly measure the motion of critical sections could have significant impact on both design and monitoring of large constructed facilities. In particular, a set of velocity transducers or accelerometers, located at critical points, could provide information on the response of the structure to various types of loads. Data obtained from such measurements on models could be used to set standards and specifications for use in design and construction. Devices mounted on actual bridges, used to monitor the structure over time, could provide information on how the response of critical parts changes as the structure ages. A set of these sensors (combined with a special-purpose integrated circuit for interfacing) would share a simple serial communication network linked to a computer for data analysis. A recent review cited measurement ranges from 10 μg to 200,000 g, with up to 10^7 dynamic range (Motamedi 1987).

The efficiency of each procedure or technique must be evaluated. This can be accomplished using several criteria including accuracy, effort, cost and duration. The most efficient methods should be adjusted to bridges. Combinations of different techniques may be also considered.

Michigan DOT has already implemented a temperature monitoring system on several state-owned bridges. Thermometers are installed in the pavement. The temperature reading is transmitted automatically by phone or radio to the computer in the MDOT headquarters in Lansing, and it is displayed on the monitor. The same line of communication can be used to transmit other information (stress, strain, deflection and so on).

Many commercially available sensors can be applied to modeling and diagnostics for large constructed facilities. It is anticipated that a small number of special-purpose VLSI chips will be designed for interfacing the sensors to data analysis computers.

6. Signal Processing

It is known that abnormalities present in a structure, such as fatigue fractures, breakdown of material, change in material properties or permanent deformation, will cause changes in the flow of acoustical and vibrational energy through the structure. The detection of such phenomena may be carried out by using various sensors, including accelerometers, velocity transducers, displacement transducers and other means discussed in this proposal. While the problem of instrumenting adequate sensors is relatively well understood, techniques for interpreting the available data in a large structure are still lacking, and present some interesting theoretical questions.

For example, the type of input (excitation) to the system to be encountered in the case of a bridge is relatively uncontrolled (e.g., heavy load passing over the bridge) and will excite a number of vibrational modes. Principal Components Analysis (PCA) is a signal processing technique which can decompose multi-modal excitations and the subsequent response of the system dynamics. Pattern recognition and feature extraction algorithms can also be applied to the study of structure-borne vibration in order to identify unusual patterns in the flow of energy through the structure.

As an aid in the implementation of these, and other potentially available digital signal processing methods, current VLSI (very large scale integration) technology offers programmable digital signal processing IC's (integrated circuits), which can be configured for a specific on-site task, and may be interfaced through a distributed network, to streamline and localize the task of processing data from multiple locations, with inputs varying in spatial location and in time.

Transient vibrational signals can be analyzed using the exponential distribution. This new time-frequency distribution (Choi 1987), allows for the energy in a signal to be represented with high resolution in a conjoint time-frequency distribution. It is expected to reveal exquisite details of vibrational signal structure. This new transform may allow cataloging of detailed signatures of various bridge responses and use of these signatures in an information theoretic pattern recognition scheme (Williams, Shevrin and Marshall 1987; Choi, 1987). Signatures of various fault conditions can be identified from experimental data and utilized to construct an automated computer based fault classifier.

The potential signals of interest can be evaluated using exponential distribution. The obtained time-frequency distribution of energy can be presented graphically in terms of intensity, contour or 3-D plots. Distinctive features or signatures in these plots, which are indicative of faults or classes of faults, can be used to build up a catalog of fault features. This catalog of features may be obtained by visual inspection of the results or by methods of statistic pattern recognition or both. Then a pattern recognition scheme can be constructed to classify groups of faults and detect where and when these patterns occur. Classifier programs can be trained on a set of known faults and subjected to a novel set of faults to test the algorithms.

There is a need for fast algorithms to compute the exponential distribution. In addition, pattern recognizing (matched) filters need to be derived for this new transform.

To recognize patterns in the signatures of various failure modes, measurements from sections of the structure can be obtained. Some of these could be simulated in a scale model.

7. Failure Detection and Isolation

Recent research has developed procedures for detecting incipient failures in dynamic systems. It has been shown that any change in the dynamics of the system can be detected using electronic signal processing in the form of a so-called failure detection and isolation (FDI) filter. The FDI can be implemented in a computer which receives inputs from a set of sensors installed on the dynamic system under study. Using specially developed algorithms and procedures, the FDI determines changes in the system dynamics which are related to the parameters in the dynamic model for the system under study.

There are procedures available for optimizing the FDI design. It can be done, for example, with respect to uncertainties in the mathematical model for the dynamic system. In certain cases only relatively crude estimates are required for some model parameters to yield a practical functioning FDI system.

The parameter changes in the dynamic system, as detected by the FDI system, can be used to assess the performance and the reliability of the dynamic system. For example, gradual changes in any parameters of the dynamic system can be used to predict failure in the later system.

The FDI concept has great potential applicability for assessing the structural reliability of bridges or other large structures. A bridge is a dynamic system which can be characterized by a mathematical model in a form which is readily incorporated in an FDI system. The parameters of this model are directly related to the load bearing capabilities of the various members of the structure. Any deterioration in a structural member due, for example, to fatigue or corrosion would be directly detected by the FDI.

The FDI system itself is potentially inexpensive and reliable. The sensors required for the FDI are deformation sensors (e.g. strain gauges) which are inexpensive and readily applied. The FDI can be implemented in a small microprocessor-based special purpose computer. Considerable self checking can be performed by this computer to access accuracy of sensor data and the signal processing procedures themselves.

The work on the failure detection method can be highly focused towards applying the existing Failure Detection and Isolation (FDI) theory (for example Min 1987) to detecting variations in the structural integrity of large dynamic structures, such as bridges. This theory already has a proven track record of detecting incipient failures in components of electronically controlled engines. Although there are obvious differences between internal combustion engines and constructed facilities, there are also many similarities between them with regard to failure detection. Among the similarities, it can be noted that both are nonlinear dynamic systems, and that a linearized state space model must be constructed for the bridge which is similar in form to the model for the engine, though of larger dimension. In addition, the degradation in performance of an engine is very analogous to a reduction in bridge structural integrity. In both cases the degradation is reflected in the model by way of parameter changes. The robustness of the model will be an issue in the applicability of FDI to detecting degradation in the bridge structure integrity.

The feasibility of applying FDI can be initially explored using simulations. Toward this end a relatively simple model can be formulated, for example a simple truss bridge structure incorporating pinned joints and elastic members. The simulation phase will then progress toward a more realistic model for a bridge incorporating such features as plastic deformation, fatigue, etc. Finally, the bridge model will be representative of a relatively simple geometric structure involving a small number of members.

The long term goal is to develop a formal mathematical structure and procedures for applying FDI theory to arbitrarily complex bridge structures. The simulation studies can be verified by tests of a scale model of a simple truss bridge, implementing the failure detection filters in a microprocessor. Structural faults can be introduced in the experimental model to evaluate the FDI performance experimentally. Finally the FDI system can be implemented on an actual bridge.

8. Conclusions

There is need for the development of efficient diagnostic procedures for evaluation of bridges. The procedures which include field data measurements, data management and evaluation, and diagnostic decisions, will be applied to evaluation of the state of the structure, development of load and response history and evaluation of the accumulated damage.

Available techniques from various branches of engineering are reviewed with regard to their use in structural engineering. In particular, the review covers signal processing, electronic sensors, microwave and acoustic methods, optical and laser techniques. The most efficient data acquisition systems, data transfer and storage, as well as data processing procedures must also be identified.

9. References

Agbabian, M.S. and Masri, S.F., ed. (1988), "Nondestructive Evaluation for Performance of Civil Structures", Proceedings of NSF Workshop, Los Angeles, CA.

Brophy, J.W. (1988), "Automated Imaging System for Bridge Inspection", Report FHWA/RD-87/090, Federal Highway Administration, Washington, DC.

Choi, H. I. (1987), "Pattern Analysis for Nonstationary Signals: Representation and Feature Selection Methods," Ph.D. Dissertation in Computer, Information and Control Engineering, University of Michigan, W.J. Williams, Chairman.

Clifton, J.R., Carino, N.J. and Howdyshell, P. (1982), "In-Place Nondestructive Evaluation Methods for Quality Assurance of Building Materials," Technical Report M-305, Construction Engineering Research Laboratory, US Army Corps of Engineers, CERL, March.

Collacott, R.A. (1985), "Structural Integrity Monitoring", Chapman and Hall, New York. Crist, R.A. and Kechter, G.E. (1987), "In-Service Measurements of Structural Performance", Buildings, ASCE Structures Congress, Orlando, pp. 258-274.

Hiltunen, D.R. and Woods, R.D. (1988), "Influence of Source and Receiver Geometry on the Testing of Pavements by the Surface Waves Method", Proceedings of the First International Symposium on Non-destructive Testing of Pavements and Backcalculation of Moduli, ASTM, Baltimore, MD, June.

Limaye, H.S. (1988), "Laboratory and Field Testing of Concrete Using Impact-Echo Technique and an FFT Analyzer", International NSF Workshop on Nondestructive Evaluation for Performance of Civil Structures, San Diego.

Manning, D., and Holt, F. B. (1984), "Detecting Deterioration in Asphalt Covered Bridge Decks," Transportation Research Record No. 899.

Marshall, R.D., ed. (1989), "Sensors and Measurement Techniques for Assessing Structural Performance", Proceedings of the NSF International Workshop, Gaithersburg, MD.

Maser, K.R. (1986), "Analysis of Concrete Deterioration Using Ground Penetrating Radar," MIT, Center for Construction Research and Education, Report CCRE 86-5.

Miller, R.K., Ringermacher, H.I., Williams, R.S., and Zwicke, P.E. (1984), "Characterization of Acoustic Emission Signals," Report No. FHWA/RD-84/025, Federal Highway Administration, Washington, D.C.

Min, P. (1987), "Detection of Incipient Failures in Dynamic Systems," Ph.D. Dissertation, (Chairman W.B. Ribbens) Department of Electrical Engineering and Computer Science, The University of Michigan, Ann Arbor.

Motamedi, M. E. (1987), "Acoustic Accelerometers," IEEE Trans. Ultrasonics, Ferroelectrics, and Frequency Control, Vol. UFFC-34, pp. 237-242.

Nazarian, S. (1984), "In Situ Determination of Elastic Moduli of Soil Deposits and Pavement Systems by Spectral-Analysis-of-Surface Waves Method", Ph.D. Dissertation, University of Texas, Austin, 453 p.

Nowak, A.S. and Absi, E., ed. (1987), "US-European Workshop on Bridge Evaluation, Repair and Rehabilitation", Proceedings, University of Michigan, Ann Arbor.

Nowak, A.S. et al. (1988), "Risk Analysis for Bridge Evaluation", Report UMCE 88-7, University of Michigan, Ann Arbor, October.

Olson, L.D. and Thompson, R.W. (1985), "Case Histories - Evaluation of Drilled Pier Integrity by the Stress Wave Propagation Methods," Proceedings of the Specialty Session on Drilled Piers and Caissons, II, ASCE, Denver, May, pp. 28-42.

Tharmabala, T. and Nowak, A.S. (1988), "Bridge Reliability Evaluation Using Load Tests", ASCE Journal of Structural Engineering, Vol. 114, No. 10, October, pp. 2268-2279.

Transportation Research Board (1985), NCHRP Synthesis 118: Detecting Defects and Deterioration in Highway Structures, National Research Council, Washington, D.C., 52 pp.

Williams, W. J., Shevrin H. and Marshall, R. E. (1987), "Information Modelling and Analysis of Event Related Potentials," IEEE Transactions on Biomedical Engineering, Vol. BME-34, No. 2, December, pp. 13-22.

Wise, K. D. (1982), "Integrated Sensors: Interfacing Electronics to a Non-Electrical World," Sensors and Actuators, Vol. 2, pp. 229-237.

Woodward, R. J. (1984), "Inspecting Concrete Bridges," Phys. Bull., Vol. 35, p. 149.

Zoldners, N. G. and Soles, J. A. (1984), "An Annotated Bibliography on NDT of Concrete, 1975-1983," In Situ/Nondestructive Testing of Concrete, V. M. Malhotra, ed., American Concrete Institute (ACI) Publication SP-82, p. 745.

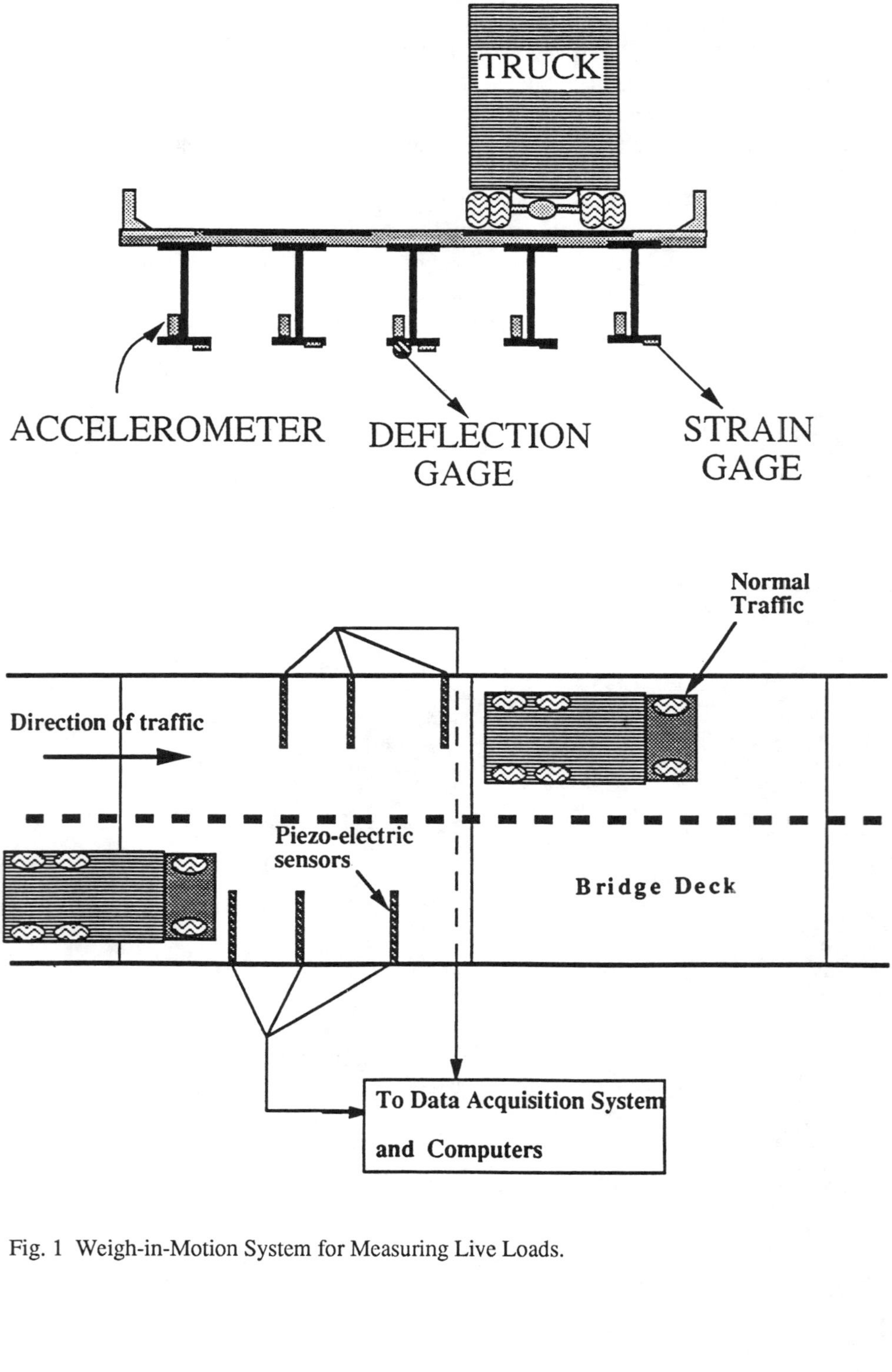

Fig. 1 Weigh-in-Motion System for Measuring Live Loads.

LONG-TERM MONITORING OF BRIDGE DEFORMATION

R. FAVRE and I. MARKEY
Institute of Reinforced and Prestressed Concrete
Department of Civil Engineering
Swiss Federal Institute of Technology
Ch-1015 Lausanne
SWITZERLAND

ABSTRACT. Some concrete bridges continue to deform even after the life normally attributed to creep. The causes and consequences of this are discussed and a possible method of calculation is proposed. The effect of the level of prestressing on the development of cracks and deflections is also mentioned. In an effort to measure these deformations and to create a possible early warning system, a hydrostatic levelling system was developed. Details of this system and actual results obtained from an excessively deformed and now repaired bridge are presented.

1. Introduction

It is evident that steel construction, in as far as it is correctly designed and calculated, has a completely elastic behaviour, i.e., deformations are entirely reversible. Deformations due to permanent loading do not increase with time. Those due to live loading such as traffic or temperature variations disappear once the action has finished.

However, it is completely different for concrete structures. Deformations due to permanent loading increase due to concrete creep and shrinkage for 5 to 10 years after the completion of the structure. Progressive cracking also occurs during this period. In fact, the more massive the concrete structure, the longer these time dependent effects occur. It is believed that the creep in concrete dams has not finished even after 30 years of service.

In order to satisfy the owner and the user of a bridge the deformations must remain limited. For example, the new Swiss Code SIA 160 limits the deflections in the following manner:

$$a_{per} \leq \ell/700$$
$$a_{inst} \leq \ell/600$$

a_{per}: long-term deflection due to the dead loading.
a_{inst}: instantaneous deflection due to the live loading.

A. S. Nowak (ed.), Bridge Evaluation, Repair and Rehabilization, 85–100.

For concrete bridges, the second condition is usually fulfilled. The first condition which is concerned with long-term deflections is not always fulfilled when the loads balanced by the prestressing are only a small fraction of the permanent dead loads or evidently if there is no prestressing at all.

Recall that a_{per} cannot be eliminated in a satisfactory manner by a camber in the construction because the evolution of the time dependent effects takes 5 to 10 years. Also note that a_{per} not only has a negative effect on the aesthetic aspect of a bridge but also on the security of the traffic (e.g. drainage of the rain water, shocks at the expansion joints). It is especially at hinges or at the simply supported ends of a bridge that the deformations provoke major inconveniences in the form of tangential discontinuities (α) of the deformed line of the structure, Figures 1 and 2.

In order to assure a high standard of concrete bridges, it appears prudent to respect generously the condition $a_{per} \leq \ell/700$. The fact that the prediction of a deflection is difficult due to the uncertainty of the actual values of many parameters, creep, relaxation, cracking and etc., encourages pessimistic assumptions in the selection of these values.

A considerable part of the long-term deformation of a bridge appears to come from the gradual propagation and evolution of cracks due to repetitious live loading, traffic loads, thermal gradients provoked by the sun, differential shrinkage due to segmental construction techniques, etc. It is being noticed in Switzerland that several prestressed bridges have ever increasing deflections after 5 years of service. These bridges require reinforcement in the form of additional prestressing.

2. The Prediction of the Long-Term Deformation of a R.C. or P.C. Bridge

There are several different methods to calculate the long-term deflection of a R.C. structure. For prestressed bridges we generally assume a State I behaviour (no cracking) in order to avoid major calculation problems.

The reality however is much more complicated. First of all, we must be aware that prestressed bridges can have cracks. This could be a result of partial prestressing where only a fraction of the dead loads are balanced. Similarly, step by step cracking could occur as a result of live loads, temperature gradients, etc., of which the effect is not entirely reversible.

To calculate the long-term deflection as best as possible, we must apply the appropriate non-linear moment-curvature relationships. Such formulae are given in the new CEB Model Code 1990 [1]. They take into account the time dependent effects of concrete (creep and shrinkage) and cracking. They define an intermediary or average curvature situated between State I (no cracking) and State II (cracking, without tension stiffening effect) [2]. These formulae are represented in Figure 3. When reloading/unloading occurs the relationship indicated in Figure 4 can be adopted.

Let us take for an example a double T-beam as represented in Figure 5. We have assumed a uniformly distributed dead load of 290 kN/m', 260 kN/m' is due to the self weight and the black top while 30 kN/m' represents the permanent live loading.

We varied the prestressing force from 0 to 30,000 kN. The amount of reinforcement was chosen is such a way that when combined with the prestressing the same ultimate resistance of the section was obtained using the theory of plasticity.

It is obvious from Figure 6 that a prestressing force of approximately 20,000 kN is required if the bridge is to have a State I behaviour (no cracking). This level of prestressing corresponds to a balanced load of 200 kN/m'. As the prestressing force decreases, the long-term deflection increases. But that which appears to be determinate in the choice of the level of prestressing is the relationship (a_{exp}/a_I).

Where

a_{exp}: expected long-term deflection taking into account the effect of cracking.
a_I: theoretical long-term deflection assuming an uncracked behaviour.

The importance of the gradual crack formation with time increases as this ratio increases. When this relationship is > 1 there is a risk that the development of cracks will continue beyond that of the evolution of creep. It is clearly evident that this bridge could not be constructed without prestressing. On the other hand, the prestressing can be used greatly to our benefit when $a_{exp}/a_I = 1$, approximately, P = 20,000 kN. To increase the standard of the structure a prestressing force of P = 29,000kN, which would balance all the permanent dead loading, could be foreseen. Recall that P is considered as the average prestressing force $P = (P_0+P_\infty)/2$ where $P_0 - P_\infty$ represents the long-term losses.

3. Some Long-Term Observations by Means of a Hydrostatic Levelling System

Hydrostatics is the study of the mechanical properties of liquids at rest. The knowledge of the laws of a hydrostatics goes back to antiquity. Archimedes of Syracuse (287-212 B.C.) started the study which was more or less completed by Pascal and Stevin in the 17th century. As for the principles of a hydrostatic levelling system, these are based on the classic laws of communicating vessels, Figure 7. These imply that the interconnected surfaces of a fluid all lie in the same horizontal plane, provided that all things are equal throughout the system. The principal parameters affecting this homogeneous system are [4]:

- gravity
- atmospheric pressure
- temperature
- variation of density
- zero flow, i.e. no loss in the elevation head due to kinetic movement.

The hydrostatic levelling system is ideally suited as a long-term monitoring system for box girder bridges, regardless of the curvature or gradient of the bridge, Figure 8. The system has the advantage that it is relatively inexpensive in comparison to other methods; it does not interfere with the traffic; measurements can be taken by trained technical staff rather than by professionals; measurements are quickly taken, 5-10 minutes per 100 m length of bridge; traffic movements (± 10 mm) do not influence the system as it is critically damped. The system can be continuously maintained in working order or can be rendered operational after a gap of a few years.

The system consists of graduated pots which are interconnected by transparent P.V.C. tubing. The pots are comprised of a metal stand with inclined funnels to receive the P.V.C. tubing and a graduated glass tube, Figure 9. These pots are mounted inside the box girder of a bridge, nominally over the supports and at mid- and quatre span. Several pots linked together form a circuit and the sum of all the circuits makes up the system which covers the entire length of the bridge. Circuits are spliced over the support or when obliged to by the geometry of the bridge. Evaporation or leakages do not create errors in the system unless they are very rapid. However, several factors can cause errors in the system. These can be classified into three categories; homogeneity of the liquid, homogeneity of the climatic conditions surrounding the system and human error.

The greatest errors are due to the homogeneity or rather the lack of homogeneity of the liquid. This results in a variation in the density of the liquid which automatically creates different hydrostatic heads and thus errors. This occurs very easily in the case of a mixed liquid as inaccurate mixing proportions and differential evaporation can quickly lead to a 5% variation in the original mixing quantities. For the case of a mixture of isopropanol and water, a 5% variation yields a 12 mm error per metre of hydrostatic head of the original mixture [4]. Thus the use of the pure liquid is advised. In addition, the liquid must be non-reactive as this causes impurities and density variations in the system.

In a similar way, non-uniform climatic conditions can create errors, Figure 10. Temperature gradients within the liquid can easily be set up by exposing the tubing to outside conditions i.e. at joints. The magnitude of these errors for water can be seen in Figure 11. Differences in atmospheric pressure can also cause errors. In fact, for an idealized case, a level head error of 26 mm could occur for a 20 m/s wind with a water filled system. Insulation against these factors is required. Finally, the human error factor is present as with all systems. Fortunately, systematic errors of observation are eliminated in calculating the deflection as they cancel out, Figure 12. However, if the system is not properly inspected before an observation, air bubbles which falsify the equipotential line can easily occur.

The system currently in use on several bridges in Switzerland uses pure demineralized water. Ispropanol was also used as it does not normally freeze, but only on a short-term basis in the winter as it degrades with time and reacts with the P.V.C. tubing. As water is used, observations have to be forsaken during the winter period, December - February, and a rubber mousse like tube is inserted into the pot to prevent the glass from breaking when the water freezes, Figure 9. The water dilates inwards towards the mousse when freezing, as the mousse has almost no resistance. When the system is correctly installed and operated the error in the deflection obtained is roughly ± 1 mm per circuit. Thus as the dimensions and geometry of a bridge grow, the number of circuits required increases. This subsequently increases the error in the system. A case study follows which shows the hydrostatic levelling system as possibly the most practical solution for monitoring a bridge during and after an additional prestressing operation.

3.1 CASE STUDY: THE LUTRIVE BRIDGE

The current hydrostatic levelling system was installed in the Lutrive bridge in December 1988. As it was necessary to have the system operational in winter, demineralized water could not be used. After the bad experiences with previous mixtures (water and alcohol) a pure liquid was sought and isopropanol (C_3H_7OH) was selected.

The Lutrive bridge is in fact two parallel twin bridges, each one containing one side of the N9, National Swiss Motorway, just outside Lausanne. Built in 1971 and 1973, the bridges cross, with central spans of 112 m and 132 m, a zone which is susceptible to a gradual land movement and two streams, the Lutrive and Macheret. The superstructure cross-section consists of a box girder of variable height and two slightly dissymmetric cantilevers, built this way to reduce the torsion in the gently curved bridges. The bridges are divided into 3 quasi-isostatic sections by 2 hinges, both of which are situated at mid-span in the central spans at the bridge deck level, Figures 10 and 13. The central spans, 112m and 132 m, underwent approximate deflections of 14 cm and 16 cm respectively. This however does not include the initial pre-camber of each span which was never actually measured but was estimated at 11 cm. The deflection was a continuous downward movement which showed no sign of stopping, Figure 14. The first dotted line represents the deflection with respect to the theoretical construction datum while the second dotted line is the assumed deflection which occurred while changing the measurement system from the traditional optical level to the hydrostatic levelling system. Note the difference in the density of the measurements taken between the two systems. This second dotted line will later be checked by measuring with both systems on the same day. The camber of the bridge is not included in these values. Extensive cracking, 0.2-0.4 mm wide inclined at 30-45° on the web could be seen.

Even though these deformations appear quite considerable, they are still relatively small in comparison with the scale of the bridge. In addition to this it must be stated that the bridge is not in any danger of collapse as the limits of the structural security are not approached. However, these deformations do concern the serviceability limit states and the public confidence in the bridge as the deflections are visible to the naked eye. Thus, it was decided to "repair" the bridge with an external additional prestressing force of roughly 1300 tonnes situated inside the box girder. The purpose of this operation is threefold:

- increase the structural security of the bridge.
- eliminate some of the obvious deflection which has already occurred.
- stop the downward deflection trend of the bridge.

The calculated instantaneous upward deflection of the bridge due to the prestress would be on the order of 50 mm for the larger of the two spans. The design engineer hopes that this value will increase with time as the action of creep will now be acting in the upward direction since all gravitational forces are equilibrated and exceeded. Relaxation will be very small as the level of stress is not very high. However, since the concrete is already old (16 years) the effect of creep will be minimal.

The additional prestressing was effectuated by using four cables per bridge, two on either side of the box girder parallel to the web, over the entire length of the bridge, 400 m. Each cable was split into three small cables at the abutments to allow for jacking as the space there was confined. Massive anchorage blocks were concreted and prestress bolted to the bridge structure at the abutments. The upward movement due to the prestressing force was conveyed to the bridge due to its varying sectional inertia permitting the use of a linear prestress cable pattern, parallel to the bridge deck, obtained by the strategically placed deviators. Figure 15 shows the cables and the deviators. In all 33 pots for 25 measurement points are used for each bridge. These are principally located over the supports, at the hinges and at mid-and quatre span.

The results of the prestressing operation for the two bridges are similar. The cables were tensioned in several stages over two days (the confined spaced required the cables to be cut every 18 cm) and measurements with the hydrostatic system were taken a total of seven times. Each measurement took 15 minutes, which was approximately equal to the time required to cut the cables. The bridge rose up in a regular manner by values which were, in total, 10% less than those calculated. Further measurements were taken after one week and subsequently at monthly intervals for five months.

During the prestressing operation a secondary control system was used to verify the results obtained by the hydrostatic system. This system, the deflection gauge method, gave similar results and is frequently used for the load testing of structures. It consists of an invar or steel wire attached to the structure and to the ground. Direct contact is made at one surface and at the other, a spring connects the wire to the surface (ground or structure). This allows the bridge to move without too much stress occurring in the wire. A pointer is attached to the wire which presses on the deflection gauge. Taking note of temperature differences during the operation and the length of the wire, the deflection of the structure can be calculated.

As can be seen in Figures 14, 16 and 17, the Lutrive bridge was subjected to a 24 hour observation. This occurred in May 1989 when the air temperature stayed roughly constant, 19°C, and the day was relatively sunny. Figure 16 is particularly interesting in that it shows the magnitude of the daily movement of the bridge and also indicates that the system is working correctly by its cyclic behavior. The deflections of the four hinges after one year indicate a stoppage in the downward deflection trend but the long-term effect of the additional prestressing operation remains to be seen.

The observations using the hydrostatic levelling system will continue to be taken three times a year on the bridges over the Lutrive and on 9 other major bridges.

4. Conclusions

The authors feel that it is essential to build concrete bridges which do not undergo excessive deformations with time. In this light, it is necessary to avoid the gradual development of cracks due to live loading, climatic effects, shrinkage, settlement, etc. This can be assured by a sufficiently high level of prestressing.

A great deal of research still remains to be done in this area. This research consists of monitoring, in a systematic manner, the deformation of bridges. Definite conclusions must be drawn from these observations. Furthermore, it is indispensable to furnish bridge designers with adequate computer applications based upon the appropriate moment-curvature relationships which were referred to in this paper.

5. References

[1] Draft Model Code CEB 1990, Comité Euro-internatinal du Béton, CEB CH-1015, Lausanne, Switzerland.

[2] Favre, R. (reporter), Beeby, A.W., Falkner, H., Koprna, M., Schiessl, P., and Jaccoud, J.P. (1985) CEB Design Manual on Cracking and Deformations, Ecole Polytechnique Fédérale de Lausanne, Switzerland.

[3] Ghali, A.,and Favre, R. (1986) Concrete structures: Stresses and deformations, Chapman and Hall Ltd., New York and London.

[4] Markey, I. (1989) Hydrostatic levelling system for the measurement of bridge deformation, Publication IBAP No. 134, Nov., Ecole Polytechnique Fédérale de Lausanne, Switzerland.

[5] Suter, R. (1989) Long term observation of concrete structures, Proc. Sym. RILEM, Lyons.

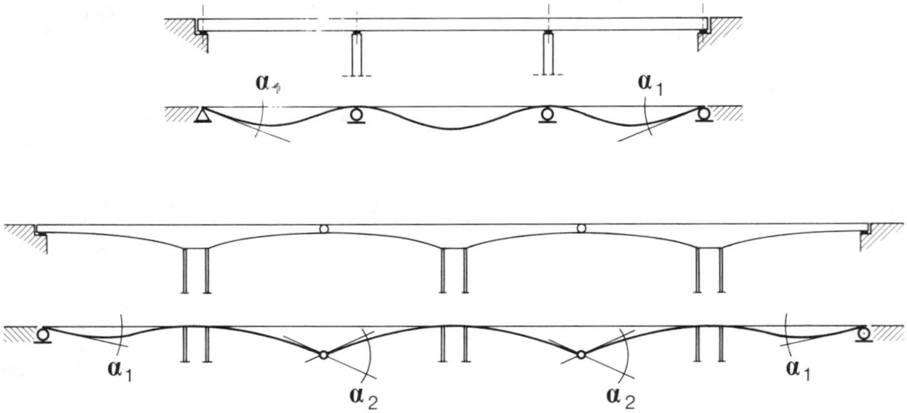

Fig. 1 Tangential Discontinuities (α).

Fig. 2 Example of a Discontinuity in a Highway Bridge.

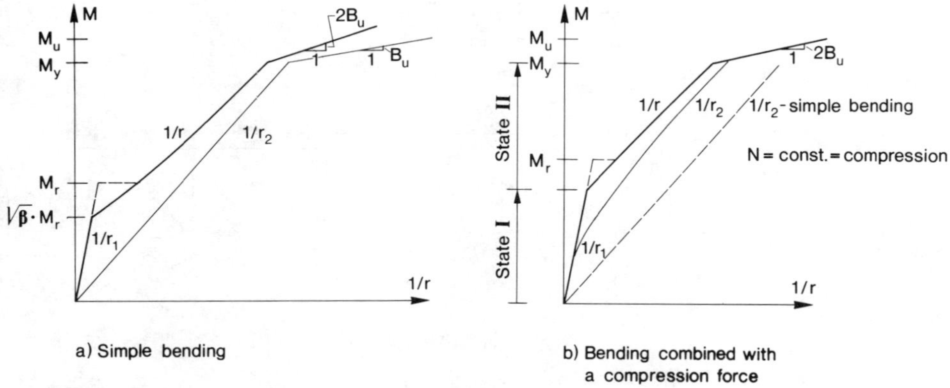

Fig. 3 Moment-Curvature Relationship.

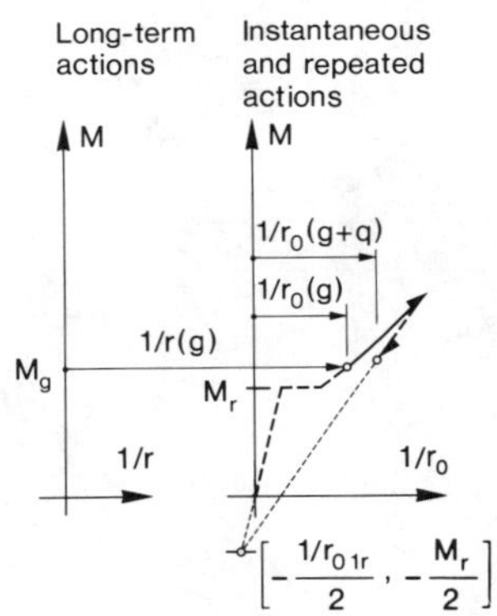

Fig. 4 Instantaneous Mean Curvature for Simple Bending: Reloading/Unloading Cycle.

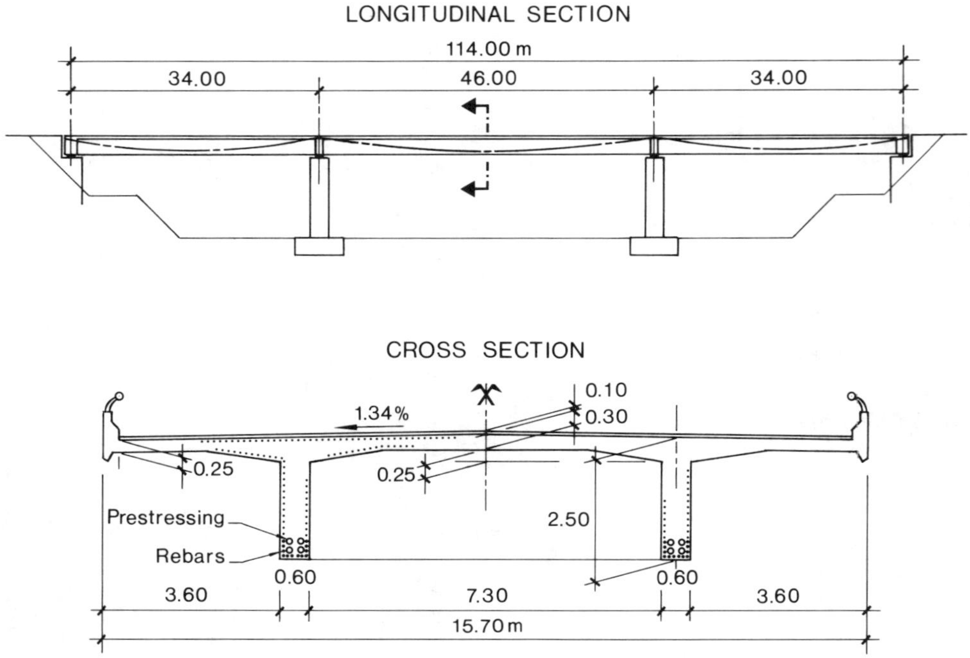

Fig. 5 Case Study of a Double T-Beam.

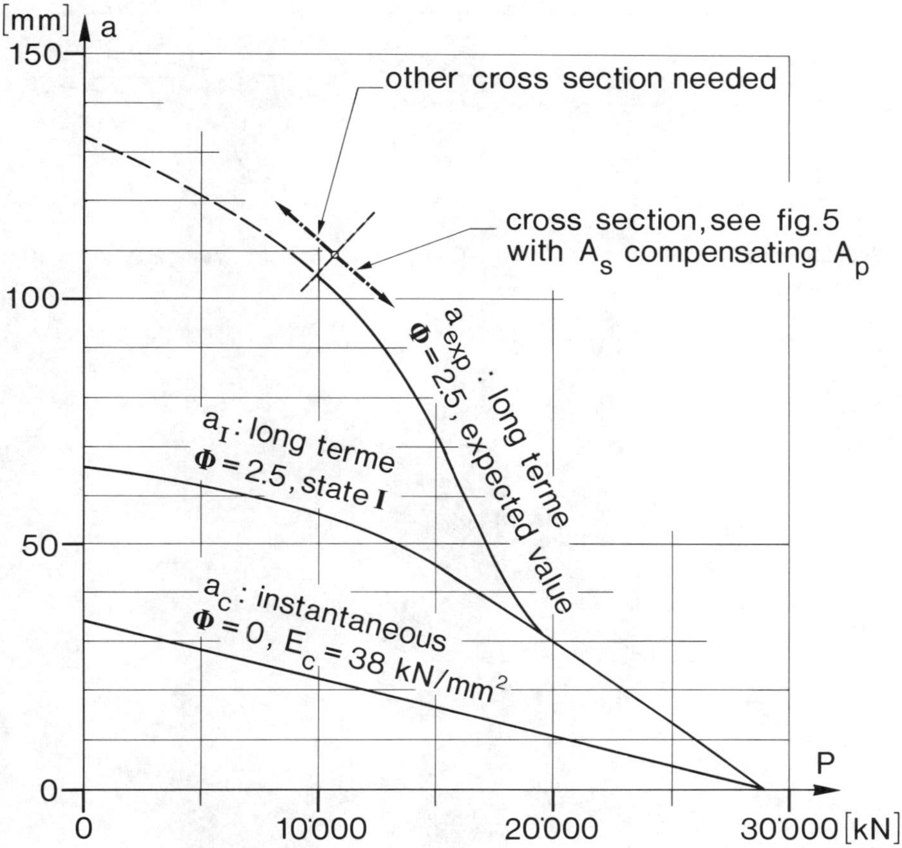

Fig. 6 Mid-span Deflection of the Central Span as a Function of the Prestressing Force P.

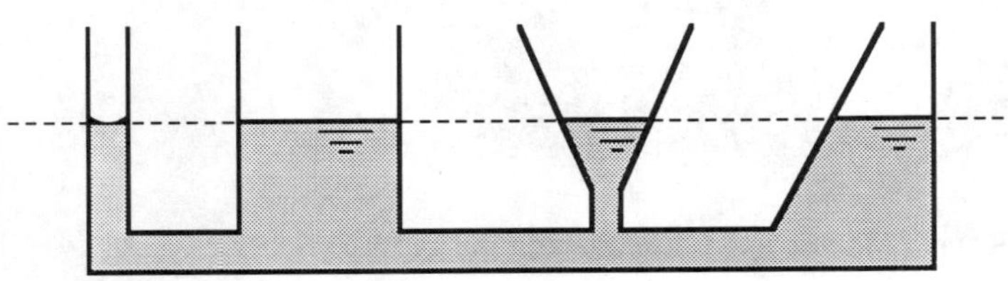

Fig. 7 Equipotential Hydrostatic Line. Principle of Communicating Vessels.

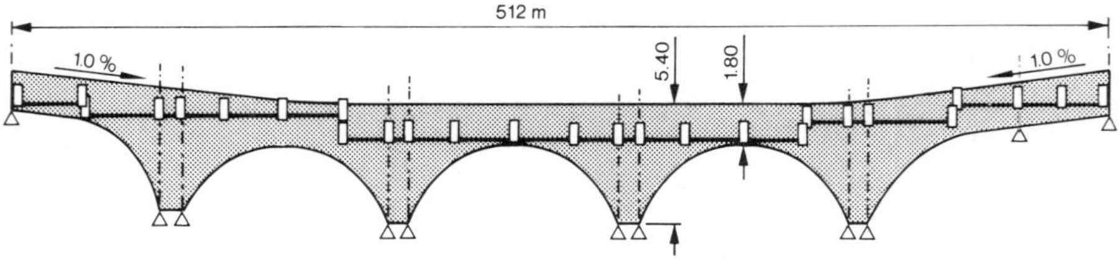

Fig. 8 Recommended Layout for the Hydrostatic Levelling System (Vertical Scale = 10 x Horizontal Scale).

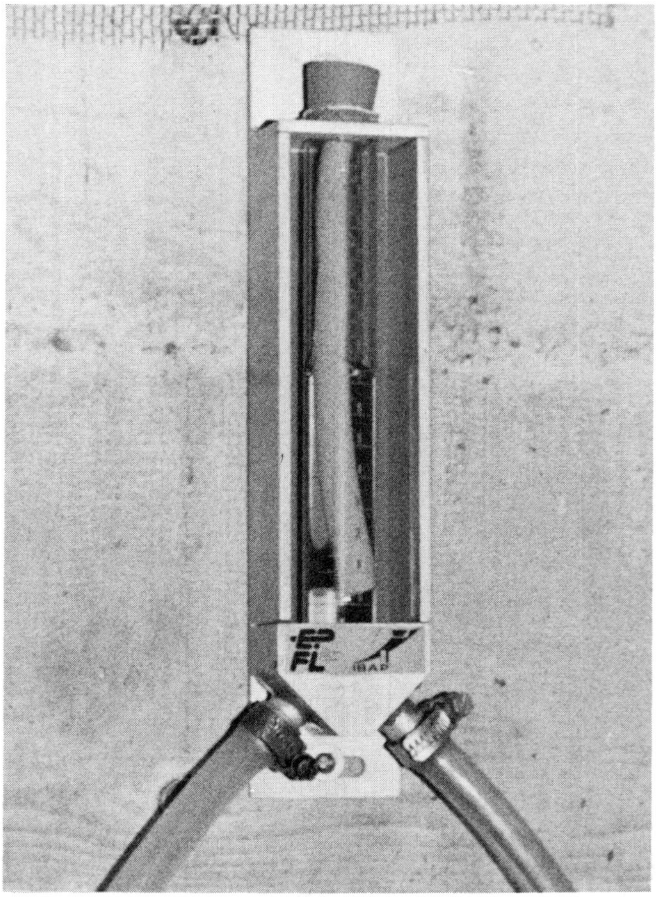

Fig. 9 Pot with Rubber Mousse and Calibration on the Glass Cylinder.

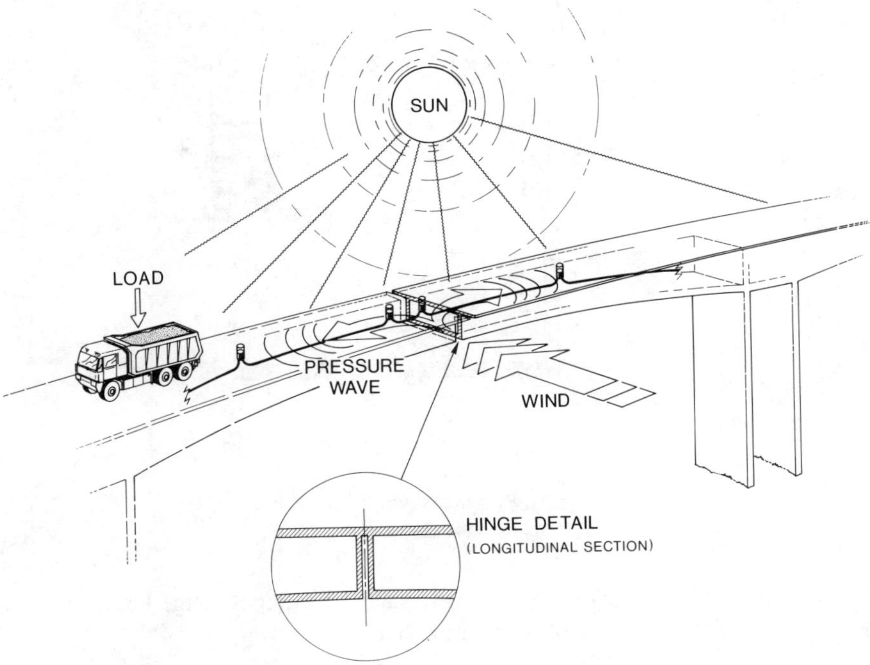

Fig. 10 Artistic Impression of the Lutrive Bridge Sowing the Hydrostatic Levelling System, Hinged Joint (Articulation) and Several Factors which Affect this System.

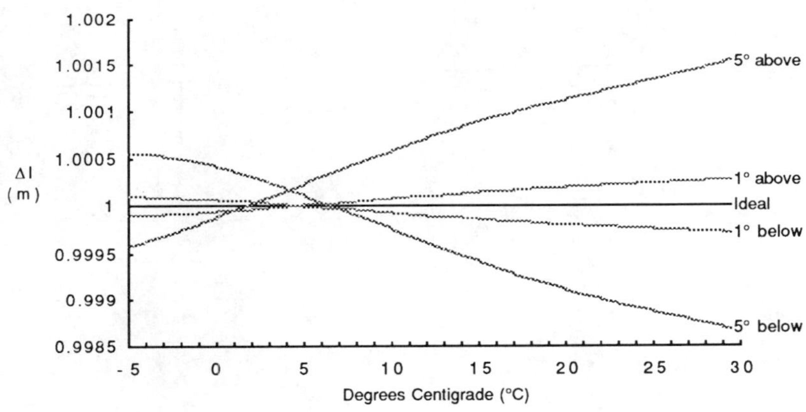

Fig. 11 Variation of the Equivalence of a One Metre Hydrostatic Head of Water with Temperature. This Graph Gives the Height of Water Necessary to Balance a One Metre Head of Water at a Specified Temperature.

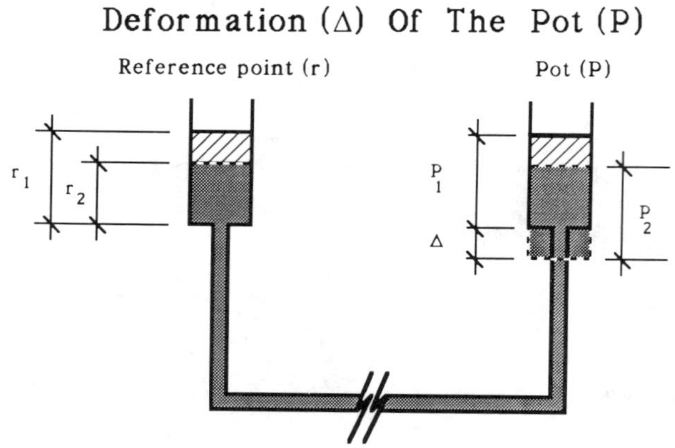

$$\Delta_{1,2} = (r_1 - P_1) - (r_2 - P_2)$$

r_1, P_1 = measurement at time t_1
r_2, P_2 = measurement at time t_2

Fig. 12 Calculation of Deformation with the Hydrostatic Levelling System. Pot (P) Undergoes a Downward Displacement of Δ Between Times t_1 and t_2.

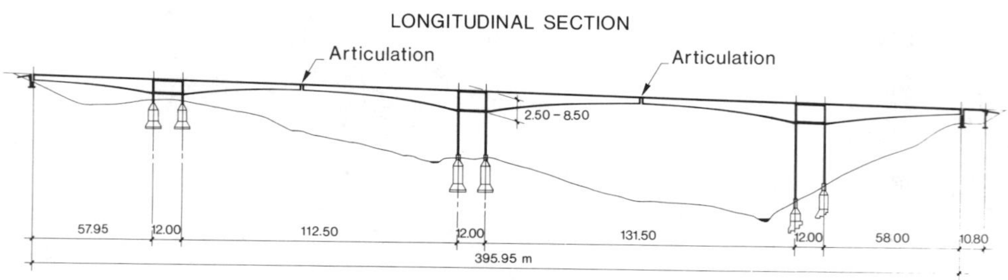

Fig. 13 Elevation of the Lutrive Bridge.

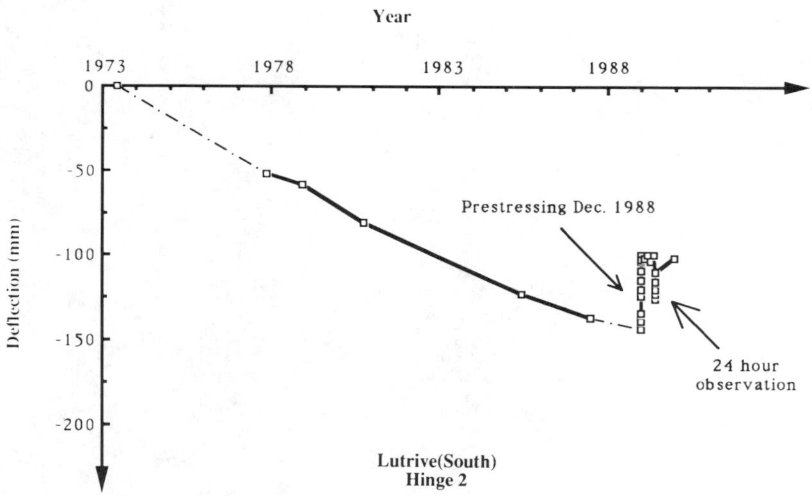

Fig. 14 Evolution of One of the Two Hinges (Articulations of the Southern Bridge).

Fig. 15 View of Prestressing System. Note that only One Cable is Acting on Each Deviator in Order to Distribute the Stresses More Evenly.

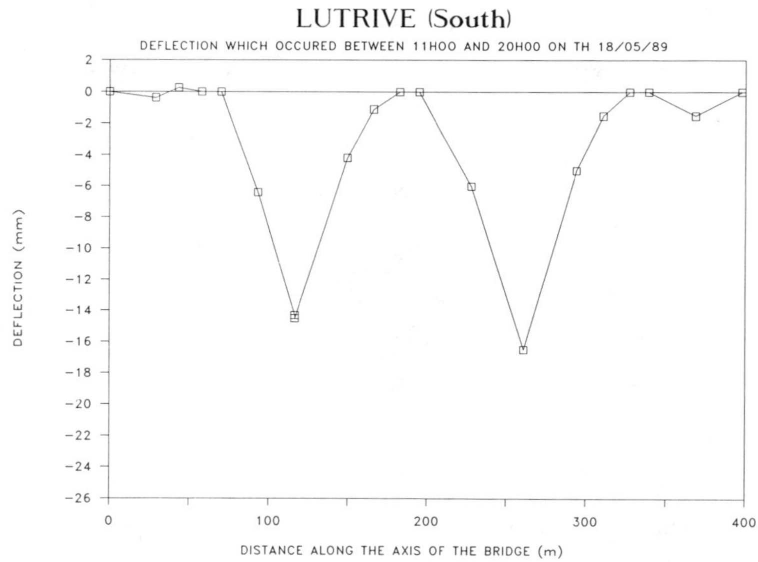

Fig. 16 With the Supports at Zero, the Maximum Deflection of the Two Hinges Occurred Between Eleven O'clock in the Morning and Eight O'clock at Night.

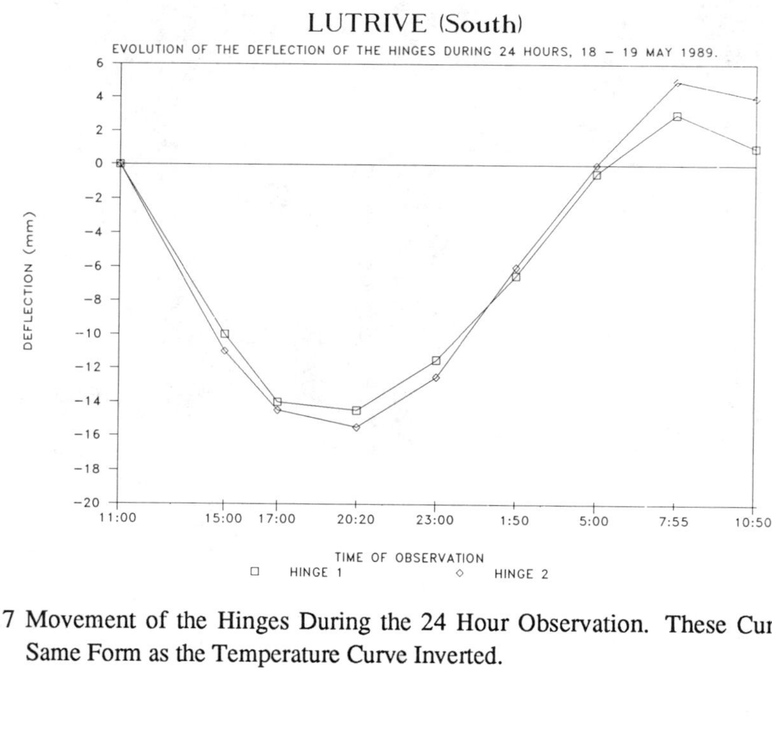

Fig. 17 Movement of the Hinges During the 24 Hour Observation. These Curves Take the Same Form as the Temperature Curve Inverted.

FLAW DETECTION IN CONCRETE USING THE IMPACT-ECHO METHOD

NICHOLAS J. CARINO
Center for Bldg Technology
National Inst. of Stands. & Tech.
Gaithersburg, MD 20899 USA

MARY SANSALONE
Civil & Environmental Engrg.
Cornell University
Ithaca, NY 14853 USA

ABSTRACT. A technique called the **impact-echo method** has been developed for flaw detection in concrete. The technique is based on a simple concept using transient stress wave propagation. Mechanical impact is used to generate a short duration stress pulse which travels into the test object as P- and S-waves. The waves are reflected by internal discontinuities or external boundaries. A displacement transducer located close to the impact point measures the time history of the surface displacement caused by the arrival of reflected waves. This paper discusses the theoretical basis of the method and gives representative examples of the results of analytical and experimental studies. A procedure called spectral peak plotting is described which simplifies interpretation of data from scans across a structure. By using this procedure a profile view of the interior of the test object can be constructed.

1. Introduction

The development of nondestructive evaluation (NDE) techniques for concrete has lagged behind the progress made in the development of techniques for relatively homogeneous materials like metals and ceramics. Most existing NDE methods are based upon monitoring the interactions of some form of radiation (electromagnetic or stress waves) with material properties or defects within a test object. The characteristics of concrete make the inspection by common forms of radiation impossible or extremely difficult. For example, concrete is a poor electrical conductor, is heterogeneous, and contains macroscopic defects which may not be detrimental to performance.

Standard test methods based upon ultrasonic stress wave propagation have been developed for nondestructive evaluation of homogeneous materials. They can be used to locate internal defects and to characterize elastic properties. The two common applications of ultrasonic waves are **through transmission** and **echo** methods. In the through transmission method, one measures the time it takes for a pulse to travel from a transmitting transducer located on one surface of the test object to a receiving transducer located on the opposite surface. If the distance between transducers is known, the wave speed can be computed, and inferences can be made about the condition of the material between the transducers. In the echo method, the transmitter and receiver are located on the same surface, and one measures the time it takes for

A. S. Nowak (ed.), Bridge Evaluation, Repair and Rehabilization, 101–118.

the pulse to travel from the transmitter to a reflecting interface (such as a crack) and back to the receiver. If the wave speed is known, the depth of the reflector can be determined.

Applications of stress wave methods for evaluation of concrete structures have recently been reviewed [1,2,3]. The implementation of ultrasonic techniques for evaluation of concrete has been hampered by the heterogeneous nature of concrete which leads to scattering of high frequency stress waves and limits their penetration into the test object. There has been success in using the through transmission method to measure ultrasonic pulse velocity. However, there has been limited success in the development of reliable echo methods for concrete structures. There have been some achievements in the testing of long slender structures, such as piles, but there is no generally accepted procedure for flaw detection in other types of members.

The National Institute of Standards and Technology (formerly the National Bureau of Standards, NBS) has sponsored basic research to develop suitable procedures for flaw detection in concrete and other construction materials. One of the techniques that has been developed is based on transient stress wave propagation and is known as **impact-echo** [4]. The research program has involved analytical studies of transient wave propagation [5-8], laboratory testing of specimens with known defects [9-12], and field studies [13]. This paper provides a brief explanation of the impact-echo method and presents results which illustrate the capabilities of the method to locate a variety of flaws in plain and reinforced concrete.

2. Principle of the Method

The principle of the impact-echo technique is illustrated in Fig. 1. A suitable stress pulse is introduced into a test object by mechanical impact on the surface. The stress pulse propagates into the object along spherical wavefronts as compression (P) and shear (S) waves [5]. In addition, a surface (R) wave travels along the surface away from the impact point. The P- and S-waves are reflected by internal cracks and voids and the boundaries of the test object. The arrival of these reflected waves at the surface where the impact was generated produces displacements which are measured by a receiving transducer. If the receiver is placed close to the impact point, the displacement waveform is dominated by the displacements caused by P-wave arrivals [5]. The displacement waveform can be used to determine the time, Δt, from the initiation of the pulse to the arrival of the reflected P-wave. If the P-wave speed in the test object is known, the distance, T, to the reflecting interface can be determined.

2.1. IMPACT RESPONSE OF INFINITE PLATE

Analytical studies have been carried out to gain an understanding of the expected response of an object subjected to impact at a point. Theoretical surface displacement waveforms have been calculated for infinite, elastic plates using the Green's function approach [4,5,10,14] and for finite plates containing defects using the finite element method [4,5-8,11]. As an example, Fig. 2(a) shows an impact-echo test on a 0.5-m thick infinite plate. The P-wave speed is 3500 m/s, and the duration of the impact is 0.25 of the time for the P-wave to travel from the top surface, to the bottom surface, and back to the top surface. Figure 2(b) shows the computed vertical displacement (using Green's function) at a point 0.10 m away from the impact point. The large downward displacement beginning at the point labeled **R** is produced by the arrival of the R-wave. After the R-wave passes by the receiver, there is no displacement until the arrival of the P-wave reflected from the bottom surface. The arrival of the reflected P-wave, which is indicated as **2P**, causes a downward displacement. The P-wave subsequently undergoes multiple reflections between the plate surfaces. The labels **2P**, **4P**, **6P**, etc. in Fig. 2(b) refer to the number of plate thicknesses that have been traversed by the P-wave when it arrives at the top surface. Note that subsequent arrivals of the P-wave also produce downward displacements. The pattern shown in Fig. 2(b) is characteristic of the impact-echo response of a wide plate or of a large planar void within a large solid. In the case of a flawed finite plate, the impact-echo response is more complex due to reflections from the plate boundaries and diffraction from the edges of flaws [7,8].

2.2. INSTRUMENTATION

An impact-echo test system is composed of three components: an impact source; a receiver; and a waveform analyzer.

2.2.1 *Impact Source.* The impact source has been found to be a critical aspect of a successful impact-echo test system. The force-time history of an elastic impact may be approximated as a half-cycle sine curve, and the duration of the impact is the **contact time** (see Fig.1). The contact time determines the frequency content of the stress pulse [14]. Impacts with short contact times contain the higher frequency (shorter wavelength) components needed for testing thin objects and for detecting small flaws. The highest frequency component with significant amplitude has a frequency value that is approximately equal to the inverse of the contact time [14]. The contact time of an impact depends primarily on the geometry and material of the impactor and on the condition of the concrete surface. An impact on a smooth, hard surface has a short contact time relative to that of the same impact on a rough surface. Small diameter (4-12 mm) steel ball bearings and spring-loaded, spherically-tipped impactors have been used successfully as impact sources. The contact times produced by these sources range from about 20 to 80 μs. The approximate value of the contact time can be determined from the recorded waveform by measuring the duration of the surface displacement due to the R-wave (see Fig. 2(b)).

2.2.2 *Receiver.* The receiver that has been found to produce successful results is a conically-tipped, broadband, displacement transducer, originally developed at the National Bureau of Standards as a secondary reference standard for calibrating acoustic emission transducers [15]. A thin lead strip is used to provide acoustic coupling between the piezoelectric tip of the transducer and the test surface. No coupling fluids or special surface preparations are required.

2.2.3 *Waveform Analyzer*. The receiver output is recorded using a digital waveform analyzer. The appropriate sampling rate depends upon the geometry of the test object and the depth of the reflecting interfaces. In current studies, slab-like specimens with thicknesses less than 1 m have been tested and a sampling rate of 500 kHz has usually been used. As will be discussed in the next section, the system should be capable of performing spectral (frequency) analysis of the recorded waveforms. An alternative to a waveform analyzer is a personal computer with an A/D (analog to digital) converter and software to perform spectral analysis.

2.3 DATA ANALYSIS

In early work [9,10], interpretation of the recorded displacement waveforms was performed in the time domain. This required establishing the time of impact initiation and the arrival time of the reflected P-wave. This was time-consuming and only feasible for simple test geometries. A more efficient approach is to analyze test results after the displacement waveforms are transformed into the frequency domain [4,14].

2.3.1 *Frequency Analysis*. The principle of frequency analysis can be understood by referring to the impact-echo response shown in Fig. 2(b). The P-wave generated by the impact propagates back and forth between the top and bottom surfaces of the plate. Each time the P-wave arrives at the top surface it produces a characteristic downward displacement. Thus the displacement waveform is periodic. For a point close to the impact point, the P-wave travel path is approximately equal to twice the thickness of the plate, and the period is equal to the travel path (2T) divided by the P-wave speed (C). Since frequency is the inverse of the period, the frequency, f, of the characteristic displacement pattern is:

$$f = \frac{C}{2T} \tag{1}$$

Equation (1) can be rewritten as follows:

$$T = \frac{C}{2f} \tag{2}$$

Thus, if the P-wave speed is known, the thickness of the plate (or the distance to a reflecting interface) can be calculated by knowing the predominant frequency in the displacement waveform. Equation (2) is the fundamental relationship for interpreting impact-echo results using frequency analysis.

In practice, the frequency content of the digitally recorded waveforms is obtained using the fast Fourier transform (FFT) technique. This technique is based on the principle that any waveform can be represented as a sum of continuous sine waves, each with a particular amplitude, frequency, and phase shift. The plot of amplitude versus frequency is called the **amplitude spectrum**. Figure 2(c) shows the amplitude spectrum corresponding to the displacement waveform shown in Fig. 2(b). There is a single large amplitude peak at a frequency of 3.42

kHz which corresponds to the frequency of multiple P-wave reflections between the bottom and top surfaces of the slab [7]. The frequency associated with reflections between the top and bottom surfaces is called the **thickness frequency**. Using Eq. (2), the calculated plate thickness would be 3500/(2• 3420) = 0.51 m, which is in agreement with the actual thickness of 0.5 m.

The resolution in the digital amplitude spectrum, i.e., the frequency difference between adjacent points, is equal to the sampling frequency divided by the number of points in the waveform record. This affects the resolution of the depth calculated using Eq. (2). For a given resolution in the amplitude spectrum, the resolution of the calculated depth improves as the depth decreases. Typically, waveforms are sampled at a rate of 500 kHz and contain 1024 points; therefore, the resulting digital spectra have a resolution of 500/1024 = 0.488 kHz.

Figure 3 illustrates the application of frequency analysis. Figure 3(a) shows the experimental amplitude spectrum for an impact-echo test performed over a solid portion of a 0.5-m thick concrete slab. The thickness frequency is 3.42 kHz. Using Eq. (1) and solving for C, the P-wave speed is calculated to be 3420 m/s. This procedure would be used to determine the P-wave speed in any test object. Figure 3(b) shows the amplitude spectrum obtained from a test over a portion of the same slab containing a disk-shaped void at a depth of 0.25 m. The peak at 7.32 kHz corresponds to multiple reflections between the top of the plate and the void. Using Eq. (2), the calculated depth of the void is 3420/(2• 7320) = 0.23 m, which compares favorably with the actual depth.

2.3.2 *Spectral Peak Plotting.* In using the impact-echo method to determine the locations of flaws within an object, tests are performed at regularly spaced points along **scan lines** marked on the surface. Examination of the amplitude spectra will reveal the presence of defects. A technique for visualizing the location of defects along a scan line is to construct a **waterfall plot** of the individual spectra. In a waterfall plot, the spectra are plotted consecutively, with a vertical offset, in the order corresponding to the test location along the scan line. By observing the location of peaks in the spectra, it is easy to identify those test points located above internal defects.

The information contained in a waterfall plot can be further processed to produce a cross-sectional view of the test object. The technique that has been developed is called **spectral peak plotting,** and it involves the following operations for each spectrum in the scan [16]:

- Locate the highest amplitude peak in the spectrum and normalize the amplitudes by dividing by the highest amplitude value.
- Select a threshold value for the relative amplitude and locate all peaks in the spectrum with relative amplitudes above the threshold value.
- For each peak above the threshold, use Eq. (2) and the P-wave speed to compute the depth (T) corresponding to the frequency value (f).
- Plot the depths at that test point along the scan line.

The above procedure is illustrated in Fig. 4. A computer program has been written which uses the individual spectra as input and generates a picture of the cross section along the scan line. Spectral peak plotting permits a rapid assessment of impact-echo results by allowing the user to "see" inside the test object. In constructing the cross-section, a line is drawn at the computed depth, and the length of the line equals the distance between test points.

3. Applications

The impact-echo method has been used successfully in the laboratory for detecting a variety of defects in concrete structures including: simulated cracks and voids in plain and reinforced concrete [4,9,13]; the depth of simulated surface-opening cracks [11]; voids in prestressing ducts [11,17]; honeycombed concrete [11]; and delaminations in slabs with and without asphalt concrete overlays [12].

Most experimental studies have been controlled-flaw studies where specimens containing flaws at known locations have been constructed and tested. However, two "blind" studies have been carried out. One study involved detecting delaminations in a reinforced concrete slab [12] and the other involved locating voids within the grout placed in post-tensioning ducts embedded in a 1-m thick, reinforced concrete specimen [17,18]. In both of these studies, the locations of the flaws were not known by the authors prior to testing. The locations of the flaws were revealed to the authors after they submitted the impact-echo results to the project leaders.

The following sections highlight the results from some of these studies; details can be found in the listed references. The objective is to acquaint the reader with the capabilities of the method.

3.1. VOIDS IN CONCRETE SLABS

One of the first laboratory specimens constructed at NIST was a 0.5-m thick, plain concrete slab having plan dimensions of 2.0 by 4.5 m. When the slab was cast, 25-mm thick polyurethane disks of various diameters were embedded at different elevations. Because of their low density, the disks simulated voids. This specimen was used to establish the basic experimental procedures for the impact-echo method and to validate the results of analytical studies [4,10]. The spectra shown in Fig. 3 are examples of the results obtained from this slab. One of the 0.5-m diameter disks embedded in the slab was inclined at an angle of 30 degrees to the bottom surface as shown in Fig. 5(a). A scan was performed over the disk using a 50-mm spacing between test points. The approximate locations of the test points with respect to the disk are shown in Fig. 5(a). The disk was located approximately between test points 5 and 14. The spectra were processed using the spectral peak plotting program, and Fig. 5(b) shows the resulting cross section along the scan line. The presence of the inclined disk is clearly indicated. Note that for the tests over the disk, the apparent thickness of the slab is greater than 0.5 m. This is due to the increased path length of the portion of the P-wave which diffracts around the disk boundary and is reflected between the slab surfaces [4,13]. The boundary of the disk is indicated by the distinctive pattern of regularly spaced lines below test points 5 and 14. This pattern is believed to be caused by diffraction effects at the boundary of the disk. Additional lines below test points 1 to 4 and 15 to 18 are associated with secondary peaks in the spectra, such as shown in Fig. 3.

3.2. DELAMINATIONS IN CONCRETE SLAB

A critical need in the nondestructive evaluation of concrete bridge decks and parking structures is a method to detect delaminations caused by the disruptive stresses arising from corrosion of reinforcing steel. While progress has been made in the development of techniques based upon acoustic impact, ground penetrating radar, and infrared thermography [12], there is still a need for a simple and highly reliable method.

The authors participated in a "blind" study designed to demonstrate the feasibility of wave propagation methods for reliable detection of delaminations in parking garage slabs [12]. A 0.20-m thick reinforced concrete slab, 3 m by 1.5 m in plan, was cast by those in charge of the study. Thin sheets of plastic were embedded in the slab to simulate delaminations. Figure 6(a) shows the planned locations of the plastic sheets. Some sheets were located on the bottom layer of wire mesh (approximately 140 mm from the top surface), and some sheets were located on the top layer of wire mesh (approximately 40 mm below the top surface). These artificial delaminations could not be detected by sounding (chain drag method). At the time of testing, the authors were advised that the delaminations were located along two lines 440 mm distant from the long sides of the slab. However, the number, sizes, or exact locations of the simulated delaminations were not disclosed.

Before describing the results, the expected response for an impact-echo test over a delaminated portion of a slab is discussed. The amplitude spectrum would be expected to contain two major frequency components: (1) a thickness frequency peak corresponding to multiple P-wave reflections between the top surface of the slab and the delamination; and (2) a low frequency peak corresponding to flexural vibration of the delaminated portion of the slab [7,12]. To measure the thickness frequency associated with the depth of the delamination, the stress pulse must contain significant components at frequencies higher than the thickness frequency. Therefore, according to Eq. (1), as the depth of a delamination decreases, shorter contact times are required for the thickness frequency peak to appear in the spectrum. If the contact time is too long, the depths of the delaminations can not be measured. However, the presence of the delaminations will be indicated in the spectra by low frequency peaks associated with flexural vibrations. These low frequency peaks would be absent in the spectra from tests over solid portions of the slab.

Initial tests were performed at 200-mm intervals along two scan lines to determine the approximate locations of delaminations. The two scan lines are labelled "A" and "B" in Fig. 6(b), and the test points (location of receiver) are numbered from 0 to 14. After the delaminations were located, subsequent tests were performed at a 100-mm spacing (or less) along two orthogonal scan lines to determine the approximate extent of each delamination.

The amplitude spectrum obtained from a test over a solid portion of the slab had a single large amplitude peak at 9.76 kHz; this was the thickness frequency [12]. Using Eq. (1) and the known slab thickness of 0.20 m, this frequency resulted in a P-wave speed of 3900 m/s.

Because the slab was tested when it was only four days old, the hardness of the surface was low and the contact times of the impacts were relatively long (about 40 to 50 μs). Therefore, the stress pulse did not contain significant frequency components above 25 kHz. Thus, according to Eq. (2), it was not possible to determine the depths of delaminations shallower than about 80 mm.

Figure 7(a) shows the cross-section obtained using the spectral peak plotting procedure for the initial scan along line B. It is evident that there is a delamination below test points 2 and 3, and there is a delamination below test point 7. There are two clues in Fig. 7(a) which indicate the presence of these delaminations: (1) lines indicating the depth of the delaminations; and (2) lines at depths greater than the slab thickness. Note that at test point 12 there is no line to indicate the depth of a delamination nor is there a line corresponding to the depth of the slab. It was concluded that there was a shallow delamination below point 12. The reasoning for this conclusion is as follows: Because of the limited frequency content of the impact, the thickness frequency associated with the shallow delamination could not be measured. However, the low frequencies associated with the flexural vibration of the delaminated section were measured, and these low frequencies resulted in fictitious depths greater than the slab thickness of 0.20 m.

Figures 7(b) shows the results of a scan orthogonal to line B, which established the approximate size of delamination number 1. It was concluded that the size of this delamination was about 300 mm in each direction [12].

Figure 6(b) shows the approximate boundaries of the delaminations detected using the impact-echo method. Delaminations numbered 1 and 2 along line B were clearly identified to be at the depth of the bottom layer of wire mesh. The spectra obtained from delamination number 3 on line B and number 5 on line A contained only very low frequency peaks indicating that these delaminations were shallow. Delamination number 4 on line A was detected, but its depth could not be determined because of its small size. The presence of delamination number 6 on line A was detected because the thickness frequency associated with the depth of the slab shifted to a slightly lower value, indicating that the stress waves were being diffracted around the defect. However, the depth of this delamination could not be determined because it was too shallow. A comparison of Figs. 6(a) and 6(b) shows that all the simulated delaminations were detected. Approximate sizes were correct, and where depths could be determined, they were accurate.

3.3. VOIDS IN GROUTED TENDON DUCTS

The authors participated in a "blind" study designed to determine whether existing test methods, based on stress wave propagation, could be used to detect voids in grouted post-tensioning tendon ducts [17,18]. The study was sponsored by the Canada Centre for Mineral and Energy Technology (CANMET), and the impact-echo method was one of the methods selected for evaluation by the principal investigator (B.H. Levelton & Associates) for the project. The feasibility of the impact-echo for locating empty tendon ducts had been demonstrated previously [11].

The test specimen, which was designed by the principal investigator, was intended to simulate a portion of a typical post-tensioned concrete ice-wall for an arctic offshore structure. During the grouting of the ducts, various artificially created voids were incorporated into the ducts. The locations of the voids within the ducts were not known by the authors prior to testing. After the wall was tested and the results presented to the principal investigator, the locations of the voids were disclosed to the authors.

Figure 8 shows an elevation and a plan of the wall specimen. The letters A through G were used to identify the seven tendon ducts: ducts A through E were placed vertically, and ducts F and G were placed horizontally. The locations of the ducts were known prior to testing. Impact-echo tests were performed at 100-mm intervals along the centerlines of each duct. Test points were identified by sequential numbers which are shown along the perimeter of the wall in Fig. 8. The receiver was located directly over the test point and the impactor was located next to the receiver as shown in the enlarged sketch in Fig. 8.

At each test point, tests were repeated until a representative amplitude spectrum was obtained. At points where the surface was in good condition, three tests were sufficient. At points where the surface was uneven, six to eight tests were required to establish a representative spectrum.

To aid in interpreting the amplitude spectra, a table was developed which listed the expected frequency values at different test points assuming there were voids in the ducts directly below the test points. Because the stress waves generated by impact propagate along spherical wavefronts, voids in adjacent ducts, as well as a void in the duct directly below the test point, can produce wave reflections. Thus frequencies corresponding to P-wave reflections from voids in adjacent ducts were also computed. The presence of peaks in the amplitude spectra at frequencies equal to these calculated frequencies was interpreted to mean that a void existed in the duct being scanned or in an adjacent duct. Because only certain frequency values were allowed in the digital amplitude spectra (multiples of 0.488 kHz in this case), the computed frequencies typically fell between frequency values permitted in the spectra.

Waterfall plots of the amplitude spectra for each scan line were plotted. For example, Fig. 9 shows waterfall plots for scan line ER (along duct E on the reinforced face) and scan line D (along duct D on the unreinforced face). If a void were present in duct E, a test from the reinforced face would result in a frequency peak at 9.84 kHz [17]. The closest allowed frequency values in the spectra are 9.77 and 10.25 kHz, and vertical lines were drawn at these frequencies as shown in Fig. 9(a). Test points whose spectra had peaks at 9.77 or 10.25 kHz were identified as possible locations for voids in duct E. The points where there was high confidence that a void was present are marked with a **Y** (for "yes") on Fig. 9 (a). For example, test points ER6 and ER11 have large frequency peaks at 9.77 kHz and are marked with a **Y**. At some points, there was a peak close to the expected frequency value or there was a series of peaks in the vicinity of the expected frequency. These points were marked as **Y?**, which signifies that there is an indication of a void, but the results are not as clear as points marked with a **Y**. For example, points ER1 and ER4 have multiple peaks in the vicinity of 9.77 kHz. It is believed that many of the points marked as **Y?** are near the boundaries of voids.

Figure 9(b) shows the waterfall plot for scan line D. Reflections from a void in duct D would be associated with a frequency of 4.55 kHz. The permitted frequency values close to this theoretical frequency are 4.39 and 4.88 kHz, which are plotted as vertical lines in the Fig. 9(b). The spectra for test points D0 and D1 had peaks at 4.88 kHz, and points D6, D7 and D12 had peaks at 4.39 kHz. Thus it was concluded that duct D had voids at its ends and center [17].

The above procedure was used to interpret the waterfall plots for all of the scan lines. The results of the tests from the unreinforced face are summarized in Fig. 10 (a), where test points identified as being over a void are shown in black. The test points where there was a question about the presence of a void are shown shaded. In drawing the extent of voids, it was assumed that, at each point where a void was detected, the void extended 50 mm to either side of the test point.

After submitting the test results to the principal investigator, the authors were given a sketch showing the actual locations of the voids. In addition, the principal investigator drilled two cores from the specimen, and provided the photographs and descriptions of the cores. Figure 10(b) shows the planned locations of the voids. A comparison with Fig. 10(a) shows that all the voids were detected. However, there are two differences between the impact-echo results and the actual void configuration: (1) the size of the voids tended to appear greater in the impact-echo results; and (2) more voids were detected by impact-echo than shown in Fig. 10(b). These differences may be attributed to the effects of diffraction by the boundaries of the voids, shrinkage of grout from the ducts, and inadequate consolidation adjacent to the ducts [17]. Diffraction of stress waves at the sharp edge of a void allows the void to be "seen" even though the test is not directly above the void. This would tend to make the voids appear to be larger than their actual sizes. Shrinkage of grout would create a thin air gap between the grout and duct, which would cause wave reflection as if a void were present. Similarly, voids adjacent to the duct, due to inadequate consolidation, would also cause wave reflections.

Based on the authors' results and the results of others who tested the wall, the principal investigators concluded that the impact-echo method "is capable of identifying relatively small, deep voids in a heavily reinforced concrete wall" and that it "holds the best potential for being adapted for field use" [18].

4. Summary

A technique called impact-echo has been developed for nondestructive testing of concrete structures. The testing technique is based on a simple concept: mechanical impact is generated on the surface of the test object, and the surface displacement close to the impact point is measured. The stress waves which propagate into the object undergo multiple reflections between the test surface and internal defects or the opposite boundary of the test object. By monitoring the surface displacements caused by the arrival of the multiple reflections (echoes), the depth of the reflecting interface can be determined.

Implementation of the impact-echo concept has been successful because of three key features: (1) the use of short duration impacts with well defined force-time histories; (2) the use of a broadband, high fidelity, displacement transducer to monitor surface displacement at a point; and (3) the use of frequency analysis for interpretation of recorded displacement waveforms.

The contact time (duration) of the impact is critical in determining the depth and size of internal defects which can be detected. As a general rule, the inverse of the contact time has to be at least as great as the thickness frequency to be measured, and the shortest wavelength (wave speed divided by maximum frequency) has to be approximately equal to or less than the dimension of the smallest flaw to be detected.

The receiving transducer must be broadband so that it will respond over a wide range of frequencies and will not contaminate the signal by its own resonant vibration. In the studies which have been performed to date, a displacement transducer has been used so that test results could be compared directly with analytical results. Broadband accelerometers or velocity transducers may also be suitable; however, these devices have not yet been investigated by the authors.

Traditionally, the results from echo testing have been analyzed in the time domain, that is, the depth of the reflecting interface has been determined by measuring the time from the start of the pulse to the arrival of the echo. The presence of surface waves and the complex displacement waveforms produced by impact-echo tests on real structures make time domain analysis impractical. Therefore, frequency analysis is the recommended method for interpreting the results of impact-echo tests.

Frequency analysis is based on the principle that the P-wave generated by the impact will undergo multiple reflection between the test surface and another reflecting interface, such as internal defect or the opposite surface. When this condition is achieved, the Fourier transform of the recorded waveform yields an amplitude spectrum with a dominant peak associated the depth of the reflector. If the P-wave speed is known, the peak frequency can be used to calculate the depth of reflecting interface. By performing tests along a scan line and using the spectral peak plotting technique, an "image" of the interior of the test object can be constructed.

The authors have conducted a series of numerical, laboratory, and field studies to develop the technical basis for the impact-method and to demonstrate its wide range of applicability. The examples presented in this paper are just a few of the many applications that have been investigated. Experimental and numerical work continues at Cornell University and NIST. Finite-element studies are underway to understand the impact-echo response of structures such as piles and beams containing diverse defects. Work is progressing on the development of a test system that will permit rapid testing and automated interpretation of test results. Efforts are also underway to develop a draft standard for using the impact-echo method for thickness measurement of concrete elements.

5. Acknowledgments

The authors thank Dr. Nelson Hsu of NIST for his invaluable contributions during the early stages of development of the impact-echo method. Gratitude is extended to Mr. John A. Bickley and Niel Cumming for arranging for our participation in the blind studies presented in the paper.

6. References

1. Carino, N.J., and Sansalone, M., "Pulse-Echo Method for Flaw Detection in Concrete," Technical Note 1199, National Bureau of Standards, July, 1984.
2. Thornton, H.T, and Alexander, A.M., "Development of Nondestructive Testing Systems for In Situ Evaluation of Concrete Structures," Tech. Report REMR-CS-10, Department of the Army, Waterways Experiment Station, Dec. 1987.
3. Sansalone, M. and Carino, N.J., "Stress Wave Propagation Methods," in Handbook on Nondestructive Testing of Concrete, Edited by V.M. Malhotra and N.J. Carino, CRC Press, in press.
4. Sansalone, M., and Carino, N.J., "Impact-Echo: A Method for Flaw Detection in Concrete Using Transient Stress Waves," NBSIR 86-3452, National Bureau of Standards, Sept., 1986, 222 pp. (NTIS PB #87-104444/AS)
5. Sansalone, M., Carino, N.J., and Hsu, N.N., "A Finite Element Study of Transient Wave Propagation in Plates," Journal of Research of the National Bureau of Standards, V. 92, No. 4, July-August, 1987, pp. 267-278.
6. Sansalone, M., Carino, N.J.,and Hsu, N.N., "A Finite Element Study of the Interaction of Transient Stress Waves with Planar Flaws," Journal of Research of the National Bureau of Standards, V. 92, No. 4, July-August, 1987, pp. 279-290.
7. Sansalone, M., and Carino, N.J., "Transient Impact Response of Thick Circular Plates," Journal of Research of the National Bureau of Standards, V. 92, No. 6, Nov.-Dec., 1987, pp. 355-367
8. Sansalone, M., and Carino, N.J., "Transient Impact Response of Plates Containing Flaws," Journal of Research of the National Bureau of Standards, V. 92, No. 6, Nov.-Dec., 1987, pp. 369-381.
9. Carino, N.J., "Laboratory Study of Flaw Detection in Concrete by the Pulse-Echo Method," in In Situ/Nondestructive Testing of Concrete, V.M. Malhotra, Editor, ACI SP-82, American Concrete Institute, 1984, pp. 557-579.
10. Carino, N.J., Sansalone, M., and Hsu, N.N., "A Point Source - Point Receiver Technique for Flaw Detection in Concrete," Journal of the American Concrete Institute, V. 83, No. 2, April, 1986, pp. 199-208.
11. Sansalone, M., and Carino, N. J., "Impact-Echo Method: Detecting Honeycombing, the Depth of Surface-opening Cracks, and Ungrouted Ducts," Concrete International, V. 10, No. 4, April 1988, pp. 38-46.
12. Sansalone, M. and Carino, N.J., "Detecting Delaminations in Concrete Slabs With and Without Overlays Using the Impact-Echo Method," ACI Materials Journal, V. 86, No. 2, March-April 1989, pp. 175-184.
13. Sansalone, M., and Carino, N.J., "Laboratory and Field Study of the Impact-Echo Method for Flaw Detection in Concrete," in Nondestructive Testing of Concrete, ACI SP-112, American Concrete Institute, 1988, pp. 1-20.

14. Carino, N.J., Sansalone, M., and Hsu, N.N., "Flaw Detection in Concrete by Frequency Spectrum Analysis of Impact-Echo Waveforms," International Advances in Nondestructive Testing, 12th Edition, W.J. McGonnagle, Ed., Gordon & Breach Science Publishers, New York, 1986, pp. 117-146.
15. Proctor, T. M., Jr., "Some Details of the NBS Conical Transducer," Journal of Acoustic Emission, V. 1, No. 3, 1982, pp. 173-178.
16. Carino, N.J. and Sansalone, M., "Impact-echo: A New Method for Inspecting Construction Materials," Nondestructive Testing and Evaluation for Manufacturing and Construction, Edited by H.L.M. dos Reis, Hemisphere Publishing Corp., 1990, pp. 209-223.
17. Carino, N.J. and Sansalone, M., "Detection of Voids in Grouted Ducts Using the Impact-Echo Method," to be published in Proceedings RILEM Workshop on Testing During Concrete Construction, Mainz, W. Germany, March 5-7, 1990, Chapman-Hall, London.
18. Cumming, N., Seabrook, P. and Malhotra, V.M., "Evaluation of Several Nondestructive Testing Techniques for Locating Voids in Grouted Tensioned Ducts," CANMET Mineral Science Laboratory Division Report 90-1, Jan 1990.

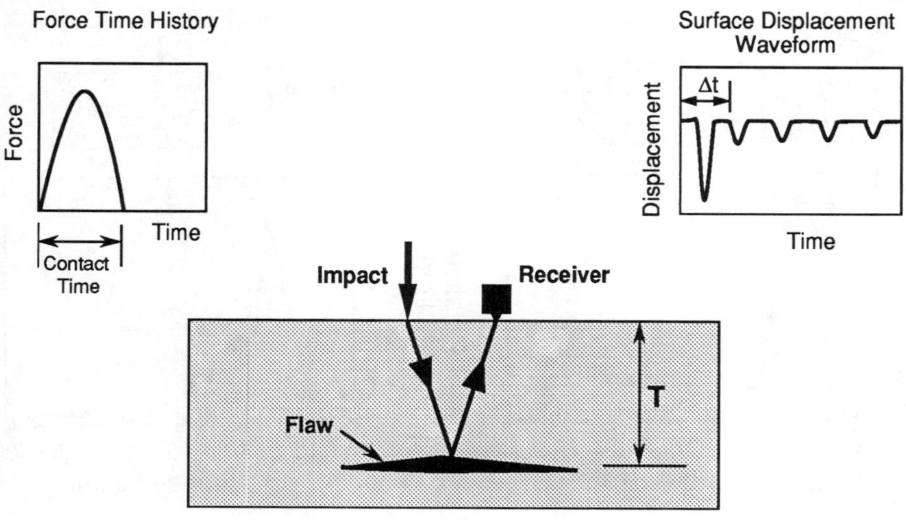

Fig. 1 Principle of the Impact-Echo Method.

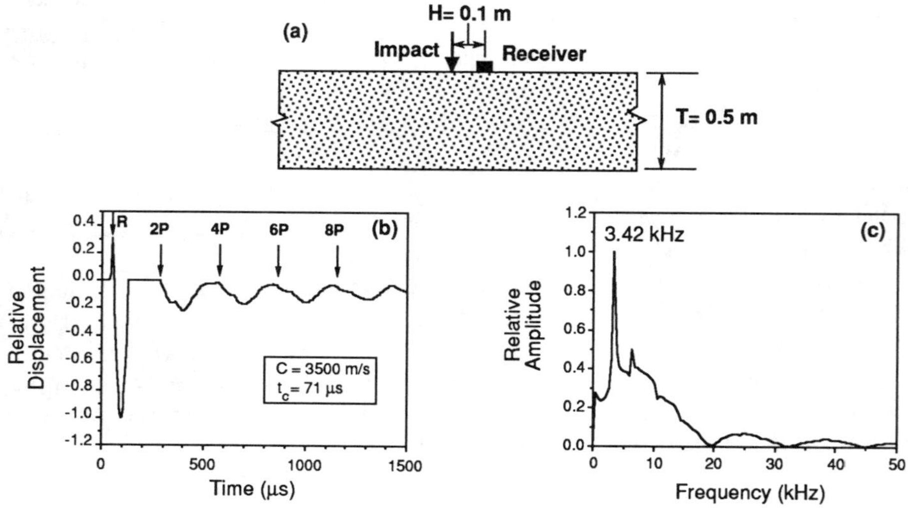

Fig. 2 (a) Impact-Echo Test on a 0.5-m Thick Infinite Plate; (b) Computed Vertical Displacement 0.1 m Away From the Point of Impact; (c) Amplitude Spectrum of Displacement Waveform.

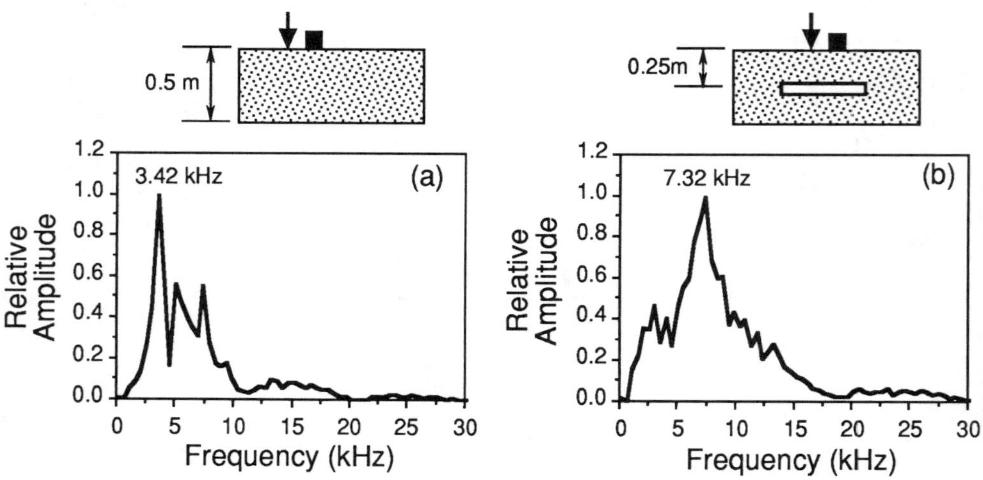

Fig. 3 Experimental Amplitude Spectra: (a) Over Solid Portion of 0.5-m Thick Slab; (b) Over a 0.25-m Deep Void in the 0.5-m Thick Slab.

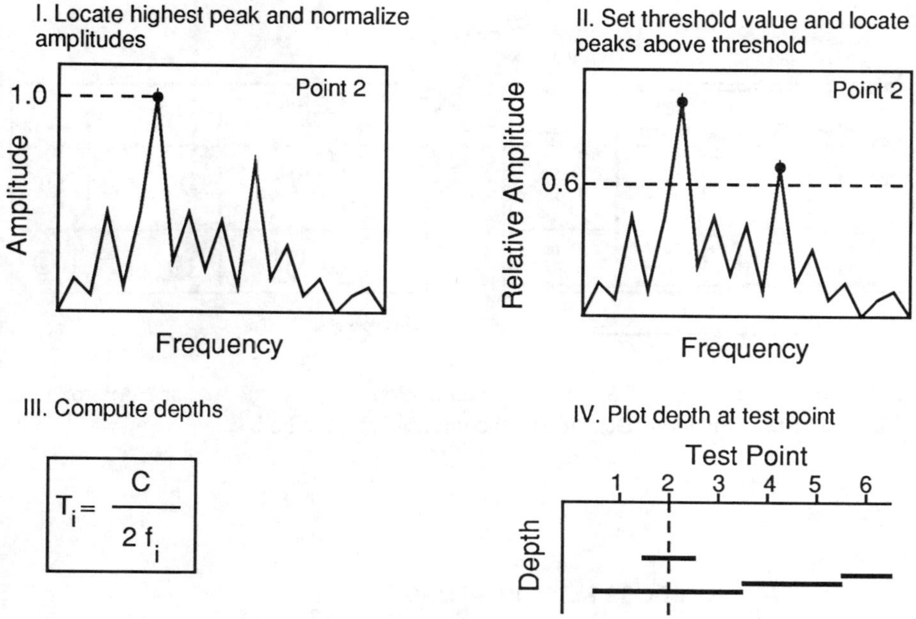

Fig. 4 Procedure Used to Construct Spectral Peak Plot from the Amplitude Spectra Along a Scan Line.

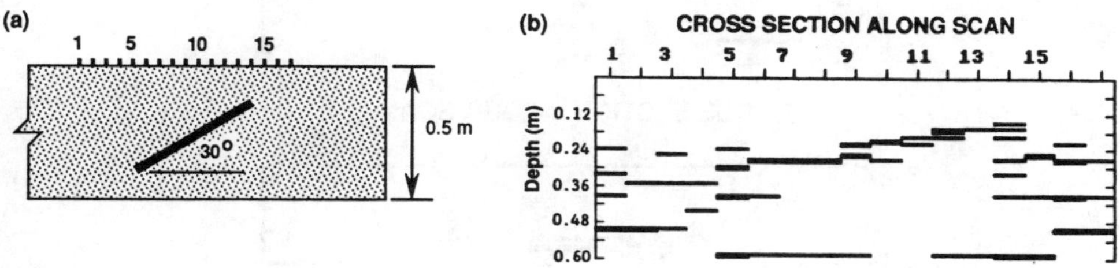

Fig. 5 (a) Cross Section of Slab Showing Inclined Disk and Test Points; (b) Cross Section Obtained Using Spectral Peak Plotting Program.

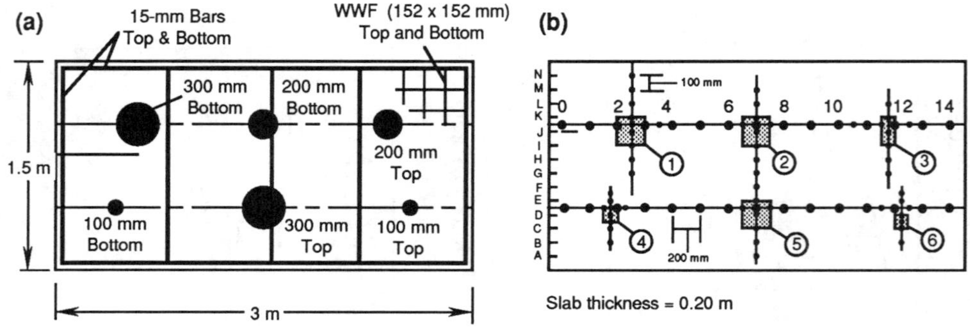

Fig. 6 (a) Planned Locations of Simulated Delaminations; (b) Test Points and Approximate Extent of Delaminations Detected by the Impact-Echo Method.

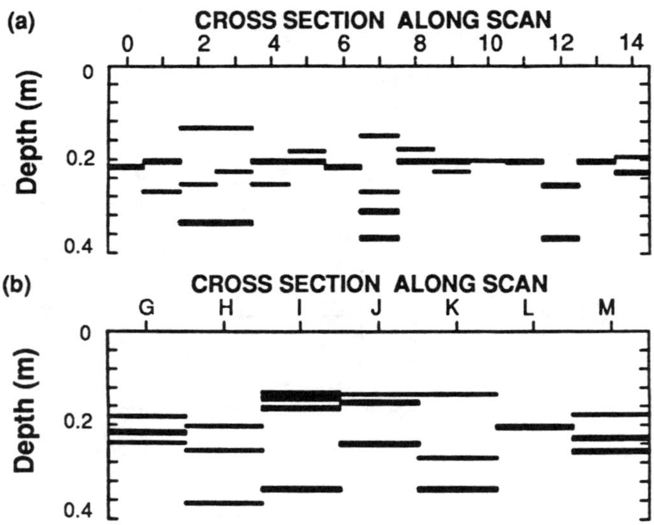

Fig. 7 (a) Cross Section Along Scan Line B; (b) Cross Section for Scan Line Orthogonal to Line B and Passing through Test Point 3.

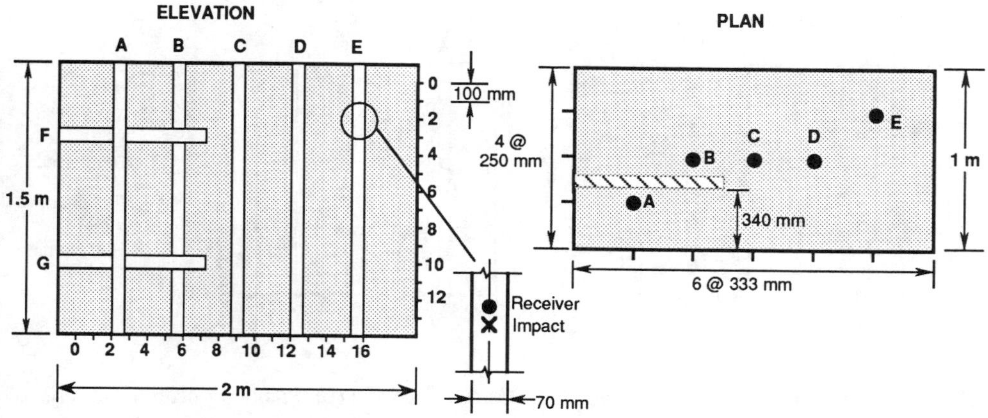

Fig. 8 Elevation and Plan of Wall Specimen; Numbers on Perimeter are the Impact-Echo Test Locations Along the Centerlines of the Ducts.

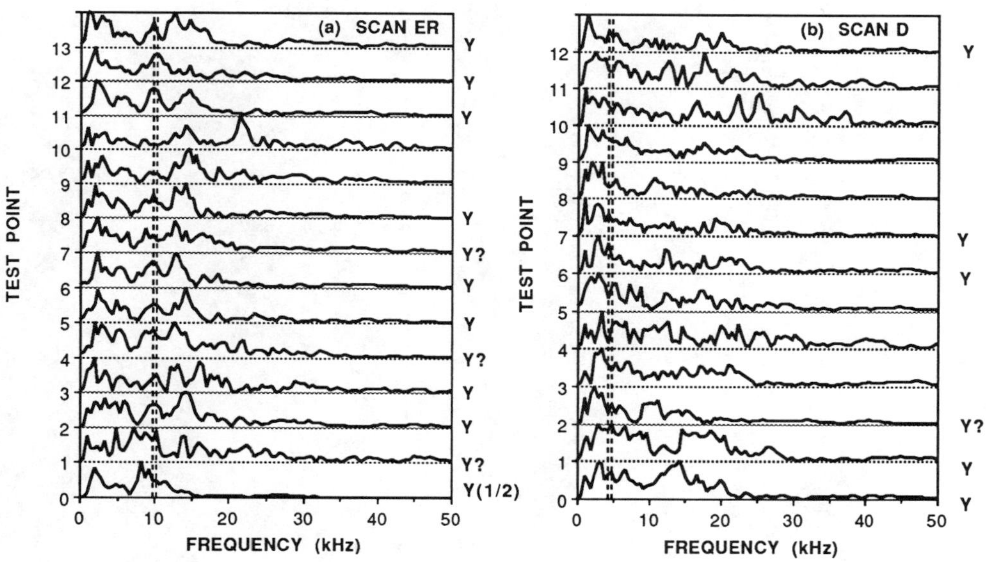

Fig. 9. Examples of Waterfall Plots for Scans Along ER (a) and Along D (b).

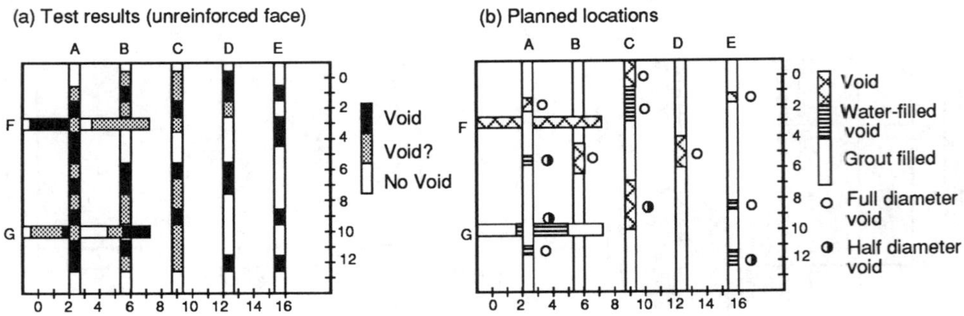

Fig. 10 (a) Impact-Echo Results from Unreinforced Face and (b) Planned Locations of Voids.

APPLICATIONS WITH OPTICAL FIBER SENSOR SYSTEM FOR MONITORING PRESTRESSED CONCRETE STRUCTURES

R. WOLFF and H.-J. MIESSELER
STRABAG Bau-AG
Postfach 21 11 20
5000 Köln 21
WEST GERMANY

ABSTRACT: Innovative use of glass fiber composites as an alternative to conventional steel reinforcement has made it possible for optical fiber sensors to be integrated into the prestressing tendons during production or directly into the concrete, thereby enabling constant monitoring of the structure to be carried out on a long-term basis.

This paper deals with the principles, development, initial applications, and future prospects of this type of sensor monitoring for concrete structures which thus become "intelligent".

1. Introduction

The demands to be met by modern man-made structures today are becoming increasingly diversified. In addition to the actual use, architectural viewpoints play an ever greater role. In the field of bridge construction these facts are nowadays demonstrated by structures which are becoming increasingly audacious. This is particularly evident in the greater spans of bridges.

Spectacular cases of deterioration in the past have shown the necessity for load bearing structures to be controlled on a permanent basis in order to guarantee their durability over a long period of time. In prestressed structures, cracks in the concrete and changes of the stress state in the tendon must be observed. Bridges, for example, can be monitored in this way by using optical fiber sensor systems which are incorporated in the high-performance fiber composite prestressing element or directly into the concrete.

2. Sensor Technology

Owing to ever increasing requirements and demands on engineering building structures, the question of durability of the materials used and life expectancy of the structure is becoming of ever greater interest. Apart from actual new construction works, investment for the preservat-

A. S. Nowak (ed.), Bridge Evaluation, Repair and Rehabilization, 119–132.

ion of building structures accounts for a considerable portion of the funds available (regular investigation of bridges in the DIN 1076 class is therefore stipulated).

2.1 THE PRINCIPLE OF MONITORING WITH OPTICAL FIBER SENSORS

In telecommunications engineering, the optical fiber sensor is first and foremost a signal transmitter. A prerequisite for its high importance is the excellent light permeability achieved. In dependence with the mechanical stresses of the optical fiber sensor, light attenuation, which is undesirable in telecommunications engineering, is used as a sensor effect in the monitoring of building structures. In contrast to telecommunications engineering, efforts made in the development of the optical fiber sensors are directed to attaining the highest possible measuring signal as a result of mechanical changes in the optical fiber sensor.

The gradient optical fiber sensor (Fig. 2), with a decreasing index of refraction from the core radius, is usually employed for such sensor application. The inner core is surrounded by a light-reflective outer sheath. It is also permeable to light but with a lower index of refraction. When a light ray is conducted though an optical fiber sensor, strew losses occur in areas of micro-bending. This resultant light loss is measured as a change in attenuation and is expressed in decibels.

As the optical fiber sensor is provided with a thin wire coil (Fig. 1), use is made of the knowledge that micro-blending can also be produced as a result of radial compression. When longitudinal tension occurs, the wire coil exerts radial pressure on the optical fiber sensor from a specific lay, thus causing micro-bending (Fig. 3). This in turn results in corresponding changes in attenuation, thereby turning the optical fiber sensor into an optical fiber extension sensor.

The optical fiber extension sensor shows changes in attenuation as a function of the applied strain. As the stresses in concrete structures do not occur evenly and localised irregularities, such as for example cracks, can also develop, local defects should also be detectable in addition to the integral attenuation measuring process. To this end, the increased attenuation signal occurring at the point of interference can be recorded with the back-scatter measuring technique known from telecommunication engineering. The type of defect and also localised changes in stress are detectable by superimposing the attenuation curves of the loaded and unloaded sensor.

Clarification of the measurements:

- the control unit
  produces electrical signals for transmitter modulation
- the transmitter
  converts the control unit's electrical signals into light signals, bunching these in the extension sensor and reference fiber
- the receiver
  receives the light signals of the extension sensor and reference fiber, converting them into electrical signals
- the measuring electronics
  process the receiver's electrical signals, recording them in an appropriate manner

3. Investigations on Structural Elements for the Testing of Optical Fiber Sensors

There are two basically different possibilities for the monitoring of building structures using optical fiber sensors. Firstly, for the monitoring of concrete structures, for example in the case of crack occurrence and its further development, and secondly, to monitor the glass fiber composite tendons. To monitor cracks in concrete building structures: in new building structures sensors specially manufactured for this purpose are embedded directly into the concrete at locations on the bearing structure which are shown by analysis to be particularly liable to stress.

This enables statements to be made at a very early stage regarding any rehabilitation measures which might required, thereby reducing the cost for such works considerably. In existing structures, particularly structures which are already defective, these optical fiber sensors are post applied to the structural elements to be monitored, so that the cracks which have already occurred can be further observed, and repairs already made can be monitored with respect to the bearing capacity of the element. In glass fiber composite prestressing tendons the optical fiber sensors are integrated directly into the fiber composite in order to show that these tendons are intact.

To prove that these optical fiber sensors fulfil the demands placed on them, comprehensive investigations were first carried out in the laboratory to determine basic applicability. The next stages were tests carried out at the University of Ghent with small prestressed beams having spans of 2.0 m and overall heights of 60 cm (Figs. 5 and 6).

All sensor applications developed to date, that is sensors integrated into prestressing tendons embedded directly in the concrete, and applied subsequently, were investigated in the tests. The beams were prestressed with two single-bar glass fiber prestressing tendons each, with and without bond.

The final practical applicability test, also executed at Ghent University, was the investigation on a prestressed beam with a length of 20.0 m and an overall height of 1.0 m (Figs. 7 and 8). The suitability of these sensors for the required applications and chosen methods of application could hereby be proven. This beam had limited prestressing comprised of a 19-bar glass fiber prestressing tendon with a working load of 660 kN. In each case a total of 3 sensors were integrated into the prestressing tendon, 8 were embedded in the concrete, and one subsequently integrated into grooves milled in the concrete.

Both the transition from a non-cracked to a cracked state and exceeding of the concrete reinforcement's yield point could be determined with the aid of optical fiber sensors (Fig. 9). Similarly, these changes could be detected by optical fiber sensors in the prestressing tendon, up to failure of the beam (Fig. 10). The conventional measuring procedures, intended to control the results of the sensor measurements, showed concurring results. The application maturity of this state-of-the-art monitoring method could therefore be impressively substantiated.

4. Applications

4.1 ULENBERGSTRASSE BRIDGE

In the Ulenbergstrasse Bridge, the world premier for a high-load road traffic bridge (bridge classification 60/30) constructed with glass fiber prestressing tendons, some prestressing tendons were provided with optical fiber sensors, in addition to those embedded directly in the concrete. This latest type of monitoring for building structures has been carried out here for the first time since the opening of the bridge to traffic in July 1986. As an example, Fig. 12 shows the results for a period of one week in August 1987.

This gives a comparison of the temperatures constantly measured within the bridge interior by means of the sensors' attenuation, as well as elongation measured at the beginning and end of the bridge.

Measurements carried our over a period of almost three years indicate no alteration of the bridge structure (Figs. 11 and 12), there by convincingly confirming a completely normal behavior for the structure.

4.2 MARIENFELDE BRIDGE, BERLIN

The pedestrian bridge at Berlin-Marienfelde, a two span T-beam concrete bridge, is the first example of a fully monitored bridge structure. Designed to have external prestressing, the bridge's prestressing is achieved my seven 19-bar glass fiber composite prestressing tendons. In addition to all the prestressing tendons with integrated optical fiber sensors, a series of these sensors were embedded into the concrete or applied subsequently.

A load investigation was carried out on the bridge in November 1989. During this test a total load of 250 concrete slabs (weight per slab 1 ton) were laid on the bridge in five layers, this load corresponding to twice the traffic load. Placing of the slabs led to an alteration in the strain state of the tendons as well as of the complete superstructure. This was in turn very impressively recorded with the aid of the optical fiber sensors

With sensors of a particularly high sensitivity, assertions could be made regarding the dynamic behavior of the bridge.

4.3 SCHIESSBERGSTRASSE BRIDGE

This three span, solid concrete slab bridge (bridge classification 60/30), with two spans of 16.30 m and one of 20.40 m, is designed with limited prestressing comprised of 27 glass fiber prestressing tendons (working load 600 kN) and post-bonding. Three glass fiber bars per tendons are provided with optical fiber sensors and there are to be four additional such sensors integrated directly into the concrete on the upper side and four on the lower side of the slab. Construction work was schedule to commence in March 1990.

5. Future Prospects

The current constantly increasing requirements and demands on heavy building structures make the question of the utilized materials' durability and the useful life of such structures in general to be of ever greater and wider interest. Compared with essentially new construction, investment for the preservation of building structures claims a not inconsiderable proportion of the funds available. The possibility of integrating sensors, by placing them directly into the bar materials during production, and by the embedding them in the concrete, opens completely new horizons for the permanent monitoring of building structures.

6. References

Rehm, G. and Franke, L. (1979) "Synthetic resin bonded glass fiber bars as reinforcement in concrete construction", German Committee for Reinforced Concrete, issue 304, pp. 3-18.

Wieser, M. and Preis, L. (1982) "Use of synthetic resin bonded glass fiber bars as reinforcement in construction engineering", Bauwirtschaft, issue 24-43, pp. 90-94.

Waaser, E. and Wolff, R. (1986) "A new material for prestressed concrete, HLV- Heavy-Duty Composite Bars comprising glass fibers," Beton 36, issue 7, pp. 245-250.

Levacher, F.K. and Miesseler, H.-J. (1988) "Monitoring tensile forces with integrated optical fiber sensors", paper at the IABSE Congress, Helsinki, pp. 313-318.

Franz, A. and Miesseler, H.-J. (1989) "Berlin-Marienfelde Research - External prestressing and monitoring of the structure with integrated sensors", paper at the German Concrete Congress, Hamburg.

Miesseler, H.-J. and Lessing, R. (1989) "Monitoring of load bearing structures by means of optical fiber sensors", paper at the IABSE Congress, Lisbon, pp. 853-858.

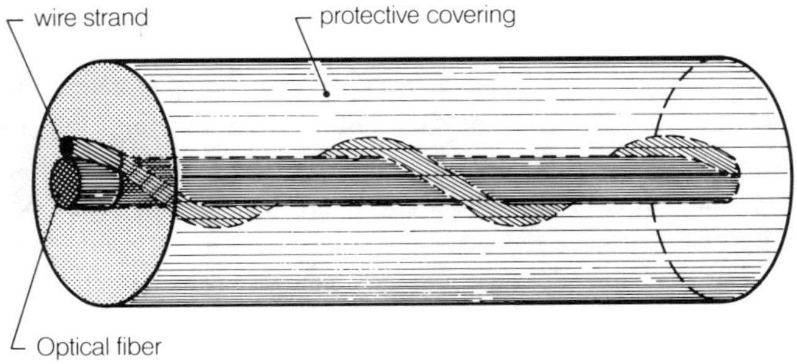

Fig. 1 Principle of the Optical Fiber Sensor.

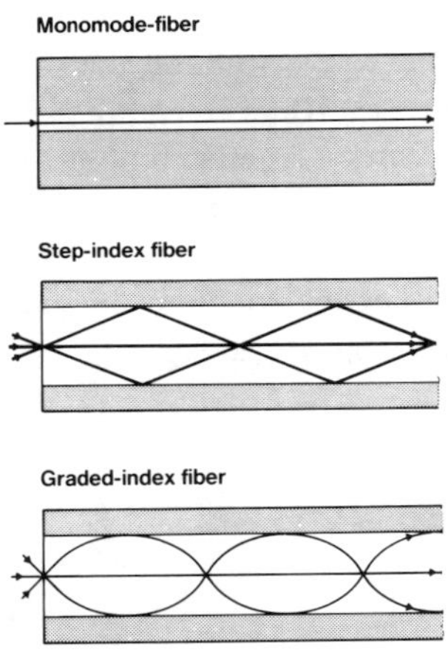

Fig. 2 Different Types of Optical Fiber Sensors.

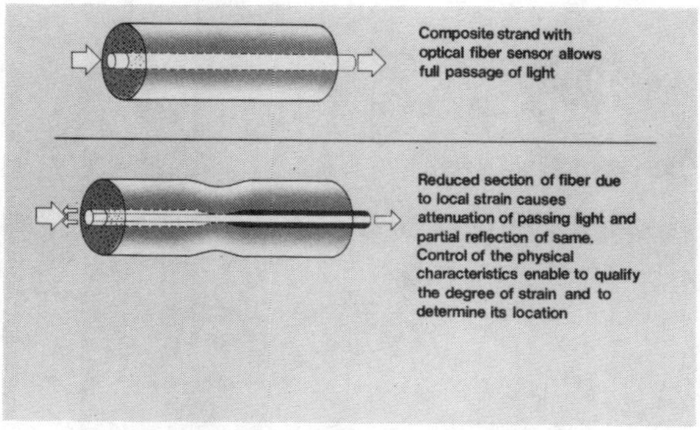

Fig. 3 Strain Control with Optical Fiber Sensors.

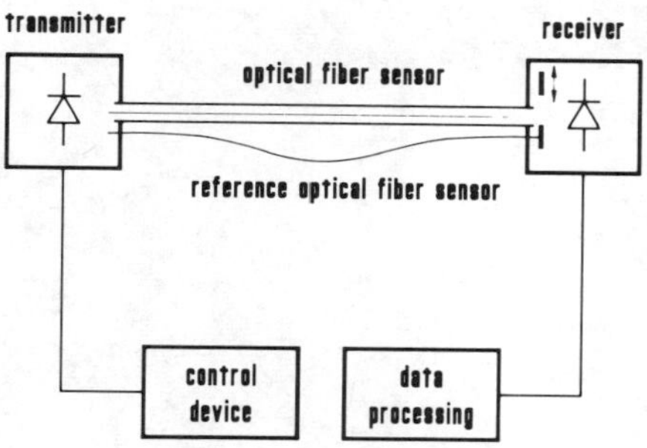

Fig. 4 Diagram of Temperature Compensated Attenuation Measuring Process.

Fig. 5 View of 2 m Beam.

Fig. 6 Cross-Section of a 2 m Beam.

Fig. 7 View of 20 m Beam.

Fig. 8 Cross-Section of 20 m Beam.

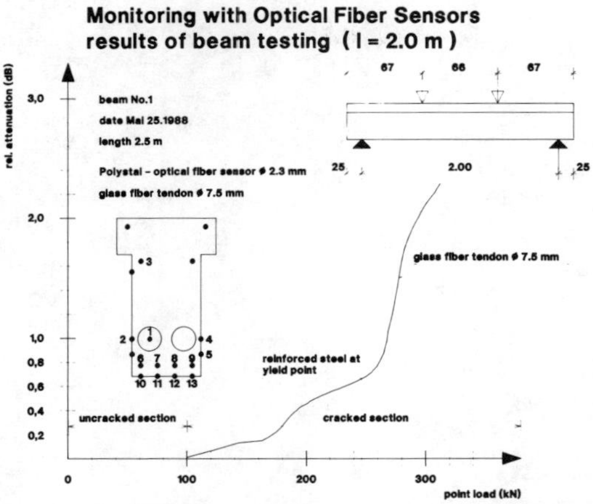

Fig. 9 Results of Beam Testing with Optical Fiber Sensors in the Tendon.

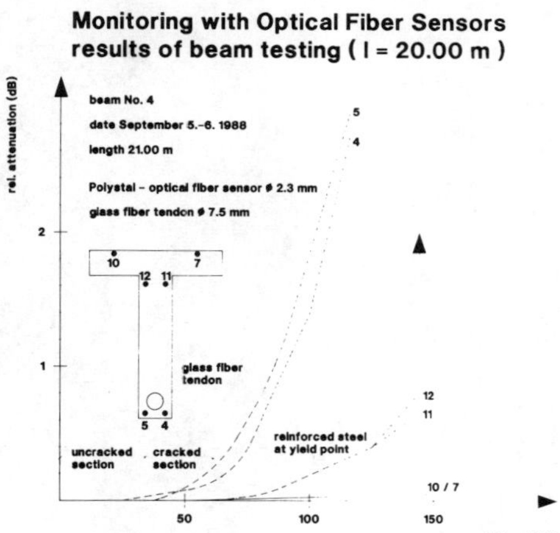

Fig. 10 Results of Beam Testing with Optical Fiber Sensors in the Concrete.

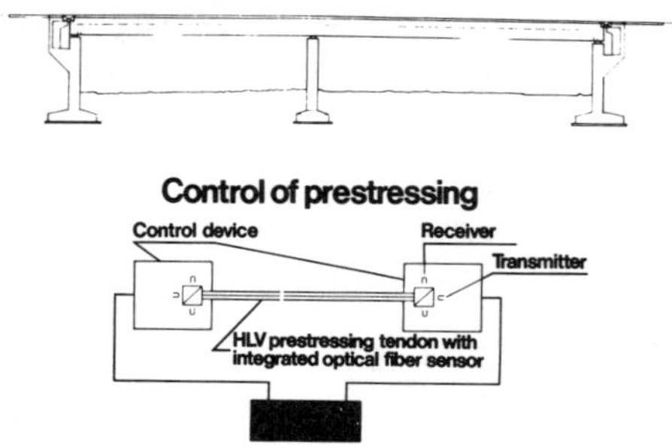

Fig. 11 Measurement Principle.

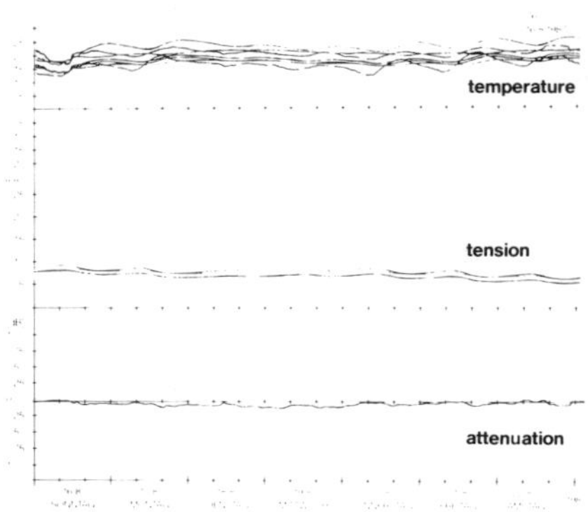

Fig. 12 Measurement Results at the Ulenbergstrasse Bridge.

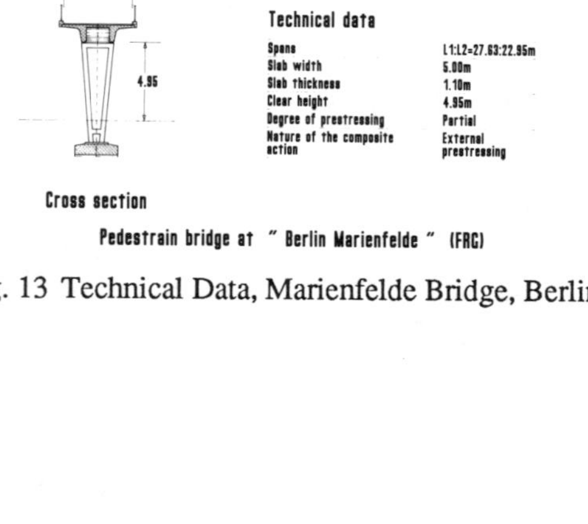

Fig. 13 Technical Data, Marienfelde Bridge, Berlin.

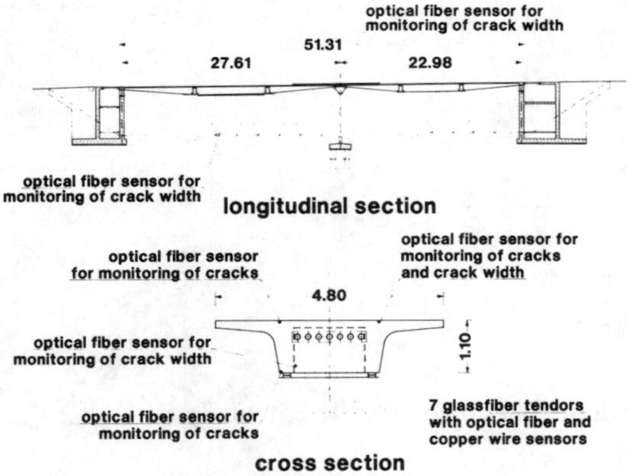

Fig. 14 Measurement Layout for the Marienfelde Bridge, Berlin.

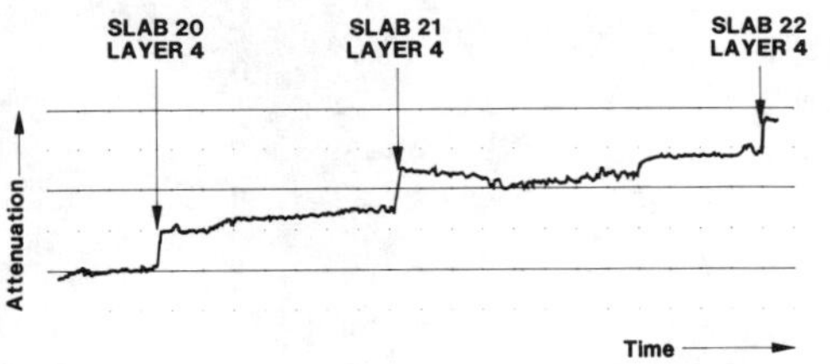

Fig. 15. Reaction of the Optical Fiber Sensors on Removal of the Concrete Slabs from the Center of the Long Span.

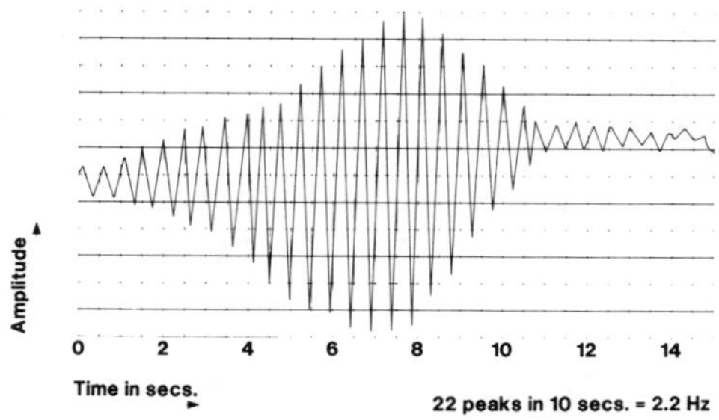

Fig. 16 Characteristic Vibration of the Bridge Caused be Specific Stimulation.

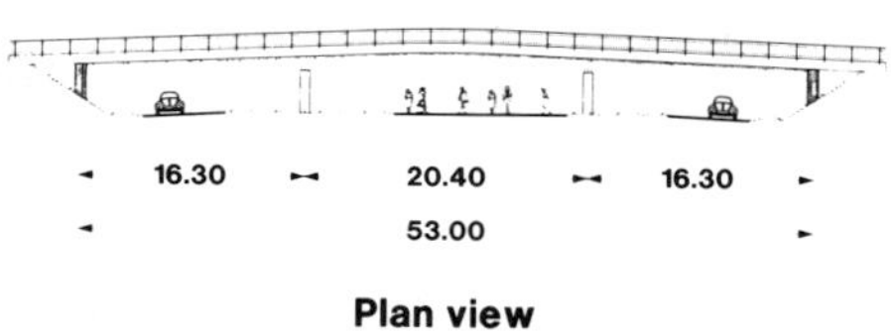

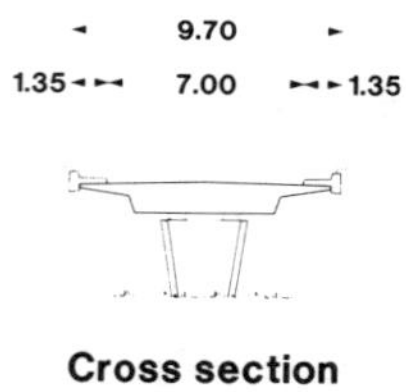

Technical data

Spans	L1=L3:L2=16.30:20.40
Slabs widht	9.70m
Slabs thickness	1.12m
Clear height	3.00m
Load class (DIN 1072)	60/30
Degree of prestressing	Limited
Nature of the composites action	post-tensioning with subsequent bond

Fig. 17 Schiessbergstrasse Bridge, Longitudinal Section.

DETECTION OF STRUCTURAL CHANGES IN A BRIDGE MODEL

M. S. AGBABIAN, S. F. MASRI, M.-I. TRAINA and O. WAQFI
Department of Civil Engineering
University of Southern California
Los Angeles, California 90089-2531 USA

ABSTRACT. This paper presents some experimental results from an exploratory study of a highly idealized physical bridge model to gauge the potential of vibrational analysis techniques to detect structural changes on the basis of response measurements and modal parameters variation.

1. Introduction

1.1 BACKGROUND

The potential of using system identification (SI) approaches for damage detection in structures has been recognized by many researchers. Similar approaches have been used extensively in other fields, including (among others) nonintrusive medical diagnostics using tomology and geophysical prospecting using seismology. Noteworthy recent efforts to apply formal SI concepts to the problem of damage detection in structures include the works of Yao (1985), Stephens (1985) and Natke (1986). A workshop exclusively devoted to structural safety evaluation based on system identification approaches was recently held in Germany, (Natke and Yao, 1988). Several studies have been performed to establish the appropriate structural parameters that are most sensitive to damage. The system's natural frequencies are easy to evaluate with a high degree of accuracy making it one of the earliest and most wide-spread tools for damage detection. Several references describe the variations of the natural frequencies due to damage, including those by Cawley and Adams (1979a, 1979b). Recent efforts include the works of Flesch (1988), who used the approach for the safety inspection of bridges, and Biswas et al.(1989) who used an experimental approach on a highway bridge.

In February 1988, an NSF-sponsored international workshop titled "Nondestructive Evaluation for Performance of Civil Structures" (Agbabian and Masri, 1988) was organized by the authors in order to (1) review the state-of-the-art of all NDE techniques, present, emerging, and future ones that offer the promise of being an effective tool for use in conjunction with civil structures, (2) identify the instrumentation requirements of the field, and (3) prioritize the research needs of the Workshop subject. Even though this Workshop encompassed all NDE methods (e.g., acoustic emission, imaging techniques, ultrasonics, x-ray, etc.), not just system

A. S. Nowak (ed.), Bridge Evaluation, Repair and Rehabilization, 133–143.

identification approaches, nevertheless the priority list of research needs gave a prominent ranking to the use of system identification techniques as a tool for global structural performance evaluation.

Several technical challenges play a major role in SI applications to structural damage detection. For example, uniqueness and observability problems are inherent in many structural systems due to the presence of redundant structural members and limited sensor locations. In addition, the available physical measurements from the sensors always contain small amounts of noise superimposed on the desired signal. The presence of such noise can lead to serious problems with the accuracy and reliability of various SI algorithms. Finally, the available response measures of some structural systems are inherently insensitive to changes in structural parameters of interest, making it difficult to devise an adequate test arrangement.

1.2 SCOPE OF PAPER

This paper explores the potential of an identification procedure to detect structural changes on the basis of response measurements.

Section 2 of this paper describes the experimental studies conducted, and Section 3 presents the formulation for the identification procedure under discussion. The usefulness of this approach for damage detection is demonstrated in Section 4 where the experimental measurements from a highly idealized model bridge are subsequently analyzed to assess the capability of the procedures under discussion for damage and deterioration monitoring.

2. Experimental Study

To illustrate the method of approach under discussion, consider the bridge model shown in Fig. 1. The model consisted of a grid of aluminum rectangular beams spaced at 5 inches center-to-center that resulted in a structure with an overall length of 45 inches and width of 10 inches.

The model design allowed the convenient response monitoring at 24 discrete locations corresponding to the physical grid points. A load cell was used to measure the applied dynamic load, and servo accelerometers were used to measure the response. The transducer-to-model connection as shown in Fig. 2 satisfied the following practical considerations: (1) the transducer was rigidly attached to the structure, (2) the structural response at discrete locations was easily measured, (3) only the vertical component, of the response was considered, and (4) the transducers did not modify the weight or the performance of the model structure.

The boundary conditions for the beam assembly simulated a "fixed" constraint. Lumped masses were placed at various locations in order to reduce the effects of the instrumentation. Selected locations allowed the introduction of adjustable levels of "reversible defects" in the bridge model as shown in Fig. 2.

The method under consideration imposes no restrictions on the nature of the excitation source to be used as a probing signal. In the present test, a location near the center of the model was subjected to a random signal of constant spectrum covering the frequency band 0 - 50 Hz. Figure 3 shows the measurement system for a typical test.

A finite element analysis of the bridge model yielded the frequencies and mode shapes shown in Fig. 4. The fundamental frequency was equal to f_1 = 8.0 Hz and was associated with the first vertical bending mode.

3. Identification Procedure

Consider a discrete MDOF linear system that is subjected to dynamic excitations. The frequency response functions (FRF) of the structure relate the response (output) of the structure to the set of excitations (input). The relationship between the response at point i and excitation at point j can be expressed in the frequency domain as:

$$Y_i(\omega) = H_{ij}\ \omega\ X_j\ (\omega) \tag{1}$$

where

$Y_i(\omega)$ = response at point i for frequency ω,

$H_{ij}(\omega)$ = frequency response function of response i with respect to excitation j for frequency ω,

$X_j(\omega)$ = excitation at point j for frequency ω.

The least-squares solution of Eq. (1) for the frequency response function H is given by (Ewins, 1984):

$$H = \frac{\Sigma_{k=1}^{p} G_{yx}}{\Sigma_{k=1}^{p} G_{xx}} \tag{2}$$

where

G_{xx} = excitation auto spectrum,

G_{yx} = response cross spectrum,

p = number of samples of spectra.

For an undamped MDOF system the eigenvalue problem is governed by: (3)

$$[K-\omega^2 M\ ]\ [\phi] = 0 \tag{3}$$

For small changes in the stiffness matrix ΔK, we expect small changes in ω^2 and ϕ. Thus, Equation 3 becomes:

$$[(K+\Delta\ K) - (\omega^2+\Delta\ \omega^2)\ M][\phi+\Delta\ \phi]= 0 \tag{4}$$

or, for a single mode i, ignoring second order terms, Equation (4) becomes:

$$[K-\omega^2\ M]\ [\phi+\Delta\ \phi]+[\Delta\ K - \Delta\ \omega^2\ M]\ [\phi] = 0 \tag{5}$$

Premultiplying Equation 41 by $[\phi]^T$, the first term of Equation 5 vanishes since the coefficient matrix in Equation 3 is symmetric, and, for a single mode i, the second term reduces to:

$$\Delta\omega^2_i = \frac{\phi_i^T \Delta K \phi_i}{\phi^T_i M \phi_i} \tag{6}$$

The system's global stiffness difference matrix ΔK entries are zero, except the entries corresponding to the damaged element local stiffness difference matrix Δk_a for element a. The numerator of Equation (5) can be expressed in terms of the strain energy (SE) of element a for mode i:

$$SE_{(ia)} = \phi_{ia}^T k_a \phi_{ia} \tag{7}$$

Consider an element in flexure; then all the stiffness elements k_{ija} are proportional to the quantity EI_a. Thus, damage in an element, through either, loss of stiffness or inertia, would influence the quantity EI and k_{ija} equally. Assume the damage is expressed as a fraction of the original stiffness:

$[\Delta k_a = p_a \bullet k_a]$

For a damage in element a, Equation (5) can be expressed as:

$$\Delta\omega^2_{ia} = p_a x \frac{\phi_{ia}^T k_a \phi_{ia}}{\phi_i^T M \phi_i} \tag{8}$$

A similar expression can be derived for damage in another element b:

$$\Delta\omega^2_{ib} = p_b x \frac{\phi_{ib}^T k_b \phi_{ib}}{\phi_i^T M \phi_i} \tag{9}$$

Combining Equations (51), (52), and (53), we get:

$$\frac{\Delta\omega^2_{ib}}{\Delta\omega^2_{ia}} = \frac{p_b \text{ x } SE_{(ib)}}{p_a \text{ x } SE_{(ia)}} \tag{10}$$

Let the dimensionless frequency variation $\Delta\omega^2$ be defined as:

$$\Delta\omega^2 = \frac{\omega^2_{damaged} - \omega^2_{undamaged}}{\Delta^2_{undamaged}} \text{ x } 100\% \tag{11}$$

Equation (54) suggest that for an equal (small) magnitude of damage, the variation in the natural frequency $\Delta\omega_{ia}^2$ for mode i depends on the relative strain energy contribution of the damaged element a for this mode.

4. Results

The analog measurements obtained as discussed above, were subsequently filtered and analyzed using conventional digital signal processing techniques to obtain some pertinent structural dynamics response measures. Among the key items obtained from this data processing phase were the model's frequency domain quantities (FDQ) that describe the structure's (output) acceleration and the applied (input) load. Figures 5 and 6 show sample FDQ plots indicating the influence of the load application position and the monitoring station location on the dynamic characteristics of the model bridge before and after the introduction of an intentional

Table 1 presents the natural frequencies of the first 4 modes for the undamaged case and all damage cases.

Define the ratio of the strain energy contents of any two elements a and b for mode i as:

$$R_{ab}^{\ i} = \frac{SE_b^{\ i}}{SE_a^{\ i}}$$

Damaged elements contributions to total strain energy for individual modes (Analytical model).

Table 2 presents the ratio $R_{ab}^{\ i}$ for the elements corresponding to the defect location.

Table 3 shows the variations of the natural frequencies ($\Delta\omega^2$) for all damaged cases, and the normalized values for variations obtained by dividing all the values by the variation of damage case 1.

5. Discussion

Subject to the limitations of the test under discussion, analysis of the results indicates the following:

- the structure's dominant frequencies can be adequately, and consistently determined using reasonable monitoring locations.
- changes in the frequency of a particular mode due to damage depends on the relative position of the damage in that mode, assuming a small change in the stiffness.
- significant damage at a sensitive location of a mode (node) may result in a major effect on the mode shape and frequency due to the sensitivity of the damaged element.
- the presence of "damage" or deterioration influences the shape (amplitude, width) of the FRF in a noticeable way; by monitoring a wide frequency range, it is easier to detect even relatively small changes/deterioration in the structure by observing the spacing between the modes.
- the input and output spectra, and the correlation function are sensitive indicators of changes in the modal parameters of the model.

Further details regarding this study are available in the work of Waqfi (1990).

6. Summary and Conclusions

This study was not meant to be a comprehensive treatment of the title subject, but rather an exploratory investigation of the usefulness of various structural parameter estimates, that can be conveniently extracted during an experimental test, to quantify changes in the characteristics of physical systems. On this basis it appears that system identification techniques may offer some useful and reliable tools for detecting the extent of changes in monitored structures.

7. Acknowledgements

This study was supported by grants from the U.S. National Science Foundation.

8. References

Agbabian, M.S. and Masri, S.F., (Editors), (1988), Proceedings of the International Workshop on Nondestructive Evaluation For Performance of Civil Structures,} Dept. of Civil Engineering, University of Southern California, Los Angeles, CA 90089-2531.

Agbabian, M.S., Masri, S.F., Miller, R.K., and Caughey, T.K., (1988), "A System Identification Approach to the Detection of Changes in Structural Parameters," Structural Safety Evaluation Based on System Identification Approaches, Edited by H.G. Natke and J.T.P. Yao, Friedr. Vieweg & Sohn, Braunschweig/Wiesbaden.

Biswas, M., Pandey, A.K., and Samman, M.M., (1989), "Diagnostic Experimental Spectral/Modal Analysis of A Highway Bridge," The International J. of Analytical and Experimental Modal Analysis, Vol. 5, No. 1, pp. 33-42.

Cawley, P., Adams, R.D., (1979), "A Vibration Technique for Non-Destructive Testing of Fibre Composite Structures," J. of Composite Materials}, Vol. 13, pp. 161-175.

Cawley, P., Adams, R.D., (1979), "The Location of Defects in Structures from Measurements of Natural Frequencies,'' J. of Strain Analysis, Inst. of Mech. Eng., Vol. 14, No. 2, pp. 49-57.

Ewins, D.J., (1984), Modal Testing: Theory and Practice, John Wiley & Sons Inc.

Flesch, R.G., and Kernbichler, K., (1988), "A Dynamic Method for the Safety Inspection of Large Prestressed Bridges," edited by Agbabian, M.S., and Masri, S.F., Proceedings of the International Workshop on Nondestructive Evaluation for Performance of Civil Structures, Univ. of Southern California, pp. 218-230.

Natke, H.G., and Yao, J.T.P., (1986), "Research Topics in Structural Identification," Proc. Third Conference on Dynamic Response of Structures, EM Div./ASCE, UCLA, 31 March - 2 April.

Natke, H.G., and Yao, J.T.P., (1988), "Structural Safety Evaluation Based on System Identification Approaches," Proceedings of the Workshop at Lambrecht/Pfalz, Friedr. Vieweg & Sohn, Braunschweig/Wiesbaden.

Stephens, J.E., (1985), "Structural Damage Assessment Using Response Measurements," Ph.D. Thesis, School of Civil Engineering, Purdue University, West Lafayette, IN.

Waqfi, O.M., (1990), "Vibrational Analysis Technique for Nondestructive Evaluation of Bridge," Ph.D. Thesis, Department of Civil Engineering, University of Southern California, Los Angeles, CA 90089-2531.

Yao, J.T.P., (1985), Safety and Reliability of Existing Structures, Pitman Advanced Publishing Program, Boston.

Table 1. Natural Frequencies for Various Damage Locations

Mode	Frequency (Hz)			
No.	Und.	D#1	D#2	D#3
1	8.03	7.96	7.71	7.93
2	20.40	20.26	19.79	18.67
3	20.90	19.86	20.19	19.83
4	40.20	38.87	37.59	36.98

Table 2. Damged Elements Contributions Total Strains Energy for Individual Modes (Analytical Model)

Mode No.	Element D#1	Element D#2	Element D#3
1	1.00	11.14	0.60
2	1.00	263.0	64.3
3	1.00	0.11	0.89
4	1.00	1.94	0.54

Table 3. Frequency Variations for Various Damage Locations (Experimental Results).

Mode	Freq. Variations, $\Delta\omega^2$			Norm. Variations		
No.	D#1	D#2	D#3	D#1	D#2	D#3
1	-1.79	- 7.84	- 2.78	1.00	4.38	1.55
2	-1.40	- 5.91	-16.30	1.00	4.22	11.64
3	-9.75	- 6.68	- 9.94	1.00	0.69	1.02
4	-6.50	-12.58	-15.40	1.00	1.94	2.37

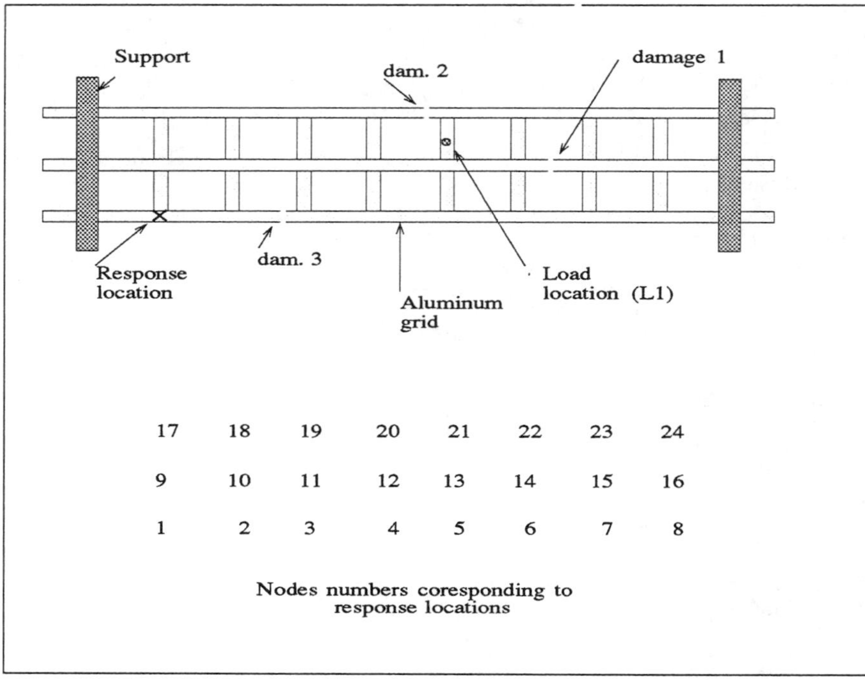

Fig. 1 Model General Shape Showing the Damage, Load and Response Locations.

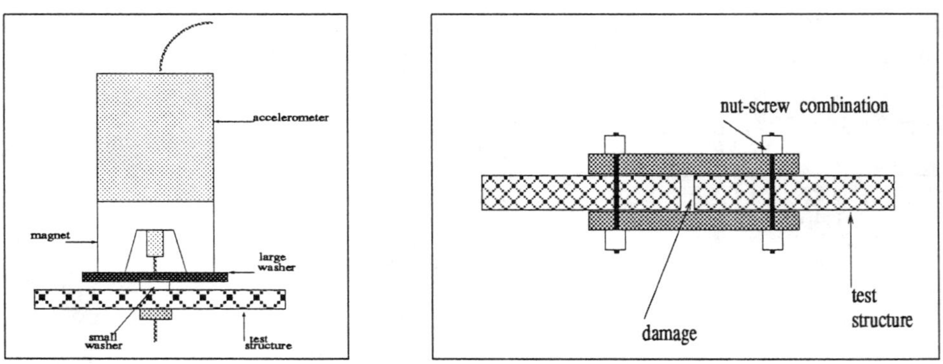

Fig. 2 Transducer Assembly, and Defect Configuration.

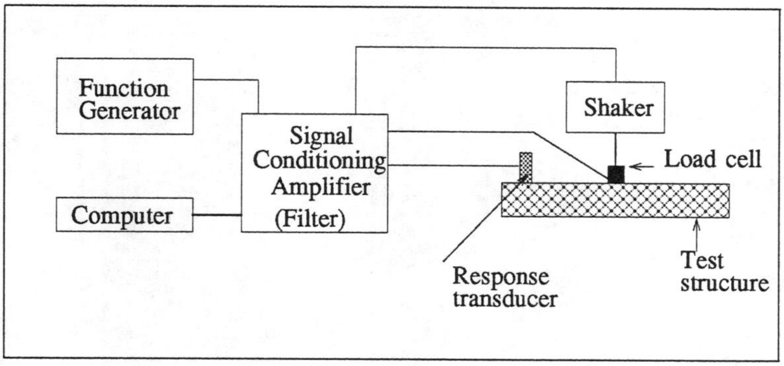

Fig. 3 Instrumentation.

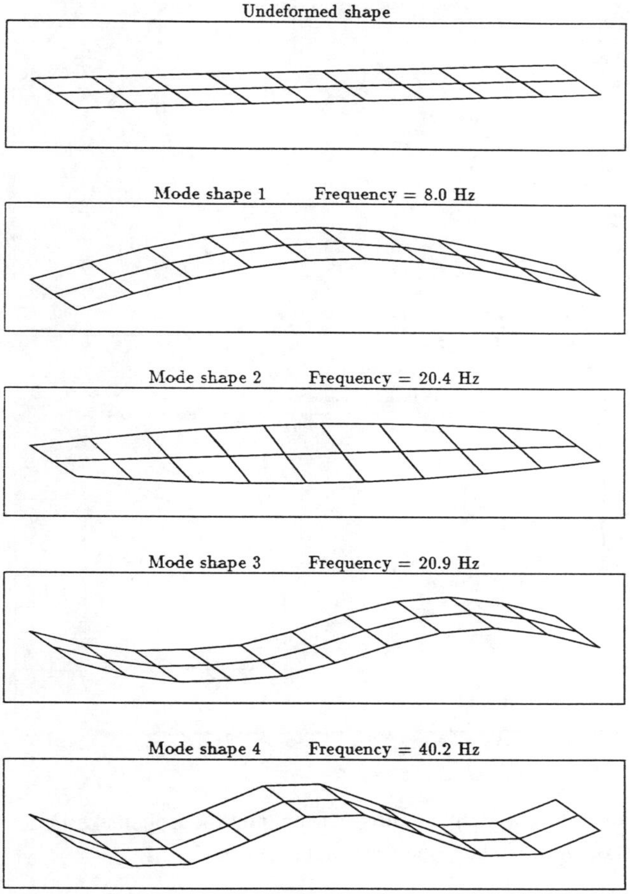

Fig. 4 Model Mode Shapes.

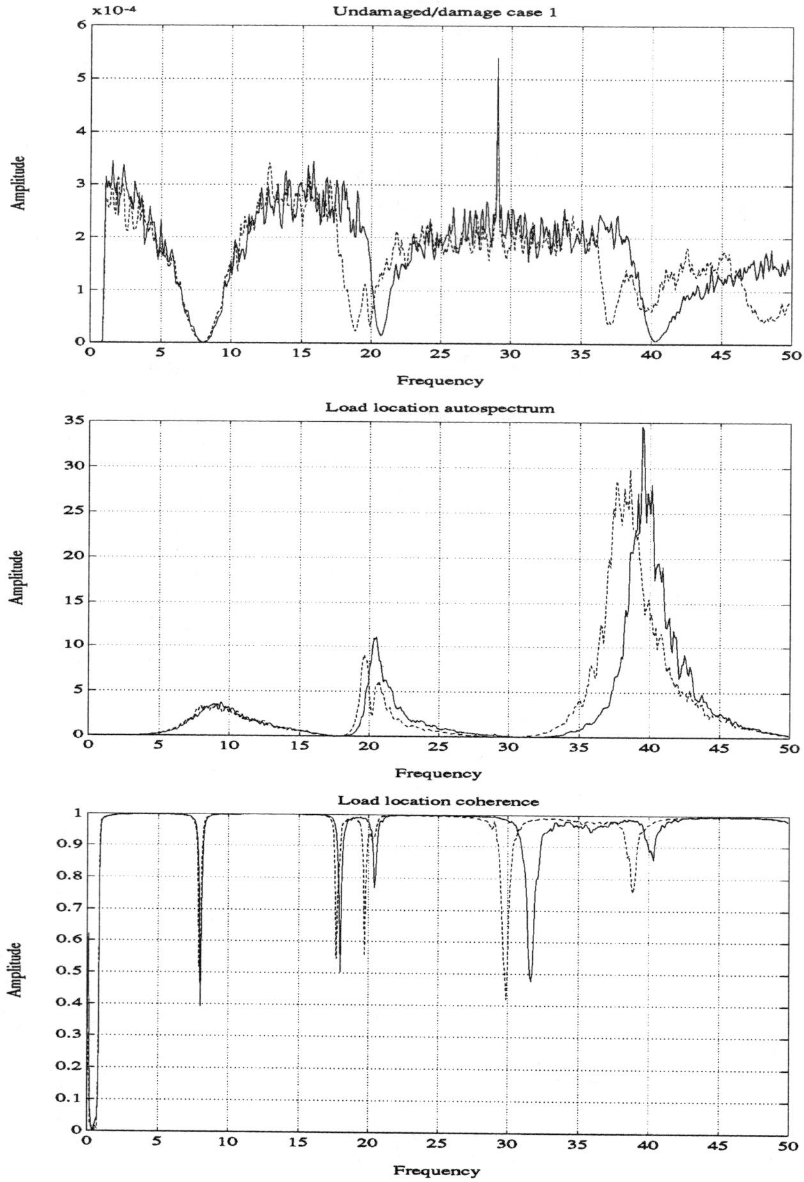

Fig. 5 Input Autospectrum, Output Autospectrum, and the Coherence Functions for the Load Locations (L1) for Undamaged Case and Damage Case 1.

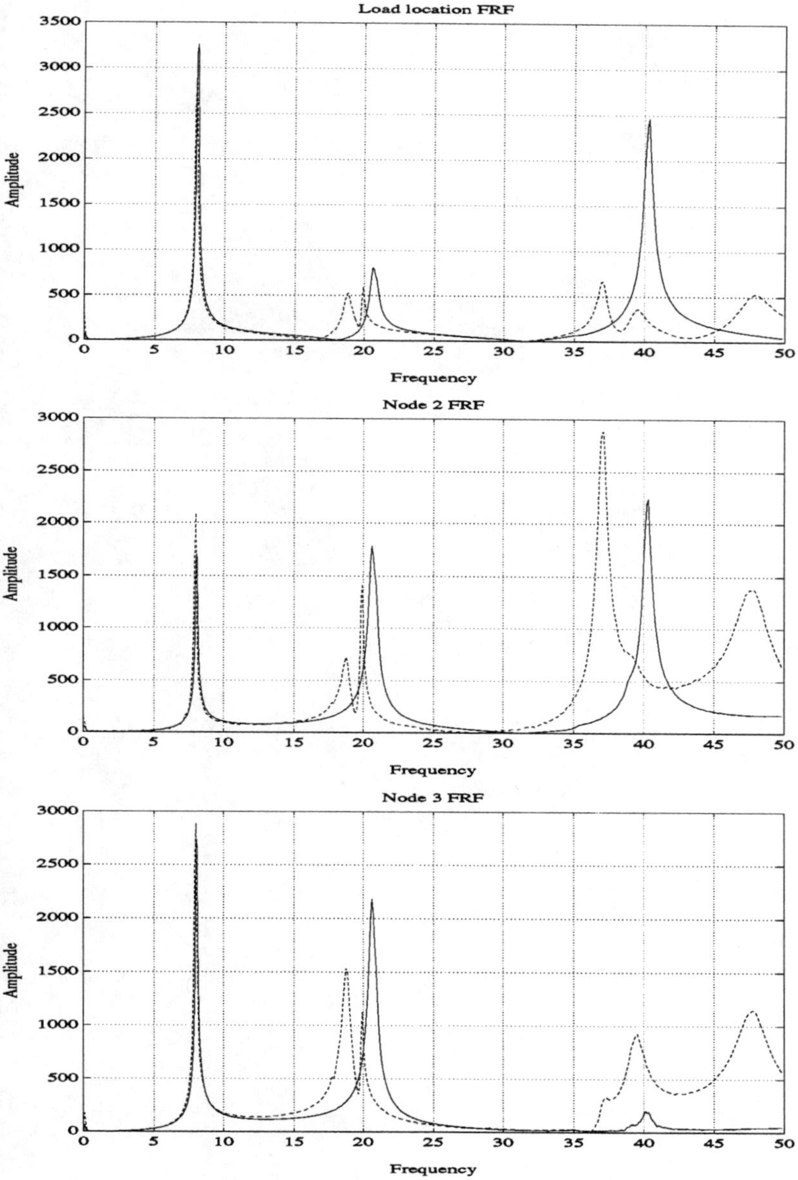

Fig. 6 Frequency Response Functions for Undamaged and Damage Case 3 (Full Damage).

DETECTION, DIAGNOSIS, AND MONITORING OF CRACKED PRESTRESSED CONCRETE BRIDGES

J. CHATELAIN and B. GODART
Ministry of Transport
Central Laboratory for
Bridges and Roads
58, Boulevard Lefebvre
75732 PARIS Cedex 15
FRANCE

J.L. DUCHENE
Ministry of Transport
Regional Laboratory for
Bridges and Roads
Rue de l'Egalité BP34
93350 LE BOURGET
FRANCE

ABSTRACT. A certain type of prestressed concrete bridge built in France form 1965 to 1975 may present a type cracking which is characteristic of insufficient bending strength. These bridges may be prematurely endangered by fatigue or corrosion of the tendons. Often these bridges need strengthening by additional prestressing. This article describes the stage that necessarily precedes such a strengthening, and that includes:

a) the detection of deficient structures among all bridges in service according to the Instruction on the Inspection and Maintenance of Bridges of the French Ministry of Transport.
b) the diagnosis based on in-situ testing, such as the weighing of bearing reactions, but mainly the decompression moment method. The principle of this method consists in progressively loading the bridges with trucks in order to find the loading case corresponding to the opening of the joint.
c) the monitoring of the bridge, which may be necessary if the rupture by fatigue of prestressing wires in the joints is of concern.

1. Introduction

The use of prestressed concrete in highway bridge construction came into widespread use in France beginning in the 1950s.

In the first generation of bridges (those built before 1960), there is reason to fear a sometimes grave pathology - two forms of corrosion of the prestressing cables: generalized corrosion and stress corrosion. At the time, the danger of such corrosion was underestimated: too little importance was attached to the proper execution and inspection of grouting, and waterproofing membranes were considered unnecessary on prestressed concrete decks.

Toward the 1960s, "balanced cantilever" construction techniques imported from Germany were used in France on prototype structures, structures with hinges in the middle of the span (Chazey bridge, destroyed and rebuilt: Beaucaire bridge, being rebuilt: and Savines bridge, in service, Figure 1).

A. S. Nowak (ed.), Bridge Evaluation, Repair and Rehabilization, 145–159.

The spans were rapidly made continuous by restraining in the middle of the span, and use of this technique was extended to the range of medium spans and finally to larger spans substantially in excess of 100 metres.

Since some effects that were especially critical in structures of this type were poorly understood (redistribution of hyperstatic reactions by creep), or even completely ignored (effects of the temperature gradient), some pathology appeared in these decks in the 1970s, a consequence of overconfidence in a technique that was still immature and incompletely mastered. This pathology takes the form of cracks in the webs and bottom slabs of the box girders.

We shall begin with a brief description of this pathology (symptoms and causes of the distress), then show how the overall mechanical condition of the deck can be assessed on the basis of measurements made on the structure.

The general principle for the strengthening of these bridges is the addition of prestressing, and it is obvious that a sound knowledge of the true mechanical condition of the structure is one of the keys to good strengthening design.

2. Nature and Causes of the Distress

In 1979, the French Ministry of Transport issued a Technical Instruction on the Surveillance and Maintenance of Bridges, intended for the ministry's operating agencies (ref. [1]). The first part of this instruction states the provisions applicable to all bridges, while the second contains a score of sections per major type of structure. Section 32.2 of ref. [1] describes the distress observed in the prestressed concrete bridges with box girders or ribs. The descriptions of distress that follow are taken from it; they are only simplified schemata of the actual signs of the distress, which are most often more or less clear superpositions of the schemata.

2.1 CRACKS REFLECTING INSUFFICIENT BENDING STRENGTH (BENDING MOMENT + SHEAR FORCE).

These cracks are most often found towards the middle of the span and in the "zero moment" zones (Figure 2).

In box girders, they most often extend completely across the bottom slab and upward in the webs, where they may be inclined in the vicinity of the bearings, because of the shearing stress (Figure 3). They most often originate from the joints between segments, a zone of reduced strength; they may be quite open at the bottom of the box girders (opening width 1 to 2 mm), but may also be very fine, even hard to see (Figure 4).

It should be noted immediately that the extent of opening of these cracks cannot be related to the degree of severity, as we shall see later. This type of cracking is often accompanied by an abnormal permanent deflection visible to the naked eye.

The cause of this insufficient bending strength are now well known:

- underestimation (or even complete ignorance) of the effects of the redistribution of the hyperstatic reactions by the creep of the concrete, which affects bridges of which the static

schema of construction differs from the final schema in service (the general case of bridges built by cantilevering and made continuous by restraining);
- neglect of the effects of exposure of the upper side of the deck to sunlight, which produces a temperature gradient in the structure in which strains are restrained, and induces large loadings (sometimes of the same order of magnitude as the maximum traffic load);
- an actual prestress in the structure that is less than the design estimates (relaxation of tendons underestimated, friction of the cables in their ducts greater than assumed in the design and highly dependent on the quality of the construction work, etc.);
- excess permanent loads (overlaying of pavements and underestimation of equipment weight, etc.);
- a prestress counted at full value starting from its anchorage, with no allowance for a diffusion length;
- deficient participation by the bottom slab because of the cracking cause by the unopposed downward thrust of the continuity cables;
- etc.

2.2 OTHER TYPES OF CRACKS

Two other types of cracks are characteristic of the pathology observed in these bridges in the 1970s:
- the diffusion and entrainment cracking that develops close to the cable anchor points when the passive reinforcements are insufficient, especially when several units are terminated in the same section; (Figures 5, 6, and 7).
- "down thrust" cracking, or longitudinal cracking of the bottom slab in the middle of variable-depth spans, caused by transverse bending of the insufficiently reinforced bottom slab by the downward thrust of the curved continuity cables (Figures 8 and 9); the wider the box girder, the greater is this bending.

2.3 MAIN CONSEQUENCE

It is obvious that a box girder with a network of cracks such as described above will not function quite as the designer intended.

As it happens, the ultimate limit state is rather little affected, and the major drawback lies in the fact that prestressing cables cross the bending cracks (which may coincide with the entrainment cracks), together with a small quantity of passive reinforcements, when the segments are cast in situ (the passive reinforcements do not cross the joints of precast segments). These reinforcements are subjected to abnormal overstresses due to traffic loads, overstresses that may be aggravated by the effects of the temperature gradient, and there is a risk of fatigue failure of these reinforcements. Right at the cracks, local phenomena in which friction and corrosion are combined may give a fatigue life shorter than that normally found on cables tested in the open air.

The main aim of the strengthening prestressing is to eliminate these tensile loads and so remove the risk of failure of the cables.

3. Detection of Distress Structures

The first step to be taken is detection of the distressed structures; surveillance of the structures in accordance with the official rules (Ref. [1]) is the means. The detailed inspection that is in principle performed every five years is the occasion for an exhaustive survey of cracking and minutious examination of the condition of the joints, and is the foundation for subsequent investigations.

4. Means of Investigation for a Sound Diagnosis

4.1 WEIGHING OF BEARING REACTIONS

Given the situation described in the detailed inspection report, in which a system of bending cracks (most often accompanied by diffusion cracks) is identified, an attempt has been first made to explain the cracking condition by a global approach:

- In all cases, a complete recalculation has been performed using modern design methods; insofar as possible, the actuals stages of construction, as reconstituted, have been taken into account and realistic data have been used (coefficients of friction and relaxation estimated as the date of recalculation, not the values imposed by the regulations at the time of construction).
- Wherever possible, the bearing reactions on abutments were measured.

The results of these investigations have invariably given measured values that are different from the design estimates, with the difference consistently in the direction of a larger than anticipated redistribution of hyperstatic forces. These measured values, put back into the recalculation, most often suffice to explain the actual cracking condition of the decks.

In reality, this measurement technique, which we shall not describe here, has occasionally proved to be inadequate, because:

- it was impossible to place a jack between the deck and the abutment;
- the indispensable temperature corrections entailed sacrificing some of the necessary precision, because the problem was too complex for a rigourous solution;
- the measurement result was compared to a recalculation that was itself full of uncertainties, and the difference found was not explained solely by an incorrectly estimated redistribution effect.

This led us to develop another method, the "method of decompression moments", which we shall now explain (Ref. [2]).

4.2 METHOD OF DECOMPRESSION MOMENTS

4.2.1 ***Theoretical behavior of a decompressed joint.*** The theoretical behaviour of a decompressed joint is illustrated in Figure 10, which shows the variation of strain $\Delta\varepsilon_b$ in the bottom fibre of the concrete and the variation of strain $\Delta\varepsilon_a$ at the joint, in the continuity cables, versus the variation of the external moment applied to the joint

Up to point I, the joint remains closed and acts like a uniform section; the strain of the prestressing tendon (and of the passive reinforcement) is equal to that of the concrete. Beyond point I, the joint opens and there is gradual decompression of the bottom slab, then of the webs.

The stress in the concrete near the crack goes to zero and remains there thereafter, while the stress in the tendons and reinforcements becomes a nonlinear function of moment.

The objective, then, is to detect the change in the behaviour of the joint through measurements of the strains of the tendons and reinforcements and of the concrete.

4.2.2 ***Principle of the decompression test.*** The aim of the decompression test is to assess the bending deficit of a prestressed concrete structure having cracked joints. Its principle consists of having a convoy of trucks move gradually across the bridge so as to produce a constant increase of the moment in the (instrumented) joints and determine at which value of the applied moment the joint is decompressed. This operation tells us how much reserve moment the joints have in the unloaded condition.

When a joint is found to be open in the unloaded condition, negative-moment loading cases (trucks on adjacent flanking spans) are used in an attempt to close the joint and determine the position of point I.

4.2.3 ***Practical execution of the decompression test.*** In practice, the decompression test is carried out after a stage of "preliminary exploration of the joints", in which limited instrumentation (basically displacement transducers spanning the joints) is used to assess the exact extent of the distress by identifying those joints that open and close under the traffic carried by the bridge. This first stage is used to select two or three representative joints (including the most distressed) for more complete instrumentation, as we shall now see.

4.2.3.1 *Use of gauge/transducer pairs.* The gauge/transducer (G/T) pair is a combination of a long-base (approximately 10cm) wire resistance strain gauge and an inductive displacement transducer, placed athwart the joint so that the two instruments are aligned.

The gauge must be far enough from the edge of the joint (at least half the wall thickness) to be unaffected by highly localized phenomena.

It is not always possible to precisely identify the moment at which the joint opens by interpreting the readings from these instruments versus the bending moment induced in the section; for this reason, it is best to use these two instruments simultaneously and plot the G/T diagram (Figure 11).

The G/T diagram is in two distinct parts. The first part, which is substantially linear and has a small slope, represents the gradual decompression of the concrete at the edge of the joint.

If the transducer is sufficiently precise (precision on the order of a micron or better), and if the decompression is ideal, the slope of this curve is determined by the functioning of the transducer as an extensometer. The second part, which is practically vertical, represents the behaviour of an open joint; the sensor indicates a large opening, while there is not further decompression of the gauge.

There are two ways of interpreting this diagram. In one, an attempt is made to identify the loading case corresponding to the intersection of the two straight lines; the equations of statics can then be used to determine the value of the bending moment at which the joint opens. In the other, an attempt is made to assess the reserve compressive strength of the concrete in the no-load condition (Figure 11). This approach uses the law of behaviour of the concrete ($\sigma=E\varepsilon$) and requires the application of strength of materials hypotheses ($M = \sigma I/v$).

We consider the first approach preferable, because it yields a result more directly by giving an estimate of the reserve bending moment at the location of the G/T pair.

The investigation of a joint by G/T pairs requires a large number of measurement points, because the interpretation of the results is often made difficult by a relatively complex working of the joint, which has contact defects created by the anchoring of the continuity cables in bottom slab, differences in shrinkage between parts of different thickness, mating faults between precast segments, etc. (Ref. [3]).

The experience acquired in investigations of the joints of many bridges enables us to classify joints into two main families, whether the segments are precast or cast in situ.

The first family consists of the joints of bridges in which the continuity cables are grouped in the two bottom gussets; faster decompression of the middle of the bottom slab and of the webs is generally found in these joints, while decompression of the bottom gussets occurs later (Figure 12).

The second family consists of the joints of bridges in which the continuity cables are distributed over the whole width of the bottom slab; contact defects are often found behind the continuity cable anchorages, reflecting entrainment cracks and perturbations of the stress field induced by diffusion cracks and down thrust cracks (Figure 13).

When an idea of the mechanical functioning of the joint has been formed the reserve bending moment at the joint can be determined by investigation, by combining all information from each G/T pair.

4.2.3.2 *Investigation of prestressing cables.* When investigation by G/T pairs fails to yield a solution to the problem, and especially when it is impossible to close a joint which is open in the unloaded condition by negative-moment loading cases, it is necessary to measure the variations of the strains of the continuity cables. This is done by opening a window, as small as possible, in the concrete, cutting the duct, removing the grout, and glueing vary small strain gauges to a few basic wires of the strands so exposed.

These strain measurements naturally make it possible to estimate the tensile loads to which the cables are subjected under gradual loading. But the interpretation of the measurements can be taken even further by comparing them to a strain calculation. This calculation, of the "reinforced concrete in compound bending" type, assumes gradual upward growth of the crack in the bottom slab, and then through the webs; correct adherence of the cables; and the theoretical prestressing force in the joint is taken as the normal force. An attempt can then be made to match the experimental and theoretical curves, as shown in the example of Figure 14, to determine the unloaded crack height and the moment needed to close the joint (in the example shown, about 9 MNm). If the joint opens in a "random" manner, the compound bending calculation is more difficult.

The greatest difficulty encountered in the investigation of cables is the impossibility of ascertaining the exact grouting quality.

Figure 15 shows the slippage lengths of three cables in the same joint, based on simultaneous tendon strain and joint opening measurements taken just above each of the cables.

4.2.3.3 *Use of clinometers*. The recent use of precision clinometers (of the order of 10-6 rd) has two advantages. First of all, when placed flanking a joint, on the top gusset, clinometers advantageously replace G/T pairs in providing information of a global character: the rotation of the joint.

Most important of all, joint rotation measurements are essential when the aim is to model the cracked structure. According to Ref. [4], two ways are possible:

- by using elastic hinges in place of the open joints, and taking into account the true coefficients of rigidity for the hinges. The use, in the calculation, of these hinges, the rotation of which is proportional to the moment applied to them ($\Delta M = K\Delta\Theta$) makes it possible to correct the scale of the moments induced by the movements of trucks, a scale which may in fact be altered by the opening of the joints of a hyperstatic structure, as we have confirmed in weighings of bearing reactions (the cracked centre span is more flexible) (Ref. [5]).
- by introducing short elements with smaller rigidities near the joints, these rigidities having characteristics derived from rotation measurements made in the cracked zone.

4.2.3.4 *Temperature conditions of structure during test.* It is always necessary to know the state of thermal loadings to which the structure is subjected at the time of testing, which must be taken into account in estimating the reserve moment in the unloaded condition.

5. Other Experiments

5.1 FATIGUE INVESTIGATION

For some structures with measured cable strain values that seem too low to require immediate strengthening, but even so are not negligible, it is appropriate to check the fatigue life of the prestressing tendons that cross the cracked joints. To carry out this check, a method of estimating fatigue damage to prestressing cables has been devised. This method, rather cumbersome to apply, is currently being simplified.

Before the method is examined, it should be pointed out that the tensile stresses recorded in the course of a day may be broken down into two signals:

- a large-amplitude, low-frequency signal produced by thermal effects;
- a weaker signal, at very high frequency, caused by traffic.

It should also be noted that the deformation of the cable produced by a given vehicle varies with the time of day at which it passes, because of the attendant variations in the thermal gradient and the nonlinearity of the relation between the moment and the tensile load of the cable.

The method consists of fitting the bridge with two LCPC piezoelectric strips connected to a fine-scale traffic analysis station to record the time history of loading for several weeks, placing a few temperature probes at judiciously chosen points to record daily variations of the temperature gradient in the structure, and recording the strains of the prestressing tendons simultaneously. The measured traffic is assumed to be steady and the moments induced by the traffic are calculated from the influence line of moments. Thermally induced moments are calculated with the help of a record of annual variations of the temperature gradient. These two moments are "added" together in a probabilistic simulation, and the tensile overloads are then calculated using the experimental moment tensile load curve. Counting the tensile loads by the "rainflow" method, applying Miner's model and referring to Woehier's S-N curves, it is then possible to assess the fatigue life of the prestressing cables of the structure.

This method, a powerful tool to assist decision-making by the operator of the structure, is however, still limited by its great sensitivity to the choice of the Woehler curve and by a lack of knowledge of the in-situ fatigue resistance of the cables (Ref. [6]).

5.2 EXPERIMENTS DURING REPAIR

We have just examined a large share of the strategy of investigation prior to repair (Ref. [7]). However, if the joints are wide open in the unloaded condition (if an opening of 1 to 2 mm is observed), major instrumentation is often unnecessary, because it is decided to inject the joints with an epoxy resin and apply additional prestress designed to take out the effects of traffic and of the temperature gradient.

If the joints are not so wide open, trucks are parked on the bridge for the injection and left in place long enough for the resin to polymerize. Strain measurements made on sections already equipped, when the trucks leave, have shown that this type of operation introduces non-negligible compressive stresses in the bottom fibre; but these are not taken into account in the calculations because of gaps in our knowledge of the long-term performance of injected resins. Strain measurements may be made again when the additional prestressing cable are tensioned, to make sure that the prestress is properly distributed in the joint and check its effectiveness on the structure.

6. Conclusions

When the bending strength of a prestressed concrete bridge is insufficient, the structure is always completely reanalyzed. It is rare for this reanalysis by itself to be sufficient for a judgement of the true mechanical condition of the cracked deck; this is why we have had to develop certain techniques of investigation, such as weighing of the bearing reactions and the "decompression moments" test. In the different versions of this test, gauge/transducer pairs or direct measurement of the tensile stresses in the cable or of section rotations are applied to a cracked joint.

Experience on about twenty instrumented bridges has shown us that the stress distribution in the vicinity of a joint may be highly complex, and that the mechanical behaviour of such a joint when it opens does not always comply with exact laws. For this reason, each operation is a special case, and the strategy used must be adapted to each new case.

This experience is part of the know-how acquired in the course of various strengthening operations carried out using additional prestressing (Figure 16), which anticipated the "external prestressing" technique now used in France for the construction of new bridges.

7. References

[1] LCPC-SETRA (1979), "Instruction Technique pour la Surveillance et l'Entretien des Ouvrages d'Art," Ministère des Transports, Direction des Routes, lère et zème parties.

[2] Chatelain, J., Bruneau, J., Duchene, J.L. (1981), "The evaluation of actual bending strength of some types of prestressed concrete decks by load tests," Int. Conf. on gestion des ouvrage d'art, Bruxelles-Paris 13-17 April, Editions ENPC, Vol. 1, pp. 15-20.

[3] Chatelain, J., Bruneau, J. (1985), "Les joints de voussoirs dans les ponts en béton précontraint," Annales ITBTP, No. 435, Série béton 230, June.

[4] Abdunur, C., Duchene, J.L. (1988) "Mesures de rotations pour le schéma statique d'un ouvrage fissuré," Proc. Int. Conf. Measurements and Testing in Civil Engineering, Lyon Villeurbanne 13-16 Sept., Vol. II, pp. 283-291.

[5] Godart, B. (1987) "Approche par l'auscultation et le calcul du fonctionnement de ponts en béton précontraint fissurés," U.S. European Workshop on Rehabilitation of Bridges, CEBTP Saint Rémy-lès-Chevreuse, 22-25 June.

[6] Jacob, B., Carracilli, J., Godart, B., Trouillet, P. (1987) "Etude en fatigue d'un pont en béton précontraint sous actions combinées du trafic et des gradients thermiques," Bulletin de liaison des laboratoires des ponts et chaussees, No. 152, Nov.-Dec., pp. 37-48.

[7] Chatelain, J., Godart, B. (1988) "Evaluation of the actual mechanical behaviour of prestressed concrete bridges," IABSE 13th Congress, Helsinki, June 6-10, pp. 325-330.

Fig. 1 SAVINES Bridge in the French Alps (1959).

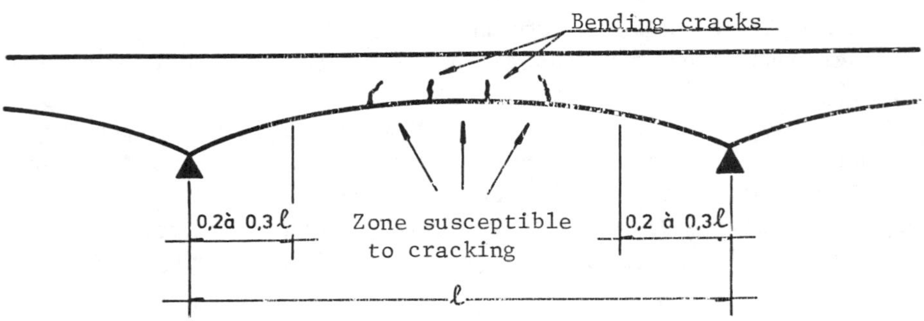

Fig. 2 Zone Susceptible to Bending Cracking.

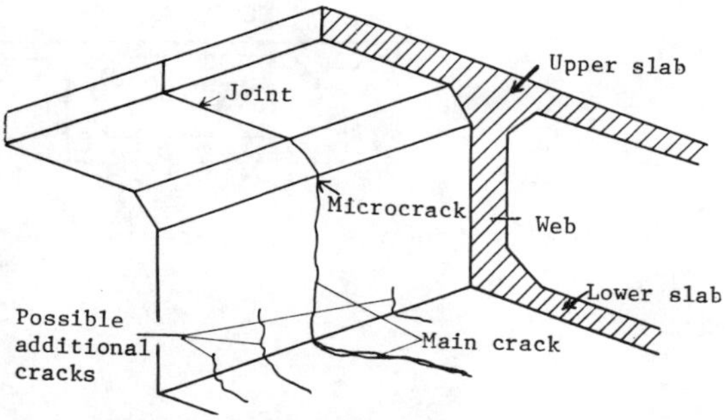

Fig. 3 Detail of Bending Cracks.

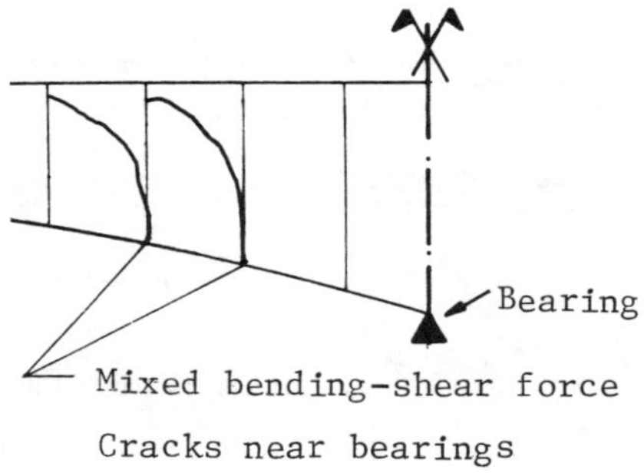

Fig. 4 Influence of the Shear Stresses on Bending Cracks.

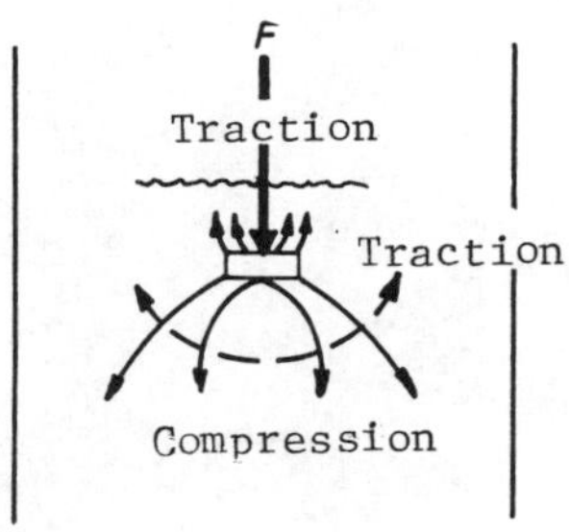

Fig. 5 Local Effects of a Concentrated Load in a Slab.

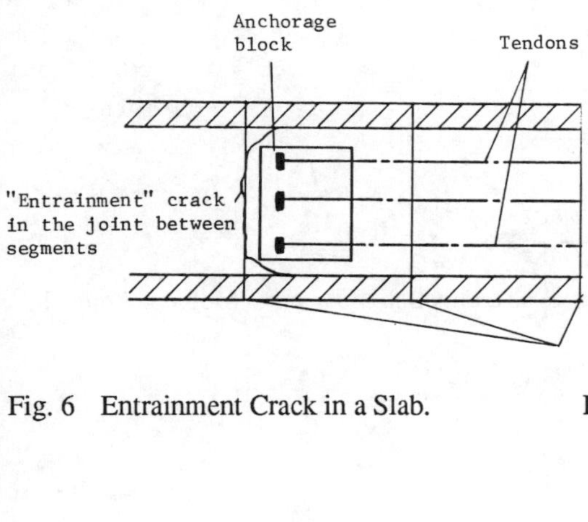

Fig. 6 Entrainment Crack in a Slab.

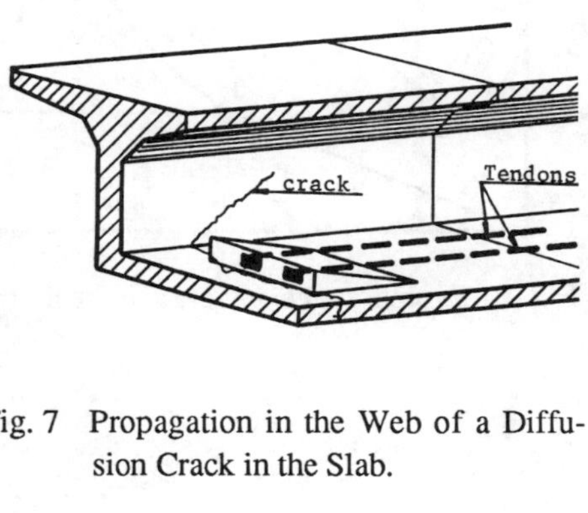

Fig. 7 Propagation in the Web of a Diffusion Crack in the Slab.

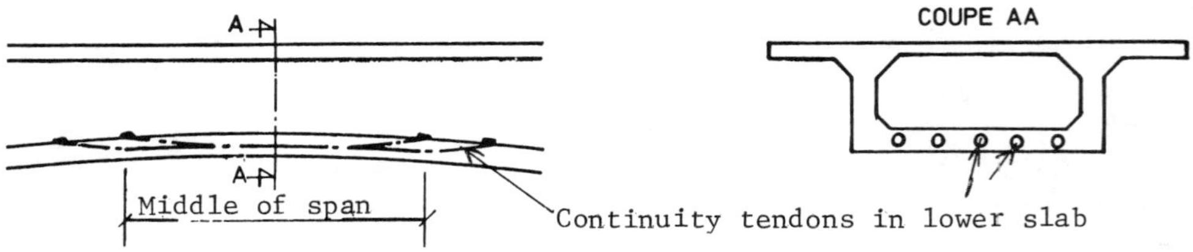

Fig. 8 "Down thrust" in the Middle of a Span.

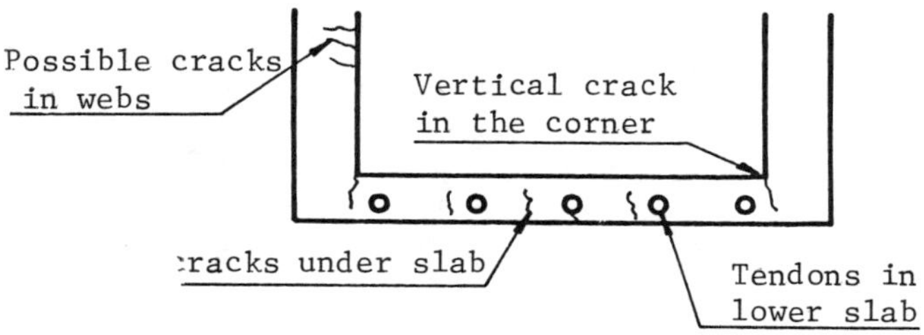

Fig. 9 Location of "Down thrust" Cracks.

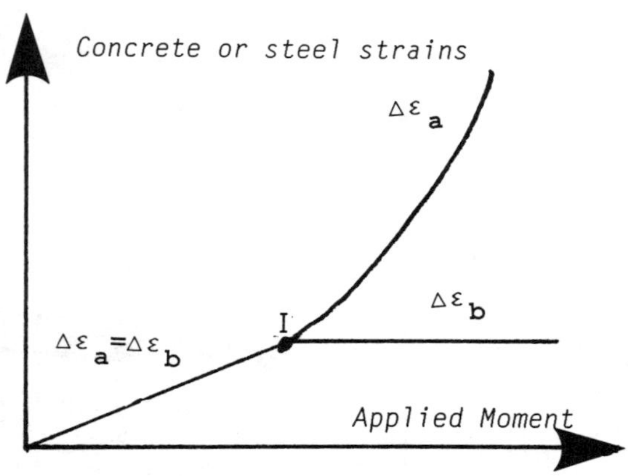

Fig. 10 Evolution of Concrete and Steel Stresses as a Function of External Moment Applied.

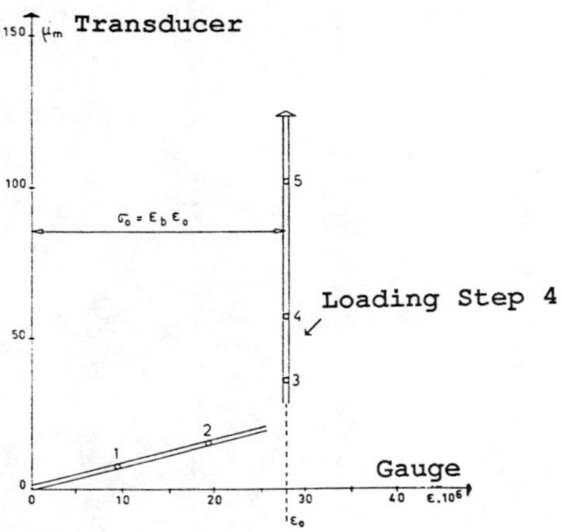

Fig. 11 Gauge/Transducer Diagram (G/T Diagram).

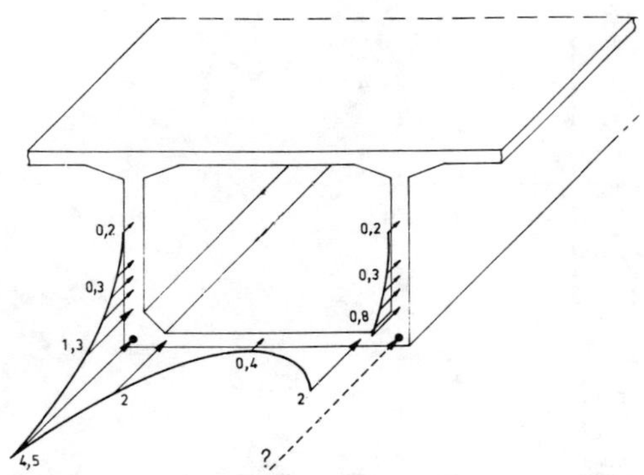

Fig. 12 Map of the Measured Stresses (MPa).

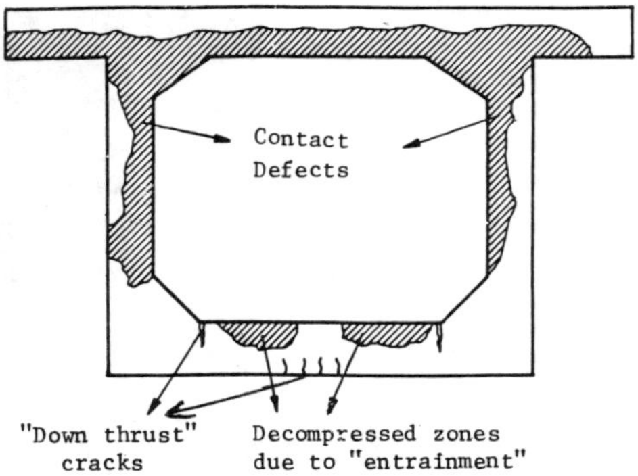

Fig. 13 Example of Contact Defects.

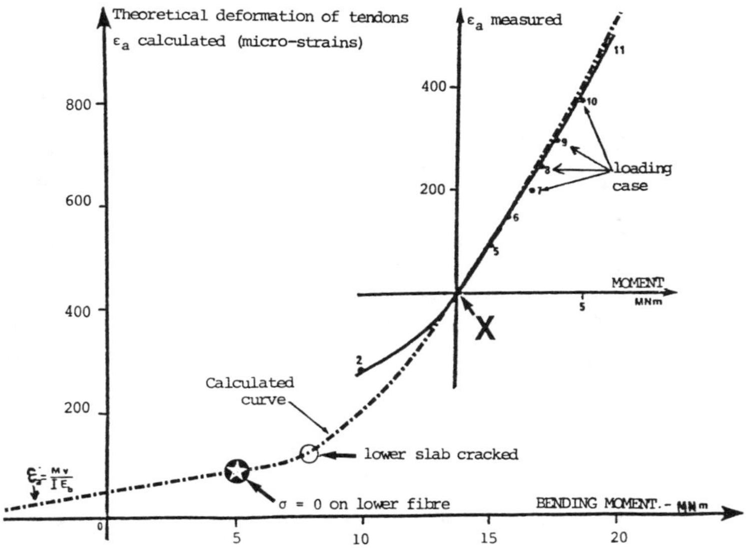

Fig. 14 Attempt to Match Experimental and Theoretical Curves.

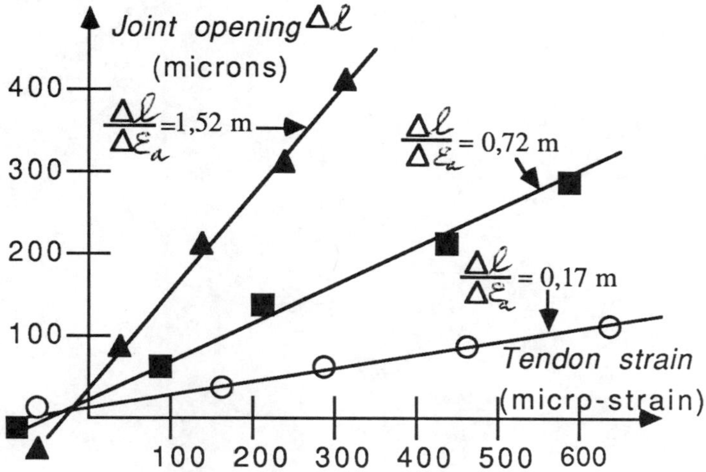

Fig. 15 Diversity of Slippage Lengths.

Fig. 16 Strengthening by Additional Prestressing.

MODAL TECHNOLOGY FOR DAMAGE DETECTION OF BRIDGES

M. BISWAS, A. K. PANDEY and M. M. SAMMAN
Duke University
Transportation and Infrastructure Research Center
School of Engineering
Durham, NC 27706 USA

ABSTRACT. Technologies of analytical and experimental modal analysis have been investigated for their application in damage detection of bridges. In the context of experiments performed on a typical highway bridge in North Carolina, various signals and signal-processed data that can be used for interrogation are presented. By introducing simulated failures in a highway bridge in Pennsylvania and in a number of laboratory bridge models and analytical models, experimental results are examined for their ability to detect faults.

1. Introduction

The two major application ares of analytical and experimental modal analysis are 1) proper diagnosis and resolution of an existing vibration problem, i.e. "trouble shooting, and 2) calibration of computational analytical model, i.e. "parameter identification", for predicting dynamic behavior (Ewings, 1984; Allemang & Brown, 1987). In addition, the use of modal technology as a way of monitoring structural integrity has been recognized in principle for some time (Adam, et al, 1978; Collacott, 1985; Stephens & Yao, 1987; Fisher & Wood, 1988).

Recognizing the need for modern, rapid and reliable non-intrusive methods of inspection of bridges, especially for timely detection of incipient structural failures, the technology of experimental modal analysis is considered a viable option.

For the purpose of using the methods of modal technology, we need not be preoccupied with finding only the modal parameters. We should investigate possible utilization of the whole range of available resources, i.e., hardware, software, and all types of signals and processed data.

The signals and data that can be generated as a result of a set of tests are classified into three main groups:

1. Non-parametric, single channel signals: this is the dynamic response signal form a accelerometer mounted on the structures. The signal is in the form of electric voltage proportional to the acceleration at the point of measurement as a function of time. Using a spectrum analyzer, the raw time domain signal is first digitized. To reduce the effect of "leak-

A. S. Nowak (ed.), Bridge Evaluation, Repair and Rehabilitation, 161–174.

age", a weighting function "window" may be applied at this time. The time record is then processed by Fast Fourier Transform (FFT), producing a frequency spectra.

2. Non-parametric, dual channel signal: this is the "Transfer Function" or the "Frequency Response Function" (FRF), a Laplace domain ratio of frequency domain signals from the accelerometer and from the force transducer of a modally controlled exciter (hammer or shaker). The transfer function can be interrogated via different views, such as 1) magnitude and phase, 2) real and imaginary parts and 3) Nyquist plots. Using standard algorithms of digital signal processing, (Elliott, 1987; Jackson, 1986; Sterns and David, 1988) all non-parametric information are generated within the same spectrum analyzer on a nearly real-time basis.

3) Parametric Data: by using different curve-fitting techniques, groups of transfer function data are processed to generate modal parameters, i.e., modal frequencies, mode shapes and modal dampings (Ewins, 1984; Allemang and Brown, 1987).

Experiments were performed on two highway bridges and a number of laboratory model bridges. The first bridge is a typical steel-girder bridge, and the tests were performed to obtain typical signals and data which would be considered "bench mark" information for subsequent experiments on other bridges. On the second bridge, a major crack was simulated by unfastening splice bolts of one of the main girder. Tests were performed and simulated by unfastening splice bolts of one of the main girders. Tests were performed and subsequently data was analyzed for both the intact and the cracked conditions. Tests on laboratory models and analytical models were performed to evaluate in detail the veracity and efficiency of different experimental techniques, analytical procedures and signal processing methods.

Failure in structure is diagnosed by detecting differences in various signals and parametric data. In general, signals at more elemental level are quicker to obtain but more difficult to evaluate quantitatively; on the other hand, parametric data takes longer to process but give a better estimation of physical changes. Changes in signals and data due to simulated failure in a real bridge as well as in models are discernible but in general they are small. Techniques of "Pattern Recognition", a field within the discipline of "Artificial Intelligence", have been investigated to identify small differences in signals or "signatures" of bridge dynamic response characteristics.

Details of experiments, typical test results and results of application pattern recognition methods are presented in the paper.

2. Typical Signal and Signal Processing - A North Carolina Bridge

Since this was the first bridge to be tested, the objective of the test was to try out different parameters involved in the data collection process. A combination of different frequency range and frequency resolution were tried. The results presented here are for a frequency range of 200 Hz and 400 lines of resolution in the frequency domain.

The two-lane bridge (I-40 West bridge over I-85 near Efland, North Carolina) is about 97.7 m (320 ft. 7 in.) long, three-span composite structure with reinforced concrete deck over structural steel welded plate girders. Each span is simply supported. The deck is 13 m (42 ft. 10 in.) wide and is supported over five girders. Figure 1 shows the typical cross section.

Experiments were performed on the 2nd girder of the middle span which is about 32.26 (42 ft. 10 in) long. Figure 2 is a schematic plan view showing locations of test points marked on the top of the deck of the test span. Thirteen points were marked on the top of the test girder and a 14th point was used to help delineate flexural modes from torsional modes. The accelerometer was mounted at the reference point (Figure 2, point 8).

To initialize the spectrum analyzer, the hammer was first struck at the reference point. Thereafter, the hammer was struck sequentially from point 1 through 14 and transfer function was measured for each point. To reduce the effect of noise on computation of transfer functions, typically, an average of readings due to five repeated hammer strikes at each point was taken.

Figure 3(a) shows the time record of a typical hammer blow and Figure 3(b) shows the corresponding frequency spectra obtained using fast fourier transform (FFT). Figure 4(a) shows a typical time record of the acceleration response at the reference point. Figure 4(b) shows the corresponding frequency spectra. Figure 5(a) shows a typical "reference point" transfer function. Figure 5(b) shows a typical "cross point" transfer function.

Using a set of transfer functions corresponding to each of the test points and utilizing procedures of curve fitting, modal parameters, namely modal frequencies and mode shapes were obtained. For comparison, finite element structural dynamic analysis was also performed. Table 1 shows a listing of frequencies obtained form finite element analysis and experimental tests. Figure 6 shows two (the 4th and 6th) of the mode shapes obtained from experimental results.

Tests performed on the bridge in North Carolina provided a bench-mark for performing experimental modal analysis on highway bridges. Experimental results also provided a good correspondence with results of finite element analysis.

3. Test Results - Pennsylvania Bridge With Simulated Failure

The object of experiments on this bridge was to examine the sensitivity of various response functions to development of cracks. The bridge was selected because of its typical composite construction, very low traffic volume and its having bolted splices, some of which could be loosened up to simulate development of cracks, without compromising the safety of integrity of the structure.

The bridge is located near the Harriburg International Airport, carrying the School House Road running North-South over the highway PA-283 running East-West between Harrisburg and Lancaster, Pennsylvania. Figure 7 shows schematic vies of the bridge. The super-structure is 71.32 m (234 ft.) long. The bridge carries two lanes and wide shoulders on each side. The reinforced concrete deck is 14.63 m (48 ft.) wide, supported by seven welded steel plate-girders @ 2.18 m (7 ft. 2 in.) on center and has 0.76 m (2 ft. 6 in.) over-hang on each side. The north and south lengths are 35.66 (117 ft.) each.

4. Test Results - Laboratory Scale Model Bridges

The objective of performing experiments on a model were to examine in detail the parametric and the non-parametric response functions for their effectiveness in detecting cracks in bridge girders. The model was constructed to allow simulation of cracks in the girders. Figure 16 shows the framing plan of the model bridge. The small bridge is a composite structure with a 44.5 mm (1.75 in.) thick ismetrically reinforced concrete slab "glued" by epoxy mortar to the top of the three W12x16 hot-rolled structural steel beams. To simulate failure, the bolts of splice near the mid-span location of the center girder were removed.

Experiments were performed with the accelerometer mounted on the top of the deck and the hammer striking on the deck, as well as with the accelerometer mounted laterally to the bottom flange of the test girder and the hammer also striking the bottom flange laterally. Results shown here correspond to latter case. Table 3 shows a comparison of frequencies and Figure 17 shows comparison of a typical mode shape for uncracked and cracked cases.

To quantitatively study the difference between the modal analysis criterion (MAC) was used (Wolff and Richardson, 1989). Given two sets of mode shapes, say set A and set B, it identifies those mode shapes from set A which are different from mode shapes of set B. MAC between two mode shapes, $V_{A,i}$ and $V_{B,j}$, can be defined as

$$MAC\,(V_{A,i}, V_{B,j}) = \frac{|V_{A,i}, V_{B,j}|^2}{V_{A,i}^T V_{a,i} V_{B,j.}^T V_{B,j}} \qquad (1)$$

MAC has a value between 0 and 1. A value close to 1 indicates that the two mode shapes are almost identical.

In addition, coordinate modal assurance criterion (COMAC), which is used to obtain point wise correlation between mode shapes, were evaluated (Lieven and Ewins, 1988). For two sets of mode shapes A and B, COMAC at point i as defined as

$$COMAC(i) \quad = \frac{(\Sigma_{L=1}^{Lmax} |({}_iV_{A,L}) \bullet ({}_iV_{B,L})|)^2}{\Sigma_{L=1}^{Lmax} ({}_iV_{A,L})^2 \bullet \Sigma_{L=1}^{Lmax} ({}_iV_{B,L})^2} \qquad (2)$$

where L_{max} is the number of mode shapes in each set. COMAC also has a value between 1 and 1. A value close to 1 indicates that there is hardly any differences between the modes shapes at that point.

For un-cracked and cracked cases, Table 4 shows a comparison of MAC values and Table 5 shows a comparison of COMAC values.

Figure 8 shows a schematic framing plan of the structure. The girders are continuous over the center pier and each has bolted splices at 7.62 m (25 ft.) away on each side of the center pier support. Figure 9 shows schematic view of a typical field bolted splice of atypical girder. There are four rows of bolts in the bottom and the top flange splice plates and 19 rows of bolts in the web splice plate. All bolts are 22.2 m (7/8 in.) - diameter. ASTM A325 "High Strength" fasteners.

The first interior girder next to the west fascia girder of the north span was selected as a typical girder for testing. For the purpose of simulating a realistic failure in the test girder, all four rows of bottom-flange splice bolts and the lowest three rows of web splice bolts on one side of the splice center line was disabled.

Dynamic test data were taken for 17 points on the deck of the north span shown in Figure 10.

Frequency spectra: Figure 11(a) and (b) show reference point frequency spectra corresponding to the intact and the cracked girder cases. Close observation of the two spectra overlaid indicates a consistent shift of the frequency peaks. For example, Figure 12 shows a blow up of the overlaid spectra within a selected range of frequencies between 70 and 90 Hz.

Frequency response functions (frf): the FRF data, displayed in various manners, indicated consistent change in frequencies. For example, Figure 13 shows comparison of the "reference point" FRF magnitude for the intact and cracked cases, within a selected frequency range of 70 and 90 Hz. Figure 14 shows the comparison of the sum of the squares of the "imaginary" parts of all 17 FRFs within the same frequency range.

Modal analysis: Table 2 shows a comparison of modal frequencies of the first eight flexural modes. For example, for the eight mode, it shows drop from 88.75 Hz. to 87.00 Hz. from the intact case to the cracked case. As anticipated, differences are more pronounced for the higher modes. Figure 15 shows the 9th flexural mode shape for the two cases.

Results indicate the presence of detectable changes in some of the response data due to simulated physical failure. Biswas, et al (1990) recently presented a detailed description of the experiments and results of the Pennsylvania bridge. A list of equipments used is also provided in the same paper. Non-parametric information, i.e., time records, frequency spectra (Fourier domain), transfer functions (Laplace domain) as well as parametric information, i.e., modal frequencies and mode shapes, have been examined. Changes in the spectra are detectable but are difficult to quantify. Frequency response functions are compared also by graphically displaying superimposed plots within selected frequency ranges. Changes in frequency response functions are detectable and quantifiable. Amongst various displays of frequency response functions, i.e. functions of real and imaginary parts of FRF, the display of the sum of the squares of the imaginary part of the composite FRF appears to show the differences best. Regarding parametric quantities, modal frequencies show small but consistent drops due to the presence of the simulated crack.

In order to interrogate non-parametric response functions, a number of techniques of pattern recognition were investigated. One of the candidate methods is the so called "Freeman chain code" and the generalized thereof (Freeman, 1974; Pavlidis, 1977; Saghri and Freeman, 1981). This method was modified to specifically interrogate typical bridge FRFs which are digitized wave forms. Figure 18 shows modified chain-code comparison of two FRF wave forms of the same un-cracked girders from readings taken at two different times. Figure 19 shows modified chain-code comparison of FRFs of uncracked and cracked girder. Appearance of numerous spikes indicate pressure of crack.

5. Conclusions

Application of technologies of theoretical and experimental modal analysis appears to be a feasible tool for detecting faults in bridge structures. Non-parametric response functions such as frequency spectra and frequency response functions, and parametric response functions such as frequencies and mode shapes are only marginally sensitive to appearance of faults in a structure. Enhancement to parametric quantities, such as MAC and COMAC and application of pattern recognition techniques to frequency response function appear to be more effective in detecting changes in a structure.

6. Acknowledgements

The work reported here is a part of the research project sponsored by the Pennsylvania Department of Transportation (PennDOT) and the U.S. Department of Transportation (USDOT), Federal Highway Administration (FHWA). G.L. Hoffman of PennDOT is the Project Manager.

J.R. Widler of North Carolina Department of Transportation provided us with the opportunity to perform experiments on the bridge in North Carolina.

The contents of this paper reflect the views of the authors who are responsible for the facts and the accuracy of the data presented herein. The contents do not necessarily reflect the official views or policies of either the USDOT-FHWA, or the Commonwealth of Pennsylvania at the time of publication. This paper does not constitute a standard, specification, or regulation.

7. References

Adams, R. D., Cawley, P., Pye, C.J. and Stone, B.J. (1978) "A Vibration Technique for Non-destructively Assessing the Integrity of Structures," J. Mechanical Engineering Science, 20(2), IMechE., 93-100.

Allemang, R.J. and Brown, D.L. (1987) "Experimental Modal Analysis," Handbook of Experimental Mechanics, Kobayashi, A.S. (ed.), SEM, Prentice Hall.

Biswas, M., Pandey, K. and Samman, M.M. (1990) "Diagnostic Experimental Spectral/Modal Analysis of a Highway Bridge," The Intl. J. of Analytical and Experimental Modal Analysis, 5(1), Society of Experimental Mechanics, 33-42.

Collacott, R.A. (1985) "Structural Vibrations," Structural Integrity Monitoring, Chapman and Hall, Ch. 7, 234-285.

Elliott, D.F. (ed) (1987) Handbook of Digital Signal Processing, Engineering Applications, Academic Press.

Ewins. D.J. (1984) Modal Testing: Theory and Practice, John Wiley & Sons.

Fisher, J.W. and Wood, J.D. (1988) "Experience with Bridge and Building Structures and Opportunities for NDE," in Proc. Intl. Workshop on Nondestructive Evaluation for Performance of Civil Struct., Agbabian, M.S. and Masri, S.F., (eds.), Dept. of Civil Engrg., Univ. of S. California, 101-120.

Freeman, H. (1974) "Computer Processing of Line-drawing Data," Computing Surveys, 6(1), 57-96.

Jackson, L.B. (1986) Digital Filters and Singal Processing, Fluwer Academic Publishers.

Lieven, N.A.J. and Ewins, D.J. (1988) "Spatial Correlation of Mode shapes, the Coordinate Modal Assurance Criterion (COMAC), in Proc. 6th International Modal Analysis Conference, SEM 690-695.

Pavlidis, T. (1977) Structural Pattern Recognition, Springer-Verlag, 161-164.

Saghri, J.A. and Freeman, H. (1981) Analysis of the Precision of Generalized Chain Codes for the Representation of Planer Curves, Transactions of Pattern Analysis and Machine Intelligence, v. PAMI-3, No. 5, 533-539.

Sterns, S.D. and David, R.A. (1988) Singal Processing Algorithms, Prentice Hall.

Stephens, J.E. and Yao, J.T.P. (1987) Damage Assessment Using Response Measurements," J. Structural Engineering, ASCE, 113(4), April, 787-801.

Wolff, T. and Richardson, M.H. (1989) "Fault Detection in Structures Form Changes in Their Modal Parameter," in Proc. 7th International Modal Analysis Conference,. SEM, 87-94.

Table 1 Modal Frequencies for North Carolina Bridge.

Mode No.	Frequency in Hz	
	Analytical	Experimental
1	2.11	2.75
2	8.25	11.25
3	17.96	----
4	30.56	31.25
5	45.37	45.00
6	61.77	60.00
7	79.24	70.00
8	97.42	76.25

Table 2 Modal Frequencies for the Pennsylvania Bridge.

MODE	FREQUENCY(Hz)		FREQUENCY
	INTACT	CRACKED	CHANGE (Hz)
1	25.25	25.00	-0.25
2	30.75	30.25	-0.50
3	35.25	35.00	-0.25
4	44.25	44.25	0.00
5	53.25	53.25	0.00
6	65.00	64.75	-0.25
7	71.25	70.75	-0.50
8	88.75	87.00	-1.75

Table 3 Comparison of the Natural Frequencies for the Intact and the Cracked Case.

Mode No.	Frequency (Hz)		Frequency Change
	Intact	Cracked	
1	85.00	72.50	-12.5
2	137.50	135.00	-2.50
3	217.50	195.00	-22.50
4	445.00	442.50	-2.50
5	605.00	575.00	-30.0
6	615.00	595.00	-20.0
7	655.00	625.00	-30.0

Table 4 MAC Values for the Cracked and Uncracked Case.

Mode No.	MAC	Mode No.	MAC
1	0.98	6	0.84
2	0.97	7	0.52
3	0.90		
4	0.84		
5	0.92		

Table 5 COMAC Values for the Cracked and Uncracked Case.

Point No.	COMAC	Point No.	COMAC	Point No.	COMAC
1	0.67	12	0.92	23	0.93
2	0.77	13	0.75	24	0.97
3	0.96	14	0.79	25	0.97
4	0.92	15	0.88	26	0.95
5	0.90	16	0.96	27	0.95
6	0.96	17	0.97	28	0.95
7	0.93	18	0.90	29	0.96
8	0.93	19	0.85	30	0.91
9	0.95	20	0.85	31	0.82
10	0.94	21	0.86	32	0.76
11	0.89	22	0.89	33	0.74

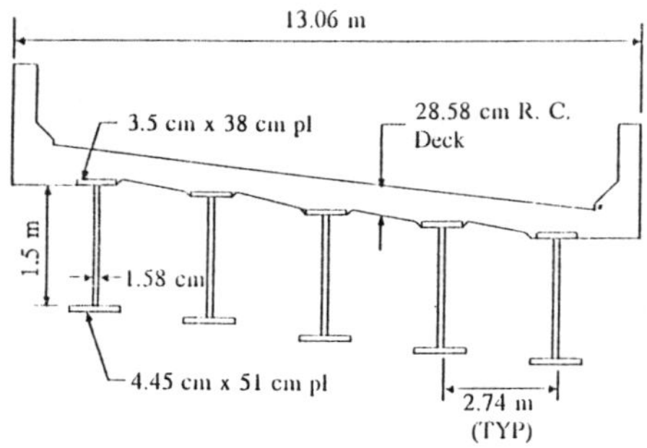

Fig. 1 Cross-Section of the North Carolina Bridge.

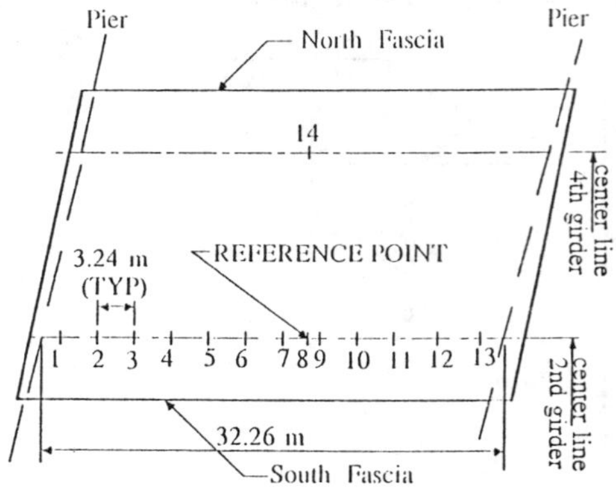

Fig. 2 Location of Test Points for the North Carolina Bridge.

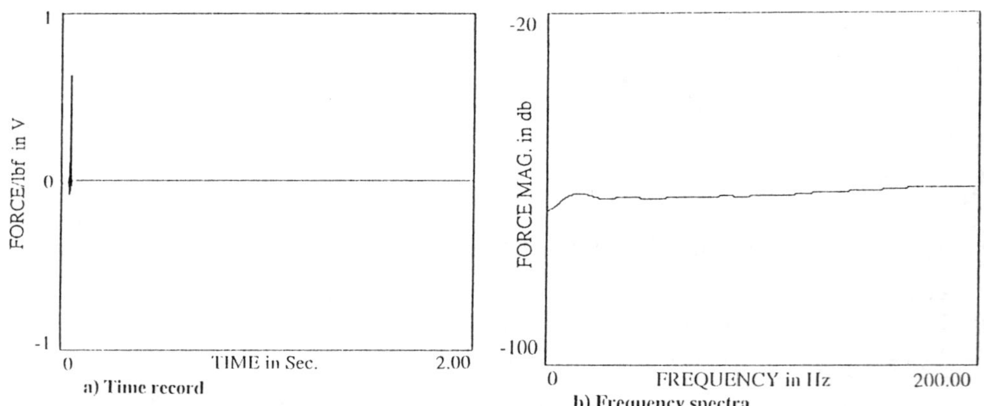

Fig. 3 A Typical Hammer Blow.

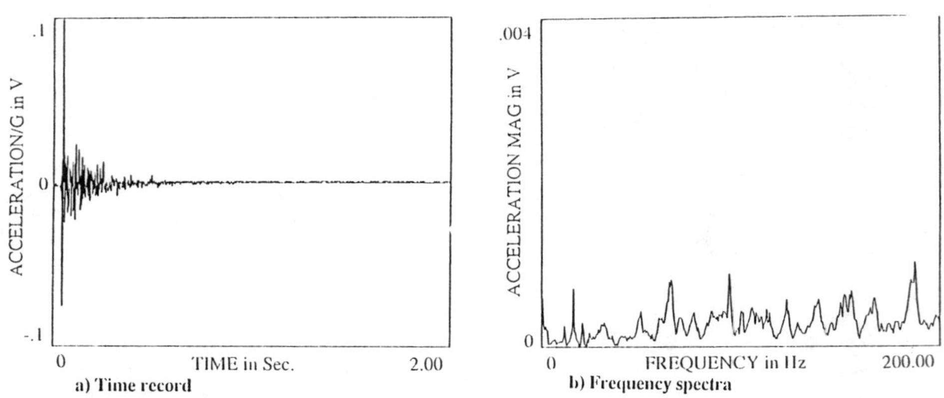

Fig. 4 A Typical Acceleration Response.

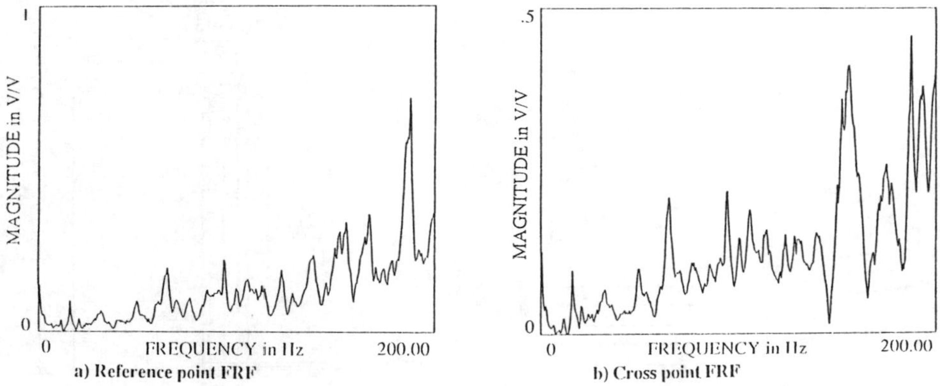

Fig. 5 A Typical Transfer Function.

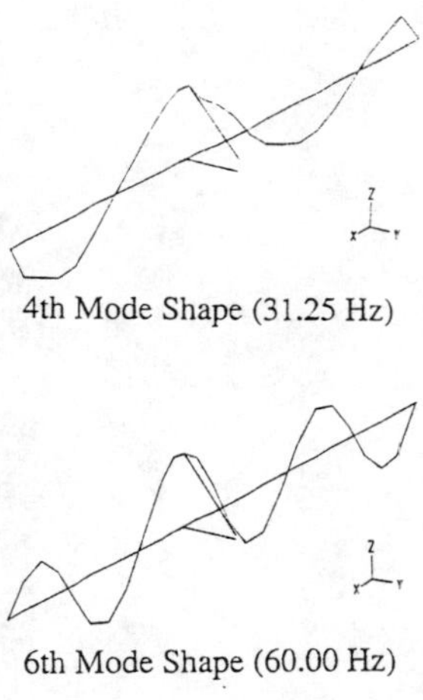

Fig. 6 Typical Mode Shapes for North Carolina Bridge

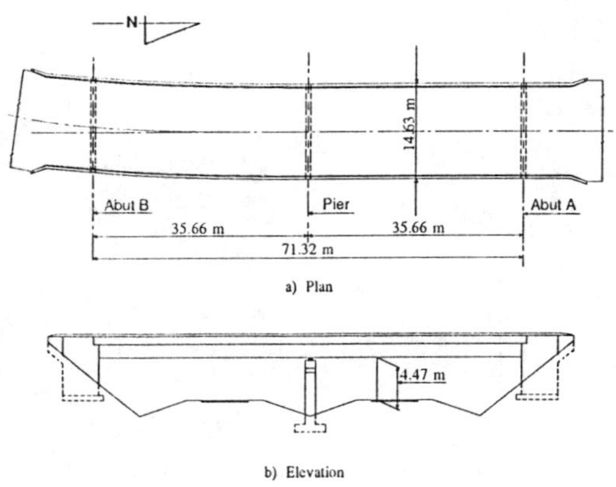

Fig. 7 Schematic View of the Pennsylvania Bridge

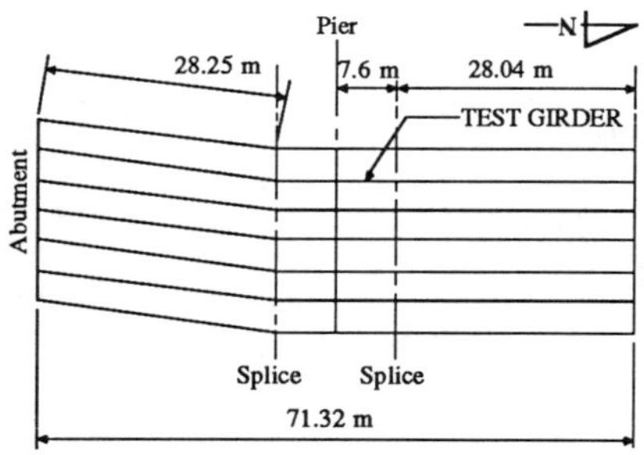

Fig. 8 Schematic Framing Plan of the Pennsylvania Bridge.

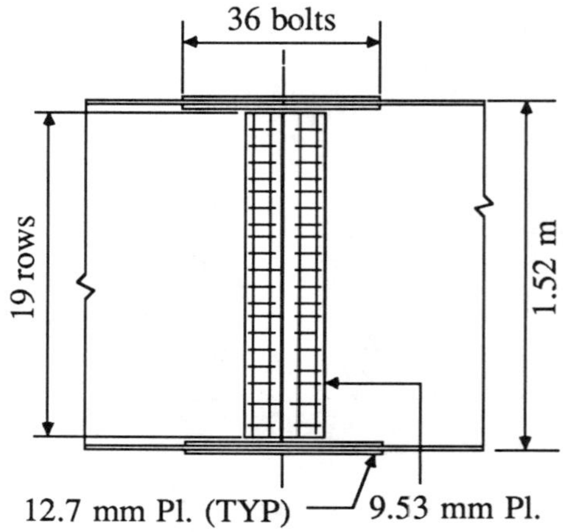

Fig. 9 Schematic Elevation of a Typical Field Bolted Splice.

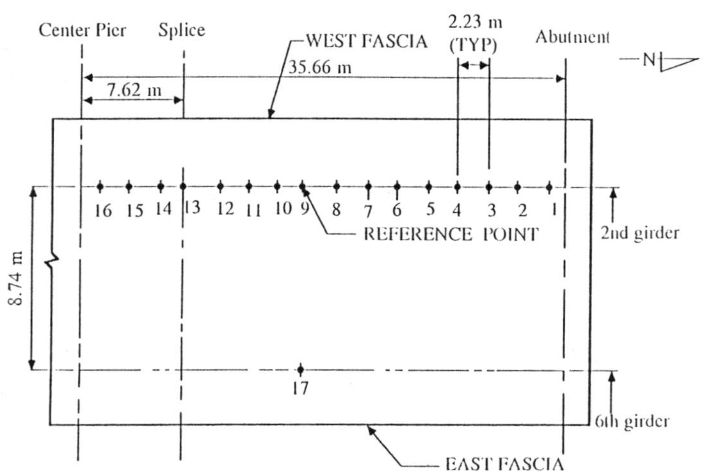

Fig. 10 Test Locations for the Pennsylvania Bridge.

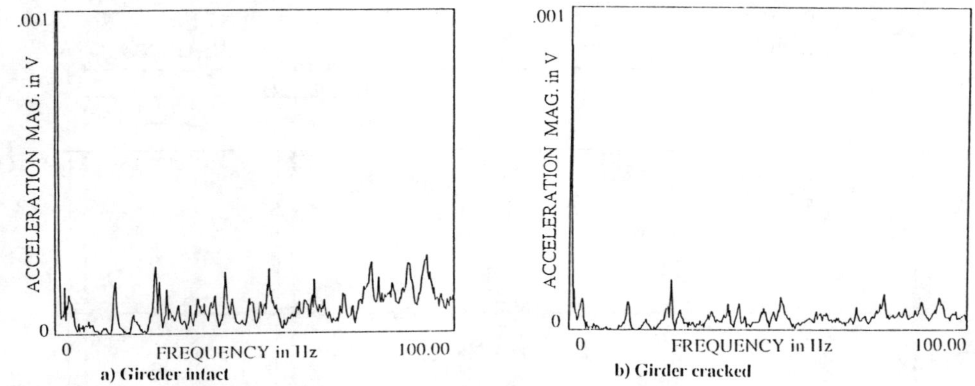

Fig. 11 Reference Point Frequency Spectra for the Pennsylvania Bridge.

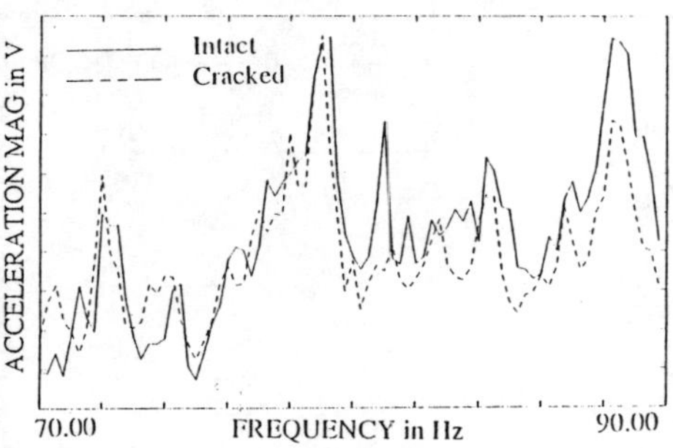

Fig. 12 Blow Up of the Overlayed Spectra.

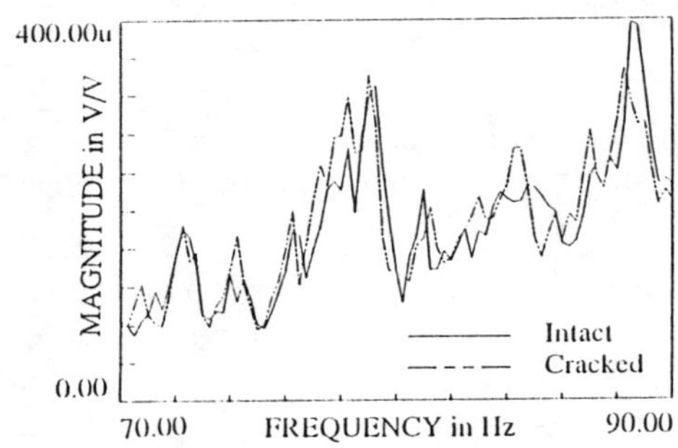

Fig. 13 Comparison of the Referenced Point FRF Magnitudes.

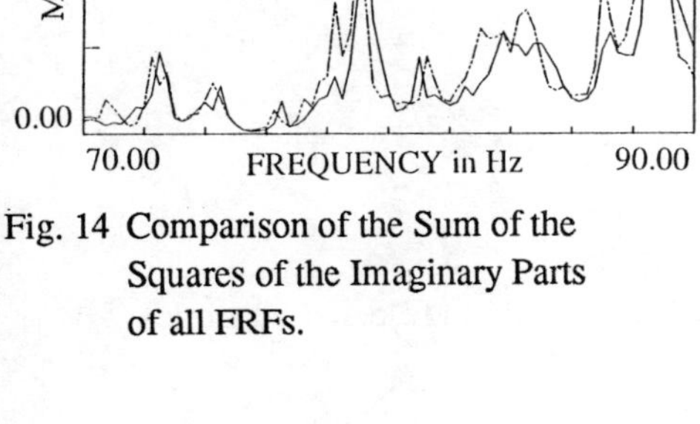

Fig. 14 Comparison of the Sum of the Squares of the Imaginary Parts of all FRFs.

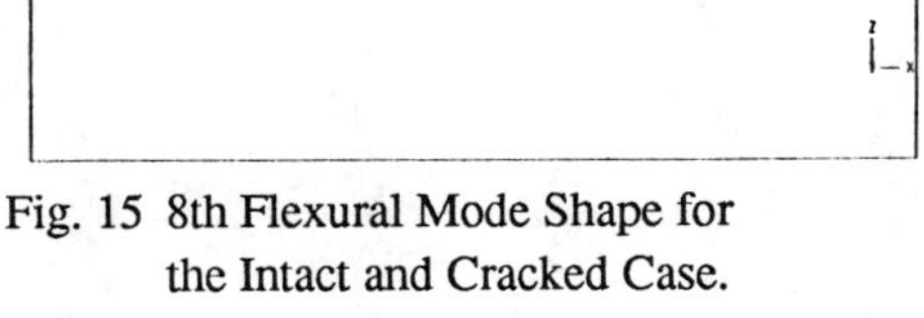

Fig. 15 8th Flexural Mode Shape for the Intact and Cracked Case.

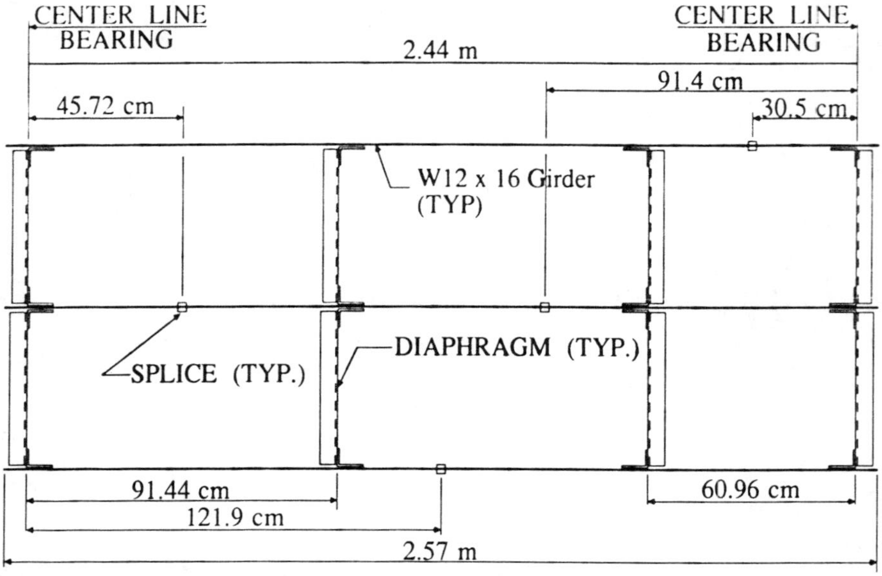

Fig. 16 Framing Plan of the Laboratory Bridge Model.

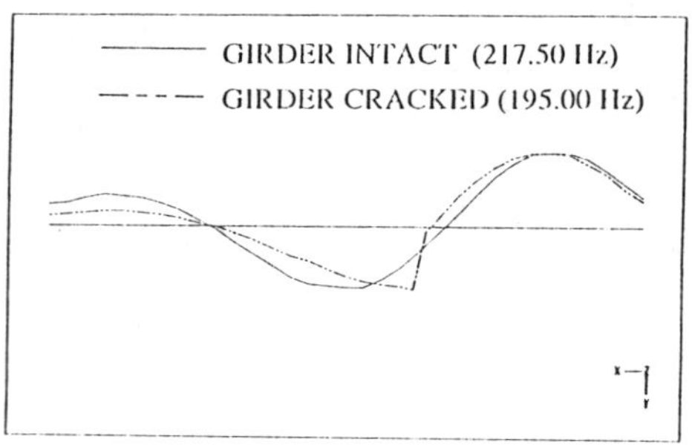

Fig. 17 Typical Mode Shape for the Cracked and the Uncracked Case.

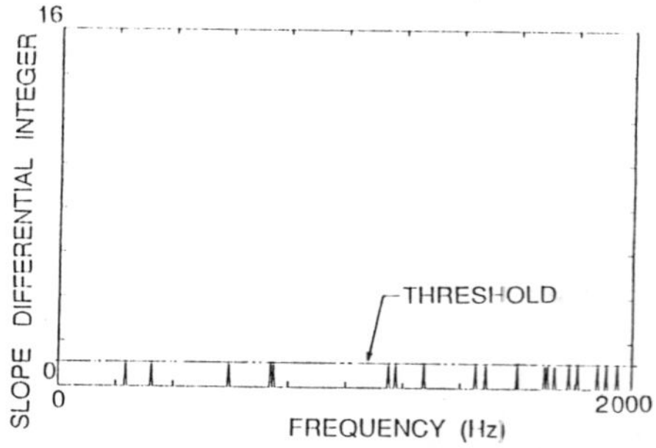

Fig. 18 Modified Chain-Code Comparison of Two FRF Wave Forms of the Uncracked Girder.

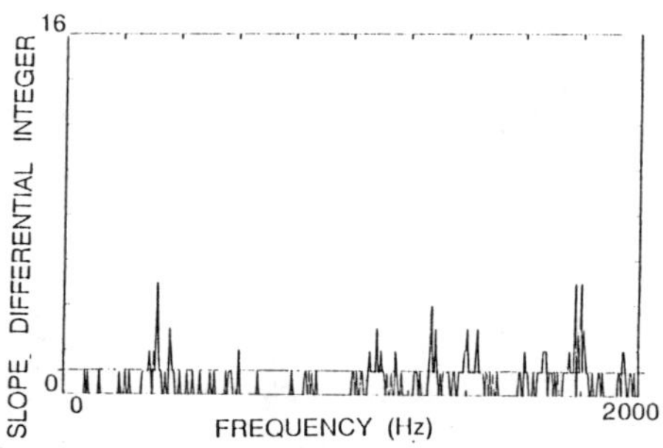

Fig. 19 Modified Chain-Code Comparison of FRFs of Cracked and Uncracked Girder.

A DYNAMIC METHOD FOR THE SAFETY INSPECTION OF LARGE PRE-STRESSED BRIDGES

R. G. FLESCH
Structural Dynamics/BVFA Arsenal
A-1030 Wien
Faradaygasse 3, AUSTRIA

K. KERNBICHLER
Inst. for Reinforced Concrete
Technical University Graz
A-8010 Graz
Technikerstraβe 4, AUSTRIA

ABSTRACT. A dynamic method for the damage evaluation of large bridges was developed by BVFA and Technical University Graz. The method is a combination of dynamic insitu-tests and dynamic calculations. The basic idea is to detect damages of bridges via changes of the dynamic parameters. The steps of application are dynamic insitu-testing, mathematical modelling, fitting of the mathematical model to the test results and sensitivity investigations. The development of the method is presented by discussing the history of application. Further references for detailed information and some examples from past projects are given. The latest investigations were started in Oct. 1988 on five bridges on Brenner motorways. The tests were finished in April 1989, analysis of measurements is still in process. More detailed results will be given in forthcoming papers.

1. Object

Due to the drastic increase in the number of vehicles passing over bridges, especially of heavy trucks, methods for bridge inspection become more and more important. The main aims are

- safety inspection of the bridge
- detection of damages at an early stage for minimizing the expense of repair

At the moment, mainly visual inspection techniques with a limited possibility for the quantification of damages, are used. This situation is unsatisfactory in our world of high-technology methods. Hence, in 1981 BVFA and Technical University Graz started the development of a global method for the inspection of structures using vibrations. The method was strongly improved over the years using the still increasing capacity of computers and measuring equipment. Mainly experience with prestressed bridges was obtained in the past. The basic concept of the method is to elaborate a dynamic model in the virgin state of the structure. In general, damages decrease the stiffness and increase damping, resulting in changes in the dynamic properties; they become different for different modes of vibration. The dynamic properties are modal frequencies (eigenfrequencies) modeshapes and modal damping ratios. Hence, damages

A. S. Nowak (ed.), Bridge Evaluation, Repair and Rehabilization, 175–185.

will lead to a certain pattern of deviations of dynamic parameters and can be used for localization and quantification of damages in a global manner.

The following steps are carried out:

- elaboration of a structural dynamic model for vertical direction (FE-model)
- experimental determination of modal parameters by dynamic insitu-tests
- indirect system identification (fitting of the structural model to the modal test model).

The procedure described above is the baseline evaluation of the dynamic properties. It should be done after the construction and before opening the bridge for traffic.

After a certain time of use (e.g. 2-5 years) the investigations should be repeated. If changes of modal frequencies, modeshapes and damping ratios are found in the experiment, the structural model must be fitted again to the test results. By comparing the old and new mathematical models, changes of stiffness in certain parts of the structure could be detected and quantified in a global sense (smeared over the FE-element in which the actual damage occurs).

If changes exceed a certain limit, which has to be fixed for any particular case, the damage area found by the global method should be investigated in detail by any local method.

With the refined information (e.g. exact location of cracks, crack-width, depth of cracks, etc., the fitting of the structural model to the experimental results can be further improved. With the resulting optimum model, the redistribution of internal moments and forces for design loads due to the stiffness change can be calculated. Based on the results, in any particular case it must be decided whether the structure is still safe or not.

The concept described above can be strictly followed only for new bridges, when the baseline is evaluated after finishing the construction and before opening of the bridge to traffic in order to have no disturbances by the traffic. For the greater part of existing bridges there is no baseline available for the virgin state. For some bridges used for several years, first measurements were carried out during the last few years giving the baseline for that certain point in time. Often the evaluation was done after some main rebuilding or visual inspection. It is argued, that severe damages which are not visible at the moment will increase until the next routine measurement to such an extent that they can be detected via the change of dynamic parameters.

For bridges without an evaluated baseline and also for special tasks, a variant of the dynamic method would be desirable, allowing the possibility of being able to detect damages from a single insitu-test. The theory for a "direct stiffness evaluation method" is discussed in [12]. Using a modified test procedure, parameters such as EI can be evaluated for any chosen element of a bending structure. Spring constants can be evaluated in a similar way. The theory is based on the fact that for $\omega \rightarrow o$ the transfer function model used for experimental modal analysis gives the stiffness coefficients. It is assumed that any disturbance during the measurement will raise problems. The theoretical concept needs practical verification.

In detail the method consists of the following parts (see Figure 1):

- structural (mathematical) modelling
- insitu-test technique
- analysis of test results (first step of system identification: elaboration of modal test model)
- indirect system identification (fitting of mathematical model to the modal test model)

Indirect identification is a still developing field with many unsolved problems. Hence, additional approaches which can help to localize and quantify damages are most desirable. One possibility is the use of a structural dynamics modification software for sensitivity investigations. Using only the modal test model, a decrease in the bending stiffness is applied systematically to the structure, giving the change in the dynamic properties due to the local stiffness change. This fruitful approach will be discussed later in more detail.

Depending on the modeshapes, the single modes have a different local sensitivity to stiffness changes. The more that modes with different shapes become available, the easier will be the localization and quantification of damages. In the past the investigations were focused on the changes in modal frequencies, although this parameter varies only proportionally to the square root of the stiffness change. In the future the changes in the modeshapes, which are much more pronounced, will be used for interpretation. Before changes in the damping ratios can be interpreted, the damping mechanisms of R/C structures must be understood better.

2. The History of Application

The development of the method is discussed in detail in [8,11]. Some important projects are mentioned in what follows.

For the Gänstorbrücke in Ulm, FRG, tests were carried out before and after some main repair work. Measurements were carried out for three states. Changes in modal bending frequencies and modal torsional frequencies were detected. The increase in stiffness resulting from the repair work was elaborated by comparing the structural models for the different states. The results are given in [3,4,5,8].

In November 1985 the investigations of Lavant bridge were started. Lavant bridge is one of the greatest prestressed frame bridges in Europe. The maximum span is 160 m and the maximum pier height is 130.35 m. The tests could be carried out in the virgin state before the bridge was opened for traffic. For the analysis of the measurements a Modal Analysis Software Package from SMS were carried out using the structural dynamics modification software. The results are given in [6,7,8].

Starting in October 1988 a baseline evaluation was carried out for five bridges on Brenner motorways. The bridge investigated first was bridge Obernberg after widening of the cross section (adding of one lane). In the latest test series a new reaction mass exciter driven by a hydraulic actuator was used. For all previous projects an eccentric mass exciter had to be used.

To obtain experience with other types of bridges a R/C arch bridge (Äußere Nößlachbrücke) and a composite bridge with steel girders and a concrete slab (Gschnitztalbrücke) was included in the program. Further, for one mushroom-slab bridge (3. Nößlachbrücke) tests were carried out to study the behavior of gaps and bearings between six substructures of the bridge. Finally another R/C frame bridge was tested to have a baseline, after a reconstruction in areas of sliding foundations due to hillside creep.

For the bearings between the six substructures of 3. Nößlachbrücke no visual inspection is possible. Corrosion could damage the gaps or the elastomere element. Hence, the dynamic method, especially the direct stiffness evaluation method, seems to be a powerful tool for the inspection. The investigations are still in progress. Using a first FE-modal, some preliminary sensitivity investigations were carried out. Figure 2 gives the sensitivity of the modal frequencies to changes in the stiffness coefficients of the bearings.

Hence, the latest series could be carried out under excellent conditions concerning test equipment and means of analysis, but with disturbances by traffic. As routine tests have to be carried out normally under traffic disturbances, one important aim is to find procedures to eliminate the disturbances. The problem will be discussed in more detail later on. The analysis for all five bridges is still in progress.

3. Focus and Limitations for the Application

All bridges tested before the latest series were large prestressed bridges. Our investigations focused on that type, because the most frequent damages of prestressed bridges are bending cracks with areas of distorted bond at both sides of the cracks giving some kind of smeared stiffness decrease, which is also the assumption in our structural model. Hence, the method seems to be suitable for prestressed bridges. A more detailed discussion of different damages and their expected effect on dynamic properties is given in [6,8].

In principle the method can also be used for R/C bridges, steel bridges or composite bridges, especially if changes in boundary areas (foundation of piers, bearing, etc.) should be detected. In the literature some cases of loss of bearings are reported, a damage which could be quickly identified by the dynamic method.

As it is not possible to close a bridge completely for testing, the measurements need to be carried out under traffic disturbances. Hence, procedures to eliminate disturbances are necessary. Initial results are discussed at the end of this paper. It is only possible to close one lane in each direction for about 2-5 days, which is the same time span it must be closed for visual inspection using the bridge inspection truck.

Next, influences from environmental conditions on the measured results must be considered. We have experience with the influence of temperature. There is an influence, if a stiffness change (e.g. due to an elongation in longitudinal direction) occurs in the direction of pronounced modal movement. The experiences from Gänstorbrücke are reported in [3,4,5,8]. It can be concluded that temperature has an influence, but the single modes are influenced to a different extent. It is assumed that after exclusion of modes which are very sensitive to temperature changes, enough modes are still available to give the expected information.

4. Insitu Tests

For a large number of measurement points it is impossible to obtain all measurements at once. In general, two basic concepts can be used for dynamic insitu tests.

- fixed position of the exciter and roving of the transducers to the next measurement points after one frequency sweep. At lest one transducer should be used as a reference and also have a fixed position.
- fixed position of transducers and roving of the exciter from one measurement point to the next point after one sweep.

In the latest bridge test series, the first concept was selected because of the following reasons:
- the heavy exciter cannot be placed in any position within a cross-section.
- the elimination of traffic disturbances seems to be easier with the first concept.

For the "direct-stiffness evaluation method", it is necessary to have the excitation and all transducers (at least 3-6) in the area of interest.

As frequency changes due to damages are often small, a precise test technique is necessary. In the past an eccentric mass exciter was used. Using a static frequency changer, the frequency could be controlled with an accuracy of 0.003 Hz. The disadvantage of the eccentric mass exciter is the quadratic force and the very low force amplitude in the low frequency range. From our experience the minimum force should be 1kN to obtain good results. Recently a reaction mass exciter driven by a hydraulic actuator was developed by BVFA. The excitation force can be kept constant during the frequency sweep using a certain control program. The exciter, the hydraulic pump and the control equipment are mounted on a flat bed lorry which is used for transportation of the heavy equipment to any excitation point. The excitation can be carried out in a vertical, longitudinal or transversal direction. The bridges are mainly excited in a vertical direction.

The response is measured by the velocity transducers Hottinger SMU 30 A. The signals are integrated to obtain the displacement response. As there are often long distances between the transducer and magnetic tape recorder, the amplifiers are mounted in boxes together with the transducers. The frequency response function for each transducer was elaborated on the BVFA-shaking table. The function is approximated analytically and used for the correction of the measured transfer functions via a complex division. To have a mesh of measurement points adequate for sensitivity investigations, the distance between measurement profiles should be about 5-10 m. Normally 4-6 transducers are distributed over the cross section in each profile. First, 2-3 positions for the exciter, adequate to excite as many modes as possible, are selected. Then the transducers are placed in one profile after the other. For each configuration a frequency sweep is carried out in the frequency range 0-10 Hz lasting about 10 minutes. As higher modes with frequencies of 10 Hz are both very sensitive to local damages and less disturbed by traffic, the frequency range will be increased in the future. The time histories of the force of excitation and of the responses are recorded on magnetic tape.

From these records transfer functions are calculated using a FFT-analyzer. In the latest series about 1000 transfer functions were obtained per bridge within one week.

From 2-3 positions of the exciter 40-60 modes were obtained. More detailed information is given in [6,8].

5. Analysis of Measurements

From the records of bridge response and of the excitation force, transfer functions are obtained by FFT-Analysis. Before the Lavant project, peak picking and a circle fit algorithm were used to determine the eigenfrequencies, modeshapes and damping ratios.

For analysis of the Lavant bridge measurements, the SMS modal analysis software MODAL 3.0 could be used for the first time. The software is very modular and provides a good fitting algorithm and very comfortable handling possibilities. The total concept seems to be trimmed towards mechanical engineering problems. Some special features of tests of very large civil engineering structures had to be met by additional software. Detailed information is given in [6,8].

6. Dynamic Calculations

The dynamic calculations were carried out mainly by Technical University Graz. SAP IV, FLASH and DYNA were used for mathematical modelling. During every project, attempts were made to find the model with the adequate level of accuracy. Only beam elements and additional spring elements were used. In our opinion the inclusion of more complicated elements will not improve the quality of modelling since material parameters of concrete structures are often quite uncertain (deviations from plan-cross sections, variances of mass density etc.). these uncertainties must be met using correction factors.

About 1200 DOF's were used for each model. In general, the bending stiffness can be modelled well but problems can arise for torsional stiffness and cross section deformation for open cross sections.

In order to get precise results, important details must be modelled well. In some cases the transversal coupling of adjacent bridges via the carriageway-slab had to be modelled. The problem was solved by eccentric connected beam elements [2,4]. The influence of the foundation was sometimes modelled well by including additional springs [2].

7. Indirect System Identification

Indirect system identification means the fitting of structural models (FE-models) to test results. The state of the art is given in [9,10] and in the Proceedings of the workshop held in Lambrecht, FRG in 1987 (see [6]).

In the projects of the past, the systematic system identification was carried out in two relatively simple steps:

- in a global manner using the modulus of elasticity. Hence, the modulus was used as a central fit parameter, representing also deviations of cross section, mass densities and boundary condition etc.

- by trial and error in the areas of structural repair, especially in the case of Gänstorbrücke [3,4,5].

For routine inspection systematic methods are necessary. Methods are under development at Curt Risch Institute, University Hannover, FRG but they are limited to models with about 30 DOF's at the moment.

8. Sensitivity Investigations

It was shown before that systematic system identification can be quite difficult for large systems. Hence, the application of the structural dynamics modification (SDM) software provides a powerful tool for interpretation of changes of modal properties. For the investigations only the modal test model is necessary, hence, no FE-model has to be fitted to test results. The software is applied to the baseline test model. To investigate the influence of local cracks on the modal parameters, the decrease in bending stiffness has to be modelled. A 3-DOF beam element (negative rib stiffener) is "roved" over the bridge, being implemented in three adjacent DOF's at one time. In that way, the influence of a local stiffness decrease (smeared over two adjacent distances between measurement profiles) on the modal parameters can be determined. The basic equations used for the software are given in [6,8] together with all references. Changes in modal frequencies due to a local stiffness decrease are presented in [6,7], changes in modeshapes are given in [7,8].

In the future, SDM will also be used for indirect system identification.

9. Disturbances by Traffic

To elaborate a practicable routine test technique, the method must be applicable during traffic. The influence of traffic was studied in detail in the latest test series. The results are still being processed. The first lesson learned is that at least one fixed transducer should be provided, which remains in the same position during all frequency sweeps (e.g. in the profile of excitation). Comparing the resulting transfer functions, disturbances can be detected. Probably a procedure for data correction can be developed. The imaginary part of the transfer function in the driving point must always have positive values, hence, negative values at certain frequencies are critera for disturbances during that sweep.

In general, traffic has the following influences on the results:

- additional excitation. As for the calculation of the transfer functions a cross spectrum is used, the results are only distorted if the disturbance has a momentary frequency component equal to the actual sweep frequency. As the sweep has to be repeated for every measurement profile, there is a low probability that disturbances will occur at the same modal frequencies in each profile. Due to the large number of profiles, distorted values can be recognized and eliminated, with enough information remaining to determine the modeshapes. In addition, if the baseline was evaluated for the virgin state without traffic, the original modeshapes are known, making the analysis of routine measurements much easier.

- additional damping, as the additional excitation is nonsynchronous with the artificial excitation.
- additional moving mass. Especially the influence from heavy trucks could be important.

The influences from additional moving masses seem to be most upsetting but it is assumed that modal frequencies are always decreased. Hence, using the large number of results, the most probable values for each modal frequency and damping ratio can be determined. During the analysis of measurements additional modes (computational modes) are used during curve fitting to take into account the disturbances. These additional modes, often having negative damping values, are removed finally.

Initial results from "Gschnitztalbrücke" are presented in Figure 3, which shows average, standard deviation and maximum value of the imaginary part of the transfer function for one fixed reference point (averaged over 18 sweeps). It is concluded that traffic disturbances are small above 3 Hz. Based on the average transfer functions for the reference points, all other transfer functions can be corrected.

10. Conclusions

The state of the art of a dynamic method for the safety inspection of structures is presented in this paper. For five bridges a large amount of data was obtained. Analysis is still in progress. Unfortunately, only a few examples are ready for presentation at the moment. The delay of analysis is mainly caused by necessary hardware and software changes.

The problem of traffic disturbances seems to be solved. Progress was made in identification of the most probable dynamic properties.

For the dynamic calculations program, DYNA (running on PC's) will be used more frequently in addition to mainframe calculations. Analysis will focus on the sensitivity of the dynamic parameters on local stiffness changes. These investigations are the basis for a new indirect system identification method and for the study of influences of damages on the dynamic properties.

To check the reliability of the SDM procedure, parallel sensitivity investigations were carried out using a FE-model and the measured model (test model) of Äußere Nößlachbrücke. A 10% stiffness decrease of the arch-elements is assumed. Besides relatively small changes in modal frequencies, constitutive changes in modeshapes were observed (e.g. a certain mode number changed from vertical bending to horizontal bending or torsion).

At the end of 1989 the theory for a "direct stiffness evaluation method" was presented in [12]. This fruitful approach will be applied to structures in the near future.

11. References

1. Kernbichler, K. and Flesch, R. (1984) "Static and Dynamic Tests, Their Qualification for Bridge Inspection and Long-Term-Observations of Bridge Structures," RILEM Symposium, Budapest.
2. Flesch, R., Kernbichler, K. and Dürnsteiner, Ch. (1984) "Dynamic testing and Modelling of Obernberg-Bridge", Proc. of Int. Conf. on Num. Meth. for Transient and Coupled Problems, Venice.
3. Flesch, R., Kernbichler, K. and Grübl, P. (1984) "Brückeninspektion mittels dynamischer Untersuchungen," 8.GESA-Symp., VDI-Berichte Nr. 514.
4. Kernbichler, K., Flesch, R. and Rauscher, G. (1984) "Dynamische Untersuchungen von Großbrücken (Massivbrücken), in-situ Versuche und Rechenmodelle," Tagung Dynamische Probleme, Univ. Hannover.
5. Flesch, R., Kernbichler, K. and Rauscher, G. (1986) "Dynamische in-situ Versuche und Rechenmodelle -Praktische Anwendung auf Großbrücken/Massivbrücken," ÖIAZ 131, Jg. Heft 10.
6. Flesch, R., Kernbichler, K. (1988) "Bridge Inspection by Dynamic Tests and Calculations - Dynamic Investigations of Lavant Bridge Proc. of Workshop on Struct. Safety Evaluation Based on System Identification Approaches, Friedr. Vieweg & Sohn, Braunschweig, Wiesbaden.
7. Flesch, R., Kernbichler, K. (1988)"A Dynamic Method for the Safety Inspection of Large Prestressed Bridges," Proc. of Int. Workshop on Nondestructive EvaluAtion for Performance in Civil Structures, Dept. of Civil Eng., Univ. of Southern California, Los Angeles.
8. Flesch, R. (1988) "Die Methoden der Baudynamik mit spezieller Berücksichtigung ihrer Anwendbarkeit zur Bauwerksinspektion," Habilitationsschrift, Teil 2, Techn. Univ. Graz.
9. Natke, H.G. (1982) "Die systematische Anpassung von Rechenmodellen an Versuchswerte als Verfahren zum Nachweis des dynamischen Systemverhaltens," Bauingenieur 57.
10. Natke, H.G. (1983) "Einführung in Theorie und Praxis der Zeitreihen- und Modalanalyse," Vieweg & Sohn, Braunschweig.
11. Flesch, R. and Kernbichler, K. (1990) "Diagnostic Dynamic Testing of Bridges on Brenner Motorway," Int. Conf. on Bridge Management, Univ of Surrey, UK, March.
12. Flesch, R. (1989) "Baudynamik," Habilitationsschrift, Teil 3, Techn. Univ. Graz.

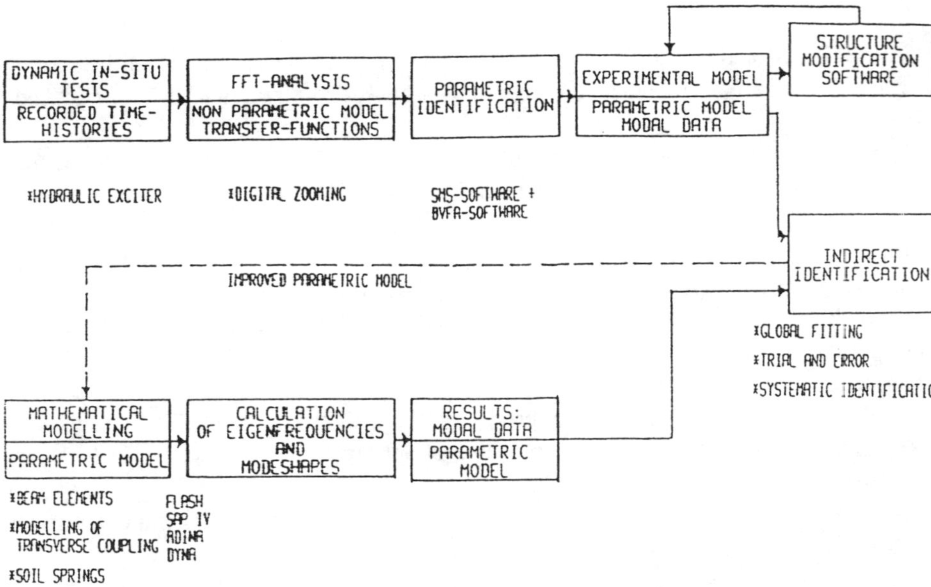

Fig. 1 Steps of the Dynamic Method for Safety Inspection.

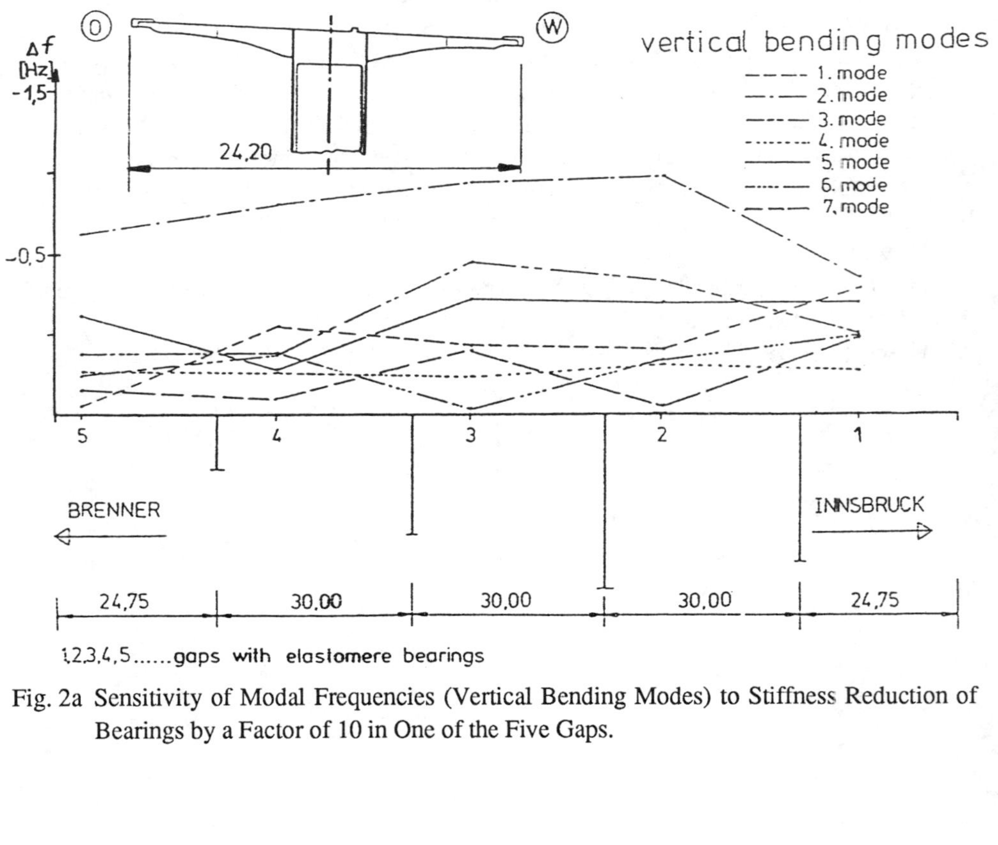

Fig. 2a Sensitivity of Modal Frequencies (Vertical Bending Modes) to Stiffness Reduction of Bearings by a Factor of 10 in One of the Five Gaps.

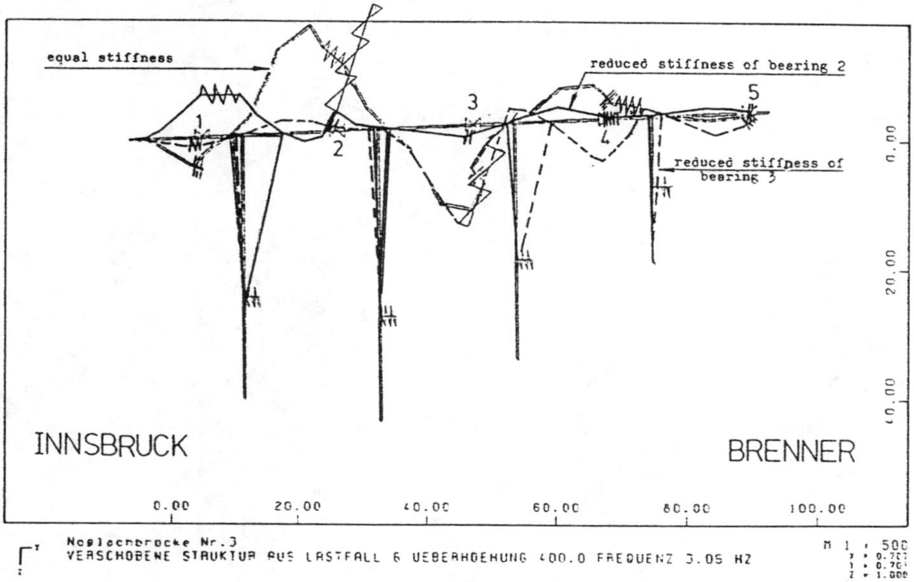

Fig. 2b Sensitivity of Mode Shape no. 6 to Stiffness Reduction of Bearings in Gap no. 2, Respectively no. 3 by Factor of 10.

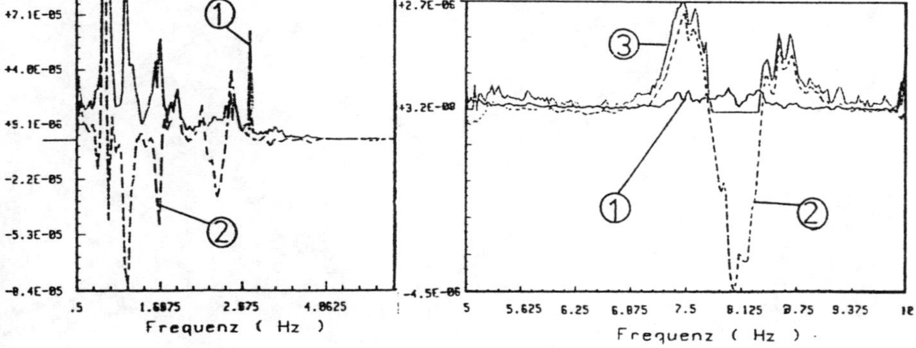

Fig. 3 Statistical Analysis of Transfer Functions in One Reference Point of Gschnitztalbrücke.

IMPROVED STRUCTURAL MONITORING WITH ACOUSTIC EMISSION PATTERN RECOGNITION

DAVID W. PRINE
GARD, Inc.
Niles, IL 60648 USA

THEODORE HOPWOOD II
Kentucky Trans. Res. Program
Lexington, KY 40506 USA

ABSTRACT. A unique acoustic emission monitoring system originally developed for in process weld monitoring has been used to monitor fatigue crack growth in a highway bridge during normal traffic loading. The system was able to clearly and reliably detect the presence of fatigue cracks that were adjacent to a row of bolts. The results of the brief experiment show that the signal processing used in this AE system may allow drastic improvements in the ability of acoustic emission to reliably detect propagating bridge flaws under adverse conditions.

1. Introduction

There is a growing need to insure the safety of the motoring public by performing periodic nondestructive evaluation (NDE) of in-service bridges. Older bridges are being subjected to loadings in excess of their original design capacity. Owing to the combined effects of age, atmosphere, and loading, these structures are prone to sub-critical, growing cracks. Newer bridges have employed ambitious designs, which often incorporate "fracture-critical" structural members. These bridges also rely on high strength steels and welding fabrication processes to provide economical structures. These bridges have demonstrated a susceptibility to crack problems caused by fabrication-related defects. Unfortunately the quality-assurance/ quality-control measures used in recently constructed bridges have not precluded fracture problems.

A promising NDE technique entails monitoring of acoustic emissions (AE) from bridges and correlation of these emissions with the integrity of the structural members. However, the user of the AE methodology faces many problems due to the environment, complexity, and size of bridges. The uncertainty of internal defect excitation places even further limitations on the AE technique. However, the main limitation with present AE technology is the ability to relate AE signals to specific source events.

Currently available AE instrumentation can perform standard signal analysis, with well defined firmware using the signal in digital form. However, this equipment and methodology does not exploit currently developed signal-processing techniques to characterize defects and their locations. Recent laboratory signal-characterization tests have yielded promising results, but it is not known whether these techniques are viable in field applications. Also, little work has

A. S. Nowak (ed.), Bridge Evaluation, Repair and Rehabilization, 187–199.

been done to optimize equipment, test methods and data analysis from bridge AE sources. The problem of positively identifying, locating, and assessing flaws is presently an active area of research in AE technology.

2. Background

GARD, INC. has been actively pursuing the application of acoustic emission monitoring to the in-process NDT of welds for over eleven years. These efforts have culminated in the development of a microprocessor-based acoustic emission monitoring system that can detect, locate, and characterize flaws in welds during the welding process. The system has been evaluated and optimized for a wide range of commonly utilized welding processes and materials. An extensive evaluation program sponsored by FHWA (Contract No. DTFH6l-80-C-00083) has just been completed in which the Acoustic Emission Weld Monitor (AEWM) was successfully tested on typical highway bridge materials and weld methods. The detailed results of this program are presented in Report No. FHWA/RD-83-006 to be published shortly. In over 350 feet of laboratory controlled welding in both A-36 and A-514 steel, using both GMAW and SAW welding methods, the GARD monitor detected 97% of the flaws (missing only one porosity). Furthermore it correctly characterized 92% of the cracks. No unconfirmed AE indications were produced, however 14 flaws which were predominantly cracks and lack of fusion were either undetected or marginally detected by radiography and ultrasonics but were easily detected by the AEWM and confirmed by metallography. The key to this success is the AE signal processing methods developed by GARD. The welding environment is a very difficult area for application of acoustic emission monitoring due to the extremely high acoustic noise levels that exist. These high noise levels preclude the application of conventional acoustic emission monitoring techniques because of the high incidence of AE signals that result from benign or non-flaw related sources. The GARD signal processing technique uses an empirically developed pattern recognition method that keys on the AE signal characteristics resulting from flaw growth and allows the vast amount of non-flaw related AE to be rejected.

The problem of detection of flaw related AE that is imbedded in a high noise background is not unique to the in-process weld monitoring application. Rather, it is typical of a wide range of potential AE applications, one of which is the in-service monitoring of bridges for flaw growth. The belief that the AEWM might prove effective for this application led to the work described in this paper.

3. Experimental Description

The tests described herein were conducted on the I-24 Twin-Arch Bridges over the Tennessee River, near Paducah, Kentucky. The tests were performed by GARD with the assistance of both Kentucky Transportation Research Pro ram (KTRP) and Kentucky Department of Highways KYDOH) personnel.

KYDOH had previously determined that the bridge had suffered out-of-plane bending cracks near the connections in the deck cross beams in the vicinity of the upper flanges. This type of cracking is caused by design and construction problems and is somewhat generic for this bridge type. The cracking is fatigue-related and is not due to any fabrication defects in the steel weldments.

In an inspection performed just prior to the AE monitoring tests KTRP confirmed several crack sites in the cross beams over the piers. The cracks were located at the termini of the upper flanges, usually at the toes of the web-to-flange fillet weld. A typical site is shown in Figure 1 . Dye penetrant and in some cases surface rust made the cracks easily visible.

The flaw sites chosen for testing purposes were all located near bolted angle splice plates that connected the cross beams to the tie-girders. It was felt that these locations would be very difficult to monitor because of large amounts of fretting noise resulting from the bolted connections. This assumption was borne out in the tests in that typically over 1000 AE events occurred per hour during the tests. The activity was in all cases associated with the passage of traffic over the portion of the bridge being tested.

Standard resonant AE sensors were used for these tests. The sensors had internal permanent magnets for attachment purposes. Silicone grease was used as a couplant. The sensors were mounted 64" apart along the edge of the angle splice plate. They were acoustically coupled to the cross beam, and the upper flange (which was the site of the crack) was located about 16" down from the top AE sensor (channel 1). The AEWM was used in a two channel linear source location mode. A third sensor was attached as near as practical to the crack site. This 3rd sensor was driven by a high power pulser and was periodically pulsed to produce a simulated AE burst to allow checkout of the AE systems performance. A typical sensor set up is shown in the lower portion of Figure 5. Each receiving sensor had a unity gain preamplifier attached to it. This pre-amp matches the high impedance of the sensor to the 50Ω cable so that cable length is not a factor. Signal cables were fed across the pier and up to the bridge deck where the AE equipment was mounted in a self-contained motor home that acted as a mobile laboratory. Figure 2 shows the mobile lab in place on the bridge.

The AEWM performs the required pattern recognition for flaw detection in real-time. The process consists of subjecting each AE event to a series of tests performed in sequence. The current flaw detection model is a three step Process. A flow chart of the process is shown in Figure 3. After computing the ringdown count (RDC) for each event the first test is applied. If the RDC lies within pre-set limits the event is passed on to the next test which is rate. This test requires that there be some number N of events (that have passed the RDC test) within some pre-set Δt or time interval. The final test is a test to see if all of the events that passed the previous two tests originate from the same location, or at least within some pre-set locational tolerance. The combination of the rate and location tests provides very high discrimination against interfering background acoustic signals, the assumption being that a growing flaw will produce higher rates of AE burst emission than other processes and that the flaw, being a localized phenomena will produce the high rate from a specific well defined location. Our use of source location as a flaw detection criteria differs radically from the traditional use of source location information. In conventional AE monitoring equipment, source location may be used to lock out given areas or regions of the structure under test, in other words, the system may be made to listen only to a specific location. This approach requires prior knowledge of the probable location of a flaw, and its degree of success depends on the flaws being locationally isolatable from potentially interfering sources (a condition that is seldom met in typical bridge structures). GARD's approach to the use of source location does not limit the monitor to a specific location. Any source location lying between the transducers is monitored. When a group of AE events has satisfied the first two criteria (i.e., ringdown count and event rate) a test is made to see that all of the group of events lie within preset locational limits of each other. For example, if a 1 inch tolerance is used, then the events that satisfied the - first two tests must have the same order of receipt at transducers 1 and 2 and their locational clock indications must not differ by more than 16 counts (16μ seconds). If this criteria is met, then a flaw indication is shown at the appropriate location. Later in this paper, the importance of all three tests will be shown.

In addition to the detection of the flaw related AE, the AEHM applies an adaptive frequency analysis model to the flaw-related events and provides a 2 category classification of the source, crack or non-crack.

For these tests a floppy disc system was used in addition to the AEWM to allow post test analysis of data. In this mode of operation, the limits for the flaw detection criteria can be varied and raw AE data can be played back through the modified model thus allowing optimal monitoring parameters to be obtained.

A photograph of the AEWM in operation is shown in Figure 4.

4. Results

A total of five sites were monitored over a three day period. Only one of these actually produced AE indications. These indications - repeated on two consecutive days and were properly located in the known crack-region. The first two sites tested were above the West Pier on the eastbound span. The first area (Location 1) had a 1 1/2" long crack at the flange terminus. Considerable AE activity resulted during the two hour test. The activity occurred in conjunction with traffic and highest amounts of activity correlated with large heavy vehicles. None of the resulting activity produced any AE indications (valid AE). This test constituted monitoring a small crack under light loading conditions. After two hours, the sensors were moved to the passing lane side of the bridge in an attempt - to get higher loading on a flaw (Location 2). This site had two 1" cracks visible in a location similar to the first. No valid indications resulted in a two hour test at this site, however, a relaxing of the flaw detection requirements (lowering rate from 4 to 2 Hz) did show some clustering of activity from the crack site along with some scattered back-ground activity.

The next site tested was located on the west bound span (Location 3). This was the most severe flaw tested. There was a 2" long through crack at the toe of the web to flange fillet weld in addition to a second crack emanating from under the angle splice plate directly above the same region. This flaw was positioned under the passing lane of the bridge and so it received the maximum possible loading during the test. Figure 5 shows the results of two separate tests performed on this site over two successive days. Each test was approximately 2 1/2 hours long. The model used for flaw detection had limits as follows:

RDC- 16 to 4000

Rate- 4 events in one second

Location-1" tolerance

In the upper right corner of the printout, the total number of received AE events is shown. For these tests we see that the totals were 2130 for the first 24 hour monitoring period and 818 for the second. These differences reflect the difference in the amount of traffic for the two monitoring periods. The AEWM Display prints sets of rectangular brackets to represent the two sensor positions with channel 1 at the left and channel 2 at the right. In this display, flaw indications will be shown at any location when the detection criteria are met. The edge of the angle splice runs along the line between the two sensors. The character, C, p, in the upper display indicates that at this location the flaw detection criterion was satisfied. Furthermore, the characterization model decided that the AE was crack related. The "P" following the comma is the truncated average of the ringdown counts for the four or more events that satisfied the detection model. In this case P signifies that the average ringdown count was between p and 99. Additional groups of events that satisfied the model are presented below the "C,P". Time of occurrence proceeds in a downward direction. The "S,:" indication is produced by our calibration pulser which was located adjacent to the bottom edge of the flange. The cracks extended around the end of the flange and above the end of the flange off toward the angle splice plate. The "S,3" indication occurs at about the end of the flange (S signifies noncrack related). One additional S,2 indication occurs near the midpoint of the monitoring region. Confirmation of a flaw in this region was not possible at the time the test was run.

The lower display was the result of another 2 1/2 hour monitoring period the following day. There was considerably less traffic during this period which is reflected in the lowered AE event count (818). One indication (S,3) occurs from the lower edge region of the flange. The photograph below the printouts shows the sensors in place. The actual orientation was vertical, however we rotated the picture 90° to place the significant features in approximately the same orientation as the printouts.

To summarize results for this site, all of the indications were grouped around the region of the known crack with the exception of one which occurred at just above the midpoint of the sensor array.

To further test the reliability of the AEWM to discriminate between the fastener noise and the crack emissions we positioned the sensors on an adjacent plate where the same pattern of fasteners existed, but no flaws were visible (Location 3A). A 2 1/2 hour monitoring period from this site produced 700 AE events but no flaw indications. The final site tested contained a fillet weld with a longitudinal crack visible (Location 4). This crack was evidently a product of the fabrication shop and produced no AE activity since no growth was occurring.

To further illustrate the power of the three test flaw detection model, the illustrations in Figure 6, 7 and 8 show what happens when we relax the rate criteria. These figures utilize a different display mode feature of the AEWM. The monitor has provision for connection of an x, y, oscilloscope for the purpose of presenting source location information. In the figures shown, the sensor positions are represented by squares at each side of the display. A bright dot signifies an AE source location that satisfies the flaw detection model. Additional successive indications at the same location further brightens the spot.

In Figure 6 we see two scope displays produced by the same data that produced the printouts in Figure 5. Here the model still requires a rate of 4 Hz or higher from any source. In - Figure 7 we see the result of playing back the, same data as in Figures 5 and 6, however, the rate criteria has been defeated. The only restriction on data is the use of a ringdown count window (equivalent to an acceptance threshold). Here we see a great deal of clutter on the displays, most of which occurs in locations where no known flaw exists. Figure 8 shows similar results for the unflawed test area. The top display is produced with a normal 4 Hz rate criterion and a very wide open ringdown count window. The next two displays are the results for the same data with no rate test and varied RDC windows. Even though this area is probably flaw free, a multitude of AE sources results. The use of source location and fixed thresholds is obviously not sufficient to eliminate the fastener noise and allow reliable flaw detection.

5. Summary

The AEWM was used to monitor several sites on a bridge where known fatigue cracks exist. The two channel linear location system sup-pressed the acoustic noise from fastener fretting and reliably and clearly detected crack-related activity even though the cracks were either immediately adjacent to or coincident with the bolt holes.

Besides providing that flaw growth related activity can be detected from noisy structural details, this work showed that very small amounts of fatigue crack growth can be detected when a bridge is subject to routine loadings. The volume of traffic on this bridge was not significant compared to bridges in more urban locations. Nor was the magnitude of the bridge loadings unusual. While the valid AE activity was usually correlated with one or two heavily loaded semi-trailer trucks, this type of traffic was very infrequent. Yet, it took no more than two and a half hours to excite the expected AE activity at flaw sites.

This work indicates that unusual load procedures such as heavy proofing loads are not necessary to excite AE activity from a crack which is already experiencing sub-critical growth due to service loading. In-service AE monitoring of a suspect area for a period of less than four hours should be sufficient to detect fatigue crack on steel bridges subjected to normal structural loading patterns. Structural discontinuities which are harmless will not generate any AE activity and will be ignored.

AE testing, incorporating the equipment and techniques described in this paper, has demonstrated three attributes which make it a desirable NDE method for inspection structures such as bridges;

(1) Good operator productivity.
(2) The ability to detect and define bridge defects (cracks).
(3) The ability to compliment (confirm) other NDE methods.

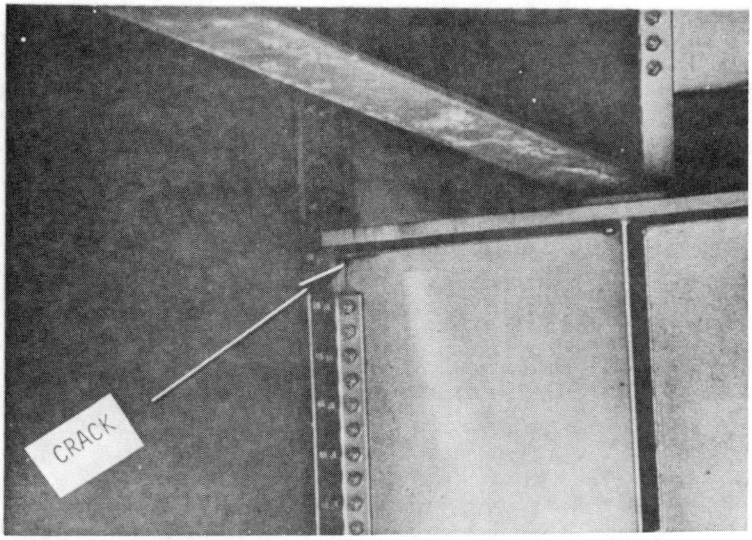

Fig. 1 Photograph Shows Typical Crack Site in Cross Beam of I-24 Bridge.

Fig. 2 Mobile AE Test Laboratory in Operation of I-24 Bridge.

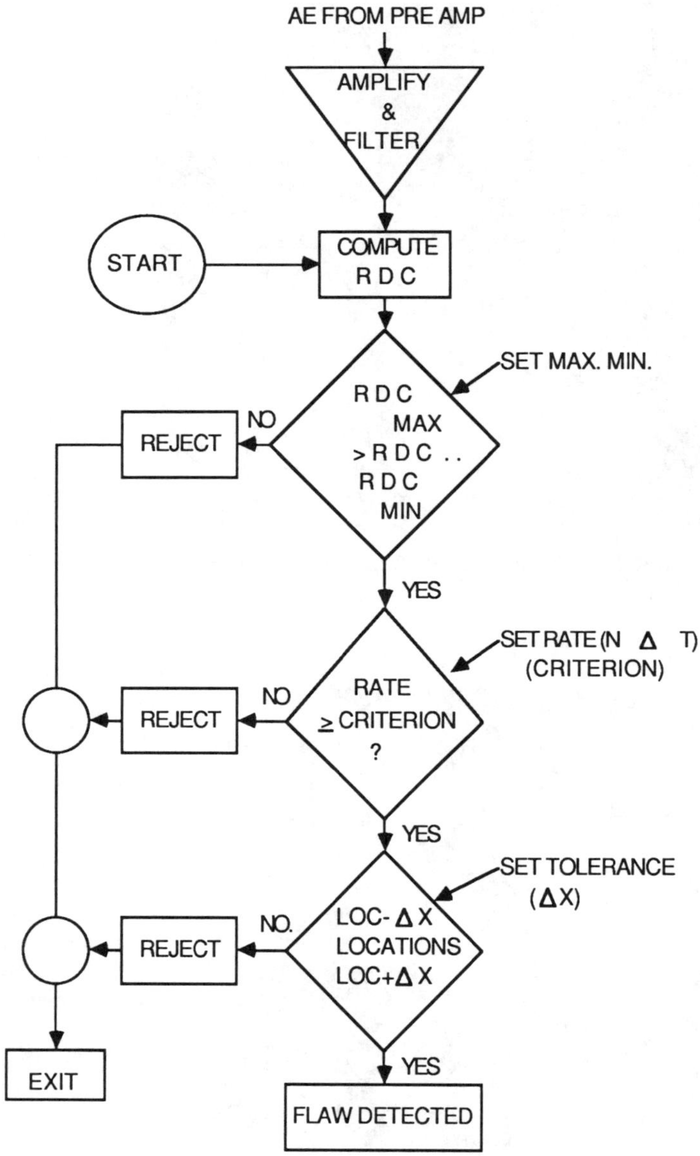

Fig. 3 AEWM Processing Flow Chart for Flaw Detection.

Fig. 4 AEWM in Operation During Bridge Monitoring Test.

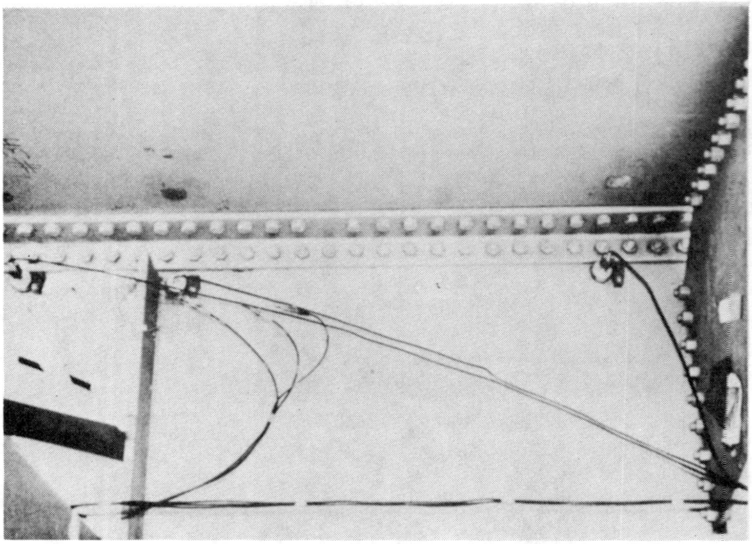

Fig. 5 AE Results for Location #1 Westbound.

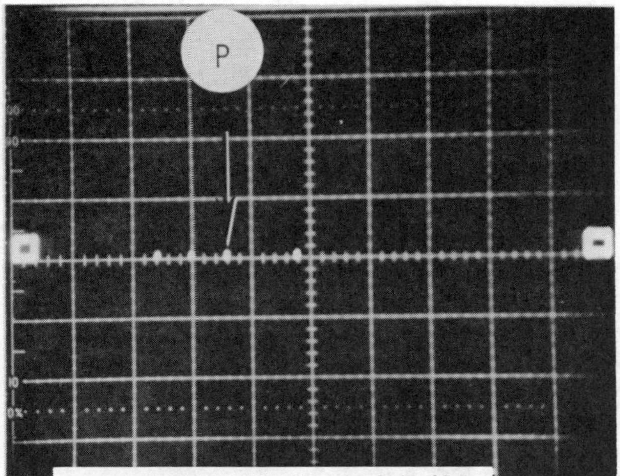

I-24 WEST LOC #1 12/9/82

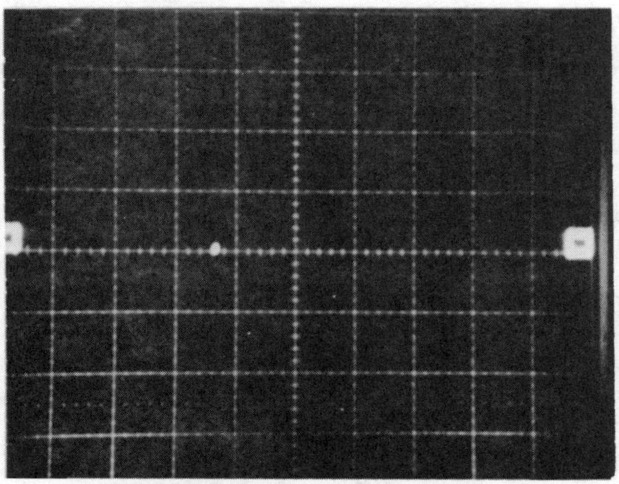

I-24 WEST LOC #1 12/10/82

Fig. 6 AE Source Location Plots.

1:1, 100-1000

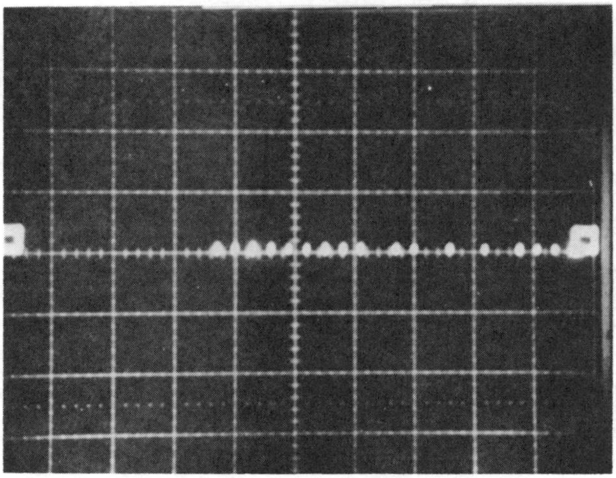

I-24 WEST LOC #1 12/10/82

1:1, 500-5000

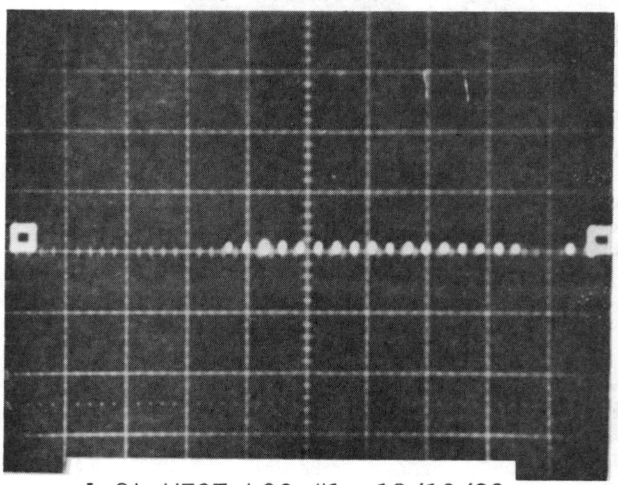

I-24 WEST LOC #1 12/10/82

Fig. 7 AE Source Location Plots.

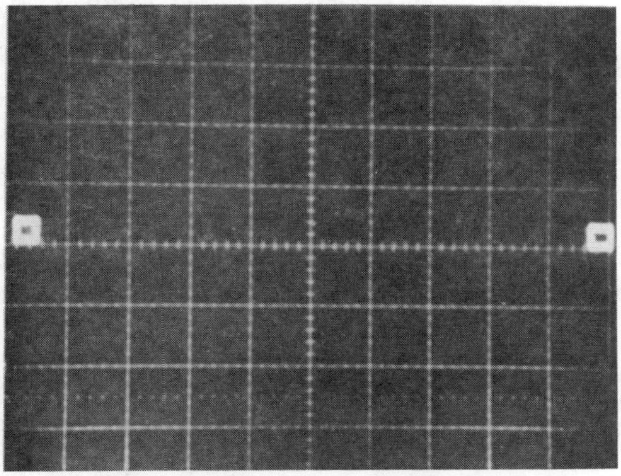

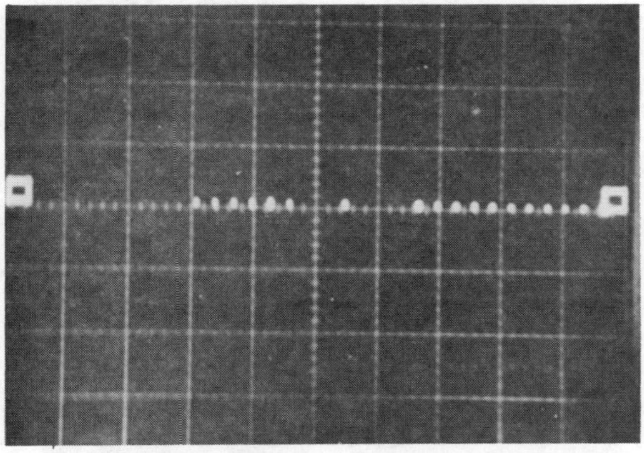

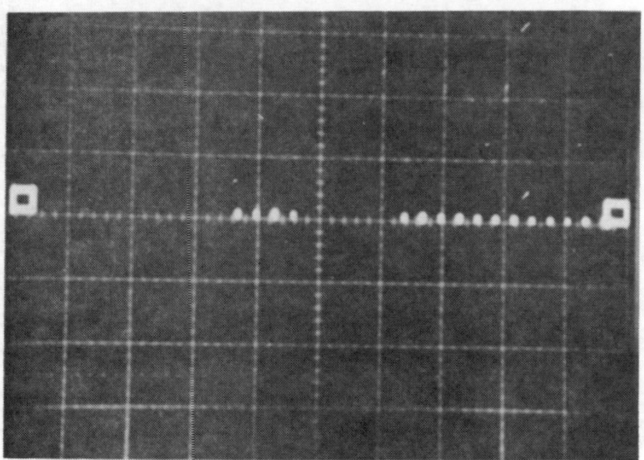

Fig. 8 AE Source Location Plots.

DIAGNOSIS OF REINFORCED STRUCTURES: AN ACTIVE MICROWAVE IMAGING SYSTEM

CH. PICHOT
Labortoire des Signaux et Systems
C.N.R.S.IE.S.E., Plateau de Moulon
91192 Gif-sur-Yvette Cedex, FRANCE

P. TROUILLET
Public Works Res. Lab.
B.P.8
54510 Tomblaine, FRANCE

ABSTRACT. This paper is devoted to a new imagining technique for the Non-Destructive Testing (NDT) of civil engineering structures such as bridges. This technique uses microwave radiation to investigate the concrete medium in a reflection mode configuration. The backscattered electromagnetic field, generated by the presence of the rebars, is measured with a linear sensor probe array and processed numerically in order to provide tomographic images of the volume under investigation. A prototype of a so-called *Microwave Camera* operating in the microwave X-band region between 7-13 GHz has been especially designed and constructed for assessing the potentials of this technique. Frequency diversity in the X-band is used to obtain a spatial resolution in the images of less than one centimeter for the localization and diameter evaluation of rebars. Performance evaluation of the prototype has been conducted on reinforced concrete slabs with practical configurations and on a bridge beam. The performance has been compared with that of magnetic metal detectors which are available on the market.

1. Introduction

The development of tomographic processes for X-rays, Gamma-rays, ultrasonics and nuclear magnetic resonance has evolved a high degree of sophistication, however, active microwave imaging is just beginning to be realized as a viable approach in certain applications as Non Destructive testing (NDT).

As with every civil engineering structure, bridges fall into disrepair over a certain period of time because of aggressions due to heavier loads and increasing traffic, to atmospheric pollution, temperature variations and water effects (moisture, frost, etc.). These structures must be checked more and more frequently by means of calculations, for supervision, maintenance and repairs, or to allow the safe passage of exceptional convoys or to take account of increasing loads, certainly greater than those expected at the time of construction. Accurate knowledge of reinforcement (number, position, and diameter of the different rebars) is essential for such verifications. Corresponding draught plans do not always exist or cannot be relied on. For these

A. S. Nowak (ed.), Bridge Evaluation, Repair and Rehabilization, 201–215.

reasons it is necessary to have in-situ recognition methods which establish mapping of reinforcement structures such as bridge beams.

The simplest method, consisting of making holes in the concrete in order to uncover the reinforcements, has major drawbacks apart from its cost: it can only be used punctually and in some cases its destructive nature prohibits its utilization. NDT techniques have made great strides in these last few years, with the multiplication of different fields of application and the introduction of new methods and techniques. But in spite of all the progress made in this domain there is, at present, no one method which gives unanimous satisfaction in the examination of reinforced concrete structures.

There is a NDT technique used with probe instruments such as the pachometer or profometer. These are based on magnetic detection (comparison of electromotive forces between a reference transformer and the transformer formed by the probe and the buried rebar). This kind of probe gives an indication of the position of rebars closest to the walls. From diagrams, it should be possible, in theory, to determine the diameter of a rebar if the concrete cover is known. The action is sensitive to a cover of less than 10 cm. In practice, these devices are proving to be too inaccurate for use, especially in the presence of numerous rebars.

Intensive research has been done on methods based on X-or Gamma-rays at the French Public Works Research Laboratory. These methods are probably the most interesting since they allow an accurate verification of the conformity and state of the rebars. Nevertheless, they have the following drawbacks: They are slow (except for the radioscopic technique) but above all very burdensome, heavy to handle, and require qualified personnel (if only with regard to radioprotection safety standards). Moreover, it is absolutely essential to have access to both sides of the investigation structure (measurement by transmission).

Within this context, it appears interesting to investigate a new method which as far as possible will not have the drawbacks evoked in the existing methods and techniques.

The Microwave Camera prototype developed at the Laboratoire des Signaux et Systemes has been patented by the French National Center for Scientific Research (CNRS) and the French Public Works Research Laboratory (LCPC) under no. 8810479. This paper deals with an experimental study to determine the Microwave Camera performance in the case of practical rebar configurations. In that way, we have studied the spatial resolution in the images, the penetration depth, the ability to detect butts (end of rebars), horizontal and vertical overlaps, the influence of concrete and, finally, the study of a complex case (bottom flange of a bridge beam). Different concrete slabs with dimensions 1m x 1m (except for the flange of dimensions 0.7 m x 0.75m), representative of practical construction configurations, were especially built.

The investigated parameters on the different slabs and on the bottom flange of a bridge beam are reported in Table 1.

In the first part, we present the physical and technological aspects of the Microwave Camera. In the second one, or show some typical experimental images obtained with the Microwave Camera. In the third part, we present the results with magnetic metal detectors available on the market. Finally, we compare the performances of these probing techniques and draw conclusions on their respective interest for the civil engineer.

2. Complex Permittivity of Concrete in the Microwave X-band Region

An important point deals with the behavior of concrete in the Microwave X-band region. In such a way, we are interested in the values of the complex permittivity ε^*. The nature of the aggregates used can be a factor in variation of the electromagnetic characteristics. Two different natures of aggregates were used in the test slabs. The first ones are *Diorite* aggregates which are composed of 80-90% silica and the other ones are *Moselle* aggregates containing less than 10% silica. Beside these different natures, the moisture or water content is an important factor which affects the complex permittivity and the propagation of electromagnetic waves through the concrete by increasing the attenuation. Complex permittivity measurements at 10 GHz have been performed at the Public Works Research Laboratory in Rouen, France on various concrete samples from the test slabs. The samples were first dried in an oven and then moistened little by little. The degree of moisture was determined by weighing. The volumetric degree of moisture M_V is defined by:

$$M_V = (W_W - W_D)/V$$

where W_W represents the weight of the wet sample, W_D the weight of the dry sample and V the volume of the dry sample.

From the measurements, the values of the real part (ε') and the imaginary part (ε'') of the complex permitivity ε^*, as a function of M_V (%), can be given in a first approximation (up to 8%) as linear with the following formulas:

Diorite:

$$\varepsilon' = 0.307\, M_V + 4.9$$
$$\varepsilon'' = 0.117\, M_V + 0.11$$

Moselle:

$$\varepsilon' = 0.335\, M_V + 4.8$$
$$\varepsilon'' = 0.135\, M_V + 0.14$$

3. Microwave Camera System

The experimental set-up of the Microwave Camera is shown in Figure 1. The Microwave Camera is composed of a microwave source (with frequency ranging from 7 GHz to 13 GHz) and one or two transmitting antennas (sectoral horns of dimensions 60 cm x 60 cm) radiating an electromagnetic incident field into the concrete. The incident field excites the rebars inducing a surface current on them. This current generates a microwave scattered field. The backscattered field is measured with a linear sensor array (length 32 cm) placed in the aperture of a receiving antenna (of the same kind of transmitting antennas). The principle of the measuring technique is based on the Modulated Scattering Technique (Bolomey (1982)). According to this technique, the sensor array consists of 64 diode loaded dipoles. By modulating each diode successively, with a multiplexer at a low frequency rate, a modulated scattered field proportional to the local field at the corresponding dipole location is reconstituted at the output of the receiving antenna. After double synchronous detection, one gets the modulus and phase of the microwave scattered field at the array location. This solution avoids any individual microwave connection between the probes and the receiver. No microwave multiplexer is needed, reducing the cost and complexity of the array structure.

The acquisition time for the signal (microwave scattered field in modulus and phase) is at present about 7 seconds per frequency. The number of frequencies generally used is about 13, which gives a total amount for the acquisition time on the order of one and a half minutes.

The imaging algorithms take into account the diffraction effects (Diffraction Tomography) (Chommeloux et al. (1986), Tabbara et al. (1988)) which are important due to the rebar dimensions versus frequency (resonance domain). They allow the display of tomographic images of the volume of concrete under investigation. The frequency diversity technique is used in order to improve the spatial resolution in the image. The total processing takes about 4 minutes for about 13 frequencies with a minicomputer PDP 11/23 in order to display the tomographic image on a color TV screen.

The innovation of the system lies in its conception of the measurement technique and the development of imaging algorithms based on Diffraction Tomography (Chommeloux et al. (1986), Tabbara et al. (1988)). The receiving antenna is placed between the two transmitting antennas. The source and microwave detection are rigidly locked with the antennas, which allows the system to be moved without perturbations in the measurements.

4. Microwave Images and Performance of the Microwave Camera

The physical parameter contained in microwave images and displayed on the screen is the induced current generated by the electromagnetic incident wave at the surface of the rebar. But the quality of any imaging system is not perfect and is determined by its impulse response, which is the resulting image when the object is a point. The spatial resolution of the system is then the dimensions of the spot. The resolution of the system is dependant on the incident wave frequency. In principle, the use of higher frequencies leads to better resolution but, in the microwave domain and for most materials like concrete this leads also to higher attenuation of the incident electromagnetic field. So the frequency range for a good compromise between resolution and penetration must be found.

Tests which have been performed on the different slabs lead to the following conclusions:

- Maximum investigation depth is actually about 7 or 8 cm, which corresponds to the zone where, in general, the rebars are concentrated. For example, in the case of the bottom flange of a bridge beam, the Microwave Camera is capable of detecting 14 rebars out of a total of 19. The use of smaller transmitting and receiving antennas should detect 16 rebars out of 19.

- Diameter evaluation is generally underestimated for large diameters and overestimated for small diameters. For example, for a diameter of $\Phi = 30$ mm, an average error of 6 mm is found on the images, and for a diameter of $\Phi = 10$ mm, an average error is 8 mm.
- Error in the location (center of the rebar) varies, on the average, from 4 mm for large diameters (Φ=30 mm), 75 mm for middle diameters (Φ =20 mm) and 90 mm for small diameters (Φ=10 mm).
- The horizontal overlaps are detected if the rebar diameter is greater than 1 cm.
- The vertical overlaps are detected. Nevertheless, due to a non-uniform incident field generated by the transmitting antenna with frequency variation, a considerable error in the vertical position in the case of several parallel overlaps may occur.
- The complete detection of overlaps could be possible with an exploration in both perpendicular directions. The intersection of perpendicular rebars does not constitute a peculiarity for the system.
- Reinforcement in a corner does not pose any particular problem for the system with the use of a concrete plate or, even better, a lighter material with the same electromagnetic characteristics in the X-band.
- Water is a highly dissipative medium for microwaves, particularly in the X-band. For concrete structures subject to dripping, it is necessary to dry the surface, and for fresh concrete structures (up to one month old), no measurement is possible.
- Concerning the type of the concrete, we found no strong variations in this frequency domain in the electromagnetic characteristics for various kinds of concrete.
- If the wall surface contains irregularities, this may lead to additional reflections which may deteriorate the images. It would be necessary, in this case, to use an intermediate medium in order to better adapt the antennas to concrete.

5. Results with Magnetic Metal Detectors

Existing light equipment essentially gives an analog response which includes both the dimension effects of the embedded rebar and the distance effect between the rebar and the probe.

In an attempt to separate these two areas of information, a prototype has been realized and developed since 1983 at the French Public Works Research Laboratory (LRN-CECP) in Rouen (Nerrieres and Faucon), starting from a metal detector available on the market (profometer) in order to dispose of two types of digital information:

- signal amplitude
- position on the wall being investigated

The equipment has been tested by experimental design on the same slabs as those used for the Microwave Camera.

The different configurations investigated are defined in Table 2.

At the center of the investigated domain (* in Table 2), 10 tests were performed in order to evaluate the variability of the measurements.

Figures 10 and 11 show the curves for signal amplitude in terms of the position along the surface of the concrete exhibiting the respective influence of cover and diameter.

Peak value and bandwidth are each connected to the diameter Φ and to the cover C by a relation of the form:

$$f(\Phi,C) = \alpha + \beta\,\Phi + \chi E + \delta\,\Phi C$$

From this preliminary calibration the diameter and the cover can be determined separately with a respective standard deviation of 1.5 mm and 4 mm.

Tests conducted on the different slabs lead to the following conclusions:

- The vertical location of rebars could, in theory, be determined by the corresponding cover variations. Unfortunately, the upper rebars hide the lower ones in the overlapping zones.
- The horizontal overlap (juxtaposition of 2 rebars) could be displayed by means of an equivalent diameter, but this would require complementary studies.
- Butt detection is possible if the space between rebars is greater than 1 cm.
- Intersections of rebars are detected by differences in the cover along the two perpendicular rebars.
- Detection of a single rebar among a group of rebars (case of bottom flange of a bridge beam) requires a complex numerical code which has not yet been developed.

In conclusion, it is possible to determine diameter and concrete cover for an <u>isolated</u> rebar with satisfactory precision for the civil engineer with the appropriate equipment. Nevertheless, this calibration and the difficulty in distinguishing a single rebar among a group of rebars at present constitute two serious limitations to the development of the system.

6. Conclusion

A Microwave Imaging system (so-called Microwave Camera) for Non-Destructive Testing (NDT) of reinforced structures has been presented and discussed. Using frequency ranging, and measuring the microwave backscattered field, this system allows us to obtain mappings (tomographic images) of concrete under investigation and to detect the presence of rebars. The maximum auscultation depth is 8 cm. The diameter and location of the different rebars are given with an accuracy of less than 1 cm. The original technique provides fast and accurate measurement of the microwave scattered field at different discrete frequencies. The imaging algorithms used to form the images allow the display of pseudo-color images within a few minutes with a microcomputer.

Experimental studies on test slabs and on the bottom flange of a bridge beam are reported in order to determine the performance of the Microwave Camera. A comparison between the performance of the Microwave Camera and those given by magnetic metal detectors which are available on the market has been made. The results are summarized in Table 3.

The Microwave Imaging Technique is bound to face swift development owing to the richness of information contained in the images, which are obtained rapidly and allow the participation of the operator's knowledge.

7. References

Bolomey, J.Ch. (1982) "La méthode de diffusion modulé : une approche au relevé des cartes de champs microondes en temps réel," Onde Electrique, 65, 73-78.

Chommeloux, L., Pichot, Ch. and Bolomey, J.Ch. (1986) "Electromagnectic modeling for microwave imaging of cylindrical buried inhomogeneities," I.E.E.E. Trans. Microwave Theory Tech., MTT-34, 1064-1076.

Tabbara, W., Duchene, B., Pichot, Ch., Lesselier, D., Chommeloux, L., and Joachimowicz, N. (1988) "Diffraction tomography: contribution to the analysis of some applications in microwaves and ultrasonics," Inverse Problems, 4, 305-351.

Nerrieres, F. and Faucon, J. (1983) "Numérisation d'un profomètre, notice, Prototype No. 643-501," Centre d'Etudes et de Construction de Prototypes de Rouen (CECP), LCPC, Paris, France.

Trou illet, P., Schwartz, J. and Truck, J.P. (1988) "Détection d'aciers dans le béton armé à l'aide d'un montage prototype," Report LCPC, Paris, France, AER 65-14.

Table 1 Investigated Parameters.
((*) : $\Sigma_{yp} = 280$; (**) : $\Sigma_{yp} = 280$ and 420)

Investigated Parameter	Rebar Diameter (mm)	Concrete Cover (mm)
butt location (end of bar)	10, 30 (*)	30
horizontal overlap	10, 12, 30 (*)	30
vertical overlap	10, 12, 30 (**)	30
spatial resolution of the system	20 (*)	30
investigation depth	10, 16, 20 (*)	30 to 350
bottom flange of a bridge beam	30 (19 rebars) 12 (frame) 18(stirrups) (*)	30 minimum

Table 2 Investigated Covers and Rebar Diameters.

Concrete cover C (mm)	Rebar diameter Φ (mm) 10	16	20	30
36	+	+	+	+
57	+	*	+	+
78	+	+	+	

Table 3 Comparative Table of Microwave Imaging and Magnetic Metal Detector.

Investigated Parameter	Microwave Imaging	Magnetic Metal Detector
Rebar Diameter Φ (error ΔΦ)	Detected Average ΔΦ: 6 mm (large Φ) Average ΔΦ: 8 mm (small Φ)	Detected Average ΔΦ: 0.3 mm standard deviation: 1.5 mm
Rebar depth D (error ΔD) or Concrete Cover C (error ΔC)	Detected Average ΔP: 4 mm (large Φ) Average ΔP: 7.5 mm (mean Φ) Average ΔP: 9 mm (small Φ)	Detected Average ΔC: 2 mm Standard deviation: 4 mm
Butt (end of rebar) Minimum interval L between 2 butts	Detected L=75 mm with present system	Detected L=10 mm
Horizontal overlap	Detected if Φ≥10 mm	Detected but numerical code required for detection
Vertical overlap	Detected	Not detected because lower rebar hidden by upper rebar
Concrete nature *Diorite/Moselle* aggregates	No incidence	No incidence
Horizontal resolution R. (error ΔR)	Average ΔR: 5 mm if Φ≥10 mm	Impossible to measure with present system
Rebar in a corner	No incidence if medium of adaptation	No incidence but requires complementary tests
Maximum auscultation depth	70 - 80 mm	tested up to 80 mm limitation at 90 mm
Measurements on bottom flange of a bridge beam	Detection of longitudinal rebars on the two lower laps	No measurement

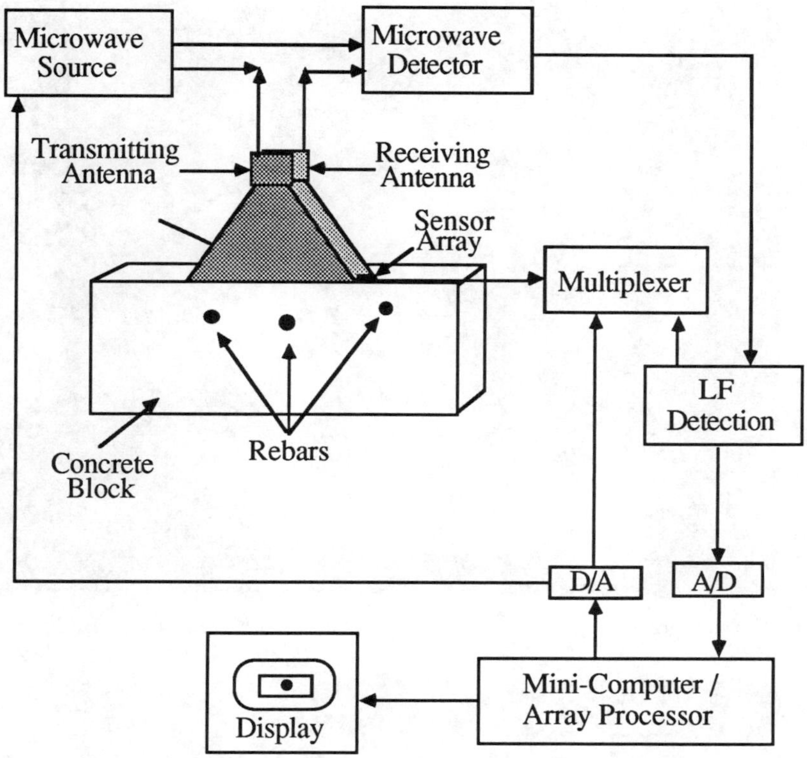

Fig. 1 Microwave Camera, Block Diagram of Prototype

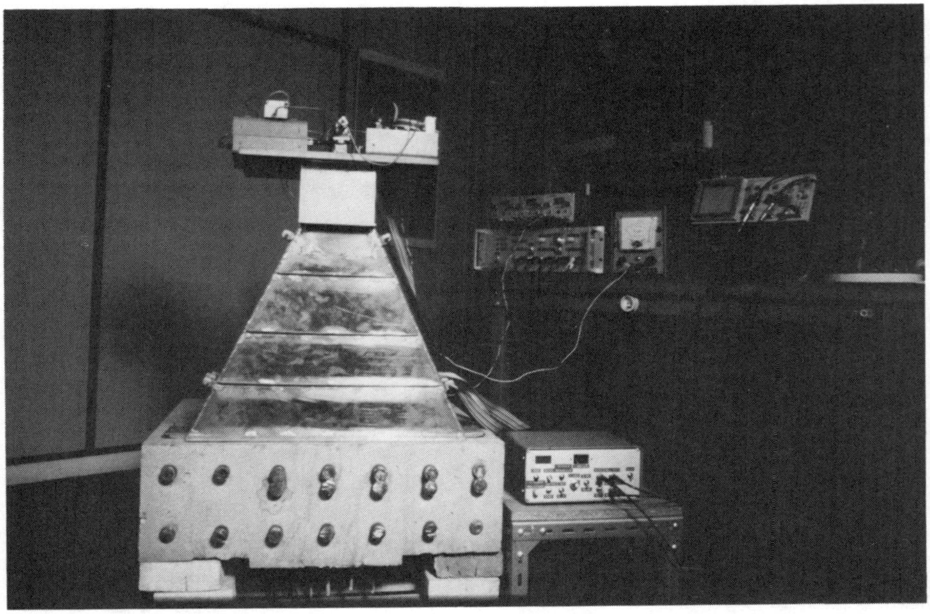

Fig. 2 Photograph of the Microwave Camera on a Bottom Flange of a Bridge Beam.

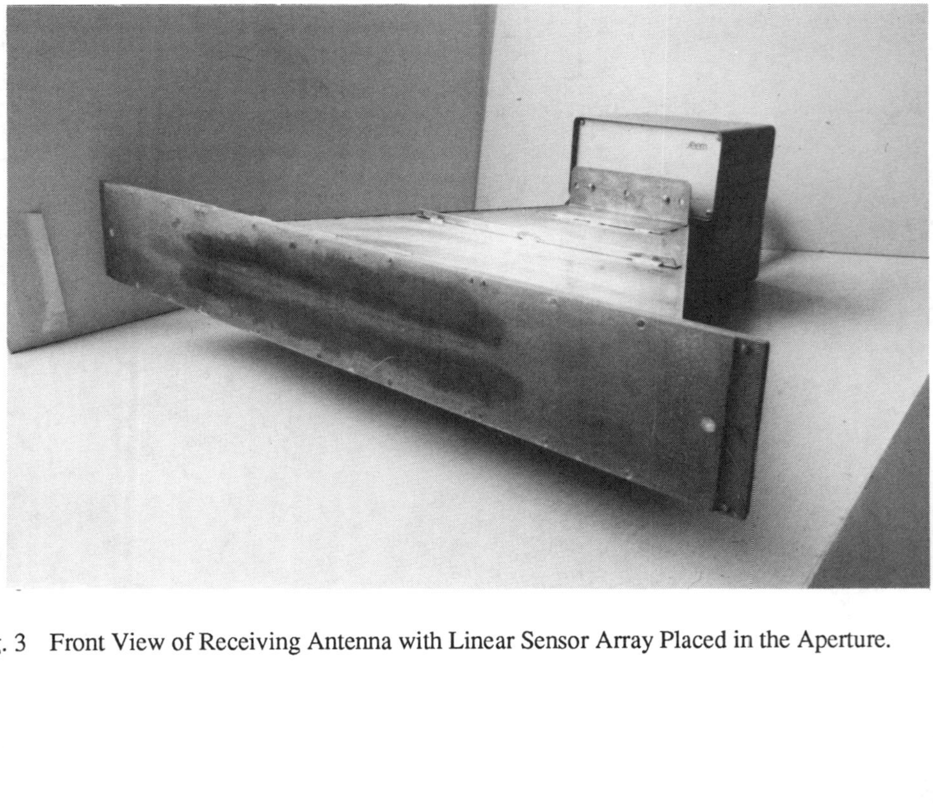

Fig. 3 Front View of Receiving Antenna with Linear Sensor Array Placed in the Aperture.

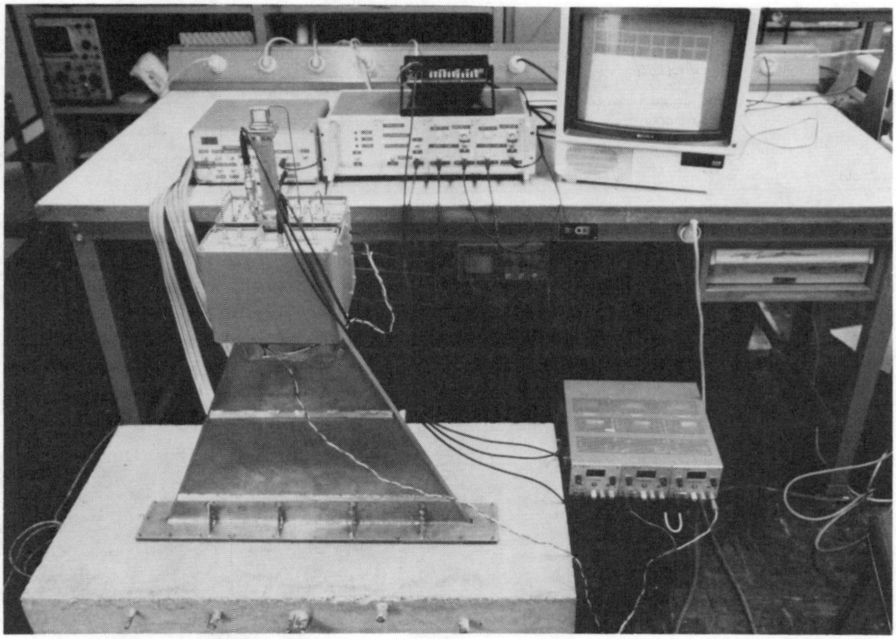

Fig. 4 Photograph of Receiving Antenna on a Test Slab with Electronic Equipment.

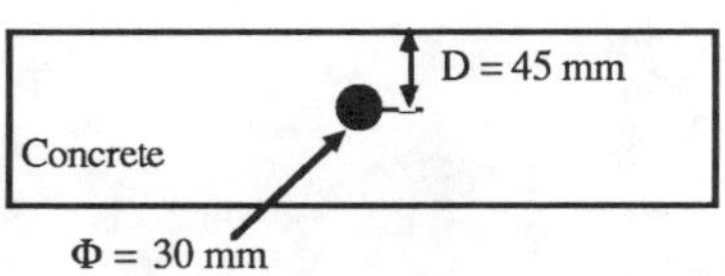

Investigated case

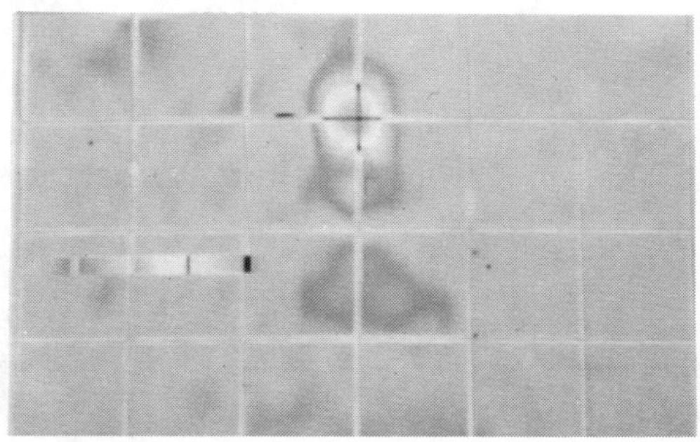

Microwave image (40 mm /division)

Fig. 5 Microwave Image of a Single Rebar of Diameter Φ=30 mm.

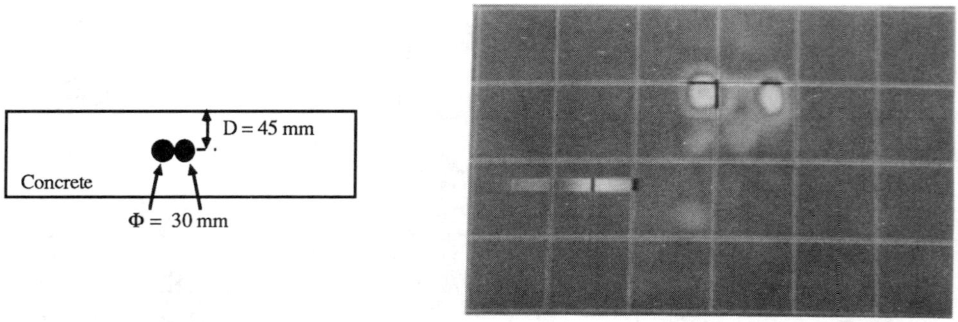

Investigated case | Microwave image (40 mm/division)

Fig. 6 Microwave Image of Two Rebars of Diameter Φ=30 mm.

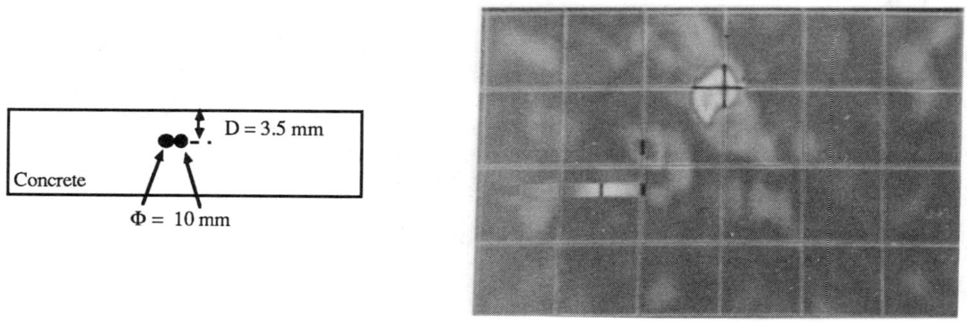

Investigated case | Microwave image (40 mm/division)

Fig. 7 Microwave Image of Two Rebars of DIameter Φ= 10 mm.

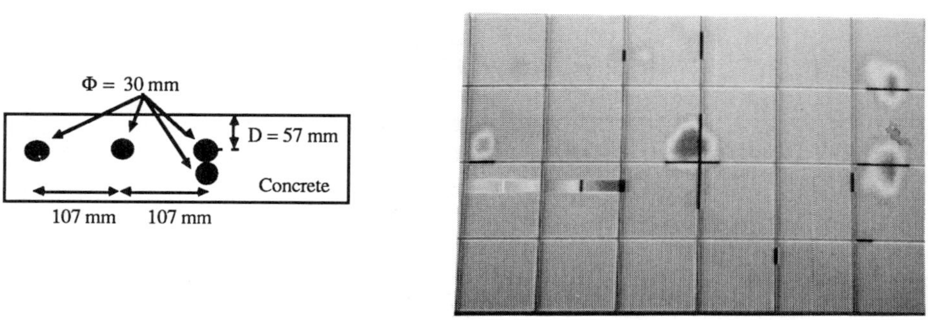

Investigated case on a bottom flange of a bridge beam (the 4 upper left rebars in Fig. 2) | Microwave image (40 mm/division)

Fig. 8 Microwave Image of Four Rebars of Diameter Φ=30 mm on the Bottom of a Bridge Beam.

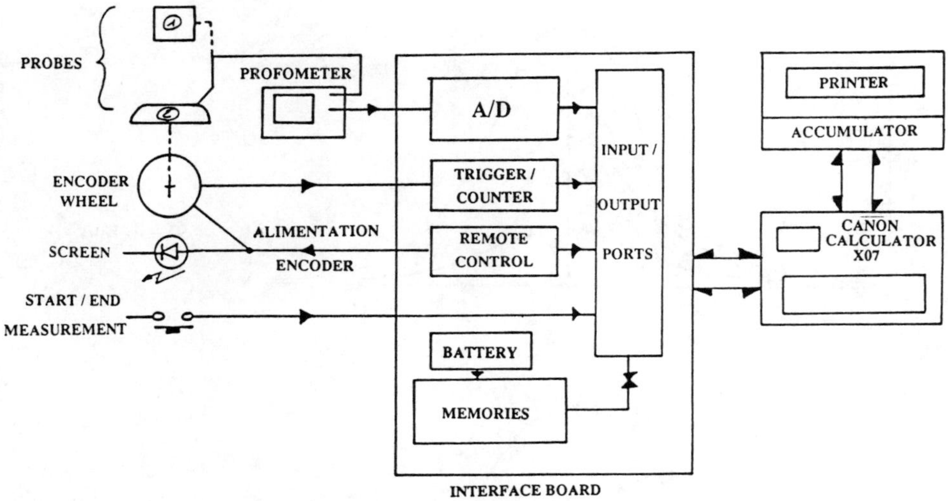

Fig. 9 Prototype Assembly at LRN-CECP Block Diagram.

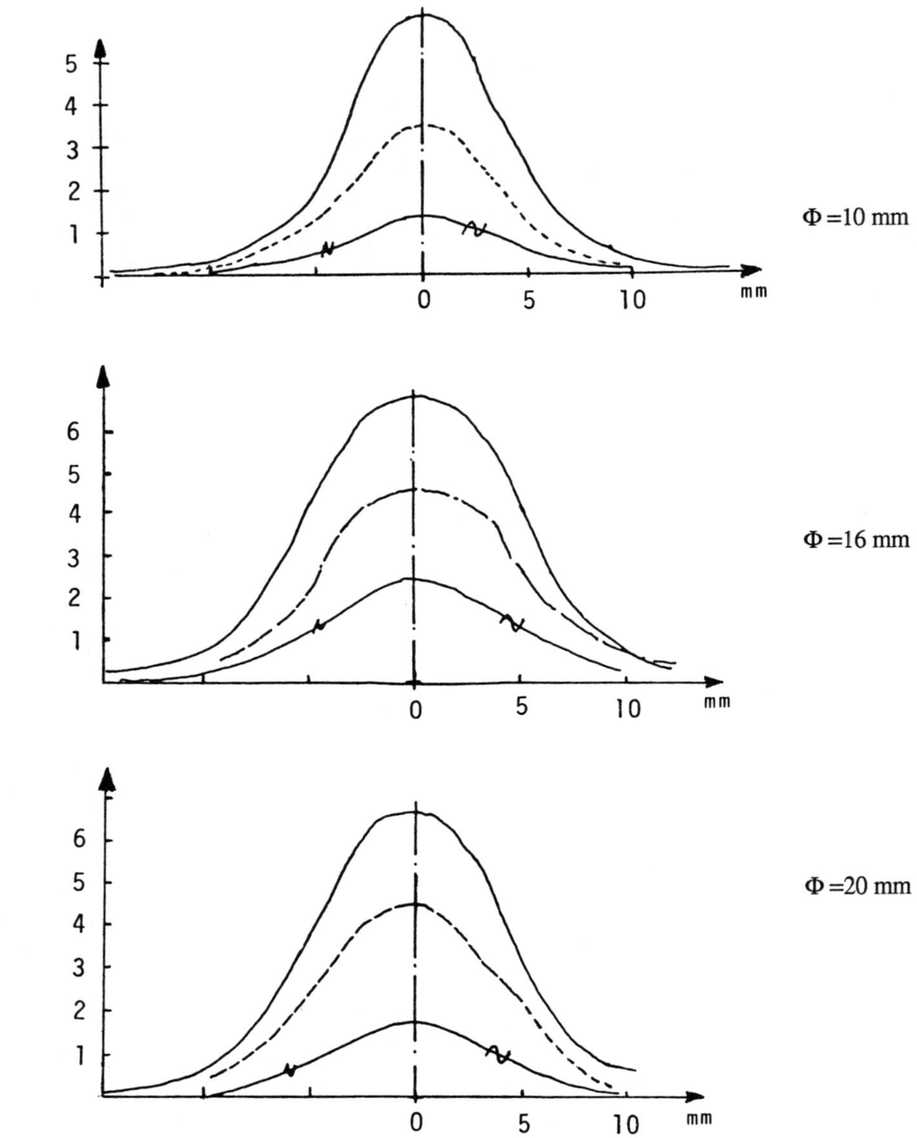

Fig. 10 Variable Cover C, Constant Diameter Φ (Signal Amplitude Versus Position).
(_______ C=36 mm, ------ C=57 mm, _______ C=78 mm)

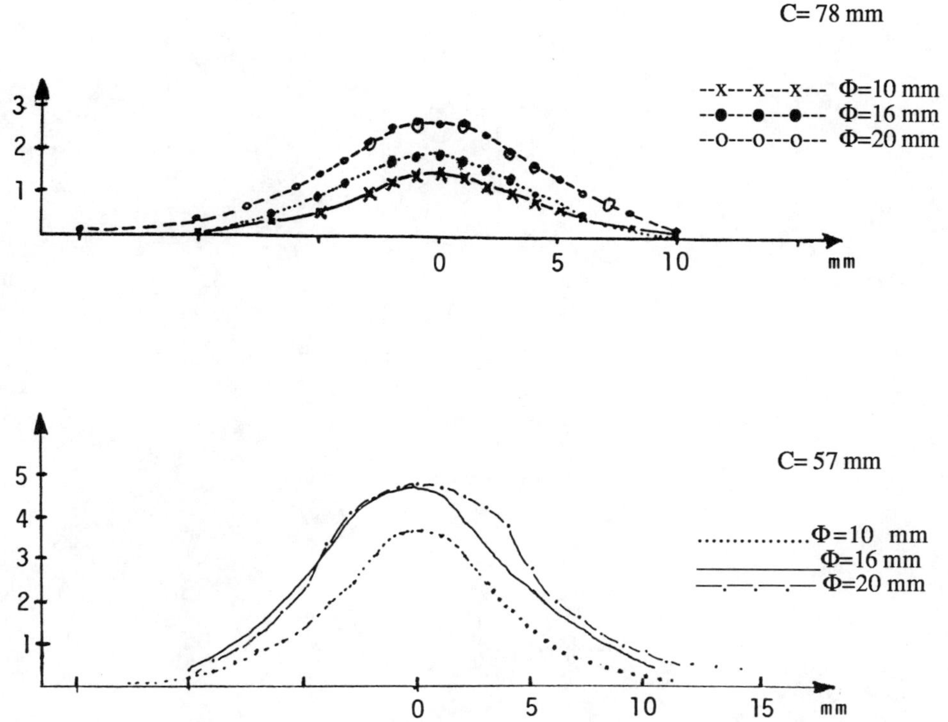

Fig. 11 Variable Diameter Φ, Constant Cover C (Signal Amplitude Versus Position).

MONITORING PRESTRESS FORCES IN A BOX GIRDER BRIDGE

M. "SAIID" SAIIDI
Civil Engrg. Dept.
University of Nevada
Reno, NV 89557 USA

JOSEPH SHIELDS
Ferrari & Associates
1000 Bible Way
Reno, NV 89502 USA

RODNEY JOHNSON
Nevada Dept. of Trans.
Carson City, NV 89712 USA

ABSTRACT. This article presents a study of prestress losses measured directly in the field by instrumenting several tendons of an actual post-tensioned box girder bridge. The measured losses are found to be in general agreement with theoretical value but deviate from them afterward. Possible explanations for the differences are discussed.

1. Introduction

Numerous studies have been conducted on different aspects of prestress loss prediction in pre-tensioned and post-tensioned concrete members. A large number of studies have involved laboratory measurement of one or more components of prestress losses [1,2,3]. Other studies have been aimed at developing analytical procedures to predict losses [4,5,6]. Only a few of the reported studies have incorporated field measurement as a means of investigating prestress force changes as a function of time [7].

Prestress losses consist of six components in post tensioned bridges, namely, elastic shortening, friction, anchorage set, relaxation, creep, and shrinkage. The most common method of prestress loss prediction is the method developed by the American Association of State Highway Transportation Officials (AASHTO) [11], in which procedures for computing each stress loss component are outlined. The losses determined based on the AASHTO method are the maximum values expected during the life of the structure. A potentially more accurate method is the time-step method in which the interactions among different components are taken into account [8]. Two of the major sources of time-dependent prestress loss in structures are creep and shrinkage of concrete. The available procedures to account for these effects are based on research data for climates which are different from the type of climates encountered in the State of Nevada. The purpose of this article is to report on a study of prestress losses in an actual bridge (called the Golden Valley Interchange) which was instrumented and monitored for a 15 month period.

A. S. Nowak (ed.), Bridge Evaluation, Repair and Rehabilization, 217–228.

2. Description of the Bridge

The Golden Valley Interchange (Fig. 1) consists of two parallel 155 ft.(47.3 m), simple-span, post-tensioned concrete box girder bridges. Both structures are 45 ft. wide on a 30o22'16" skew. The interchange is located approximately 5 miles (8 km) north of Reno, Nevada. Each bridge is 7 ft. (2.13 m) deep with six 12 in. (305 mm) wide girders throughout (Fig. 2). An 8 in. (203 mm) thick vertical diaphragm separates the cells at midspan. Each girder has three grout ducts containing thirty-one, nineteen, or eighteen 0.5 in. (12.7 mm) diameter 7-wire strands. Only the two westmost girders on the southbound structure were instrumented. A total of 484 strands were used in the bridge. The tendons were placed in a parabolic profile with eccentricities of 2.5 in. (64 mm) at the ends and 37.5 in. (953 mm) at the midspan of the bridge.

The strands are of Grade 270, Low-Relaxation (ASTM A-416) prestressing steel. The tendons containing 31 strands inside of 4 1/4 in. (108 mm) o.d. spiro-type, semi-rigid, galvanized grout ducts have the following VSL anchorage assembly: EC5-31 bearing plates, E5-31 anchor-heads, Type 1.3 wedges, and SP-31 spirals. The tendons with 18 and 19 strands have EC5-19 bearing plates, E5-19 anchor-heads, Type 1.3 wedges, SP-19 spirals, and 3-3/4 in. (95 mm) od. spiro-type, semi-rigid, galvanized grout ducts. The mild steel is of ASTM A-615 Grade 60 reinforcing steel. Based on the test data supplied by the manufacturer on the prestressing steel, the yield stress was 258 ksi (1,779 MPa) and the ultimate strength was 276 ksi (1903 MPa). The measured modulus of elasticity was 29,900 ksi (206,160 MPa).

The Golden Valley Interchange had Class EA Modified Major concrete in the top slab and Class AA Modified Major in the webs and the bottom slab. The design concrete compressive strength for the post-tensioned concrete in the bridge was 3500 psi (24.1 MPa) at the time of post-tensioning and 4500 psi (31.0 MPa) at 28 days. The measured concrete cylinder strength was 5200 psi (35.9 MPa) and 6100 psi (42.1 MPa), at post-tensioning time and at 28 days, respectively. The concrete unit weight was measured at 139 pcf (2,224 kg/cubic m).

The concrete placement sequence was as follows: bottom slab, webs, and the top slab. The webs were placed with the grout ducts and the anchors in place. The prestressing steel was threaded through the grout ducts after the top slab was cast. The tendons were initially stressed to 189 ksi (1303 MPa), 70% of the specified ultimate strength, using a hydraulic ram and a jack chair centering device.

Some special modifications were made to the bridges to accommodate this study. A 3 ft. (0.91 m) square hole through the bottom slab on the south end of westmost cell was installed to facilitate our access after the bridge is opened to traffic. Also, a 2 ft. (0.61 m) diameter manholes was built on the top slab near midspan to provide access to the exterior cell. Blockouts on both girders inside the cell were provided for direct access to the prestressing steel near midspan (Fig. 3).

3. Instrumentation

Electrical strain gage impulses were detected using a Hewlett Packard 9000 series microcomputer interfaced with a 3054 data acquisition system. Software developed for this project was used to collect, manipulate, and store strain gage data.

Dial gage assemblies with a 30 in. (762 mm) gage length were used in the Golden Valley Bridge (Fig. 3). These dial gage assemblies had a 0.4 in. (10.2 mm) range and a sensitivity of 0.0001 in. (0.0025 mm).

A total of four tendons, two on the interior girder and two on the exterior girder, were instrumented with electrical resistance strain gages. Three strain gages were bonded to three different wires in one strand in each of the instrumented tendons. Therefore, a total of 12 strain gages were bonded to the prestressing steel to allow for redundancy of measurements. Two strain gages were bonded to unstrained pieces of strand to act as temperature reference gages. The strain gages were installed after fifty percent of the initial prestressing force had been applied. The instrumented tendons were stressed to the full prestressing force after all the other strands had been stressed.

The electrical resistance strain gages were bonded to the prestressing steel using M-Bond 200 adhesive and 200 Catalyst. The prestressing strand was thoroughly cleaned and degreased before bonding. After the wire leads were soldered, teflon tape and a butyl rubber compound was applied over the gage for cushioning and to protect against moisture. A combination of expanding foam and partial blockouts were used to keep grout away from the strain gages. In addition, the strain gages were left open to the air so that the blockout would not provide an environment in which pressure could build up and destroy them. This proved successful, and only three strain gages out of twelve and one reference gage were damaged during the grouting process.

Mechanical Strain Gage supports were connected to concrete using an epoxy resin. The locations of mechanical strain gage points and electrical strain gages are shown in Fig. 3.

Because the bulk of prestress losses tend to occur early in the life of a structure, the measurements were taken on small intervals immediately after prestressing and gradually increasing intervals thereafter. Measurements were taken 1 day, 2 days, 3 days, 1 week, 2 weeks, 4 weeks, 6 weeks, 13 weeks, 25 weeks, 12 months, and 15 months from the time of initial prestressing. On each day, many measurements were taken to allow averaging of hourly data throughout the day and to account for temperature effects on the structure.

As prestress forces change, a structure will experience time-dependent deflections. Therefore, midspan deflection relative to the abutment was measured on each measurement day using a surveying level and a Philadelphia Rod with a sensitivity of 0.01 ft. (3 mm). The data were collected on the east and the west edge of the bridge.

4. Measured Data

4.1 DIRECT PRESTRESS LOSS MEASUREMENTS

As was pointed out in Section 2, three of the twelve strain gages installed directly on the tendons were damaged due to the grouting process. The strain readings from Most of the other nine gages were erratic subsequent to the grouting process, but stabilized after a week. The measurements after grouting were referenced back to the seven-day data and a best fit curve was placed between time zero and seven days.

The measured time-dependent prestress loss values are shown in Fig. 4 in solid line with "+" sign marks. The data for the first 91 days follow an expected trend. The tendon stresses between the day 91 and day 175 dropped only by a very small amount. This time window spanned early December 1988 to early March 1989. This period in the Reno area happened to be associated with unusually large amount of precipitation. It is possible for the bridge to absorb some moisture during this period and expand, thus compensation for the expected losses due to steel relaxation. Between day 175 and the anniversary of prestressing, the tendon stresses dropped quite notably. This period covered several dry summer months, and the significant stress loss is somewhat anticipated. The change in the subsequent three months was less drastic. If the 175-day data point is ignored, the prestress losses would follow the dashed line in Fig. 4 which indicates a smoother trend. The kink in the solid measured curve appears to be due to seasonal changes. It is planned to collect six more data points during 1990 to capture such changes.

4.2 CONCRETE STRAIN MEASUREMENTS

The dial gage assemblies had a 30 in. (762 mm) gage length and were mounted three wall thicknesses away from the blockouts to avoid any influence by the opening. The concrete strain data were converted to a prestress loss by first fitting a least-squares line to the measured strains and then finding the strain at the level of the tendon from this line. Figure 5 shows a sample of the measured time-dependent concrete strain distributions. It can be seen that some readings showed a fairly linear distribution along the section depth while others deviated from a straight line significantly. Since the tendons are bonded, the strain change in the concrete at the level of the tendon was assumed to be the strain change in the tendon. Figure 4 compares the time-dependent prestress losses measured from concrete strain data with the electrical strain gage losses. As can be seen from Fig. 4, the strain measurements on the concrete surface using a 30-inch (762-mm) gage length provided rather inconsistent method for predicting prestress losses. It is possible that a longer gage length, say approximately 40 inches (1.0 m) which is about one-half of the box girder depth, would have led to more consistent data.

4.3 ANCHORAGE SET LOSS

The anchorage set prestress losses were measured. The tendons were stressed and strain gage data were taken before and after the wedges were set. The difference between the two measurements represents the effect of anchorage set on the prestressing force at midspan. The data were taken on three of the four instrumented tendons. The average measured loss was 4.2 ksi (28.96 MPa).

4.4 TIME-DEPENDENT DEFLECTIONS

The relative time-dependent deflections were measured on the west and east sides of the Golden Valley Interchange and are presented in Fig. 6. A negative camber indicates a downward deflection. The slight increase in camber after the first month is attributed to the effect of the compacted backfill.

5. Analysis

Time-dependent prestress loss components, namely those due to creep, shrinkage, and tendon relaxation were calculated using a time-step method and the AASHTO procedure [11] for lump-sum estimate of different losses. In addition, the prestress loss due to the anchorage set was determined using a procedure recommended by the Prestressed Concrete Institute [12].

5.1 THE TIME-STEP METHOD

The time-dependent prestress losses were calculated using the ACI Committee 209 recommendations [9] with minor modifications described in Ref. 8. Only the important aspects of the application of the method are presented in this article. More detailed information can be found in Ref. 10.

Generally, shrinkage losses are predicted by multiplying the prestressing steel modulus of elasticity by the ultimate shrinkage strain of concrete and several correction factors. Shrinkage correction factors are based on relative humidity, temperature, member size, slump of concrete, fine aggregate percentage, cement content, and air content. Creep losses are estimated using a similar approach. The ultimate creep coefficient of concrete is multiplied by the modular ratio, a time-function, the stress at the centroid of the section, and several creep correction factors.

Reference 9 suggests a value of 2.35 for the ultimate creep coefficient and 780 micro-strains for the ultimate shrinkage strain in the absence of experimental data. The average annual relative humidity was taken at 50 percent and the average temperature was assumed to be 50 Fahrenheit degrees (10 Celsius degrees). These figures are based on the climatic data collected at the Reno International Airport from 1948 to 1987. The airport is located at approximately 8 miles (12.9 km) south of the Golden Valley Interchange.

The time-step method recommended in Ref. 9 was implemented in computer program "PLOSS". The program was developed in BASIC on a personal computer. All parameters required that were unknown or not tested for were estimated using values suggested by ACI Committee 209 [9]. The computed results are shown in Fig. 4. The values at time zero in this figure are after the instantaneous losses have taken place. Note how the measured and predicted curves in Fig. 4 have roughly the same shape and have approximately the same value up to the end of the third month (90-day reading). Between this time and the next reading which was at 175 days, the actual losses were very small compared to the theoretical value. This is attributed to the relatively humid and unusually moist weather between early December 1988 and March 1989 in the Reno area. Between March and September 1989 (a relatively dry season), a significant amount of creep and shrinkage took place in the bridge, thus reducing the prestress forces. Because a constant relative humidity is assumed in program "PLOSS", these seasonal changes were not predicted in the time-step method. To determine if the sharp drop in the prestress forces are continuing, more data were obtained on December 7, 1989. It was noted that the rate of stress loss had slowed down between September and December due to the drop in temperature relative to the summer temperature.

5.2 LUMP-SUM ESTIMATE OF SEPARATE LOSSES

The lump-sum method of estimating separate prestress losses is widely used in design because it is simple and it involves a relatively small number of parameters. This method has been adopted by AASHTO [11]. Using the bridge material and geometric properties and a relative humidity of 50 percent, the total time-dependent prestress loss components for the Golden Valley Interchange was found at 25.1 ksi (173.1 MPa) based on the lump-sum method.

5.3 INSTANTANEOUS LOSSES

Of the instantaneous prestress losses, only the effect of anchorage set at midspan was determined. The losses due to anchorage set may be calculated by a method recommended by the Prestressed Concrete Institute (PCI) [12]. According to this method the prestress loss due to anchorage set is computed at the jacking end. The stress change at the end affects the tendon stress over a length of member which is determined based on friction and other factors. Assuming a uniform distribution of friction forces over this length, the effect of anchorage set can be determined at any point. For the Golden Valley Bridge, the calculated prestress loss at midspan caused by the anchorage set was 4.4 ksi (30.34 MPa) which was in close agreement with the measured data of 4.2 ksi (28.96 MPa).

5.4 DISCUSSION

Table 1 compares time-dependent prestress losses, as calculated using the time-step method [9] and the AASHTO Method [11], with the measured prestress losses at the end of the design life of 40 years. The measured prestress loss value after 40 years was obtained by fitting a semilogarithmic regression curve to the measured data and extrapolating to 40 years. This approach was used for both the data for the first 175 days and the data for the first 455 days. Note how the AASHTO and time-step methods yield practically the same value in Table 1, but the extrapolated measured data are quite larger. This is particularly true when the one year and 15 month data are included. Because the measured data are affected by seasonal changes, it may well be that the extrapolated values are unrealistically high. Plans are underway to collect more data throughout 1990 to capture seasonal changes.

6. Summary and Conclusions

This article presents a study of the prestress losses measured in the field on a post-tensioned box girder bridge at the Golden Valley Interchange near Reno, Nevada. The primary objective of the study was to measure time-dependent prestress losses directly on tendons and compare the results with losses obtained from concrete surface strain data and theoretical values. The bridge was instrumented with electrical resistance strain gages at midspan on one interior and one exterior girder. The change in surface concrete strain as a result of prestress losses was also measured on the two westmost girders. In addition, the time-dependent deflection of the bridge was measured. The data have been collected for 15 months thus far.

The measured prestress losses in the Golden Valley Interchange were in close correlation with the theoretical time-step values (Fig. 4) during the first 175 days. The 12-and 15- month data showed larger stress loss than predicted. The most plausible explanation at this time is the relatively dry summer climate in the Reno area and the resulting large shrinkage. The prestress losses based on concrete surface strains did not agree with the tendon stresses or the theoretical time-step results, although they formed a useful parameter for checking the overall trend.

The calculated anchorage set loss which was based on the PCI [12] method was in close agreement with the measured value.

7. Acknowledgements

The study presented in this paper was sponsored by the Bridge Division of the Nevada Department of Transportation (NDOT). Mr. Dean Weitzel, a staff engineer, and other staff of NDOT are thanked for their extraordinary cooperation in the course of the study. Dr. Antoine Naaman, a professor of civil engineering at the University of Michigan, is thanked for his advice on the instrumentation and helpful comments on the final report. The authors are also indebted to Mr. Jesus Pedroareana, civil engineering technician, for his careful field work and to Messrs. David Straw and Mike Burgess, civil engineering students, for their assistance in the field.

8. References

1. Bandyopadhyay, T.K. and B. Sengupta, (1988) "Effect of Deferred Initial Loading and Percentage of Steel on Time-Dependent Losses and Deformations of Prestressed Members," ACI Structural Journal, pp. 190-205.
2. Branson, D.E., Meyers, B.L., and Kripanarayanan, K.M., (1970) "Loss of Prestress, Camber, and Deflection of Noncomposite and Composite Structures Using Different Weight Concretes," Report No. 70-6, Department of Civil Engineering, University of Iowa.
3. Huang, T. (1973), "Prestress Losses in Pretensioned Concrete Structural Members." Lehigh University, Fritz Engineering Report No. 339.9.
4. Chadha, G., (1982) "Design Aids for Prediction of Long-Term Prestress Losses Due to Creep and Shrinkage," ACI Publication SP-76, Designing for Creep & Shrinkage in Concrete Structures, pp. 109-124.
5. Ghali, Amin, Sisodiya, Romy G., and Gamil S. Tadros, (1974) "Displacements and Losses in Multistage Prestressed Members," ASCE Journal of the Structural Division, Vol. 100, pp. 2307-2322.
6. Sinno, A.M. and Howard L. Furr, (1970) "Hyperbolic Functions For Prestress Losses and Camber," ASCE Journal of the Structural Division, pp. 803-821.
7. Richardson, James E., (1984) "Field-Measured Post-Tension Prestress Loss in Stress-Relieved Strands," California Department of Transportation, Report No. FHWA/CA/SD-84/02.
8. Naaman, A.E., (1982), "Prestressed Concrete Analysis and Design, McGraw-Hill Book Company, New York, 664 pp.
9. ACI Committee 209, (1971) "Effects of Concrete Constituents, Environment, and Stress on Creep and Shrinkage of Concrete," ACI Publication SP-27, Designing for Effects of Creep, Shrinkage, and Temperature in Concrete Structures, pp. 1-42.
10. Shields, J., Saiidi, M., (1989), "Direct Field Measurement of Prestress Losses in Box Girder Bridges," Report No. CCEER-89-4, Civil Engineering Department, University of Nevada, Reno.
11. American Association of State Highway Transportation Officials (AASHTO), (1989) "Standard Specifications for Highway Bridges," Washington, D. C.
12. PCI Committee on Prestress Losses, (1975), "Recommendations for Estimating Prestress Losses," PCI Journal, Vol. 20, No. 4,pp. 43-75.

Table 1 Predicted Prestress Losses Based on Different Methods.

METHOD	PRESTRESS LOSS IN 40 YEARS (KSI)
AASHTO	25.1
Time-step method	25.3
Extrapolated measured data based on the first 175 days	35.9
Extrapolated measured data based on the first 455 days	47.

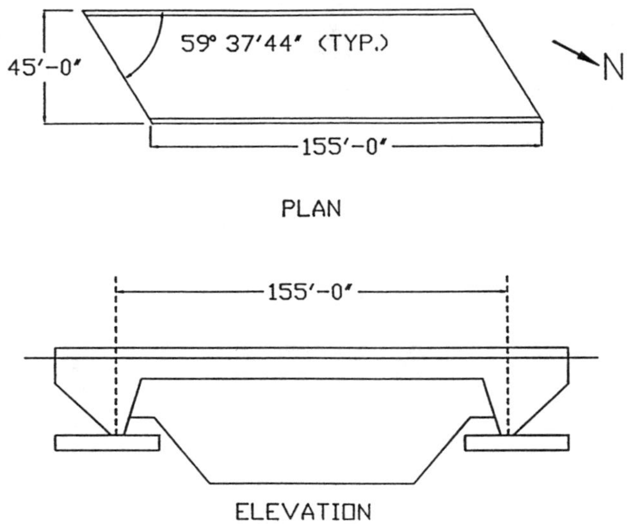

Fig. 1 Golden Valley Interchange Plan and Elevation.

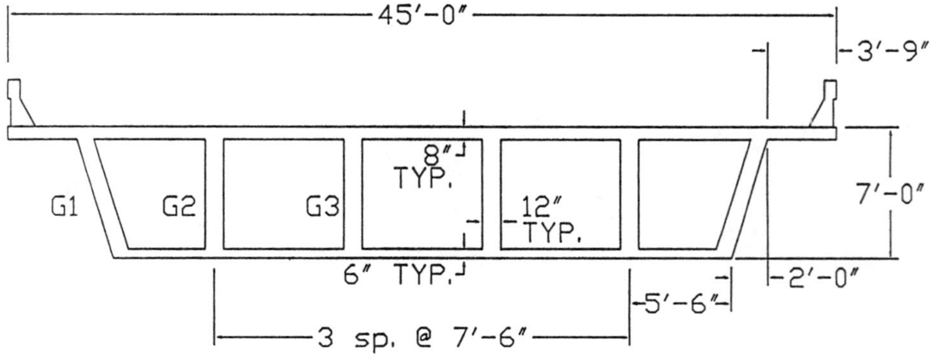

Fig. 2 Bridge Cross Section.

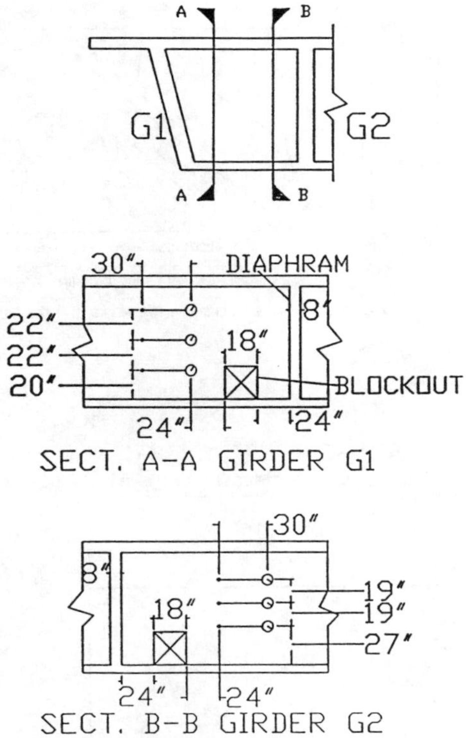

Fig. 3 Location of Access Blocks and Mechanical Gages.

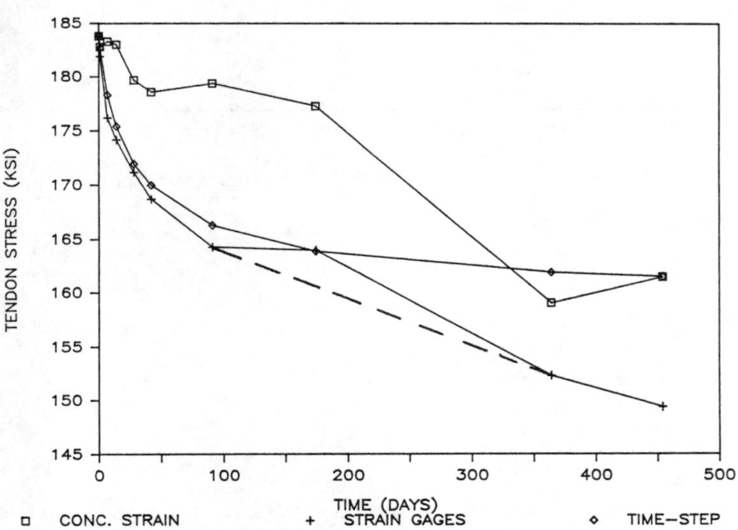

Fig. 4 Measured and Predicted Prestress Force Changes.

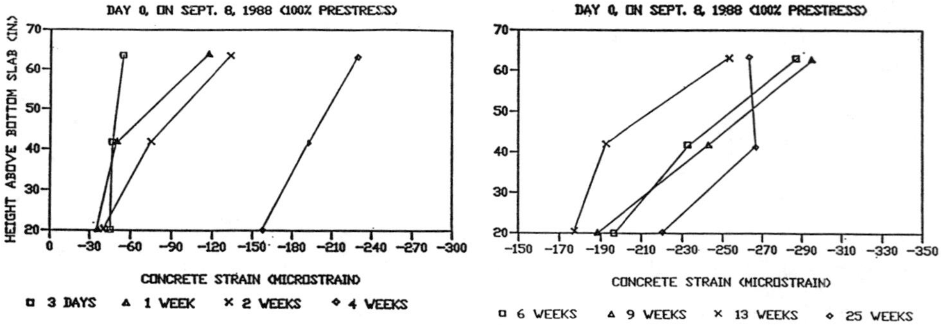

Fig. 5 Concrete Surface Strains Data for the West Wall.

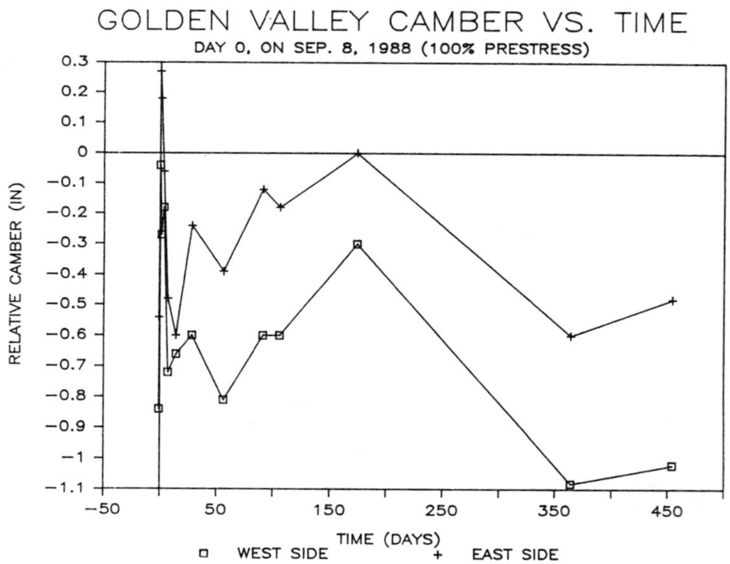

Fig. 6 Measured Deflection of the Bridge.

ASSESSMENT OF BRIDGE DECK REPAIRS BY A NON-DESTRUCTIVE TECHNIQUE

A.G. DAVIS
Testconsult
Warrington, Cheshire, WA3 6QT
UNITED KINGDOM

B.H.E. HERTLEIN
ESSI - Testconsult
P.O. Box 6098
Asheville, NC 28816 USA

1. Introduction

At present, the extent of delamination or debonding of concrete and concrete repairs in bridge decks is assessed by such techniques as sounding with a steel rod or hammer, or chain drag. While useful for quick assessment these methods are subjective, and depend largely on the acuity of the operators' hearing. It is necessary to repeat the test several times, each time with a different operator, and average the results to reduce the subjective influence. The CEBTP Paris have developed a test known as Transient Dynamic Response (TDR), in which the equipment duplicates the ear of the operator.

A trial of the method was arranged by the Ohio State Department of Transportation in October 1989 on a two lane concrete/steel bridge over Interstate Highway 270 near Columbus, Ohio. This bridge deck is being used to evaluate the performance of 67 repair patches of various sizes cut out of the original deck concrete, and replaced with differing proprietary repair mixes.

2. Methodology

Originally developed for the non-destructive control of deep foundations (Ref. 1,2,3), this method has been increasingly applied to concrete slab testing over the last eight years (Ref. 4,5).

The Transient Dynamic Response method is based on the measurement of the dynamic impedance of the concrete unit and of its support to transient force.

The deck is excited by a blow from an instrumented hammer containing a load cell. The response is monitored by a velocity transducer, and the two signals, force and velocity, are recorded on a data acquisition system. Digital treatment of the signals by Fast Fourier Transform (FFT) converts the original time-based data to frequency responses; the velocity spectrum is then divided by the force spectrum, giving a transfer function or "mobility" which is the characteristic signature of the deck at the point of test. This frequency response gives several useful parameters:

A. S. Nowak (ed.), Bridge Evaluation, Repair and Rehabilization, 229–233.

a) The average impedance over the frequency range 100-1000 Hz. This is a function of concrete thickness mainly.
b) The presence or absence of high mobility at 100 Hz approximately. This shows the presence of voiding or debonding beneath the concrete layer, and its degree.
c) The degree of damping over 100-1000 Hz, which is an indicator of continuity within the concrete unit.

3. Trials Results

In the trials carried out on 18 October 1989 only the East-bound lane was tested, and time constraints limited the testing to 29 patches together with 5 intermediate tests on the original deck concrete for correlation purposes. A total of 74 test results were recorded in 1.5 hours.

In assessing the test results, a distinction was made between debonding and delamination as follows:

Debonding - loss of adhesion between repair material and substrate, but both parts remain in contact

Delaminated - separation of repair material and substrate; typically a fine, shallow void, often containing loose debris, silty material or water.

Figures 1 and 2 show examples of typical test results and Table 1 gives the results obtained for these tests, with descriptions of the faults detected.

4. Discussion

The repair patches varied considerably in size and shape from 18" x 18" (450 mm x 450 mm) to 3 ft x 6 ft (900 mm x 1.8 m). The smaller patches require one test per patch, increasing to 3 tests per patch for the largest.

Several of the test results showed evidence of distress within the repair material, and typically this is a result of low density caused by inadequate compaction, or very fine micro-cracking in the material itself, probably caused by overstress.

Figure 1 gives examples of two sound responses; one from the original bridge deck and one on patch #3a. Typical average mobilities of between 9 and 18 x 10^{-7} m/N/s are measured, with a constant average value over the total frequency range.

Figure 2 shows one response with debonding starting (patch #5), indicated by a very sharp rise in average mobility from 14 to > 100 x 10^{-7} m/N/s at 350 Hz.

The second response from patch #6 is typical of delamination, with a high peak at frequencies below 100 Hz.

When in normal use on bridge decks, the method can identify not only debonded and delaminated ares, but also those parts of the deck which are distressed by cracking or by corrosion damage. The method is also used as a quality control check in new construction to detect such problems as honeycombing and poor compaction.

5. References

1. Davis, A.G. and Dunn, C.S., (1974) From theory to experience with the non-destructive vibration testing of piles, Proc. Inst. Civ. Engrs., Part 2, 57, pp. 571-593.
2. Stain, R.T. (1982) Integrity Testing, Civil Engineering (England), April/May, 1982.
3. Williams, H. and Stain, R.T., (1987) Pile Integrity Testing - Horses for Courses, Proc. Int. Conf. on Foundations and Tunnels, London, England, pp. 184-191.
4. Pederson, C.M. and Senkowski, L. J. (1986) Slab Stabilization of PCC Pavements, Transportation Research Board Meeting, January, 1986.
5. Hertlein, B.H.E. and Davis, A.G., (1987) Non-destructive testing of concrete pavement slabs and floors with the Transient Dynamic Response Method, Proc. Int. Conf. on Structural Faults and Repair, London, England, Vol. 2, pp. 239-246.

Table 1

Test No and Location Location	Observations
d 1-4 original deck between patches 6 and 23	Sound concrete
P 3a -3	Patch material sound and properly bonded
P 5-2 centre of patch	Patch debonded at centre
P 6-1 nearside of patch	Patch delaminated at nearside

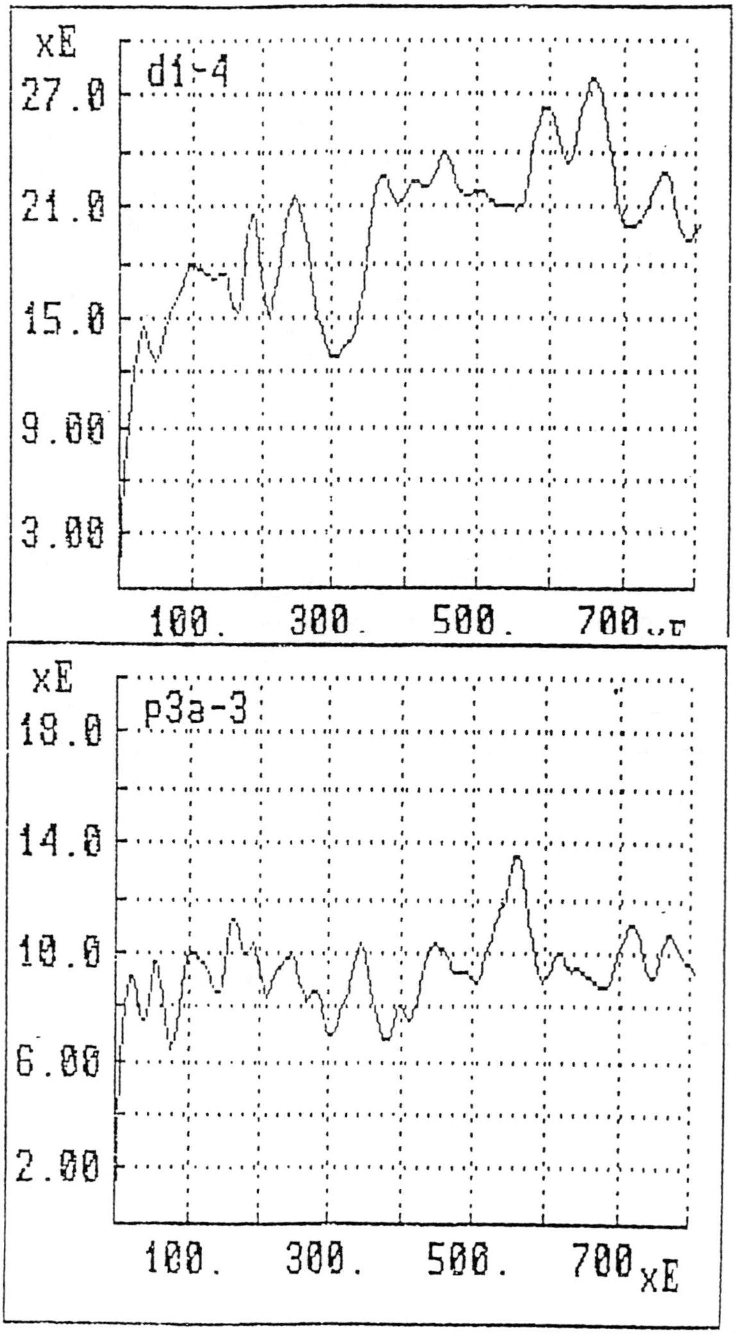

Fig. 1 Typical Response Curves. Sound Deck and Patch.

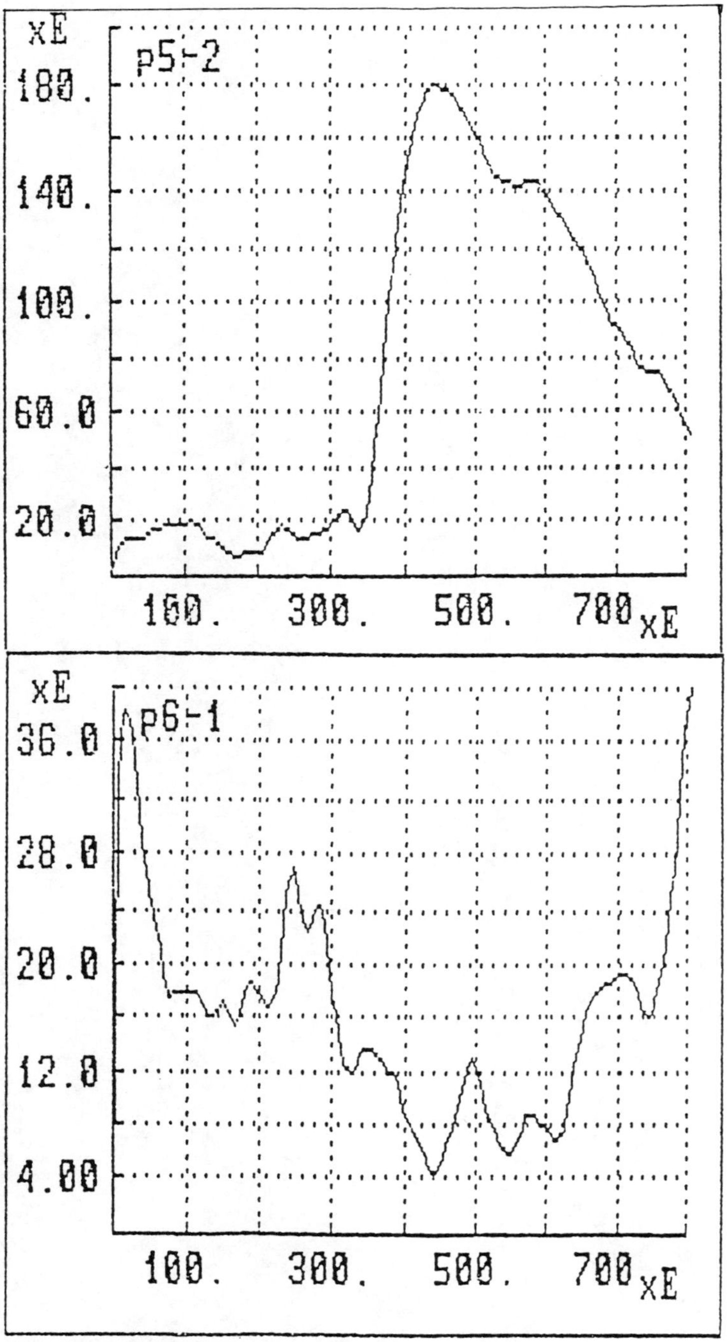

Fig. 2 Typical Response Curves. Defective Patches

DYNAMIC CHARACTERIZATION OF POSTTENSIONED RODS IN STRESSED TIMBER SYSTEMS

H. L. CHEN, G. VENKATESH and
H. V. S. GANGARAO
West Virginia University
Dept. of Civil Engineering
Morgantown, WV 26505-6101 USA

J. E. STEPHENS
Montana State University
Department of Civil Engineering
Bozerman, Montana 59717 USA

ABSTRACT. An analytical and experimental approach is presented to monitor the tension levels in posttensioned high strength rods of stressed timber systems. The response of rods under an impact load is characterized and then correlated with tension in the rod. This approach is simpler and faster than the existing techniques and is intended for use during routine structural inspection. The dominant frequency of the rod has distinguishable variation for varying tension levels.

1. Introduction

A simple technique to monitor the stresses in, and the physical condition of, posttensioning rods in stressed timber bridge systems is investigated. An estimated eighty stressed timber bridges are proposed to be built nationwide, thirty of which will be in West Virginia. In these structures, the local rigidity of the deck and global rigidity of the deck/girder are dependent on the level of posttensioning in the system. The prestressing forces have to be monitored more frequently in timber systems because of high creep losses than in concrete. Only limited information has been collected on the loss of posttensioning forces in stressed timber bridges under in-service conditions, and even less is known concerning the physical deterioration of the posttensioned rods over time.

Initial work at WVU has focused on developing a technique to determine rod tension forces that can be readily used during routine inspections. Following the technique proposed herein, the acceleration response of the rod in the transverse direction generated by an impact load is measured and this response is then correlated with the tension levels. This approach is considerably simpler, faster, and less expensive than the present approach of successively inserting a load cell under each anchorage point. After the development work is completed, the technique should be applicable for any posttensioned timber system.

A. S. Nowak (ed.), Bridge Evaluation, Repair and Rehabilization, 235–248.

2. Description of Stressed Timber Bridge System

The stressed timber bridge system of immediate concern in this investigation consists of longitudinal deck planks sandwiched between wood girders, as shown in (Figure 1). The planks are 9 inch oak decking; the girders, 6 inch by 45 inch manufactured wood beams. The planks and girders are clamped together by transverse, posttensioned steel rods to form a composite load carrying system. The bridge configuration shown in (Figure 1) typical for a single lane with a span of approximately 75 feet, has 45 tensioning rods spaced at 24 inches. A more complete description of this particular structure, located in Charleston, West Virginia, is given by (Dickson and GangaRao 1988). The posttensioned rods used in the bridge shown in (Figure 1) are Dywidag Threadbars (Dywidag 1988). These rods have been historically used in posttensioning concrete, soil, and rock applications (Naaman, 1982). The bars have a deformed surface, with the deformations acting as coarse threads. The bars are 5/8 inches in diameter and were initially tensioned to a load of 20 kips (approximately 56% of yield strength of the rod). This tension is held in the rods by nuts and washers on each end that react against the timber bridge deck. Rod tensioning is accomplished using a hydraulic jack. Retensioning is required a few weeks after construction, and a third tensioning may also be necessary. The loss of posttensioning forces can be dramatic (e. g. up to 50% in the first 30 days, Dickson and GangaRao 1988). The tension levels in the rods can be checked by reinstalling the jack and load cell on each successive rod which is time consuming. Currently, routine tension checks are recommended at two week intervals for the first two months after construction; every six months for the next two years; and annually thereafter. This inspection schedule may be modified as additional data is collected.

3. Approach For Determining Rod Tensions

The technique proposed herein for determining the forces in the posttensioning rods consists of first characterizing the transverse response of the rod under an impact load and then correlating this response with rod tension level. Following this approach, an accelerometer is mounted on the rod, the structure is tapped some distance from the accelerometer, and the resulting response is measured. The location of the accelerometer and impact point are selected such that the response can be measured. This approach to determining rod tension, while indirect, is simpler and/or less expensive than the current technique.

Several measurement locations and impact scenarios could be used in implementing this dynamic characterization approach to rod tension measurement (Stephens and GangaRao 1989). Candidate configurations were selected primarily based on the desirable characteristics of routine inspection technique. That is, such techniques should be simple, inexpensive, and efficient. Based on these criteria, two test configurations were selected in continuation of the study.

1) transverse acceleration is measured impacting the rod at mid span and measuring a distance away (Figure 2). Two holes were drilled in the model deck to access the rod and rigid bodies were glued to the rod for the convenience of impacting and measurement.
2) transverse acceleration is measured at one end of the rod while the other end is transversely impacted. The idea is to excite transverse modes of vibration impacting the cantilevered free ends of the rod. Techniques are available that measure rod tension more directly than those proposed herein. These techniques include strain gauge on each rod and active or passive load cells under anchorage point. The directness of these approaches is attractive in that the physical relationship between the measurement process and the desired information is more easily understood and more reliable results may possibly be obtained than from a dynamic characterization approach. Practical implementation of these techniques is, however, difficult. The costs of placing active load cells on each rod would be expensive. These load cells would be exposed to harsh environments and have expensive maintenance costs. Monitoring the tension in t he rods using passive load cells merits further investigation. Existing techniques used to measure residual tension in bolts were also researched for approaches that could be adapted to the rod tension problem.

4. Instrumentation

1) IBM industrial computer.
2) Rapid System R360, Real time spectrum analyzer.

The analyzer consists of the R300 digital processing (DSP) interface board, the Rapid System 4 by 4 digital oscilloscope peripheral and the turnekey R360 Real Time Spectrum Analyzer software. The spectrum analyzer utilizes four simultaneous channel acquisition. The R360 software provides convenient methods for selecting data acquisition modes, processing display and data storage options.

3) The measuring device is a PCB accelerometer with a magnetic base and a mountable base. Model 348A09, sensitivity 100mV/g, and frequency range 1 to 2000 Hz.
4) PCB Impact Hammer, Model 086B03, Range 0-500 lb. The hammer sensitivity for the Plastic head used is 9.2 mV/lb.

5. Simplified Analytical Model

(a) Case1

$$EI\ d4y/\ dx4 - T\ d2y/\ dx2 + Ky + C\ dy/\ dt + rA\ d2y/\ dt2 = P(t)\ d(x-x) \qquad (1)$$

where C and K represent damping and stiffness contributions from wood

E is the Youngs modulus.

A is the cross sectional area.

I is the moment of inertia.

Initial conditions $y(x,0) = 0$ $y'(x,0) = 0$

Boundary conditions $y(0,t) = 0$ $\quad$ $y'(0,t) = 0$

$y(L,t) = 0$ $\quad$ $y'(L,t) = 0$

Greens function's solution

$$y(x,t) = \quad G(x,t-t)\,P(t)\,dt \tag{2}$$

where G(x,t) is the Greens's function which represents the solution of the system due to harmonically concentrated load.

The convolution integral can be expressed in frequency domain as

$$Y(x,w) = G(x,w)\,P(w) \tag{3}$$

(b) Case 2.
Similar mathematical formulation as Case 1 can be developed for Case 2. The solutions can be derived incorporating the changes in boundary conditions due to K.

6. Experimental Procedure

Preliminary evaluations of the approaches proposed herein to determine rod tension using dynamic characterization are being accomplished on a full scale model of a section of a stressed timber bridge. The two approaches described above, involving en d impact on the rod while the acceleration response is measured at the other end of the rod and impact on the rod and transverse response measurement on the rod have been studied. For this evaluation, the rod in the model was tensioned at different levels and subjected to standard impact at one end with the hammer while the acceleration response was recorded at the other end. Similarly for varying tension levels the rod was impacted at the center and acceleration response recorded 5 inches away from the point of impact.

The experimental model is of span 45.5 inches and 24 inches in width. The model consists of 26 numbers, 1.75 inch thick red oak deck planks. The rod is a 5/8 inch diameter Dywidag Threadbar with nuts and washers on both ends. The end forces placed in the rod for this investigation are shown in Table 1.

Transverse impact through the impact hammer were delivered to the rod during testing. Acceleration was measured on the rod at one location (Figure 2). The impact of the hammer was also recorded. Acceleration and impact data was collected on a P/C based digital oscilloscope and were stored on diskette for later analysis. Testing was done on the rod alone for varying levels of end forces, impacting the rod with the hammer and acceleration data being measured on the rod (Figure 3).

7. Data Processing

The processing involved reformatting the data by the Rapid System Utilities. DAIDSP a dynamic analysis software was used for processing. The accelerometer being sensitive was capable of acquiring minor disturbances. A linear regression was perform ed on the data to remove any discrepancies (noise). Transfer Functions were obtained for various tests. The results were spectral magnitudes and frequencies.

8. Results

The acceleration response measured during each impact test was processed and collective results scrutinized for repeatable relationships between characteristics in the observed response and the corresponding tension levels in the rod. The data was relatively clean, and processing consisted of obtaining power spectral density(PSD)/magnitude plots in the frequency domain using a FFT algorithm. The power spectral density is the magnitude squared of the first half of the FFT and the factored out length of the signal. This aids in comparison of frequency spectra of different signals and considered to be the power of a signal at a particular frequency. Specific response features that are investigated are the magnitude of the acceleration response and the dominant frequencies of the response. The response of the rod stressed at a level was consistently distinguishable from the response of the same rod tensioned at other levels. The dominant frequency of the measured response appeared to be related to rod tension levels.

Typical acceleration responses measured on the rod tensioned at 2 and 20 kips are shown in (Figure 5). These plots indicate time versus acceleration. The collected data appears clean. A typical impact input by the impact hammer is also shown in the time domain (Figure 6). Spectral magnitude(PSD) plots were generated from these measured data.

Power spectrum Density (PSD)

$$PSD\,[f(t)] = |f(w)|^2 \qquad (5)$$

f (w) is the FFT of f (t)

Transfer functions were obtained as a ratio of the measured output and their corresponding impact input.

$$\text{Transfer function} = \frac{\text{PSD [measured acceleration time history]}}{\text{PSD [Input impact time history]}}$$

Spectral magnitudes indicating the dominant response alone are shown in (Figure 7). There was increase in frequency at increased tension levels. These tests were conducted on the model deck with variation tension. The deviation from the trend in frequency versus tension may be attributed to some of the deck planks impinging the rod at different locations. This trend was not observed in measurements done on the cantilevered portion of the rod. The specific nature of these components of the response is probably driven more by peculiarities of end anchorage.

a) Effect of prestress

In case of C = 0 and K = 0, the solution of the homogeneous equations of (1) was derived as

$$\cos aL \cosh bL = 1 \qquad (4)$$

where,

$a = [- x +(x2 + w2/a2)\ 0.5]\ 0.5$

$b = [\ x +(x2 + w2/a2)\ 0.5]\ 0.5$

$x = T/2EI$ and $a2 = EI/rA$

This solution gives the normal modes of transverse vibration of the prestressed beam. The effect of the prestress(tension T) on the vibration frequency is shown in (Figure 8), where the frequency of the first mode of a fixed - fixed finite beam subjected to axial tension is plotted.

For a simply supported beam, the modes of vibration due to the effects of prestress are given in standard references (Graff, 1975). In (Figure 8) the experimental result is compared with the analysis. The change in frequency due to tension are calculated for both fixed-fixed and pinned-pinned conditions. The experimental results are closer to the fixed end conditions. The rod is 40 inches long and 5/8 inches in diameter. The frequencies are found to increase with tension.

b) Effect of wood

As shown in (Figure 9), the variation of frequency due to increase in tension is plotted for experimental measurements and analytical results. Experiments were conducted on the model timber deck. The analytical results were obtained from solutions in section (a) where C = K = 0 were assumed. The comparison shows a positive influence of tension on frequency Wood planks do not seem to have pronounced effect on the vibrating frequency of the rod. The experimental results are closer to the fix ed-fixed conditions. Equation (1) is being analyzed incorporating numerical values of C and K to understand the influence of wood on the behavior of stressed timber decks. These values are not available yet.

9. Summary

A technique is proposed herein for monitoring the tension levels in posttensioning rods of stressed timber bridge system. Such bridges are gaining popularity nationwide; hence efficient approaches are needed to monitor their pretension levels. The basic approach presented herein for monitoring rod tensions involves the dynamic response of the rods under impact loads. The change in the natural frequencies due to tension in the rod was calculated and measured. The results show a distinguishable variation of the tension level in the rod from the frequency response. The technique discussed herein can be an efficient tool in monitoring the tension levels of rods in stressed timber deck systems. Further investigations are being carried out for possible application on a full scale model (Figure 1) and in the field.

10. Acknowledgements

This work was supported by the National Science Foundation Fund NoMSM-8802265 and West Virginia Department of Highways through the Constructed Facilities Center at West Virginia University.

11. References

Bickford, J. H. (1988) "Using Ultrasonics to Measure the Residual Tension in Bolts", Experimental Techniques, Nov. 1988, pp. 3-5s.

Collacott, R. A., Structural Integrity Monitoring, Chapman and Hall Ltd., New York, 1985.

Dickson, B. (1988), "Notes on Timber Stressing", Timber Bridge Short Course, Nov. 10, 1988, West Virginia University, Morgantown, WV.

Dickson, B., and GangRao, H. V. S. (1988), "Design of Stressed Timber T-Beams for Highway Bridges", Proceedings, Symposium on Bridge Research, National Science Foundation and Iowa State University, Sept. 26-27, 1988, Des Moines, Iowa, pp. 59-63.

Dywidag Systems International, Inc. (1988), Lincoln Park, N. J.

Graff, K. F. (1975) Wave Motion in Elastic Solids, Ohio State University Press.

Naaman, A. E. (1982), Prestressed Concrete Analysis and Design Fundamentals, McGraw-Hill, Inc.

Stephens, J . E., GangaRao, V . H . S and Govindaraj, . V., (1989), Dynamic Characterization of Posttensioned Rods in Stresses Timber Systems, proceedings, ASCE Structures Division, May 1 - 5, 1989 San Francisco, CA.

Table 1 Experimental results of dynamic characterization tests on model deck and simple tension tests on rod.

TEST ON ROD

TENSION(Kips)	FREQUENCY (Hz)
0	39.06
2	73.24
4	83.01
6	87.89
8	92.77
10	107.42

TEST ON MODEL TIMBER DECK

TENSION (Kips)	FREQUENCY (Hz)
2	73.24
4	78.13
6	87.89
8	97.66
10.5	107.42
12	112.31
14.5	117.19
16.5	122.07
18	126.95
20.5	131.84

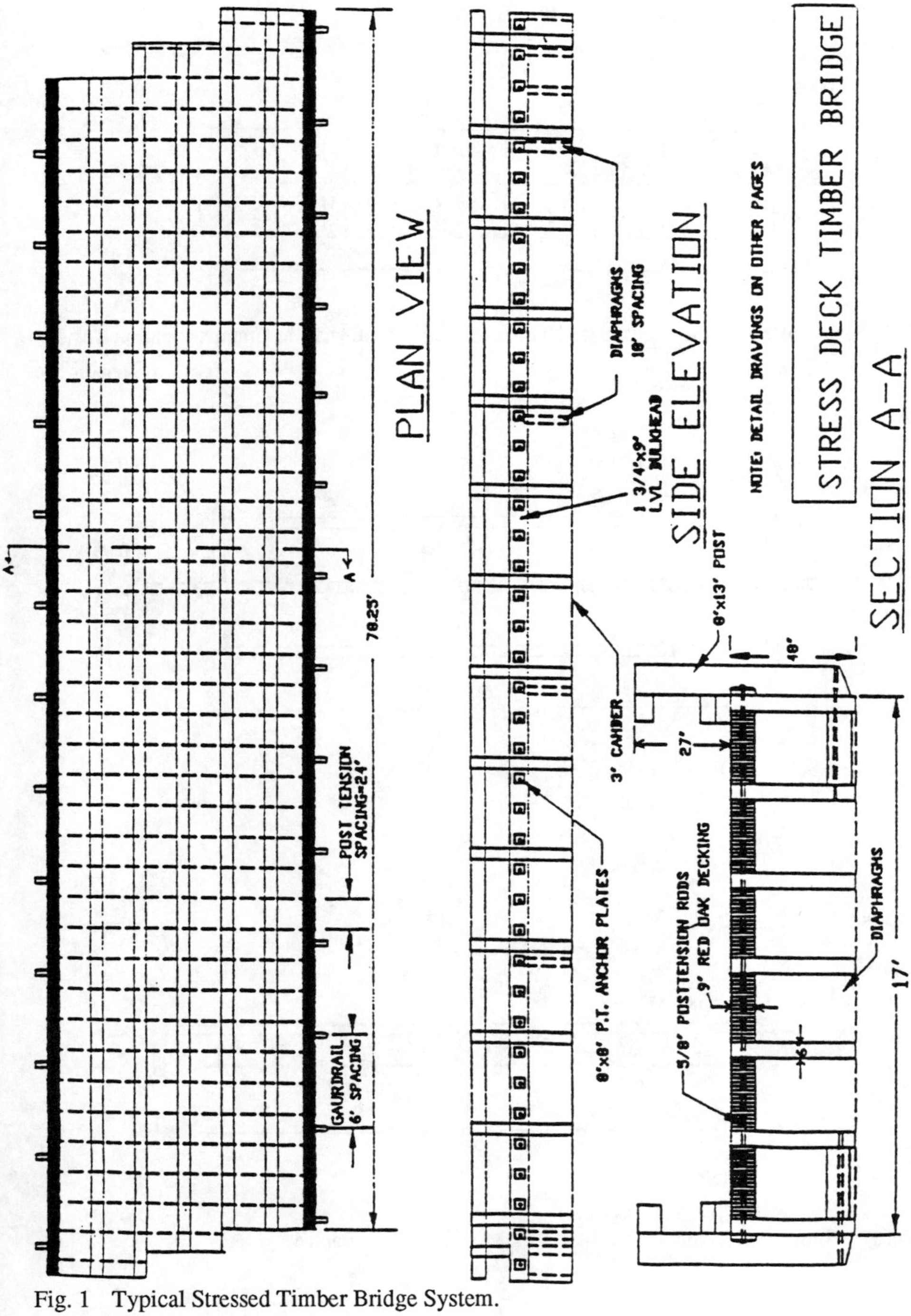

Fig. 1 Typical Stressed Timber Bridge System.

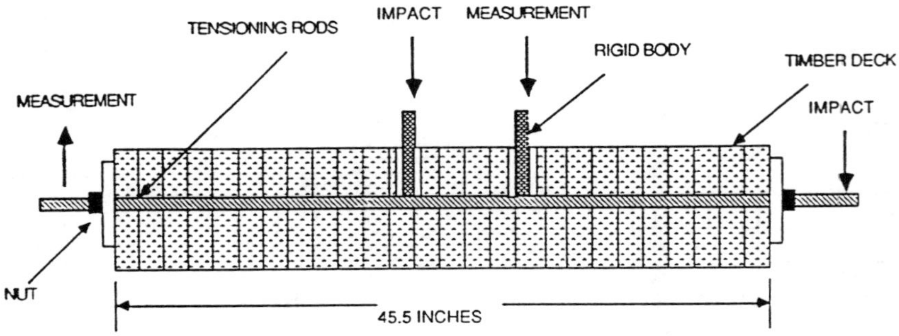

Fig. 2 Model of a Section of a Stressed Timber Bridge and Dynamic Characterization Tests.

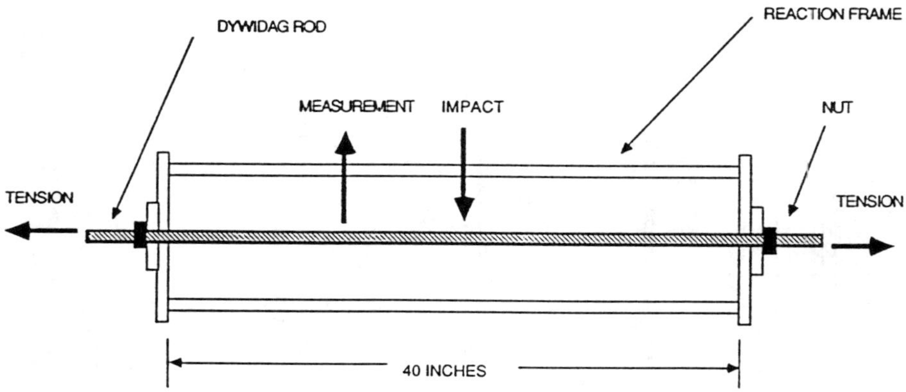

Fig. 3 Test Setup for Rod Tension Tests.

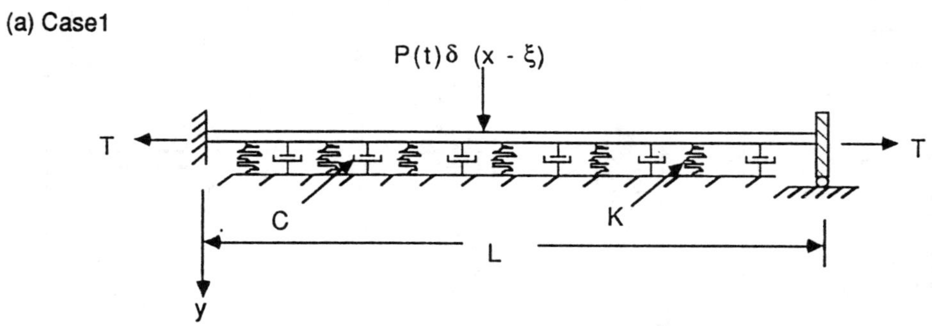

Fig. 4(a) Simplified Model of Timber Deck Model for Approach 1.

(b) Case 2.

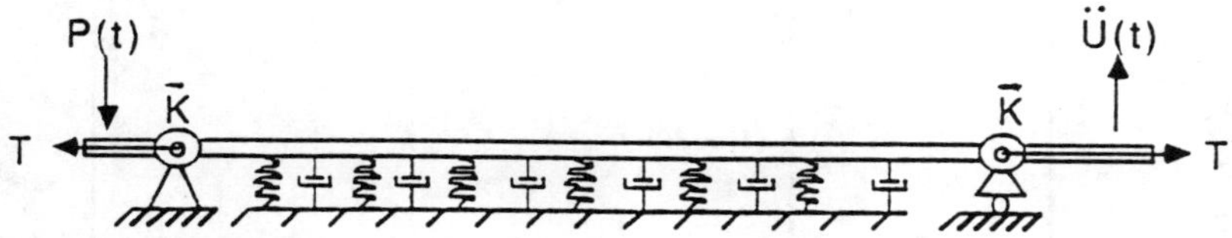

Fig. 4(b) Simplified Model of Timber Deck Model for Approach 2.

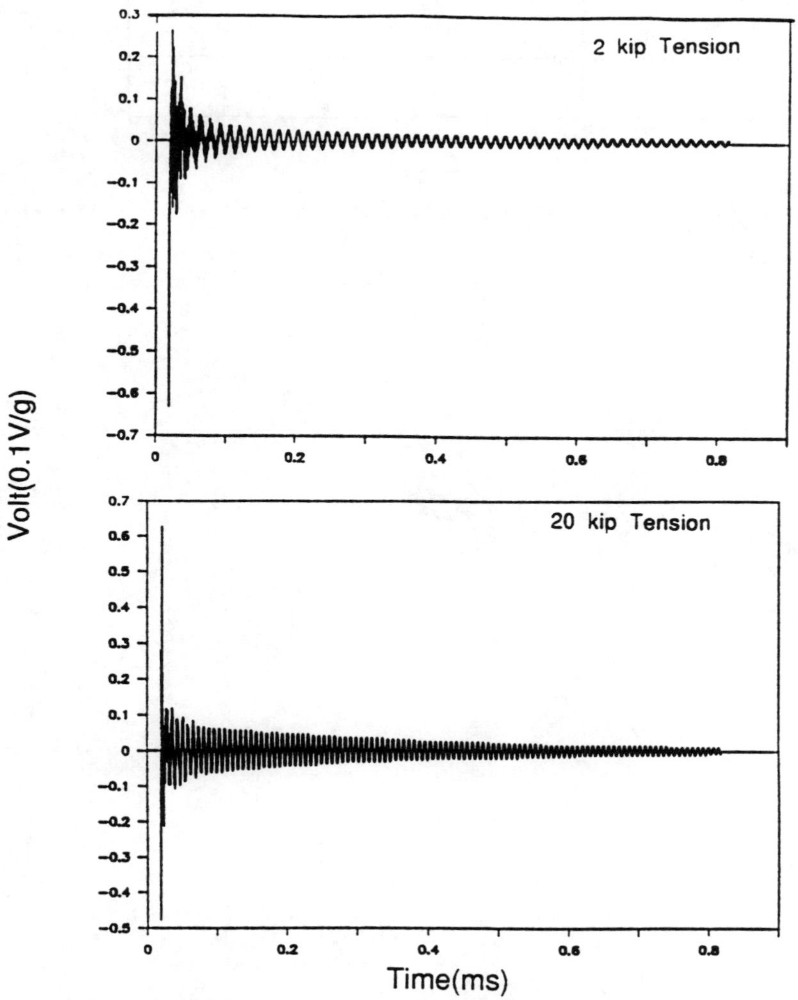

Fig. 5 Acceleration Time Histories for Model Deck Approach 1.

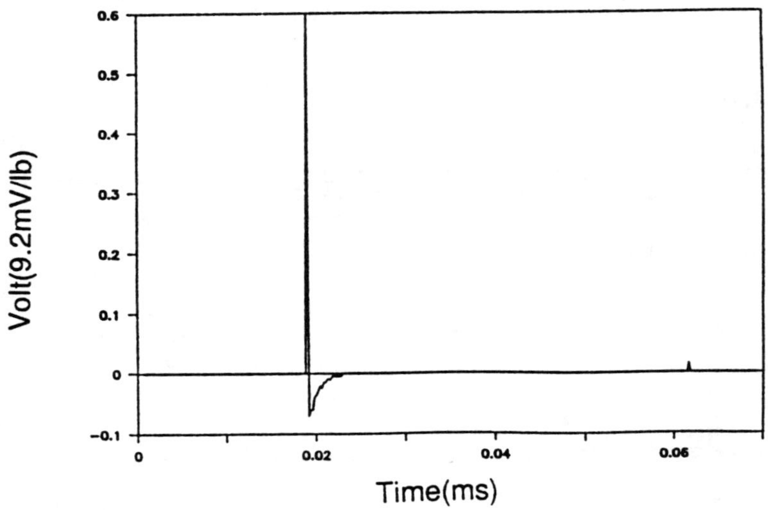

Fig. 6 Time History of Impact Input by Impact Hammer.

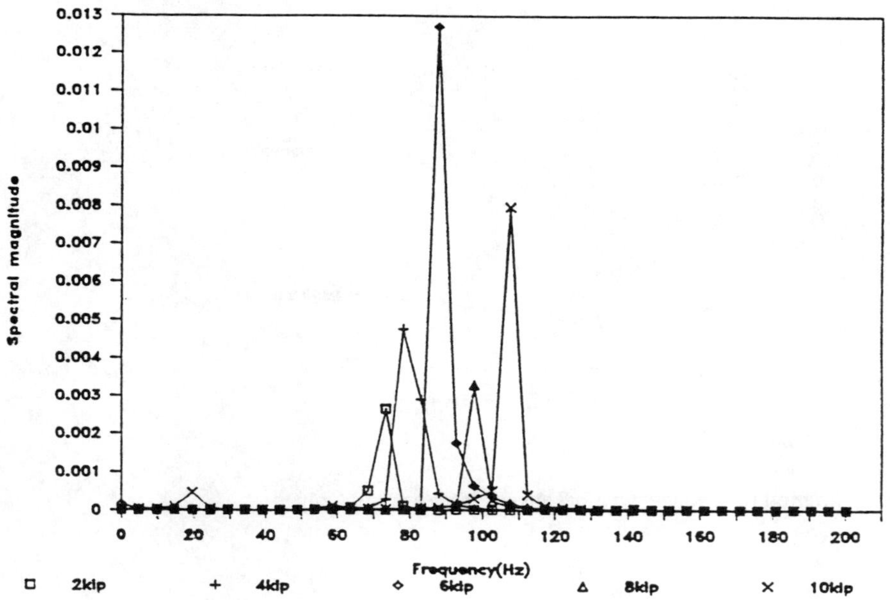

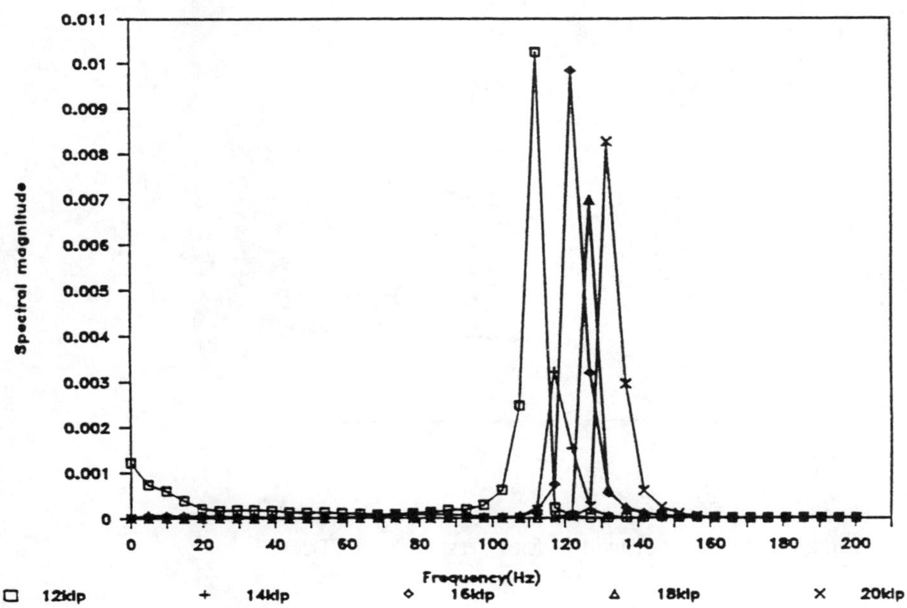

Fig. 7 Power Spectral Magnitudes for Different Levels of Tensions.

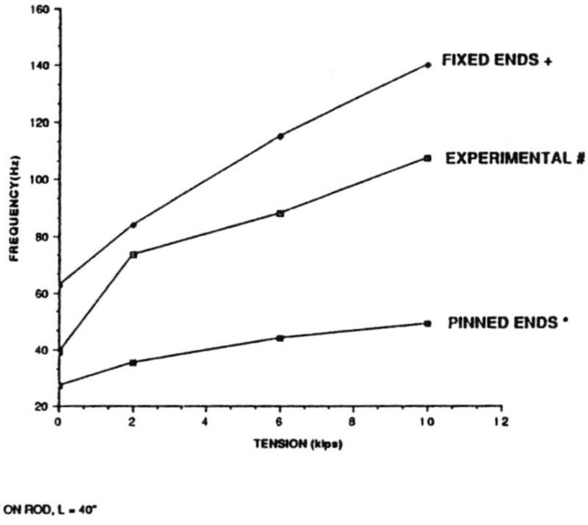

Fig. 8 Variation of Frequency with Tension, Test on Rod Only.

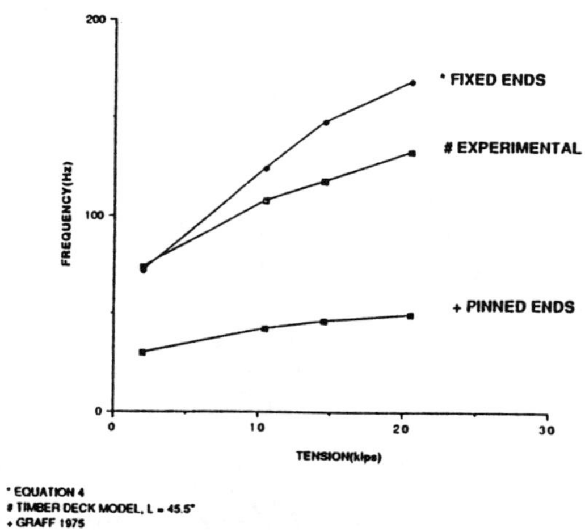

Fig. 9 Variation of Frequency with Tension, Test on Model Deck.

TECHNIQUES FOR ASSESSING THE CONCRETE (STATE) CONDITION IN BRIDGES

G. GRIMALDI[1], R. BERISSI[2], P. BREVET[3], A. RAHARINAIVO[4]
French Laboratories for Bridges and Pavements
1. Regional Laboratory, Melun
2. Regional Laboratory, Aix-en-Provence
3. Central Laboratory (LCPC) Division Metal Structures and Cables
4. Central Laboratory (LCPC) Scientific Directorate
58, boulevard Lefebvre, F-75732 Paris Cedex 15, FRANCE

ABSTRACT. The controls of the (state) condition of concrete and rebars are carried out by using non-destructive methods. A set of methodologies and facilities has been developed for determining the damaged areas to be repaired in a bridge.

- Measuring the surface concrete permeability: the principle is to check the pressure in a jar, previously fixed on the concrete surface and in which vacuum has been created. The pressure increase is due to the air flux through the concrete pores.
- Measuring the carbonation depth: the pH (colour) indicator is applied at the bottom of a small slot, correctly sawn, in the concrete.
- Measuring the half-cell potential of rebars at points which are close to one another for determining the areas where rebars are most probably corroded.

1. Scope

Repairs are less expensive for reinforced concrete bridges when deteriorations are monitored during the early stage. The damaged areas are indeed still small and the bearing capacities of rebars are not yet decreased.

The techniques used for assessing states (or conditions of embedded rebars must not be destructive. A set of such methods has been developed in the French Laboratories of Bridges and Pavements. They deal with plain concrete and embedded rebars.

The control results make it possible to determine the bridge areas which are to be repaired.

A. S. Nowak (ed.), Bridge Evaluation, Repair and Rehabilization, 249–258.

2. Concrete Permeability

2.1 PRINCIPLES

The concrete porosity near free surfaces is complex. It changes with time, due to water exchanges between the concrete and its environment, and to the flux of materials coming from the environment. Pores are between 0.1 and 100 μm in size. They are open, meaning that they communicate with the external medium.

Porosity is the outstanding factor controlling the rate of concrete change. It can be characterized with a test [1] involving permeability to gas fluxes.

A metal jar is applied on the concrete surface. At its contact with this surface, an appropriate putty roll is installed for tightening. The air in the jar is pumped out (Figure 1). When the pressured obtained is low enough, the pumping is stopped. Atmospheric air comes back into the jar through the concrete pores. The concrete involved is a circular area, with a diameter two or three times greater than the jar diameter. Its thickness is on the order of 10^{-3} or 10^{-2} m.

Figure 2 shows (p - t) curves, where p is the pressure in the evacuated jar and t is time. As for "lack fluxes", this curve is described by the equation

$$p = p_0[1 - \exp(-t/T)] \qquad (1)$$

where T is a parameter which characterizes the concrete permeability: T increases as the permeability decreases.

2.2 TEST PROCEDURE

The air flux in concrete strongly depends on its water content. So, for measuring parameter T (or permeability) the surface concrete humidity is to be known and controlled.

At each checking point on the concrete surface, a series of four tests is carried out, as follows, for measuring parameter T (or permeability):

- Condition A: the concrete surface is in its natural, present condition, ("as encountered state"), the value (T_A) of parameter T is determined,
- Condition B: some water is sprayed on the concrete surface for 15 minutes, the value (T_B) depends on the rate of water ingress into the concrete (sorption process)
- Condition C: after this wetting, the concrete surface is dried with warm air for 15 minutes, the value (T_C) of parameter T depends on the rate of water flow out of the concrete (desorption process)
- Condition D: after the previous test (condition C), the concrete surface is dried for 15 minutes more. The value (T_D) of parameter T is then determined. It corresponds to a conventionally "dry state".

2.3 INTERPRETATION

For usual dry concretes, parameter T ranges between 50 and 10 000 seconds. So the permeability of a naturally dry concrete is said to be

- low, if $T \geq 5\ 000$ sec,
- fair, if $100 \text{ sec} \leq T < 5\ 000$ sec and
- high, if $T \leq 100$ sec.

In fact, for real bridges, concrete is more or less dry and its permeability should be determined under these known conditions. In other words, the concrete permeability must be characterized with T_D and not with T_A. According to the obtained T_D value, the permeability class is determined as noted above.

3. Carbonation Depth Measurement

3.1 PRINCIPLES

The carbon dioxide coming from the environment can react with the hardened cement components. The concrete pH can be decreased by this reaction from about 12 (sound concrete) to 9 (highly carbonated concrete). It is usually checked by pouring a colour indicator on freshly broken concrete. If the indicator is phenolphthalein, the tested concrete is grey (as natural) if $pH \leq 9.1$ and it is pink if the concrete is not carbonated. The carbonation depth is usually determined by using a concrete core: this sample is broken and the indicator is poured on its fracture surface.

A new test principle is to check the concrete pH without taking cores.

3.2 TEST PROCEDURE

A slot is obtained in the concrete by sawing it with a rotating disk (Figure 3). Under appropriate conditions and with skilled labour, the bottom of the slot is not covered by cement powder coming from the various concrete depths.

As the notch bottom is of uneven depth, the colour of the pH indicator sprayed on it, shows up the carbonated concrete layer and its thickness (carbonation depth).

3.3 INTERPRETATION

Concrete carbonation can result in rebar corrosion. Therefore, carbonation depth is to be compared with concrete cover on rebars.

The carbonation grade is said to be

- low, if the sound concrete (not carbonated) thickness is greater than 18 mm,
- medium, if the sound concrete thickness ranges between 10 and 13 mm,
- high, if the sound concrete cover is less than 10 mm thick.

4. Half-cell Potential

4.1 PRINCIPLES

Measuring the half-cell potential of rebars in concrete makes it possible to know the probability of corrosion. The half-cell potential (E_C) measurement is carried out following rules, such as ASTM C876-87 recommendations.

It has been shown that E_C values depend on the concrete cover condition [2] [3]. But, some "critical" values of E_C can be determined. They depend on the oxygen access into the concrete: in parts of a bridge which are in contact with the atmosphere, these "critical" E_C values usually indicate that:

- if $E_C < -350$ mV (reference electrode Cu - Cu SO_4 satur.) rebars are probably corroded (class R),
- if $E_C > -200$ mV, rebars are probably sound (class S),
- if $-350 \leq E_C \leq -200$ mV, rebar corrosion is about 50% probable (class M).

4.2 PROCEDURE

The E_C measurements are made for determining where the bridge areas to be repaired are. Therefore, the checking points should be as close to one another as possible. The distance between two points must be less than 0.2 m, which is far shorter than recommended by ASTM C876-87 (about 1 m).

4.3 INTERPRETATION

The half-cell potential(E_C) values at various points of the bridge are used for drawing up a "potential map". Three types of maps are drawn, as follows:

a) "Mesh values": E_C values are indicated at their checking points on the map,
b) "Equipotential curves": curves corresponding to given E_C values are drawn up on the map (Figure 4.a).
c) "E_C classes": only two equipotential curves are drawn up, they correspond to E_C critical values. So they are frontiers separating areas of different E_C classes (Figure 4.b). Class R area includes rebars which have the highest probability of being corroded. Two or three holes should be bored into such an area to verify that rebars are actually corroded.

In fact, the previous E_C "critical" value (i.e. - 200 mV and - 350 mV/Cu - Cu SO_4) are sometimes questionable. It is generally agreed that they are not valid for bridge piles which are immersed in water.

Newer questions have been asked. First: are these "critical" E_C values related to some physical condition of the rebars? Then: what are the E_C "critical" values for rebars in unusual concrete, e.g. containing silica fume?

An answer to both questions has been given previously when corrosion is induced by chloride ions [4]. It can be recalled as follows. Rebars corrode when the chloride content [CL^-] in concrete (at their contact) is higher than a threshold which depends on the concrete pH ($[CL^-]_c$ = 0.6 $[CL^-]$ and $[CL^-]_c = [OH^-]$). This result has a physical meaning [5][6]. So, if a correlation is pointed out between E_C "critical" values and $[CL^-]$ critical values, then the half-cell potential E_C classes also have a physical meaning.

Figures 5.a and 5.b give results obtained with reinforced mortar specimens. They confirm that E_C "critical" value (E_C = - 350 mV) corresponds to chloride content critical value ($[CL^-] > [OH^-]$), for usual ordinary Portland mortar (Figure 6.a). But when silica fume is added to mortar, the E_C critical value is lower than - 350 mV, even when $[CL^-]$ critical value ($[CL^-]_c$ = 0.6 $[OH^-]$) holds.

5. Applications - Conclusions

5.1 APPLICATIONS

The measurements of permeability, carbonation depth and half-cell potential are used for determining the damaged areas in a reinforced concrete bridge.

It is of interest to apply the three controls because their results are different and complementary. If only the concrete permeability and carbonation depths are measured, the results are interpreted according to Figure 6. If the concrete cover is sound (carbonation grade is low) and impermeable, the controlled area is durable. On the contrary, if the carbonation grade is high and the concrete cover is of high permeability, rebars are or will be corroded.

In any other case, the half-cell potential E_C should be measured.

5.2 CONCLUSIONS

Three techniques have been developed for assessing the concrete condition in bridges. The objective is not to give a global assessment of the bridge parts, but to determine the bridge areas where reinforced concrete needs to be repaired, not withstanding the mechanical problems.

These methods are, first, the permeability measurement, then, the carbonation depth determination and, last, the half-cell potential check. The results of these controls are to be combined [7] for locating needed repairs in reinforced concrete bridges.

References

[1] Berissi R., Bonnet G. and Grimaldi, G., (1987) "Mesure de la porosite ouverte des betons hydrauliques," Proc. 1st International Congress From materials science to construction materials engineering," RILEM, 7-11, p. 119-122.

[2] Raharinaivo, A., Brevet, P., Grimaldi, G. and Pannier, G., (1986) Relationships between concrete deterioration and reinforcing-steel corrosion. Durability of Building Materials, Vol. 4, p. 97-112.

[3] Grimaldi, G., Brevet, P., Pannier, G. and Raharinaivo, A., (1986) Factors influencing electrode potential of steel in concrete,"British Corrosion Journal, Vol. 21, no. 1, p. 55-62.

[4] Raharinaivo, A., Brevet, P., Grimaldi, G. and Carpio, J., (1989) Techniques for assessing the residual lifetime of reinforced concrete civil works," 9th European Congress on corrosion Utrecht, N.L., Paper BU-097.

[5] Raharinaivo, A. and Genin, J.M.R. (1986) "On the corrosion of reinforcing steels in concrete, in the presence of chlorides," Materiales de construccion, Vol. 36, no. 204, p. 5-16.

[6] Raharinaivo, A., (1987) "Forecasting the reinforcing steel corrosion in concrete, under the action of chlorides," US-European Workshop on Bridge Evaluation, Repair and Rehabilitation, Nowak, A.S., Absi, E., editors, St. Remy les Chevreuse, F., Juin, 22-25, p. 459-464.

[7] Grimaldi, G., Berissi, R., Bonnet, G., Brevet, P. and Raharinaivo, A., (1989) Assessing the conservation state of concrete and its reinforcement," 11th International Congress CIB (International Council for Building Research...) Paris, F., Theme II, Vol. I, p. 229-238.

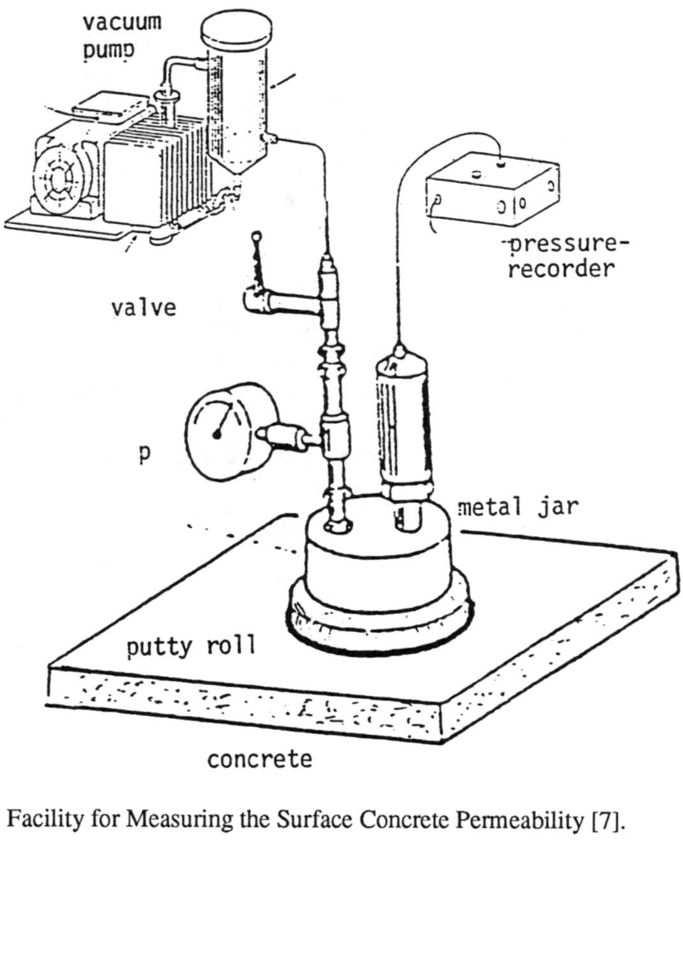

Fig. 1 Facility for Measuring the Surface Concrete Permeability [7].

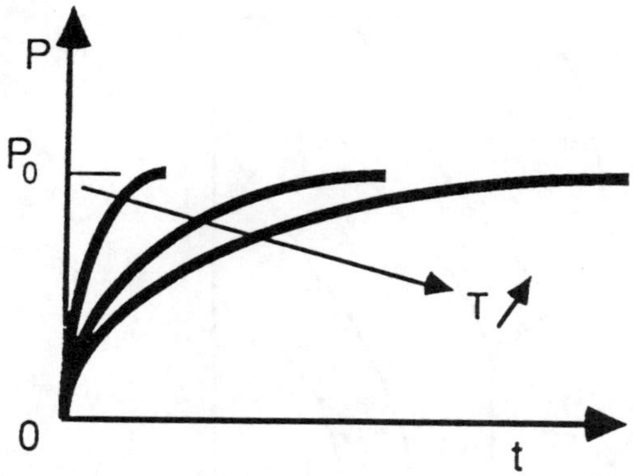

Fig. 2 Pressure p in a Previously Evacuated Jar on a Concrete Surface, Versus Time.

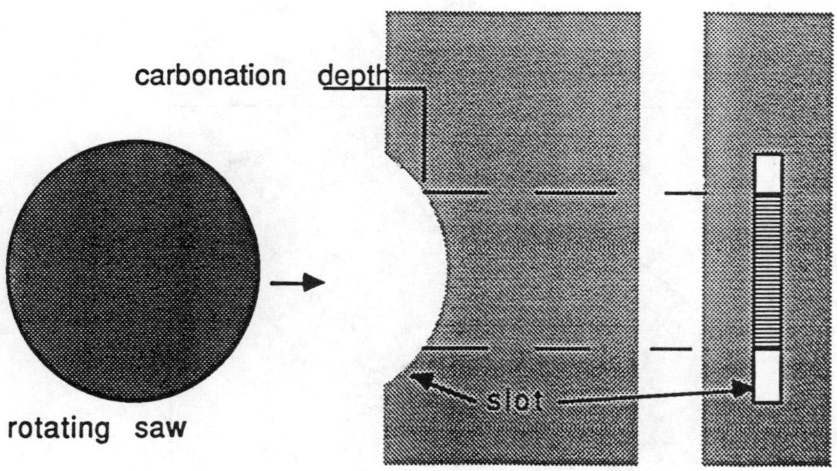

Fig. 3 Slot Sawn in Concrete for Measuring the Carbonation Depth.

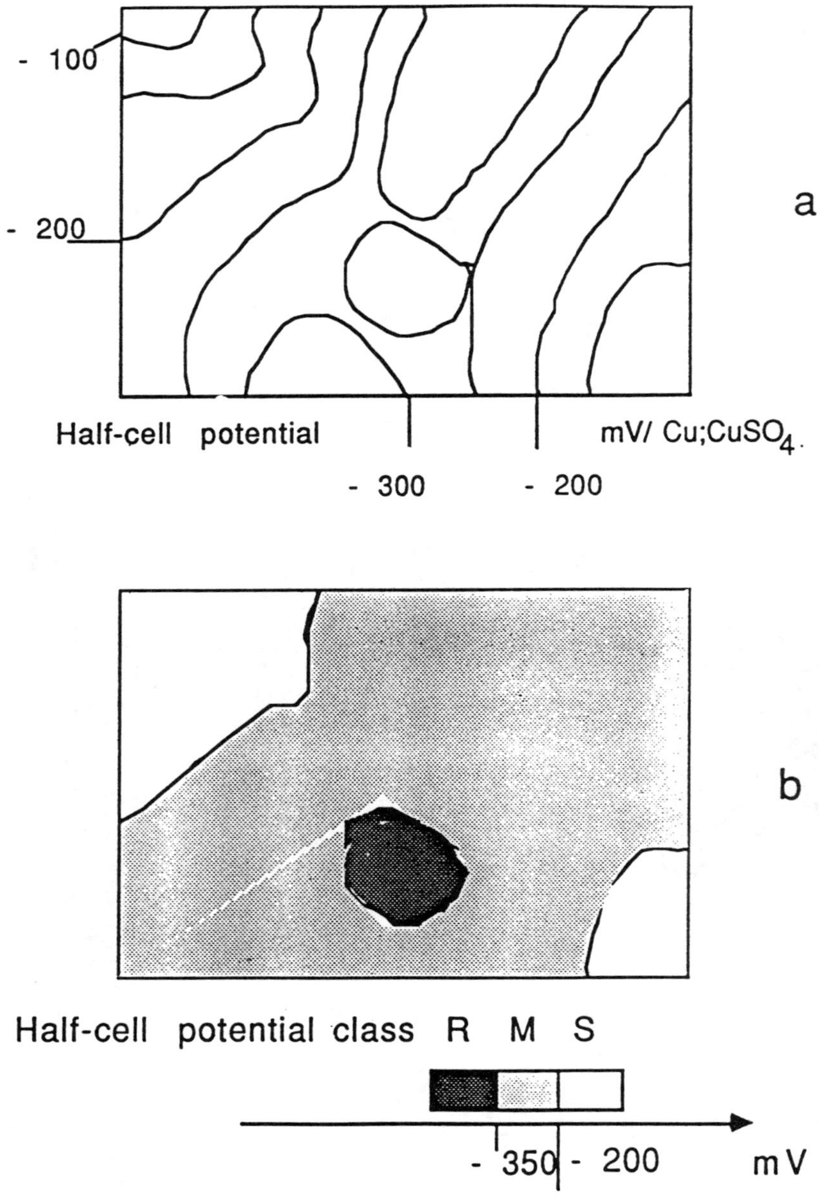

Fig. 4 Half-Cell Potential E_C Mapping; a) Equipotential Curves; b) Areas of Various E_C classes.

5.a.

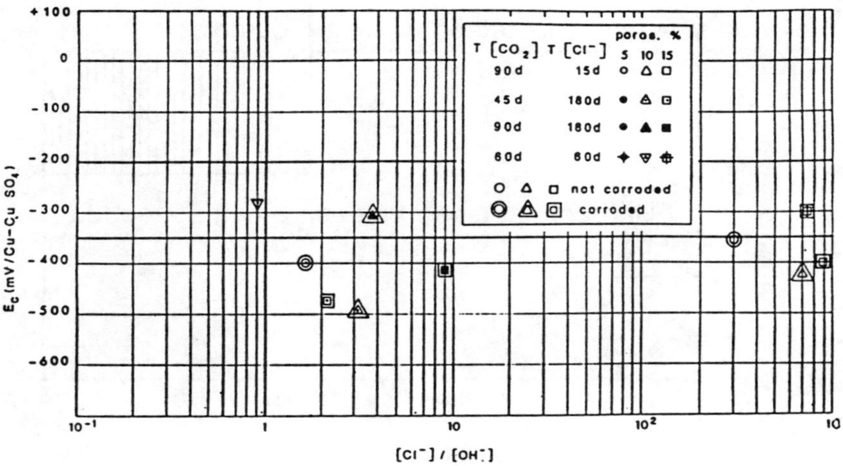

5.b.

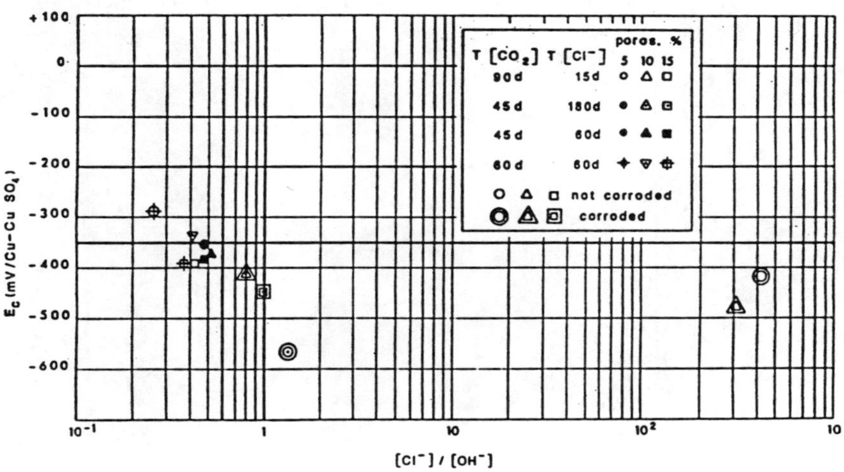

Fig. 5 Half-Cell Potential E_C of a Steel Wire Embedded in a Mortar Specimen, After Curing in CO_2 Atmosphere (During T[CO_2] and Chloride Penetration (During T[C1-], Versus [C1-]/[OH-] Were [Cl-] is the Chloride Content. Poros = Mortar (Volume) Porosity; a) Mortar of Ordinary Portland Cement; b) Mortar of Ordinary Portland Cement with Silica Fume Added [4].

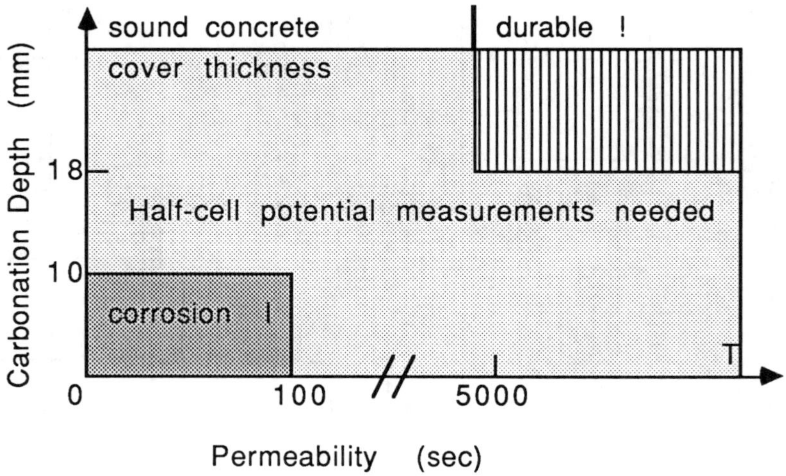

Fig. 6 Interpretation of Test Results Concerning Concrete Permeability and Carbonation Depth.

III- Loads and Analysis

FATIGUE LOADING AND DESIGN FOR ROAD BRIDGES

GERT KONIG and PETER SEIFERT
Institut für Massivbau
Alexanderstr. 5
D-6100 Darmstadt
WEST GERMANY

1. Introduction

Due to the observed fatigue damages at coupling joints of prestressed concrete road bridges and the increased application of partial prestressing, the interest in an appropriate fatigue loading and design procedure has grown. In recent years European codes of practice made no or only very simple provisions for the fatigue design of concrete road bridges. This paper presents a new design concept on the basis of traffic measurements and numerical simulation. The static and the dynamic problems are treated separately.

2. Method for the Simplified Fatigue Design Procedure, Based on Standard Spectra

In [9] the following general design equation for fatigue is proposed:

$$\gamma_{sd} \cdot \Delta\sigma(\Delta M_{EC}) \cdot \alpha_p \cdot \bar{\phi}_{fat} \leq r_{e,1,design} = r_{e,1,characteristic}/\gamma r \qquad (1)$$

with

γ_{sd} = 1.10; safety factor for model uncertainties [1],
γ_r = 1.15; material safety factor for steel [2],
$\Delta\sigma(\Delta M_{EC})$ = stress range due to the bending moment caused by the EC-load model [3],
α_p = conversion factor between the stress range due to the EC-load model and the constant amplitude spectrum with 10^6 cycles, resulting in the same damage as the real traffic,
$\bar{\phi}_{fat}$ = mean impact factor,
$r_{e,1,design}$ = design value of resistance at 10^6 cycles,
$r_{e,1,characteristic}$ = characteristic point on the S-N-curve at 10^6 cycles.

A. S. Nowak (ed.), Bridge Evaluation, Repair and Rehabilization, 261–275.

The following calculations are based on a linear relation between stress range $\Delta\sigma$ and range of effects (e.g. bending moment ΔM).

The constant $(d\sigma/dM)$ is determined for the maximum stress range. This results in exact stress ranges for the large bending moment ranges and slightly overestimated stress ranges for the small bending moment ranges. Hence, the following damage calculation is slightly on the safe side.

The equivalent stress range S_e is determined as follows,

$$S_e = \alpha_p \cdot \Delta M_{EC} \tag{2}$$

with

S_e = amplitude of the constant amplitude spectrum, resulting in the same damage as the real traffic,

ΔM_{EC} = range of effects due to the EC-load model, determined as follows:

- load is applied to the positive influence surface adjacent to the point of interest,
- load is applied to the negative influence surface with the greatest absolute values, adjacent to the positive surface,
- the range of effects is calculated by adding both absolute values.

The spectra of effects are determined by the rain flow counting method and converted to standard spectra (see Fig. 1). In Fig. 1, $\bar{\bar{S}}$ = shape parameter of the spectrum; max n_{tr} = total number of trucks; maxS = maximum effect (14 days - maximum); n_p = number of cycles with at least the amplitude max S; n(S) = number of cycles with at least the amplitude S. Applying the Miner's rule, results in equivalent constant amplitude spectra [9],

$$S_e = \left[\int_0^{\max n} S^*(n)^{k_2} \left(\frac{dn}{\max n}\right)\right]^{1/k_2} \left(\frac{\max n}{n_1}\right)^{1/k_2} \tag{3}$$

with:

$S^*(n)$ = range of effects, exceeded by an cycles,

n_1 = 10^6,

k_2 = decisive slope of the S-N-curve.

The results are presented in Fig. 2.

Increasing the shape parameter $\bar{\bar{S}}$ and the stress exponent k_1 results in a larger damage.

3. Numerical Results on the Basis of the Proposed EC-Load Model and Measured Traffic Data

The fatigue load model, proposed by Germany and Belgium in the EC 9-12-committee, is used [1] (see Fig. 3).

Various locations including the upper and lower limits within Europe, are considered as shown in Fig. 4 and Fig. 5. Lorry flows up to 7000 trucks per day were observed per direction. Various simulations are carried out for representative static systems like main girders and deck plates. The results are calibrated using the load model given in Fig. 3. The HCL is positioned in the major traffic lane on the bridge. Interestingly, the most stable results are obtained, using the same model for local and global effects. This is due to the fact that the HCL-vehicle is a much more realistic vehicle representation than any other artificial axle combination. Hence, the calibration result is independent of the shape of the influence line considered.

The following simple rule for practical application can be given:

$$\max S_o = \Delta M_{EC}\,(100\%\ \text{HCL - fatigue}) \cdot 1.05 \quad \left\{ \begin{array}{l} \text{multispan girder (all sections)} \\ \text{local effects } (L < 15\ \text{m}) \end{array} \right. \tag{4}$$

with
1.0 heavy traffic (Auxerre, F);
0.6 light traffic (Doxey, GB);
0.74 medium traffic (Brohltal, D).

$\alpha_{p,o}$ is determined using Fig. 2 with $\bar{S} = 2.5$ for urban traffic and $\bar{S} = 4.0$ for long distance traffic,

$$\alpha_{p,o} = f(\bar{S}, k_2, n_1, \max n)$$

$$= \left(\frac{\max S_o}{\Delta M_{EC}}\right) \cdot \left[\left(\frac{S_e}{\max S_o}\right) \cdot \left(\frac{n_1}{\max_n}\right)\right]^{1/k_2} \left(\frac{\max n}{n_1}\right)^{1/k_2} \tag{5}$$

where

$\max n$ = number of trucks during lifetime, e.g. 100 years,
n_1 = number of cycles at the characteristic value on S-N-curve ($n_1 = 10^6$),
k_2 = $2\,k_1 - 1$ = decisive shape of the S-N-curve,
$\alpha_{p,o}$ = present α_p - value for the given bridge,
$\max S_o$ = present maximum 14 day range of effects.

According to European truck manufacturers, there is no tendency to increase the legal limits and, consequently, the mean values of the truck weights. Nevertheless, there is a strong trend to replace light two-axle and three axle vehicles by heavy articulated vehicles. There is also a considerable decrease in the number of empty truck runs.

An appropriate $\alpha_{p,design}$ - value for a 100-year design period can be found as follows:

$$\alpha_{p,design} = \alpha_{p,o} \cdot 0.5\,(1+\psi) \cdot \left[0.5 \cdot \left(1 + \frac{MDTT_{100}}{MDTT_0}\right)\right]^{1/k_2} \cdot 0.5 \cdot (2 + \delta P) \tag{6}$$

with

$\alpha_{p,o}$ = present α_P - value for the given bridge,

$$\psi = \frac{(S_e/\max S)_0}{(S_e/\max S)_{100}},$$

$(S_e/\max S)_0$ = equivalent range of effects, determined with present $\bar{S}$-value (Fig. 2),
$(S_e/\max S)_{100}$ = equivalent range of effects, determined with future $\bar{S}$-value,
$MDTT_0$ = present design truck flow (available for most German roads),
$MDTT_{100}$ = design truck flow in 100 years, to be defined by the authorities,
δP = increase of the truck weights in 100 years.

4. Introduction of Dynamic Effects into the Fatigue Verification

This chapter is based on intensive parameters studies [7]. The impact factor for fatigue is defined in such a way that multiplying all the stress amplitudes of the static histogram by a constant impact factor results in an increased histogram, leading to the same damage as the original dynamic histogram.

Hence, the impact factor ϕ_{fat} is computed as follows:

$$\Sigma n_{i,dyn} \cdot (S_{i,dyn})^{k_2} = \Sigma\, n_{i,stat} \cdot (\phi_{fat} \cdot S_{i,stat})^{k_2}$$

This results in

$$\phi_{fat} = \left[\frac{\Sigma n_{i,dyn} \cdot (S_{i,dyn})^{k_2}}{\Sigma n_{i,stat} \cdot (S_{i,stat})^{k_2}} \right]^{1/k_2} \qquad (7)$$

The highway bridge is represented by a finite element model with lumped masses. The dynamic bridge problem may be described by the following standard second order differential equation:

$$[M]\ddot{U} + [D]\,\dot{U} + [C]U = P(t) \qquad (8)$$

in which [M], [D], [C] and P(t) represent the mass, damping and stiffness matrices, and the traffic load vector. Bridge mass - vehicle mass interaction is neglected within this formulation. This is accurate enough for heavy concrete bridges with spans greater than 25 m [9]. The bridge damping is idealized as linear viscous Rayleigh-damping. The damping matrix [D] is a linear combination of the mass matrix [M] and the stiffness matrix [C].

$$[D] = \alpha \cdot [M] + \beta \cdot [C] \qquad (9)$$

Various tests in Switzerland [2] resulted in 1.3% mean bridge damping D_{Br}. The dynamic response of the bridge under traffic loading can be determined by using the Newmark numerical method. The truck axle loads are proportional to the surface roughness.

On the basis of numerous measurements, the surface roughness can be modeled as a normally distributed, stationary, ergodic random process. The power spectral density of the surface is approximated by straight lines in a double-log-diagram.

$$\Phi_h(\Omega) = \Phi_h(\Omega_o) \cdot \left[\frac{\Omega}{\Omega_o}\right]^{-w} \tag{10}$$

Table 1 gives the classification according to the ISO-Standard, with: σ_h = standard deviation for a speed of 80 km/h and excitation frequencies 1 Hz $\leq f \leq$ 20 Hz. In Germany 87% of the roads have good or very good surface quality.

For simplification purposes, a coupled n-axle truck is modeled by n two-degrees-of freedom partial truck models (Fig. 6). The tire damping can be neglected, since it is very small for high speeds [7]. Eq. 11 describes the partial truck model.

$$\begin{bmatrix} (c_1+c_2) & -c_2 \\ -c_2 & c_2 \end{bmatrix} \begin{bmatrix} z_1 \\ z_2 \end{bmatrix} + \begin{bmatrix} d_2 & -d_2 \\ -d_2 & d_2 \end{bmatrix} \begin{bmatrix} \dot{z}_1 \\ \dot{z}_2 \end{bmatrix} + \begin{bmatrix} m_1 & 0 \\ 0 & m_2 \end{bmatrix} \begin{bmatrix} \ddot{z}_1 \\ \ddot{z}_2 \end{bmatrix} = \begin{bmatrix} c_1 h(t) \\ 0 \end{bmatrix} \tag{11}$$

A chain of n axles, represented by n partial trucks (see Fig. 6), with constant distances and constant speed (70 to 80 km/h) represents the bridge loading. The loading is simulated by Monte Carlo method from the axle load spectra, defined as the product of the roughness spectra and the vehicle transfer functions. Truck geometries, weights and distances are taken from the locations listed in section 3. The truck frequencies are varied within the common range, given by the European truck manufactures:

$$\text{-axle frequency } f_2 = \frac{1}{2\pi} \left(\frac{c_1 + c_2}{m_1}\right)^{1/2} = 8 \text{ .... } 15\text{Hz} \tag{12}$$

$$\text{-body frequency } f_2 = \frac{1}{2\pi} \left(\frac{c_2}{m_1}\right)^{1/2} = 0.8 \text{ ... } 3\text{Hz} \tag{13}$$

$$\text{-body damping } D_2 = \frac{d_2}{2 \cdot (c_{2'lin} \cdot m_2)^{1/2}} = 0.10 \text{ ... } 0.30 \tag{14}$$

Fig. 7 shows the eigenfrequencies from tests [2] for typical concrete bridges.

The simulated bridges clearly represent an upper limit in terms of eigenfrequencies (see Fig. 7). This increases the degree of resonance between bridge and truck axles. Due to this fact the impact factor $\bar{\phi}_{fat}$ increases with decreasing span widths (see Fig. 8). The resonance between the bridge and truck bodies is negligible, since it is related to the larger spans with many simultaneous axles on the bridge. All these axle loads have different phase angles, resulting in reduced dynamic effects when multiplied by the "dynamic influence line". At 0.2 L_2 and support: L* = mean spanwidth of the two spans in the vicinity of the support; and at midspan L* = spanwidth of the considered span.

In general $\bar{\phi}_{fat}$ is larger at the support than for the span and decreases with increasing span width.

Fig. 8 is based on mean bridge damping. In general it can be written:

$$[\phi fat\ (DB_r = 0.8\%)\text{-}1] \div \bar{\phi}_{fat}\ (DB_r = 1.3\%)\text{-}1] \div [\bar{\phi}_{fat}(DB_r = 1.8\%)\text{-}1]$$

$$= 1.30 \div 1.00 \div 0.85 \qquad (15)$$

The effect of surface roughness can be expressed by the following equation:

$$[\bar{\phi}_{fat}\text{-}1] = [\Phi_h(\Omega_o)]^{1/1.9} \qquad (16)$$

In general it can be stated that $\bar{\phi}_{fat}$ depends on the surface quality and on some resonance effects for medium spans above 25 m.

5. Comparison Between the Present German Design Concept and the New Proposal for the Eurocode

For example, a coupling joint of a 4-span-bridge is investigated.

Input Parameters:

L	= 50 m, 4 spans (40/50/50/40)
k_2	= 2 k** - 1 = 5, n_i** = 10^6
$\Delta\sigma$**	= $r_{e,1,design}$ = 80 MN/m^2
$\Phi_h(\Omega_o)$	= 4 cm^3
ϕ_{fat}	= 1.36 (with L* = 45m, see Fig. 8)
γ_{sd}	= 1.10
γ_r	= 1.15
ΔM_{EC}-Bruls	= 3210 kNm (100% HCL-fatigue)

From Eq. 4,
extr. max S_o =3210 • 1.56 • 0.74 = 2494 kNm (Brohltal)
extr. max S_{100} =3210 • 1.05 • 1.00 = 3370 kNm (Auxerre)

Design lifetime: 100 years
$MDTT_0$ = 6600 trucks/day
$MDTT_{100}$ = 10000 trucks/day

$\bar{S}_o$ = 3.6 (Brohltal)
$\bar{S}_{100}$ = 4.0 (Auxerre)
max n_o = $MDTT_0 \cdot 100 \cdot 365$
(index 0 = present, index 100 = in 100 years)

From Fig. 2 and $\bar{S}_0 = 3.6$:

$$\left(\frac{S_e}{\max S_0}\right) \left(\frac{n_1^{**}}{\max n_0}\right)^{1/k_2} = .525$$

$$\alpha_{p,o} = \left(\frac{\max S_0}{\Delta M_{EC}}\right) \cdot \left(\frac{S_e}{\max S_0}\right)$$

$$= 0.77 \cdot 0.525 \ \left(\frac{100 \cdot 365 \cdot 6600}{10^6}\right)^{1/5} = 1.222$$

From Fig. 2 and $\bar{S}_{100} = 4.0$:

$$\psi = \frac{(S_e/\max S)_{100}}{(S_e/\max S)_o} = \frac{0.550}{0.525} = 1.048$$

Application of Eq. 6 leads to:

$$\delta P = \frac{\max S_{100} - \max S_0}{\max S_0} = \frac{3370 - 2494}{2494} = 0.351$$

and then,

$$\alpha_{P,design} = \alpha_{p,o} \cdot 0.5 \cdot (1 + \psi) \left[0.5 \cdot \left(1 + \frac{MDTT_{100}}{MDTT_0}\right)\right]^{1/k_2} \cdot 0.5 \cdot (2 + \delta P)$$

$$= 1.222 \cdot 0.5 \cdot (1 - 1.048) \cdot 0.5 \left[(1 + \frac{10000}{6600}\right)^{1/5} \cdot 0.5 \cdot (2 + 0.351)$$

$$= 1540$$

The α_p-value is based on the assumption that the considered bridge is presently loaded by the Brohltal traffic and in 100 years by the Auxerre spectrum with 10,000 trucks per day. Due to the expected opening of the borders within the European Community this assumption is realistic.

Using $\Delta\sigma_z \, [\frac{MN}{m^2}] = \Delta M \, [MNm] \cdot (\frac{d\sigma_z}{dM})_{secant}$

with $(\frac{d\sigma_z}{dM})_{secant} = 10.70 \; [\frac{1}{m^3}]$,

from Eq. 1,

$$\gamma_{sd} \cdot \Delta\sigma_z \, (\Delta M_{EC}) \cdot \alpha_p \cdot \bar{\phi}_{fat} \leq r_{e,1,characteristic} \, (n = 10^6)/\gamma r$$

with $r_{e,1,characteristic} \, (n = 10^6) \;\; = \; (\frac{2 \cdot 10^6}{10^6})^{1/k_2} \cdot r_{e,1,char}(n = 2 \cdot 10^6)$

$$= \; (2 \cdot 80)^{1/5} \;\; = \; 91.90 \; [\frac{MN}{m^2}]$$

$$1.10 \cdot 10.70 \, [\frac{1}{m^3}] \cdot 3.210 \, [MNm] \cdot 1.540 \cdot 1.36 \leq 91.90/1.15 \, [\frac{MN}{m^2}]$$

$$79.13 \leq 79.90 \quad [\frac{MN}{m^2}]$$

The present German design practice [5],[6],[8] result in

$$\gamma \cdot \Delta\sigma_z \, (\Delta M_{DIN}) \cdot \alpha_{P,DIN} \leq r_{e,1,characteristic} \, (n = 2 \cdot 10^6)$$

$$(\frac{1}{0.7}) \cdot (10.70 \quad [\frac{1}{m^3}] \cdot 6.300 \, [MNm] \cdot 0.5) \leq 80 \quad [\frac{MN}{m^2}]$$

$$48.15 \leq 80 \quad [\frac{MN}{m_2}]$$

6. Conclusion

The following parameters of major importance for a fatigue design are summarized:

- The minimum surface quality,
- the relation between stress and effect for the considered structural detail,

- the estimate for the most unfavourble temperature loading,
- the definition of the correct traffic density for the design period, (e.g. 100 years),
- the estimate of the characteristic S-N-curve.

On the basis of these input parameters a realistic design is possible. The presented example shows that the new design concept results in 65% increased design stress range compared with a present design practice in Germany. This is due to the fact that the dynamic effects were underestimated in the past. Anther reason is the increase of the truck weights, the degree of loading, and traffic density.

Finally, it should be stated that a refined design procedure does not eliminate the necessity for accurate structural detailing.

7. References

[1] Bruls, A., (1988) Example for a Unified European Load Model, Contribution to the Preliminary Background Report for Eurocode 9 on Actions - Part 12.

[2] Cantieni, R. (1983), Dynamic Load Tests on Highway Bridges in Switzerland, 60 years of Experience of EMPA, Report No. 211, Dübendorf.

[3] CEB Model Code 1990, first predraft (1988), CEB-Bulletin d' Information No. 1990, Lausanne.

[4] CEB-Bulletin d'Information No. 188. (1988), Fatigue of Concrete Structures. State of the Art Report, Lausanne.

[5] König, G., Gerhardt, H.-C. (1986), Beurteilung der Detriebsfestigkeit von Spannbetonbrücken im Koppelfungenbereich unter besonderer Berücksichtigung einer möglichen Rißbildung, Deutscher Ausschuß für Stalbeton, Heft 370.

[6] König, G., and Giegold, J. (1984), Zur Bemessung von Koppelfugen bei Massivbrücken, Beton-und Stahlbetonbau 79, No. 67, pp. 141-147, 191-197.

[7] Mao, Q. (1989), Parameterstudie zum ermüdungswirksamen Schwingbeiwert für Straßenbrücken, Institut für Massivbau, TH Darmstadt.

[8] Normenausschuß Bauwesen: Beiblatt 1 zu DIN 1072 (1985), Straßen-und Wegbrücken, Lastannahmen, Erläuterungen, Beuth Verlag, Berlin.

[9] Seifert, P. (1989), Vereinheitlichte Europäische Straßenverkehrsbelastung auf Brücken, geeignet für Tragfähigkeits-, Ermüdungs- und Gebrauchsfähigkeitsnachweise, Abschlußbericht zum Forschungsvorhaben FE Nr.: 15.160 R 87 F des Bundesministers für Verkehr, September.

[10] Zahnmesser, W, (1981), Auswirkungen verschiedener Lastzugkombinationen auf die Straßenbeanspruchung. Forschung Straßenbau und Straßenverkehrstechnik, Bundesminister für Verkehr, Heft 327.

Table 1 Classification of roadway roughness (ISO-TC 108).

Quality of Pavement $\Phi_h(\Omega_o)$ (cm^3) for $\Omega_o = 1m^{-1}$, w = 2

Class		Lower Limit $\Phi_h(\Omega_o)$	Mean value $\Phi_h(\Omega_o)$	Mean value σ_h(cm)	Upper Limit $\Phi_h(\Omega_o)$
very good	A	0.5	1	0.183	< 2
good	B	2	4	0.367	< 8
average	C	8	16	0.733	< 32
bad	D	32	64	1.467	< 128
very bad	E	128	256	2.933	< 512

Standard Spectrum of Effects

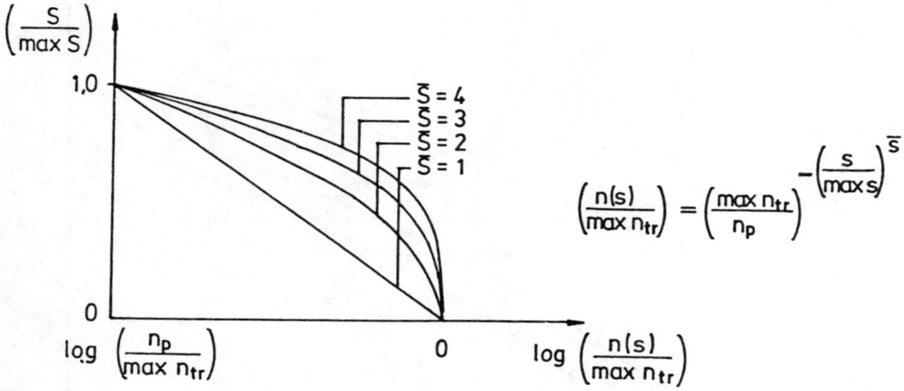

Corresponding Probability Density Function
(Continuous representation of the histogram)

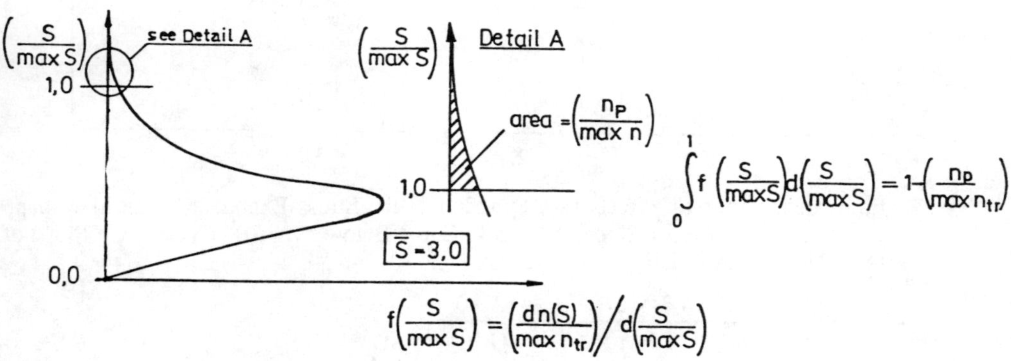

where	$\bar{S}$	=	shape parameter of the spectrum
	max n_{tr}	=	total number of trucks
	max S	=	maximum effect (14 days - maximum)
	n_p	=	number of cycles with at least the amplitude max S
	n(S)	=	number of cycles with at least the amplitude S

Fig. 1 Standard Spectrum of Effects and Corresponding Density Function.

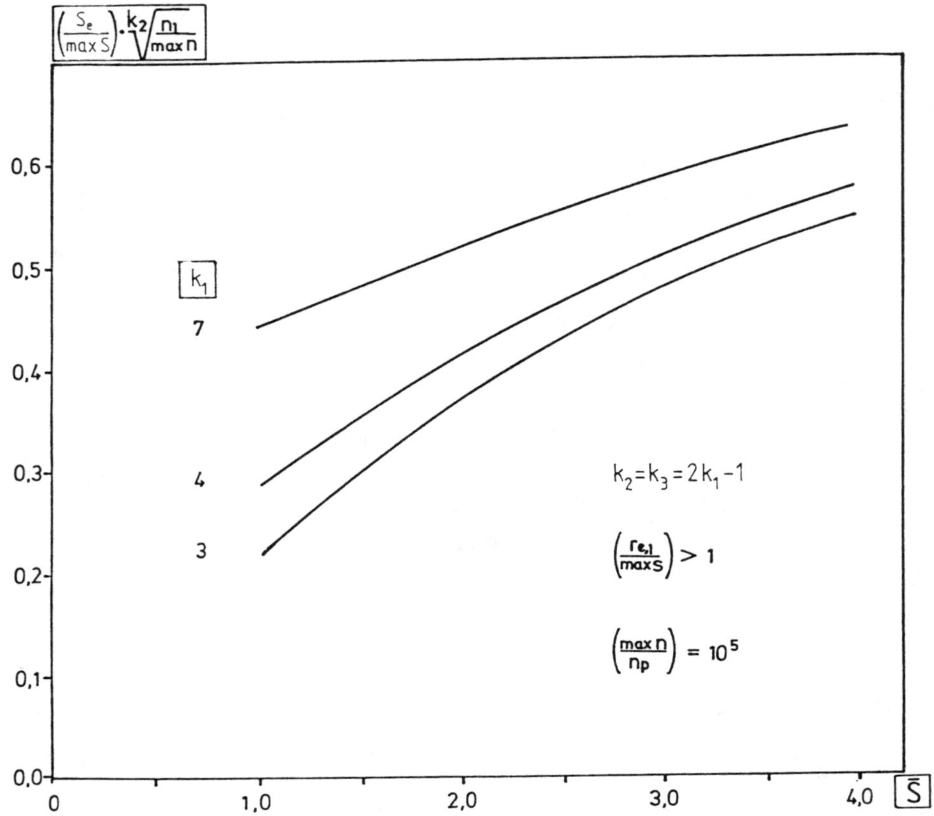

Fig. 2 Equivalent Range of Effects as a Function of the Stress Exponent k_1 and the Shape Parameter S with: k_1 = Slope of the S-N-Curve Below $n_1 = 10^6$ Cycle; k_2 = Slope of the S-N-Curve Above $n_1 = 10^6$ Cycle.

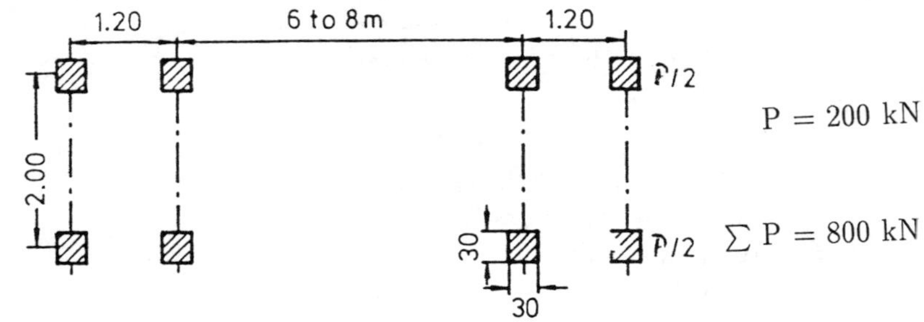

Fig. 3 EC-Bruls-Load Model, HCL-Fatigue [1].

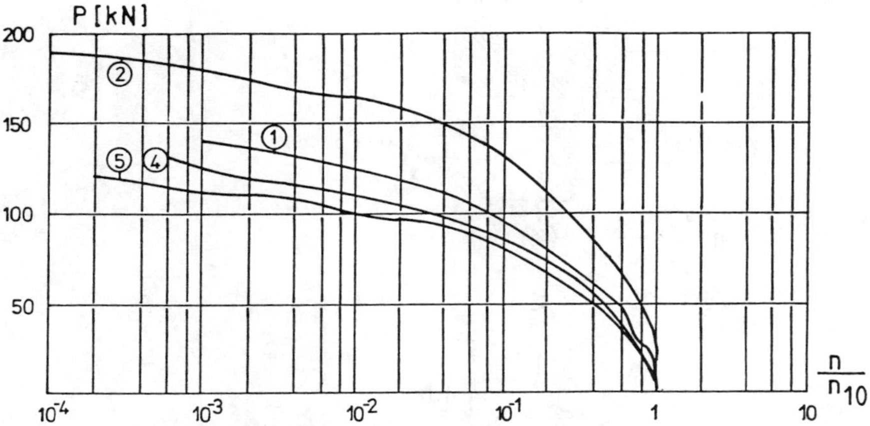

Fig. 4 Single Axle Loads for Representative Locations: 1 - Brohltal, 2 - Auxerre, 4 - Forth, 5 - Doxey, n_{10} = Number of Axles with $p \geq 10$ kN

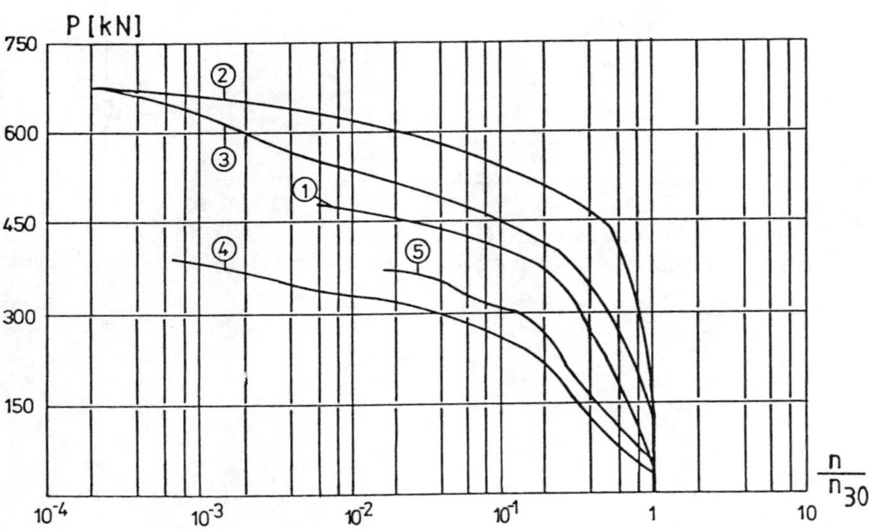

Fig. 5 Truck Weights for Representative Locations: 1- Brohltal, 2- Auxerre, 3- Périphérique, 4-Forth, 5- Doxey n_{30} = Number of Axles with $p \geq 30$ kN.

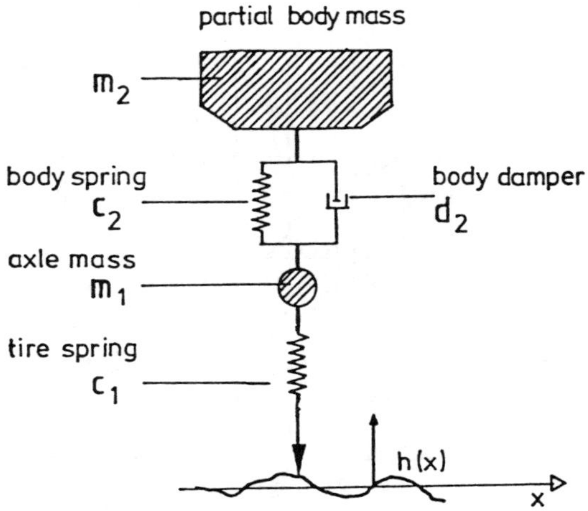

Fig. 6 Representive Partial Truck Model.

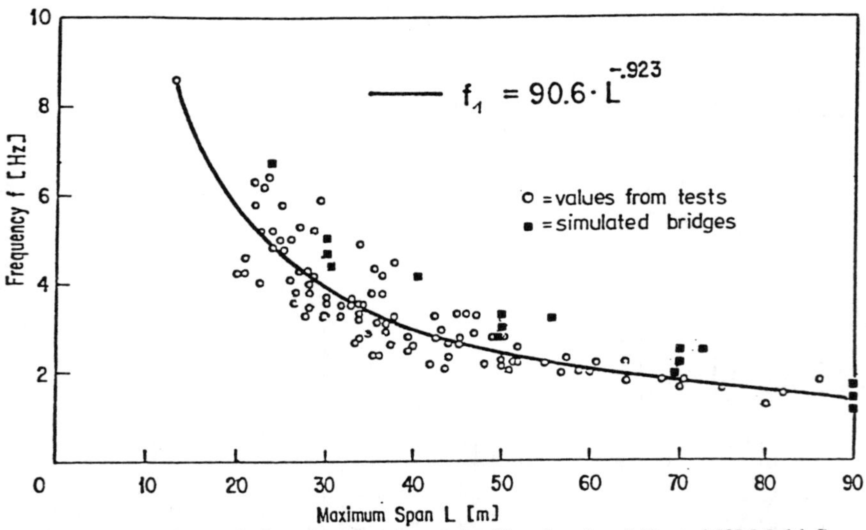

Fig. 7 Comparison of Eigenfrequencies of the Simulated and Tested [2] Multi-Span Bridges.

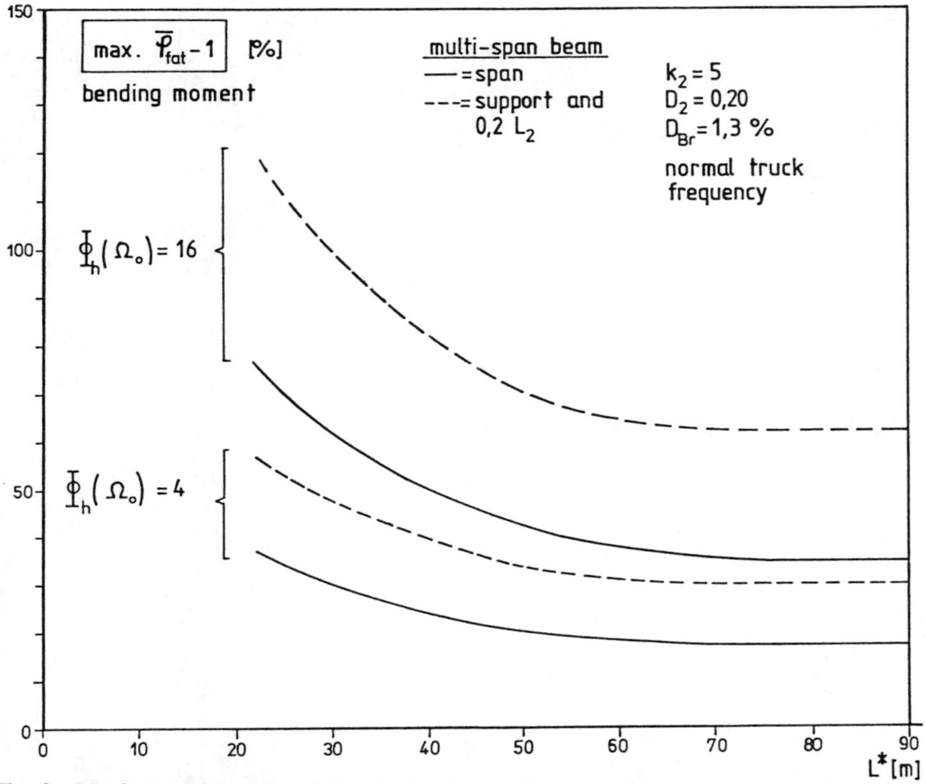

Fig. 8 Maximum Impact Factor ϕ_{fat} for Multispan Girders as a Function of the Spanwidth and the Surface Roughness (w = 2).

BRIDGE LOAD MODELS AND TRUCK WEIGHT REGULATIONS

MICHEL GHOSN
Department of Civil Engineering
The City College of New York
New York, NY 10031 USA

FRED MOSES
Department of Civil Engineering
Case Western Reserve University
Cleveland, OH 44106 USA

ABSTRACT. Historically, truck regulations have instituted control on axle and gross weights with bridge formulas based on limiting nominal stresses in certain types of bridges. These stress limitations do not usually lead to consistent or defensible safety levels and further ignore the cost impact of the weight regulation on all highway systems. This paper illustrates how new truck weight regulations can be developed which balance economy and provide acceptable safety levels. Target safety levels are derived from existing AASHTO bridge evaluation and rating procedures which have shown adequate performance safety levels. Reliability indices are used to relate the statistics on existing and proposed truck weight regulations and the dynamic behavior and resistance variables of existing bridges. bridge costs associated with new truck weight regulations are presented. These costs have been estimated by using the bridge inventory data of the Federal Highway Administration and statistics established for bridge capacity rating, span, material type and geometry.

1. Introduction

In the United States, the maximum vehicle weight allowed in the current law for general operation is 80,000 pounds. In addition, the federal limits for single axle and tandem weights are 20,000 and 34,000 pounds, respectively. The axle group weights are regulated based on what is known as the bridge formula is given by:

$$W = 500 \left(\frac{L_n}{N-1} + 12N + 36\right) \quad (1)$$

where: W is the overall gross weight in pounds on any group of two or more axles, L is the length in feet of the axle group, N is the number of axles in the axle group. The bridge formula was designed to avoid overstressing HS-20 bridges by more than 5 percent and H-15 bridges by more than 30 percent. The justification for these overstress ratios seems to be rather arbitrary.

The bridge formula has been criticized as overly conservative and critics often cite the experience of the province of Ontario, the New England states, the state of Michigan etc. that allow

A. S. Nowak (ed.), Bridge Evaluation, Repair and Rehabilization, 277–291.

higher loads than permitted by the current federal legislation for bridges designed based on the same AASHTO code (Standard Specifications 1983). Their bridges, it is claimed, do not seem to deteriorate at a faster rate than in other states which restrict their loads based on the formula in Eq. 1. A recent study by a group of researchers at Texas Transportation Institute under FHWA sponsorship, developed a new bridge formula . The new formula is based on the same overstress criteria as the existing formula but is more effective in preventing stresses from exceeding the specified overstress ratios. The proposed formula differs from the current formula by using only axle spacings to determine the maximum weight that can be carried by axle groups (James et. al. 1986):

$$W = (34 + L)\,1000 \qquad L < 56 \text{ ft.}$$
$$W = (62 + L/2)\,1000 \qquad L > 56 \text{ ft.} \qquad (2)$$

A study by the Transportation Research Board (TRB 1990) developed estimates of the impacts of various proposals for changes in truck weight regulations including bridge cost impacts. A panel of experts gathered to conduct the TRB study also recommended that the bridge formula be changed to the following:

$$W = (26 + 2L)\,1000 \qquad L < 24 \text{ ft.}$$
$$W = (62 + L/2)\,1000 \qquad L > 24 \text{ ft.} \qquad (3)$$

This formula was adopted from the TTI study by considering the stress limits on HS 20 bridges only. The panel also recommended that all states should be allowed to establish permit programs for vehicles over 80,000 lbs when the permit vehicles satisfy the current bridge formula given in Eq. (1) for trucks with up to nine axles.

The use of the arbitrary overstress ratios in the original and proposed formulas has been widely criticized. These ratios do not seem to consider increased damage due to repeated load applications, the state of deterioration of existing bridges, or the likelihood of overloads and simultaneous truck presence. It is widely accepted that a more rational approach should be developed based on structural reliability theory (Moses and Ghosn 1987). The aim is to obtain the overstress ratios using statistical data on bridge safety. The steps involved in such an analysis should be based on determining acceptable safety levels using statistics on the safety margins of typical bridges including the likelihood of overloads, simultaneous truck presence, impact allowance, girder distribution and strength deterioration or loss of serviceability. New safety criteria should then be developed on the basis of limiting the number of posted bridges based on traffic and funds available for rehabilitation.

The objectives of this paper are to present a reliability-based procedure to determine the optimal allowable overstress on highway bridges, considering both static and dynamic effects, and to review the existing bridge formulas in order to determine the impact of increasing or decreasing the legal load limit based on safety and cost criteria.

The desirable safety levels will be calculated using reliability theory that has been extensively used in recent years. In fact, structural code writing groups have found it advantageous to consider formal probabilistic techniques in assessing the safety of existing provisions and in introducing new code checking formats. The usual application of reliability theory for bridges has been in determining the risks of structures under current loads or the calibration of design or evaluation codes. The problem of establishing acceptable overstress criteria and consequently obtaining a bridge formula is an inverse problem, in the sense that, we are to determine the allowable loads over a network of existing bridges with a varying degree of deterioration and serviceability.

Proposed heavier truck weight regulation may increase the number of structures that must be load restricted (posted) or else scheduled for strengthening or replacement. In this paper, bridge costs associated with new truck weight regulations are determined using the bridge inventory of the Federal Highway Administration statistics.

2. Load Modeling and Structural Reliability

In the last few years there has been an increasing awareness of the importance of correlating design and evaluation safety factors to the uncertainties inherent in estimating the strength of members, the behavior of structural systems and the loading of the bridges. One step in that direction was the adoption by AASHTO of Load Factor Design (LFD) which specifies different factors for different loads based on the In the last few years there has been an increasing awareness of the importance of correlating design and evaluation safety factors to the uncertainties inherent in estimating the strength of members, the behavior uncertainty associated with determining these loads. The Ontario Highway bridge Design Code (OHBD 1983) went one more step in that direction by calibrating the safety factors using formal reliability theory. In addition, AASHTO is presently considering the adoption of reliability-based formulations for rating evaluation of bridges and to update the AASHTO design specifications based on reliability and LRFD principles (Moses et al. Dec. 1987 & 1987, Kulicki 1990).

In structural design, safety may be described as the situation where capacity (strength, resistance, fatigue life, etc.) exceeds demand (load, moment, stress ranges, etc.). Probability of failure, i.e., probability that capacity is less than applied load may be formally calculated; however, its accuracy depends upon detailed data on the probability distributions of loads and resistances. Such data is often not available so approximate models have been used for calculation in such codes as AISC LRFD (Load and resistance factor design) or the Ontario Highway bridge Design Code.

The most direct reliability formulation is through Safety Indices. Let the reserve margin of safety be defined as, Z:

$$Z = R - S \tag{4}$$

where R is the resistance, and S is the total load. The safety index, β, is defined as:

$$\beta = \frac{Z}{\sigma_Z} \tag{5}$$

where: Z is the mean value of Z and σZ is the standard deviation of Z. β gives the number of standard deviations that the mean margin of safety falls on the safe side. b provides an exact risk (failure probability) if R and S follow normal distributions. Reliability, R_o, is defined as the probability that the structural member survives and if Z follows a normal distribution, R_o is given by:

$$R_0 = \Phi\,(-\,\beta\,) \tag{6}$$

where Φ is the normal distribution function. In general, β from either normal or lognormal models are used as estimates of the reliability of a structural member even if its capacity and applied load are neither normal nor lognormal. To improve on these estimates, "Level II" methods have been developed (Thoft-Christensen and Baker 1982). Level II methods involve an iterative calculation to obtain an estimate to the failure probability. This is accomplished by a linearization of the reserve margin function, Z, at the point on the failure surface closest to the origin, after the surface is mapped into a normal space. In this paper, the reliability calculations are executed using a level II reliability program developed at Imperial College of London.

The reliability calculations require a complete description of the statistics of the resistance and the loads. The load S in equation (4) can be divided into two parts: The dead load and the live load. R on the other hand is determined by looking at the statistics of the resistance of the existing bridge network. For the purpose of this study concerned with the determination of acceptable truck weight regulations over the existing bridge network, the unknown variable is the live load effect.

In this study, the proposed bridge formula limiting truck weights is developed for bridges designed according to AASHTO's Working Stress Design method (WSD). The effect of the adoption of the proposed formula on the existing bridge network with different rating levels as determined by FHWA bridge inventory statistics, is investigated in a second stage in terms of the safety impact and economic costs.

AASHTO's WSD method calculates the required safe minimum member capacity of a bridge utilizing a live load envelope represented by AASHTO's truck and lane loads. An approximate analysis procedure determines a girder distribution factor based on the type of bridge, and a dynamic factor is determined based on the span length. These design or nominal values for loads and load effects determine the required nominal resistance R_n such that:

$$R_n = \frac{1}{0.55}\,(D_n + L_n\,(1 + i_n)\,g_n) \tag{7}$$

where: D_n is the nominal dead load of the member, L_n is the live load as specified by AASHTO, in is AASHTO's dynamic impact factor and g_n is the girder distribution factor. 0.55 is the limiting stress level for steel girders. Based on several experimental studies, it was found that the mean resistance R of a bridge member is higher than R_n. R can be obtained from the nominal resistance using a bias factor such that:

$$\bar{R} = b_r R_n \tag{8}$$

where: b_r is the resistance bias and R_n is the nominal resistance. To obtain the mean resistance, a bias b_r of 1.1 is assumed. The coefficient of variation associated with the resistance is 12%. D_n is usually obtained from the self-weight of the structure accounting for the weight of asphalt and other non-structural members. Hansell and Viest (1987) found that the dead load is related to the design live load and the span length by the formula:

$$D_n = 0.0132\,(\,L_n\,(1 + i_n)\,)\ \text{span length}$$

The mean dead load value D was also obtained from equation (9) ie. the dead load bias is estimated at 1.0. The dead load coefficient of variation used in these calculations is 9%. bridges are designed to safely withstand the maximum live load expected over the service lifetime of the structure. In short to medium span bridges, maximum load is usually based on the occurrence of several heavy trucks simultaneously on the bridge. Each occurrence of one or more vehicles on the bridge (herein called loading event) is characterized by the number of trucks in the event, their gross weights, axle spacings, axle weight distribution and the relative position of these trucks with respect to each other. All these factors are random variables which should be accounted for in a model to calculate the maximum loading on a bridge. AASHTO proposes an HS-20 vehicle whose effect on bridges should envelope the effect of the actual truck loads. Several studies, however, showed that the HS-20 vehicle does not provide a consistent envelope for all bridge spans under current truck traffic conditions. For this reason, the nominal live load effect L_n has to be calculated using AASHTO's HS-20 truck but the mean live load and its coefficient of variation will be calculated using a formula developed by Ghosn and Moses (1985).

A simulation program was developed to obtain the maximum live load effect in the lifetime of a bridge structure. The program uses a standard simulation truck and executes a convolution over all possible gross weight categories and slot positions behind a leading truck. This simplified simulation program was used to study the results from a large number of sites. The results were subsequently used to calibrate a simple load prediction formula for evaluating the maximum expected load over long periods of time. (Moses and Ghosn 1985). The load model estimates the actual live load moment in the lifetime of a bridge structure due to the static passage of truck loads and is given by:

$$L = a\,m\,W_{.95}\,H \tag{10}$$

a is a factor that calculates the maximum moment effect of a typical standard truck assuming a one unit gross weight. a is deterministic and its value for different span lengths is given in Table 1. For spans of 60 ft. and less a is obtained from a typical single unit truck. For spans longer than 60 ft. a typical semi-trailer truck is used as the basis for calculating a. (Moses and Ghosn 1985).

m is random factor that accounts for the variability in the axle configuration of random trucks compared to the standard trucks used for the calculation of a. Table 1 shows the different mean values of m and their coefficients of variation.

$W_{.95}$ is the value corresponding to the upper 5% of the gross weight histogram. From a study for FHWA (Snyder et. al. 1985), gross weight histograms were collected at several sites throughout the U.S. The average $W_{.95}$ obtained for semi-trailer trucks was 75 kips and the C.O.V. is 10%. For single unit trucks, the average is 47 kips and the C.O.V. is 15%.

H is a factor that relates the moment of one truck with the characteristic gross weight $W_{.95}$ to the estimated maximum lifetime load on two-lane bridges. H was obtained from the results of the simulation program with different truck histograms and is dependent on the span length as shown in Table 1.

Equation (10) gives the maximum static load effect on a bridge. Therefore, a bridge impact factor i should be included as given in eq. (11). From measurements taken at several sites, estimates of the dynamic loads as a percentage of the static load are obtained (Moses and Ghosn 1983). In this analysis the mean dynamic to static load effect ratio is assumed to be 1.17 with a coefficient of variation (C.O.V.) of 8%.

Also, the distribution of load to individual members depends on the geometric configuration of the bridge. For slabs on steel-girder bridges, field measurements yielded a bias between the measured and the AASHTO girder distribution factors equal to 0.88. Eq. (11) then gives the maximum dynamic moment on the most heavily loaded girder with g as the girder distribution factor. g has a mean obtained using AASHTO's value with a bias of 0.88 and is associated with a C.O.V. of 8%.

For a fifty-year lifetime, one needs to account for possible growth in the weights and volume of heavy trucks traveling over the bridge network. In previous reliability studies a growth factor Gr with a mean value of 1.15 and a C.O.V. of 10% was assumed such that the total truck load effect on a bridge member is given by:

$$M = a\, m\, W_{.95}\, H\, i\, g\, Gr \tag{11}$$

The safety margin for the reliability analysis can then be expressed as:

$$Z = R - D - a\, m\, W_{.95}\, H\, i\, g\, Gr \tag{12}$$

This formula accounts for the current truck load effect by considering the statistics of current truck gross weights, multiple occurrence on the bridge and truck configuration. Also indirectly considered in the H factor is the truck traffic volume. It is also assumed that changes in truck volume over the service lifetime of the bridge might produce an increase in the live load effect of up to 15% (Gr =1.15).

Using this data base, the safety index calculations are executed for steel bridges of different simply-supported spans designed using WSD criteria with HS-20 loading. The safety index inherent under current situation is plotted in Figure 1. The safety index thus obtained varies from about 2.5 to 4.2. The results indicate that the shorter spans exhibit a higher risk to failure than the longer spans. This is due to the basic procedure in WSD which applies one safety factor for all loads ie. loads that can be accurately estimated such as dead loads and loads that can be less accurately estimated such as truck loads. Shorter spans have lower dead loads and therefore they produce lower safety indices.

Figure 1 shows that the lowest safety index value accounting for the growth in traffic is about 2.5. This value corresponds to the shortest simple span considered ie. 30 ft. In structural reliability, the target safety index utilized to develop a new code is often calibrated based on the performance of structures designed to satisfy the current code. Changes in the safety index inherent in current criteria and design procedures is justified when there is a consensus among engineers that current criteria do not provide adequate safety. In the bridge engineering community, there is a general satisfaction with the general performance of simple span highway bridges including short span bridges. Because current 30 ft. span bridges provide adequate safety under current loading conditions, and since this safety level is represented by a safety index beta equal to 2.5, therefore, it was decided herein to use a target index equal to 2.5 to calibrate the new bridge formula. The target safety index measures the level of safety against failure (yielding) of the most critical bridge member. Due to structural redundancy, failure of the most critical member will not necessarily lead to the collapse of the complete structure. Another justification for using the 2.5 value as the target reliability level is due to the fact that the 30 ft. span controls the loading on short wheeled vehicles. These vehicles currently operate on the nation's bridges without any observed or documented risk for bridge safety, one cannot then justify increasing the required safety level which will unfairly mandate a reduction in current truck loadings. A separate study by Moses et al. also showed that a value of 2.5 corresponds to the average safety index for operating stresses (ie. using a safety factor of 1/0.75 instead of the 1/0.55 of equation 7) which are regularly used by a number of states for rating existing bridges.

3. Calculation of Safe Load Envelope

The steps involved in applying reliability procedures to obtain a bridge formula may be summarized as follows: 1) Choose suitable safety criteria. The safety index widely used in structural reliability theory as a measure of structural safety is used in this study as the basis for the determination of the safety of bridge members. 2) Select an acceptable reliability level. For example, a safety index of 2.5 for redundant bridges seems to provide a reasonable safety target based on the performance of existing bridge members. This safety index target of 2.5 corresponds to a probability of about 0.6% that the safety and serviceability criteria will be satisfied in any member of the bridge. 3) Choose a range of typical bridges with different design criteria, span lengths, and configurations giving a representative sample of the nation's bridges. These bridges should include simple as well as continuous spans, both steel and concrete bridges should be considered. In this study, simple span steel bridges were used to obtain a bridge formula. The implications in terms of safety and cost of other types of bridges are studied separately as part of the cost analysis. 4) Use statistics on the safety margins of these typical bridges including the likelihood of overloads and simultaneous truck occurrences to obtain the live load envelope that will produce the target safety index. It will be assumed that the uncertainties (C.O.V.) of the live load random variables will remain the same as currently observed. The live load envelope as defined herein is the mean total bridge live load required to produce the target safety index for each span length. 5) Calibrate a bridge formula that will produce the load envelope obtained in step 4). 6) Verify that the bridge formula will lead to an acceptable number of bridges that will need upgrading to support the proposed additional load. 7) Review the implications of adopting the suggested bridge formula in terms of safety of typical steel and concrete bridges of simple and continuous spans. This should include strength requirements and fatigue and serviceability. 8) Study the costs required to maintain the bridge infrastructure under the proposed bridge formula.

Changes in the legal truck weight regulations will produce changes in the characteristics of the regular truck traffic over highway bridges. Specifically the weights and the axle configurations of the trucks will be affected. According to our basic truck load model (eq. (11)) it is $W_{.95}$ that will be affected the most. The effect of the changes in the truck weight regulations on H will be secondary. If trucks are allowed to carry more loads, the total number of truck trips required to carry the total loads in a jurisdiction will probably decrease. This will affect the number of simultaneous occurrences on a given bridge site which is represented by the variable H. H was found to be sensitive to only very large changes in truck volumes, for this reason, H was assumed to remain constant for the purposes of this study. Also, H represents the shape of the weight histogram above the $W_{.95}$ value, thus we are also assuming that the percentage of overloads will remain the same after the implementation of the new weight regulations.

In this study, we are assuming that changes in the legal weight limits will produce a shift in $W_{.95}$ such that the ratio of $W_{.95}$ to the legal limit remains constant. At the same time, m is assumed to remain constant. These are pure assumptions made herein since no data is available to account for future changes in these parameters.

The next step in this analysis consists of determining the live load envelope required to produce an acceptable safety level. This is done as follows: Assuming that current bridges are designed according to AASHTO's WSD method (equation 7) an optimization program is developed to determine the mean live load L such that the target reliability of 2.5 is exactly matched for each span length considered. Table 2 gives the live load envelope required to reach a target safety index b equal to 2.5 for steel bridges designed according to equation 7. The values shown correspond to the required total moment effect for two lane bridges.

Using the load envelope developed above, a formula can be obtained using the procedure outlined by the TTI study (James et al. 1986). The bridge formula was designed to give a relationship between the weight of a truck and its length. The steps involved are summarized as follows:

- A truck satisfying the bridge formula will be assumed to have a total weight W and total truck length B.
- Assume the truck weight W is uniformly distributed over the truck length B.

- Several values of B are used such that B varies between 1 and 120 ft.
- Given a truck length B, find the moment effect Mr of a unit load uniformly distributed over B. This is done for each span length.
- For every span length, find the load envelope M_t required to obtain the target safety index 2.5 (see Table 2).
- For each span length, find $W_{.95}$ t (target weight) such that $mM_r W_{.95}$ t H is equal to M_t. a is not included herein since it is implicitly considered in M_r.
- Repeat the previous steps for every truck length B. For each truck length, several $W_{.95}t$ values are obtained corresponding to the load effect of every span considered.
- For each truck length B find the lowest $W_{.95}t$ value: $W_{.95}t$ min.
- A bias factor gives the relationship between the legal load W_t and $W_{.95}t$. A bias of 1.07 is used for spans greater than 60 ft. based on current truck weight statistics for semi-trailer trucks. The bias factor for shorter spans governed by single unit trucks is 1.09
- Plot W_t min versus B. This curve provides an envelope that the distributed load Wt should satisfy in order to insure that all span lengths will produce a safety index of at least 2.5.
- Find an algebraic expression that will fit the W_t min versus B curve as close as possible. This will be the bridge formula to be utilized.

In this approach we assume that distributing truck weights uniformly over the truck length provides a safe envelope for all typical truck configurations and axle spacings. Figure 2 shows the Wt versus B curve obtained using the calculations performed in this analysis. The curve is compared to the curve given by TTI and the one proposed by the TRB Truck Weight Study (James et al., TRB)

The proposed formula, as discussed earlier, was developed such that a simply-supported steel bridge designed to satisfy AASHTO's WSD criteria for HS-20 loading will have a safety index beta = 2.5 when subjected to the loads expected over the next fifty years if the new bridge formula is implemented. The projections of the loads is based on the assumption that a lateral shift in the gross weight histograms accompanies any shift in the legal limit. With this assumption, the number of illegal and overloaded vehicles as a percentage of the total traffic remains unchanged from the present situation. Also, to account for future increases in truck traffic and the more frequent number of multiple truck occurrences caused by that, a traffic growth factor of 1.15 is included in the maximum load model. The proposed formula obtained by fitting the Wt min vs. B results is as follows:

$$W = (1.64\,L + 30)\,1000 \qquad \text{for } L < 50 \text{ ft.}$$
$$W = (0.80\,L + 72)\,1000 \qquad \text{for } L > 50 \text{ ft} \qquad (13)$$

4. Cost Analysis

A major portion of the total cost impacts for new regulations is the effect on the existing bridge population. Some 130,000 bridges are now rated structurally deficient with an estimated $53 billion replacement or upgrading cost. If a new truck weight regulation introduced higher legal loads, a larger number of structural deficiencies will increase replacement costs. The procedure outlined herein for the cost analysis, follows the method developed by Moses for TRB's Truck Weight Study (Moses 1989).

To provide a base case for the cost analysis, bridges under current truck regulations are checked by vehicles corresponding to the current Federal bridge Formula. This means that they should provide adequate capacity under the AASHTO legal vehicles. A bridge is considered deficient under this base case scenario if these AASHTO vehicles cause stresses that exceed the operating stress level plus 5% tolerance. The operating stress level is obtained based on eq. 7 with a safety factor 1/0.75 rather than 1/0.55 and a load L_n obtained from the AASHTO rating vehicles not the HS loading. These proposed criteria are similar to the rating methods used by many state agencies.

The cost allocation study considered a large sample of bridges that represent the highway classifications and regions in the U.S. bridges of different spans, geometries, material and age were analyzed. The sample was obtained from the Federal Highway National bridge Inventory System (NBI). The predictions using the Base Case model produced a total number of deficiencies close to the 130,000 estimated deficiencies which appears in the Secretary of Transportation annual report.

The cost analysis is divided into four parts: 1) Overstress costs for new bridges, 2) Overstress costs for existing bridges, 3) Fatigue costs of new bridges, 4) Fatigue costs of existing bridges. In this paper, only the overstress costs for existing bridges are presented, these costs represent more than 80% of the bridge impact costs (Moses 1989). In addition, the methods to be followed for the other portions of the cost analysis are outlined.

4.1 OVERSTRESS COSTS - NEW BRIDGES

It is herein assumed that as a result of changing the weight regulations, new designs will follow the same checking format of AASHTO as given in eq. 7. New bridges will be designed for live loads L_n obtained using an HS type vehicle with a gross weight that will provide an upper bound to the new legal loads. Typically, an increase in HS level causes only a small percentage increase in cost for each bridge type. For example, a 25% increase in design HS level increases the cost of a new bridge construction by about 5%. 10,000 new and replacement bridges per year are assumed. This provides an estimated \$3 billion per year for the construction of new bridges under the current legal limits. Using the \$3 billion per year estimate, the additional costs implied by different weight regulations can be obtained based on their respective required HS design levels.

4.2 OVERSTRESS COSTS - EXISTING BRIDGES

The current US bridge network consists of 600,000 bridges which would need upgrading if a new legislation allowing higher truck weights is implemented. In this analysis, it is assumed that all deficient bridges under any proposed legislation will have to be replaced. The possibility of posting or closing bridges has been ignored because these options will entail economic and productivity costs to the shipping industry exceeding the cost of replacing the affected structures. Upgrading options are seldom used in practice because of Federal rules requiring that upgraded bridges should satisfy all regulations on geometry, lane widths, side barriers... Thus the cost estimates given herein are upper bounds to the values that will actually be encountered.

As previously mentioned, additional deficiencies are defined with respect to operating stress levels plus 5% load tolerance using legal vehicles satisfying the bridge formula under consideration. Replacement cost of deficient bridges do not reflect the existing condition and age of the structures. Table 3 shows the estimates of current number of deficient bridges in comparison to the expected number under the new regulations given in eq. (13). This is presented for two different highway categories: primary systems and non-primary systems. The estimate is also given separately for steel, reinforced concrete and prestressed concrete bridges. The cost implications of such number of deficiencies is given in terms of total length of deficient bridges. Construction costs are usually given for unit area and since bridges have standard widths, the total length of deficient bridges is directly related to the total cost.

The results of Table 3 show that a large increase in the total length of deficient steel bridges on primary highway systems will accompany the implementation of equation (13). The increase in the length of deficiencies is about 5.7 times the current length. The additional length of deficiencies for the other types of bridges and highway classification is in general less than 2.3 the current levels.

One reason for this large number of deficiencies is that many existing bridges do not satisfy the WSD HS-20 criteria used to determine the target reliability level.

4.3 FATIGUE COSTS - NEW BRIDGES

The fatigue cost analysis is performed based on an effective truck weight that produces a damage equivalent to the fatigue damage produced by the actual traffic (Moses 1989). The damage costs as calculated using this method reflect the effect of a change in the gross weights of trucks, the effect of change in the axle spacings and axle weight distribution, and the number of truck passages required to transport a given amount of freight. Fatigue may not always govern the design and fatigue costs are charged only to bridges designed to new HS levels that will envelope the new regulations but will be unable to sustain the higher fatigue damage.

4.4 FATIGUE COSTS - EXISTING BRIDGES

A fatigue model was developed by Moses (1989) to extrapolate the present fatigue damage costs as they increase in the future due to the passage of time and the accumulation of more cycles of loading and changes in weight regulations. Damage ratios for existing bridge members for projected truck traffic scenarios are calculated. Costs are obtained by extrapolating present estimated annual fatigue damage costs and related expenses.

5. Conclusions

A procedure to determine a bridge formula that provides acceptable safety levels over a bridge network is presented. The procedure utilizes a reliability analysis and through an optimization program calculates the required legal truck load effect that produces a uniform safety index for existing bridges designed according to current AASHTO criteria. The proposed formula is based on the reliability analysis of simple span steel bridges satisfying AASHTO's WSD specifications. The effect on the national bridge network resulting from the implementation of the proposed formula is expressed in terms of the number of additional deficient bridges and in the cost of replacing these structures. This analysis accounts for the costs and safety of steel and concrete bridges of simple and continuous spans.

6. Acknowledgement

The work described herein was funded by the Federal Highway Administration research project "bridge Overstress Criteria". Partial support was also provided by Transportation Research Board's "Truck Weight Study".

7. References

1. Transportation Research Board, 1990, Truck Weight Limits: Issues and Options, Special Report 226, Washington, D.C.
2. Hansell, W.C. and Viest, I.M., October 1971, "Load Factor Design for Steel Highway bridges", Engineering Journal AISC.
3. James, R.W., Noel, J.S., Furr, L.H., and Bonilla, F.E., 1986, "Proposed New Truck Weight Limit Formula", Journal of Structural Engineering, Vol. 112, No. 7.
4. Kulicki, J.M., 1990, "Development of Comprehensive bridge Specification and Commentary", NCHRP 12-33.
5. Manual for Maintenance Inspection of bridges, 1982, AASHTO
6. Moses F. and Ghosn M., November 1987, "Discussion on Proposed New Truck Weight Limit Formula", Journal of Structural Engineering.
7. Moses, F. and Ghosn, M., January 1985, "A Comprehensive Study of bridge Loads and Reliability", FHWA/OH-85/005.
8. Moses, F. and Ghosn, M., August 1983, "Instrumentation for Weighing Trucks-In-Motion for Highway bridge Loads", FHWA/OH-83/001.
9. Moses, F., Ghosn, M., and Gobieski, J., September 1985, "Weigh-In-Motion Applied to bridge Evaluation", FHWA/OH-85/012.
10. Moses, F., Schilling, C.G. and Raju, K.S., November 1987, "Fatigue Evaluation Procedures for Steel bridges", NCHRP Report 299.

11. Moses, F. and Verma, D., December 1987, "Load Capacity Evaluation of Existing bridges", NCHRP Report 301.
12. F., September 1989, "Effects on bridges of Alternative Truck Configurations and Weights", NCHRP HR 2-16(b).
13. Nowak, A.S. and Grouni, H.N., 1982, "Safety Criteria in Calibration of OHBD Code", Proc. Intl. Conf. on Short and Medium Span bridges, Toronto, Canada.
14. Ontario Highway bridge Design Code, 1983, Ontario Ministry of Transportation and Communications.
15. Snyder, R.E., Likins, G.E. and Moses, F., 1985, "Loading Spectrum Experienced by bridge Structures in the United States", Report FHWA /RD-85/012.
16. Standard Specifications for Highway bridges, 1983, AASHTO
17. Thoft-Christensen, P. and Baker, M.J., 1982, Structural Reliability Theory and Its Applications, New York, Springer-Verlag.

Table 1 Input Data for Reliability Analysis.

Span (ft.)	a	m		H	
		mean	C.O.V.	mean	C.O.V.
30	6.07	0.92	15%	2.63	10%
40	8.57	0.93	12%	2.69	10%
60	13.57	0.94	6%	2.75	10%
80	13.40	0.93	9%	2.78	7%
100	18.40	0.95	7%	2.80	7%
125	24.40	0.96	6%	2.86	7%
150	30.90	0.96	5%	2.87	7%
175	36.90	0.97	4%	2.98	7%
200	43.40	0.97	4%	3.05	7%

Table 2 Required Load Envelope for HS20 Steel Bridges.

Span ft.	Maximum Moment kip-ft
30	717
40	1231
60	2438
80	3902
100	5434
125	7500
150	9986
175	13820
200	18380

Table 3 Consequences of Implementation of Proposed Bridge Formula.

	Total # of Bridges	Current Deficiencies		Expected Deficiencies	
Primary Highways		Number	Lengthx10^3 ft	Number	Length.x10^3 ft
Steel	41140	1738	119.3	7248	679.5
R/C	25510	1174	43.7	2346	124.4
P/C	26474	1264	95.6	2896	249.7
Non-primary Highways					
Steel	154401	80243	3759.1	18742	1234.5
R/C	81671	21510	652.8	11957	521.0
P/C	48568	5503	247.5	6079	378.9

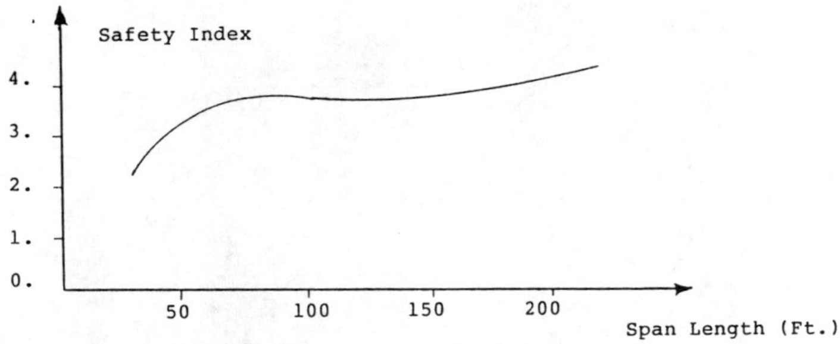

Fig. 1 Safety Indices Under Current Truck Traffic.

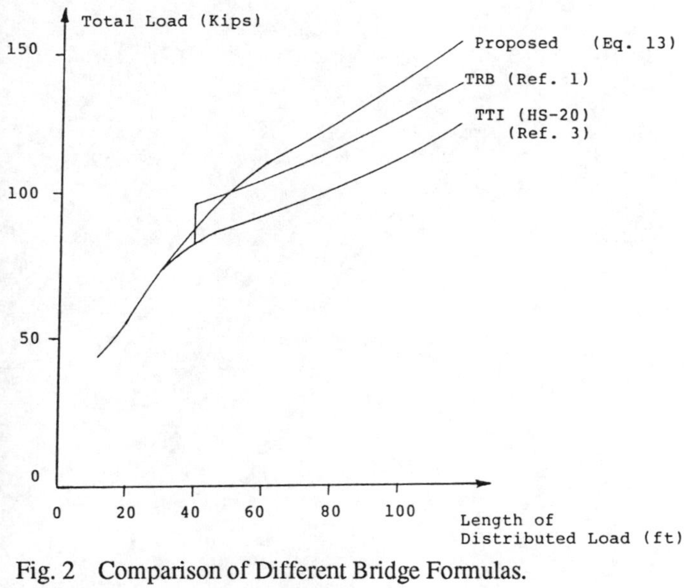

Fig. 2 Comparison of Different Bridge Formulas.

DYNAMIC BEHAVIOR OF BRIDGE UNDER FULL TRAFFIC

R. EYMARD, F. GUERRIER, B. JACOB
Laboratoire Central des Ponts et Chaussées
58 Boulevard Lefebvre
75015 Paris, FRANCE

ABSTRACT. In order to calibrate traffic load models on bridges, the dynamic effects produced by vehicles in vertical and horizontal motion on a bridge must be computed. Several physical phenomena must be taken into account, such as the dynamic behavior of the bridge, and that of the vehicle axles, which are vertically excited by the pavement roughness. Because many situations involving several trucks on a bridge must be handled, a simplified computer program has been developed; the case of simply-supported beams has been studied, under the dynamic load of real traffic. Numerical comparisons of the static and dynamic effects are presented, in different forms: the computation of local extremae gives the value of dynamic impact factors, and the decomposition of the time-history in cycles provides an evaluation of the fatigue damage.

1. Dynamic Impact Factor

The knowledge of the traffic load is essential for the design of bridges; for this reason, the Eurocode 9-12 [1] on bridge traffic loads must give the definition of extreme loading; it must include the dynamic effects which result from the vertical motion of the loads. This phenomenon is due to the excitation of the suspension of the vehicles by the pavement roughness, and by the vibration of the bridge itself.

Various models [2,3] have been developed for the dynamic study of bridges under traffic loads. Some of them contain a detailed finite element description of the structure, which leads to very unwieldy calculations: at each time step a bridge resolution must be made, coupled with the vertical motion of the vehicles; for this reason, only a short time period can be computed, involving a small number of vehicles.

For such methods, the dynamic impact factor (DIF) can only be defined as the ratio of the maximum static to the maximum dynamic loads, for identical loading cases; a calculation which is realisible for only one or two vehicles. So it is impossible to avoid a bias, when multiplying maximum static loads, obtained during the calculation of traffic composed of thousands of vehicles by a DIF computed with this method.

A. S. Nowak (ed.), Bridge Evaluation, Repair and Rehabilization, 293–306.

Our purpose was to build a model which can handle the dynamic behaviour of a bridge under a full real traffic, composed of many thousands of trucks with the speed, axle load and spacing information available with new Weigh-In-Motion systems [4].

2. Physical Modelisation

2.1 BRIDGE DEFLECTION EQUATION

In order to limit the computation time, we have only considered the case of simply supported beams (see Figure 10, from which several advantages can be drawn:

- the vibration modes are mathematically simple;
- the difference between the deformation under loading, and during vibration is small (see below).

The vibration of simply supported beams under moving loads has been previously studied by different methods. Saïd [5] used a modal approach, in which the number of computed modes can be selected. A more general beam configuration has also been solved by a modal method [6].

The partial differential equation of the deflection y(x,t) under traffic loads is

$$m_1 \frac{\partial^2 y}{\partial t^2} + CI \frac{\partial^5 y}{\partial t \partial x^4} + EI \frac{\partial^4 y}{\partial x^4} = -\sum_i p_{zi}(t)\delta_{x_i}(t) \qquad (1)$$

where the notation is:

x : abscissa of deflection,
t : time,
I : bridge inertia,
E : bridge modulus,
C : bridge damping factor,
m_1 : mass of bridge per unit length,
p_{zi} : action of axle i on the bridge,
δ_x : Dirac mass at abscissa x,
x_i : abscissa of the axle i.

Four integrations in space of δ_α, for any position α of the load (Figure 1), give the following expression for the influence line $L(\alpha,x)$ for the deflection at abscissa x:

- if $0 \le x \le \alpha \le d$,

$$L(\alpha,x) = \frac{1}{6d}\left[d^2(d-\alpha)x - x^3(d-\alpha) - x\,(d-\alpha)^3\right] \qquad (2)$$

-if $0 \le \alpha \le x \le d$,

$$L(\alpha,x) = L(d-\alpha,d-x)$$

- elsewhere, $L(\alpha,x) = 0$

where d is the beam length. The difference between $L(\alpha,x)$ and

$$\bar{L}(\alpha,x) = \frac{2d^3}{\pi^4} \sin\left(\frac{\pi}{d}\alpha\right)\sin\left(\frac{\pi}{d}x\right) \tag{3}$$

is always less that 7% of $2\,d^3/\pi^4$: this result is due to the fact that this function L is the orthogonal project of the function L on the set of sine functions of x.

This property justifies the simplification which consists of taking into account only the first eigen frequency:

$$y(x,t) = q(t)\ \sin\left(\frac{\pi}{d}x\right) \tag{4}$$

where q(t) is the middle span deflection. The equation of q(t) can be written:

$$m_1 \frac{d^4}{\pi^4} \frac{d^2q}{dt^2} + CI \frac{dq}{dt} + EI\ q = \Sigma_i P_{zi}(t) L(x_i(t),d/2) \quad (5)$$

2.2 VEHICLES VIBRATION EQUATIONS

The actual Weigh-In-Motion systems do not provide information about several topics, such as the dynamic characteristics of vehicles, or their rotary inertia. For this reason, but also for the efficiency of the algorithm, the axles of the trucks have been modeled as independent vibrating systems; each of which is composed of a suspended mass m_h connected by a spring, friction element and a damper, to a non-suspended mass m_z ;(brakes, wheels ...). The tire couples this non-suspended mass to the road: it is assumed to have a visco-elastic behaviour, also represented by a spring and damper, as shown in Figure 2.

The modelisation leads to the following equation for the action of the tyre on the road:

$$P_{zi}(t) = a_{zi} \frac{dz_i}{dt} + r_{zi} z_i \tag{6}$$

where the notation is:

a_{zi} : tyre damping coefficient,
r_{zi} : tyre elastic modulus,
z_i : tyre elastic vertical deformation.

The pavement roughness is described by the function u(x). The non-suspended mass dynamic motion can then be written as:

$$m_{zi} \frac{d^2}{dt^2} (z_i + y(x_i(t),t) - u(x_i(t))) = m_{zi}g - P_{zi}(t) + P_{hi}(t) \tag{7}$$

where the notation is:

m_{zi} : nonsuspended mass,

P_{hi}: action of suspended mass on axle i.

The action of the suspended mass is described by three terms (linear damping, elastic and friction):

$$P_{hi}(t) = a_{hi}\frac{dh_i}{dt} + r_{hi}h_i + F_{hi} \tag{8}$$

where F_{hi} is the friction force, which is related to $\frac{dh_i}{dt}$:

if $dh_i/dt < 0$ then $F_{hi} = F_t < 0$ which is the value of friction in the traction mode;

if $dh_i/dt > 0$ then $F_{hi} = F_c > 0$ which is the value of friction in the compression mode;

if $dh_i/dt = 0$ then the value of F_{hi} must be included in the interval $[F_t, F_c']$ (blocked mode).

Figure 3 represents the admissible values for the couple $(F_{hi}, dh_i/d_t)$.

The equation of the suspended mass motion is assumed to be:

$$m_{hi} \frac{d^2}{dt_2} (h_i + z_i + y(x_i(t), t) - u(x_i(t))) = m_{hi}g - P_{hi}(t) \tag{9}$$

3. Numerical methods

3.1 PAVEMENT ROUGHNESS

Axles of trucks are taken into account when their positions are in the interval $[-x_o, d+x_o]$ (the "influence area"). The length x_o is chosen sufficiently long so that the calculated vertical motion of the vehicles does not depend on x_o. The pavement roughness can either be measured, or generated by a spectral description (classification of ISO-TC108).

The function u(x) is then descretized in space with a 10 cm step. This step size is justified, because it is smaller than the tire print on the road. An example of generated pavement roughness is given in Figure 4.

3.2 TIME DISCRETIZATION

Time is discretized in constant steps Δt: let $t^n = n\,\Delta t$. At each time step n, only the N^n axles which are in the influence area are computed. For each axle i, the arrival time t_i is the time at which axle i is at position 0.

For each axle i, let $x_i^n = V_i(t^n - t_i)$, and

* $y(x_i^n, t^n) = q^n \sin(\pi/d\, x_i^n)$ if $d \geq x_i^n \geq 0$

* $y(x_i^n, t^n) = 0$ if $-x_o \leq x_i^n \leq 0$ or $d + x_o \geq x_i^n \geq d$

Equation (5) is discretized as follows:

$$\frac{m_1 d^4}{\pi^4 \Delta t^2}(q^{n+1} - 2q^n + q^{n-1}) + \frac{CI}{2\Delta t}(q^{n+1} - q^{n-1}) + \frac{EI}{2}(q^{n+1} + q^{n-1})$$

$$= \sum_{i=1}^{N^n} p_{zi}^{\ n}\, L\left(x_i^n, \frac{d}{2}\right) \tag{10}$$

where equation (6) has been transformed to:

$$P_{zi}^n = \frac{qz_i}{2\Delta t}(z_i^{\,n+1} - z_i^{\,n-1}) + \frac{r_{zi}}{2}(z_i^{\,n+1} + z_i^{\,n-1})$$

Equation (8) becomes:

$$P_{hi}^n = \frac{a_{hi}}{2\Delta t}(h_i^{n+1} - h_i^{n-1}) + \frac{r_{hi}}{2}(h_i^{n+1} + h_i^{n-1}) + F_{hi}^n$$

Let

$$Z_i^n = z_i^n + y(x_i^n, t^n) - u(x_i^n)$$

Equation (9) becomes:

$$Z_i^{n+1} - 2Z_i^n\ Z_i^{n-1} = g + \frac{\Delta t^2}{m_{zi}}(P_{hi}^n - P_{zi}^n) \tag{11}$$

Subtracting equation (7) from (9) leads to:

$$h_i^{n+1} - 2h_i^n + h_i^{n-1} = \frac{\Delta t^2}{m_{zi}} P_{zi}^n - \left(\frac{\Delta t^2}{m_{zi}} + \frac{\Delta t^2}{m_{hi}}\right) P_{hi}^n \tag{12}$$

For each time step n, the implicit equations (10), (11) and (12) are simultaneously solved for unknowns q^{n+1}, z_i^{n+1}, h_i^{n+1} or $F_{hi}^{\ n}$ for each axle i in the influence area at time t^n. This requires solving a linear system whose size $(2N^n + 1)$ depends on the number of N^n of axles in the influence area. The total number of floating point operations for the solution of the linear system is $14\ N^n + 1$, which are all vector type operations.

3.3 FRICTION

Each axle i can be either in the blocked, compression or traction mode. In the blocked mode, $h_i^{n+1} = h_i^n$. In this case, the expression for the force F_i^n can be derived from equation (12). When the force becomes greater than F_c', the axle i moves into the compression mode; if this force becomes less than F_t', the axle i moves into the traction mode.

In the compression mode, when $h_i^{n+1} < h_i^n$, or in the traction mode, when $h_i^{n+1} > h_i^n$, the axle moves into the blocked mode.

3.4 STATIC CALCULATIONS

At each time step, a "static" value $q_s(t)$ of the midspan deflection is computed from the continuous equation:

$$EI\ q_s(t) \ = \ \sum_i (m_{hi} + m_{zi}) \ gL \quad x_i(t), \quad \frac{2}{2} \tag{13}$$

4. Results

4.1 DYNAMIC IMPACT FACTOR

4.1.1 *Evaluation of the Influence of the Simplifications.* We compared our model to a previous one [5], which can take into account several vibration modes. The Table 1 provides values of dynamic impact factors obtained with three modes, and with one mode of vibration, for various span lengths, and a 2-axle truck. The difference is not significant for short span lengths (less than 35m).

4.1.2 *Two-axles Truck Simulation.* The dynamic effects of a 2-axle truck have been computed. 100m of 300m of the longitudinal profile that was used in the calculation is shown in Figure 4.

The characteristics of the truck are listed in Table 2.

The static and dynamic deflections have been computed for a 30 m long bridge, with the following numerical values:

$$EI = 34.32x10^9\ N.m^2,\ CI = 3.67x10^7\ N.m_2.s,\ m_1 = 18200\ kg/m.$$

The results are shown in Figure 5.

The dynamic impact factors with the same truck, the same pavement and different bridges have been computed. These are shown in Figure 6 as a function of the bridge is natural frequencies. High values (about 2) are obtained; notice that they have been obtained for light trucks.

4.1.3 *Results for a Real Traffic.* Traffic records of the motorway A6 at Auxerre were used (3660 trucks): they represent one day of two lane traffic. The impact factors (in the cases of the same bridges and pavements as in paragraphs 4.1.2 and 4.1.3) are plotted as functions of the span length in Figure 8: they vary from 1 to 1.3. Such small values are obtained because the configurations of two lane traffic producing extreme deflections involve at least two heavy trucks; so the vibration of the bridge with one truck is partially decreased, the second truck accentuates the damping.

4.2 FATIGUE CALCULATIONS

4.2.1 *Rain-Flow Counting.* The local extreme values of q(t) and $q_s(t)$ are used in order to count the stress cycles, using the rain-flow method [7,8]. Figure 9 gives a description of the algorithm.

4.2.2 *Fatigue Damage Computation.* In the case of metallic or composite bridges, some welded details are exposed to fatigue damage under stress variations. These stresses can have global influence lines which are linked to the bending moment. The case of the vertical stiffener to the main girder has been computed, with the pavement roughness of Figure 4, and the bridge described in paragraph 4.1.2. The rain-flow stress histograms for one day of traffic (3660 trucks), for static and dynamic calculations, are shown in Figure 10.

A fatigue calculation can then be made, using Miner's approach for the evaluation of the total damage: for this method, Woehler curves are used [9]. The class of detail which has been used is class 56, which corresponds to class 71 with reduction factor $(25/\gamma)1/4$, where τ = 70 mm is the thickness of the metallic detail. The resulting damage caused by the stresses of each class is shown in Figure 11.

The lifetime is the inverse of the integral of each damage curve. The results are 34 years and 29 years for the static and the dynamic case respectively.

Note that the large number of small stress dynamic cycles does not contribute to the fatigue calculation. However, this example shows that the difference in the number of larger stress cycles leads to a significant reduction (15%) in lifetime. This suggests that dynamic fatigue effects must be taken into consideration in future codes.

5. Conclusions

The single mode dynamic model presented in this paper:

- calculates the dynamic behavior of bridges under real traffic;
- gives frequency histograms of the deflections, moments and stress cycles of the bridge;
- calculates lifetimes of various bridge elements under specific assumptions.

The model is much less cumbersome than models that assume several modes. However, for the cases considered here, the dynamic impact factors do not differ appreciably for a wide range of parameters. It would seem useful to extend the comparison of the models to other configurations in order to determine the limit of usefulness of the single mode case.

A full traffic calculation has been performed for both dynamic and static conditions, from which lifetimes have been deduced. Dynamic effects in this case reduce the lifetime by 15%. Dynamic impact factors, however, are found to be small.

6. References

[1] Eurocode 9-12, Definition of dynamic impact factors, Project Report of subgroup 5.

[2] Drosner, S. (1987) Simulation program for dynamic effects, Lehrstuhl für Stahlbau, RWTH, Aachen, Germany.

[3] Martinez, Cabrera, Nicotera, Petrangeli, and Sanpaolesi, Dynamic local effects for R.C. slabs, Istituto di Scienza Delle Costruzioni, Pisa, Italy.

[4] Jacob, B. (1986) A high performant WIM System by piezo-electric cables and its applications, Int. Symposium on Heavy Vehicles Weights and Dimensions, Kelowna, BC, Canada, June.

[5] Saïd, M., (1984) Dynamique des ouvrages d'art sous charges mobiles, Thèse de Docteur-Ingénieur en Génie Civil, February 9th, Ecole Nationale des Ponts et Chaussées, Paris, France.

[6] Akin, J.E. and Mofid, M. (1989) Numerical Solution for response of beams with moving mass, Journal of Structural Engineering, vol. 115, No. 1, January, pp. 120-131.

[7] Downing, S.D. and Socie, D.F. (1982) Simple rainflow counting algorithms, Int. J. Fatigue, January, pp. 31-40.

[8] Brozetti, J. and Chabrolin, B. (1986) Méthodes de comptage des charges de fatigue, Construction Métallique No. 1, FAT-CAL 2-86.

[9] Conventon Européenne de la Construction Métallique, Recommandations pour la vérificiaton à la fatigue des structures en acier, CECM, avenue des Ombrages 32/36, bte 20, B. 12000 Bruxelles, Belgique.

Table 1

bridge length (m)	20	30	40	50	60	70	80	90	100
1 mode (present model)	1.21	1.15	1.08	1.05	1.04	1.02	1.02	1.03	1.03
3 modes (Saïd model)	1.21	1.15	1.12	1.08	1.04	1.06	1.05	1.03	1.04

Table 2.

The characteristics of the truck are :

axle	m_{zi} (kg)	r_{zi} (kN/m)	a_{zi} (kN.s/m)	m_{hi} (kg)	r_{hi} (kN/m)	a_{hi} (kN.s/m)	F_{hi} (N)
1	800	2000	1.6	3700	430	11.97	0
2	1200	4000	2.77	10300	1470	35.92	0

Table 3.

axle	m_{zi} (kg)	r_{zi} (kN/m)	a_{zi} (kN.s/m)	m_{hi} (kg)	r_{hi} (kN/m)	a_{hi} (kN.s/m)	F_{hi} (N)
1	800	2000	4.3	5000	400	17	0
2	800	4000	7	6800	800	30	0
3	800	4000	7	6800	1000	33	0
4	800	4000	7	6800	1000	33	0
5	1200	4000	7	6800	1000	33	0

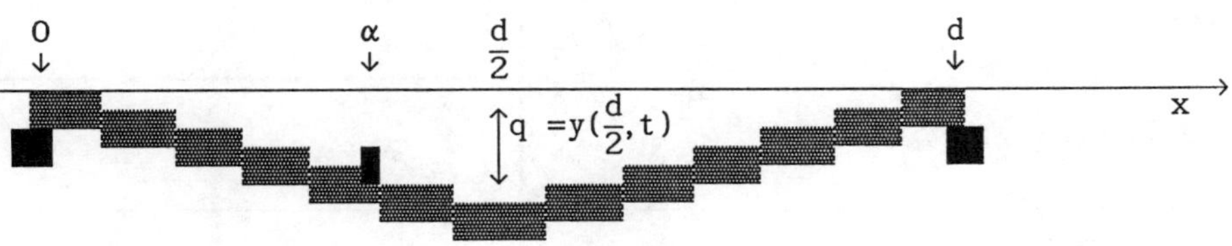

Fig. 1 Deflection as a Function of Load Location.

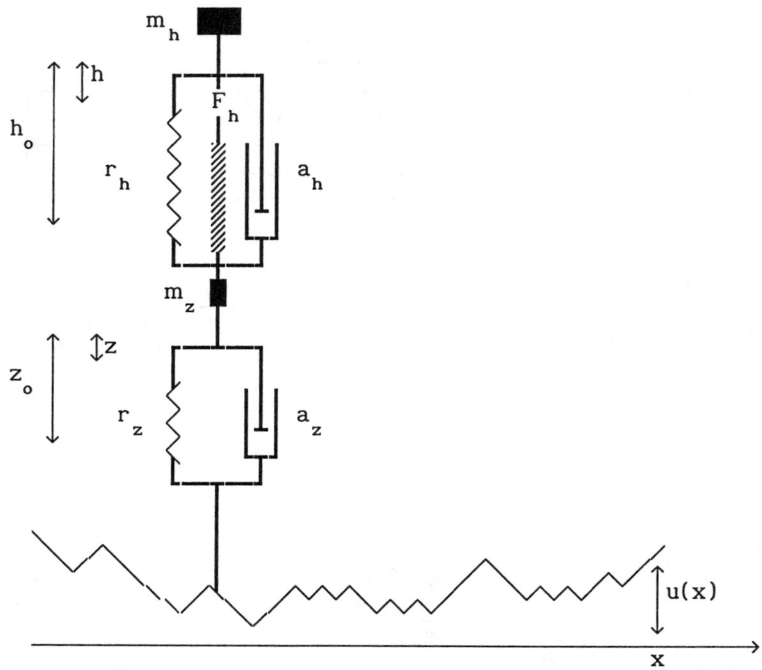

Fig. 2 Axle Description.

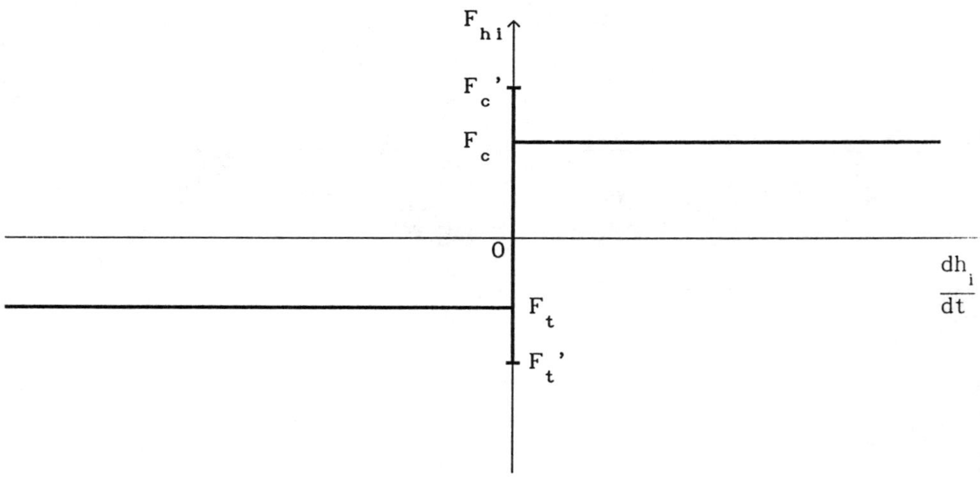

Fig. 3 Admissible Values of $(F_{hi}, \dfrac{dh_i}{dt})$.

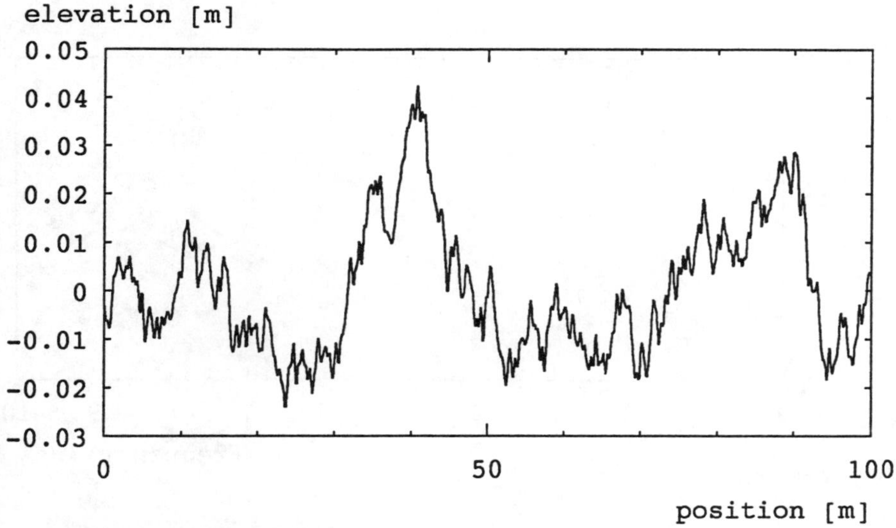

Fig. 4 Pavement Roughness Versus Position on Pavement.

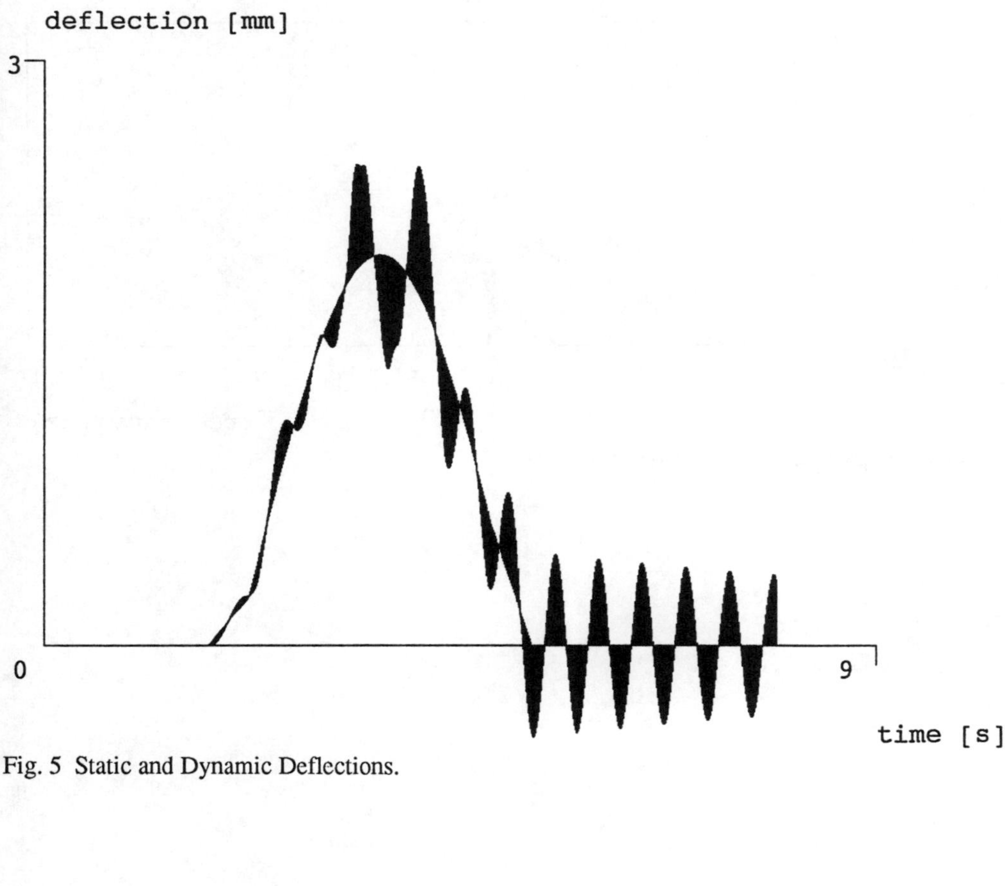

Fig. 5 Static and Dynamic Deflections.

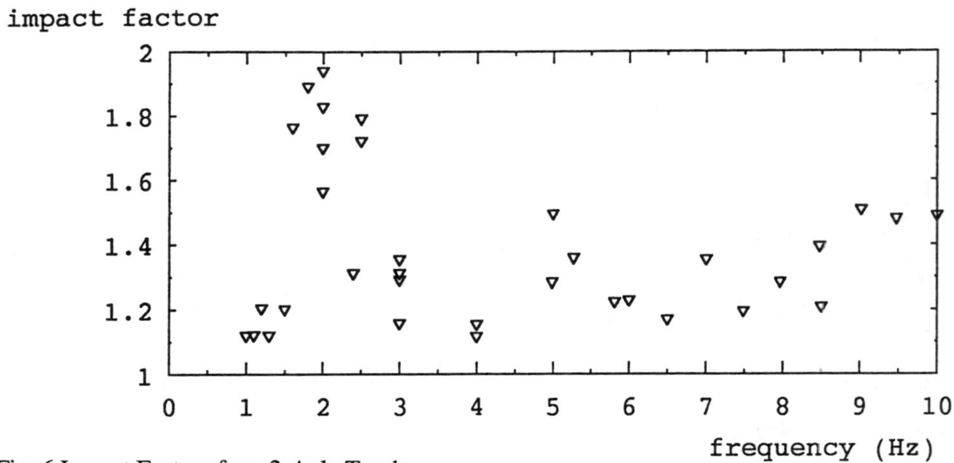

Fig. 6 Impact Factors for a 2-Axle Truck.

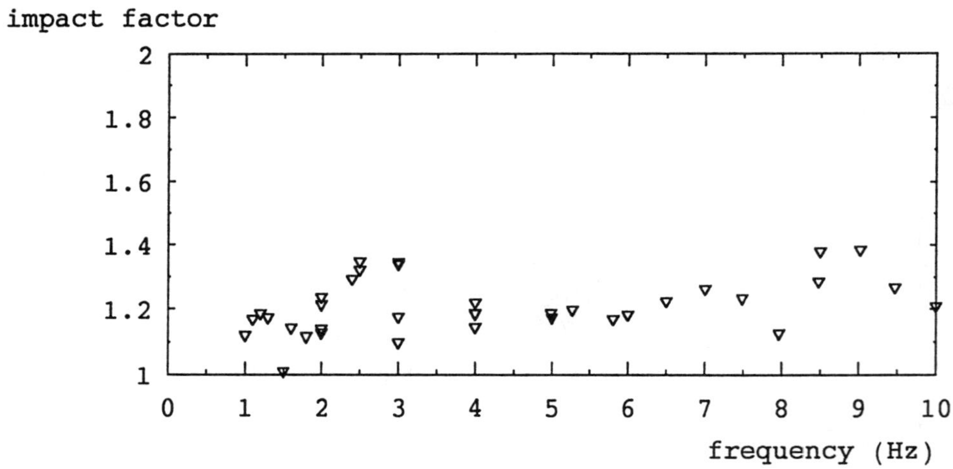

Fig. 7 Impact Factors for a 5-Axle Truck.

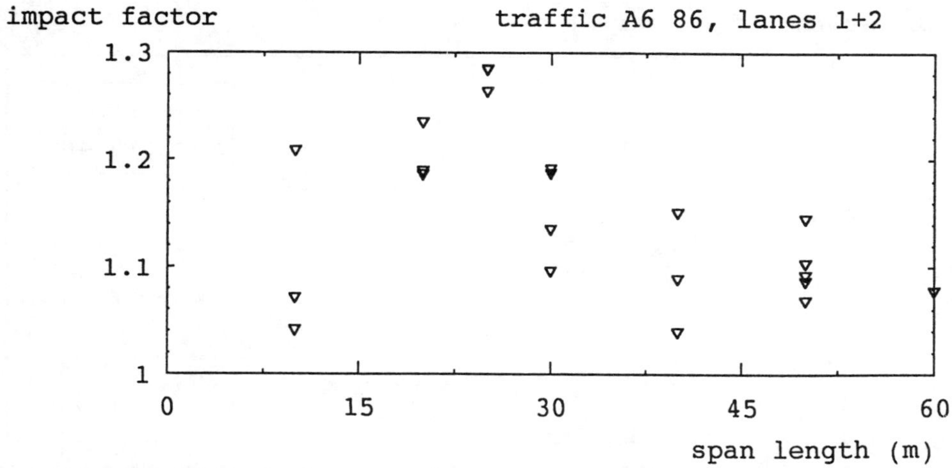

Fig. 8 IMPACT FACTOR for One Day of Two-Lane Traffic.

```
   w : beginning of cycles pile
   W : end of cycles pile
   n : size of the piles w and W (initially 0)
   ← : assignment operator

For each couple of local extrema ( v < V ) do
|    While n > 0 and w(n) > v do
|    |    Put cycle W(n)-w(n) in histogram
|    |    n ← n - 1
|    End_while
|    n ← n + 1
|    w(n) ← v
|    While n > 1 and V > W(n-1) do
|    |    Put cycle W(n-1)-w(n) in histogram
|    |    n ← n - 1
|    End_while
|    W(n) ← V
End_for
```

Fig. 9 Algorithm for Rain-Flow Cycles Counting.

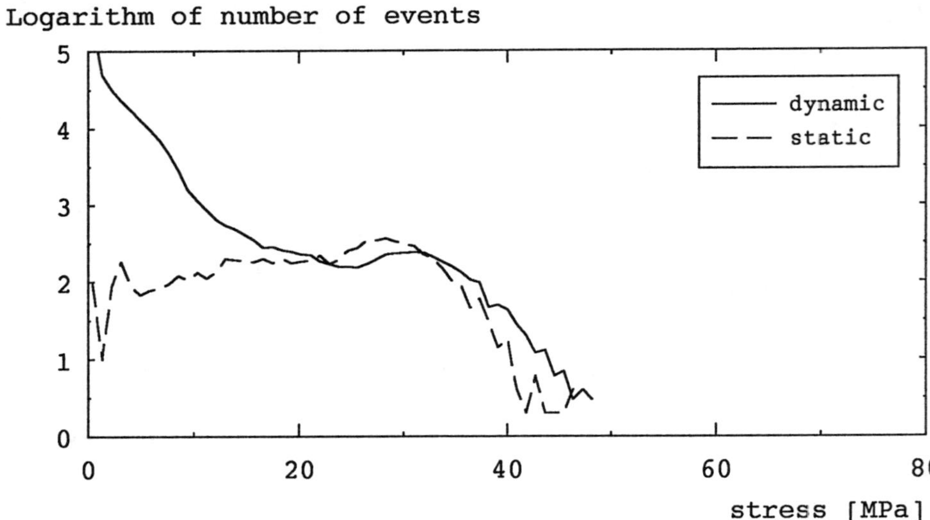

Fig. 10 Rain-Flow Stress Histograms.

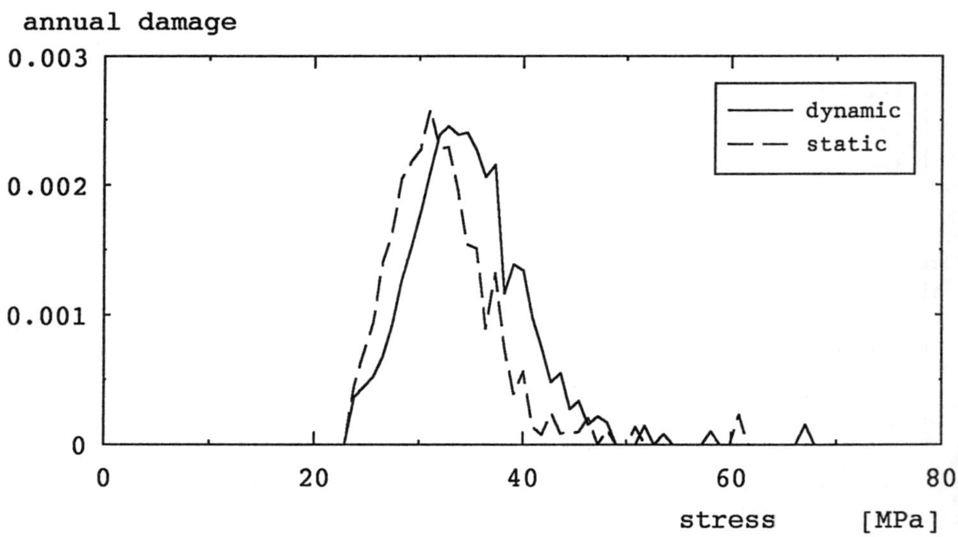

Fig. 11 Annual Damage for Each Stress Class.

FINITE ELEMENT MODELING AND BRIDGE RELIABILITY EVALUATION

JANICE J. TRAUTNER
University of Utah
Dept. of Civil Engineering
3220 Merrill Engineering Bldg.
Salt Lake City, UT 84112 USA

DAN M. FRANGOPOL
University of Colorado
Dept. of Civil, Environmental and Architectural Engineering
Boulder, CO 80309 USA

ABSTRACT. In this study an investigation is conducted on the methodology used to analytically evaluate the safety of existing bridges. This methodology is presented and applied to one of the oldest types of bridges, truss bridges. An interpretation of the AASHTO loading for the bridge is presented. Four finite element models are identified and described: two two-dimensional models, one composed entirely of truss elements and one containing beam and truss elements; and two three-dimensional models, one composed of beam and truss elements only and one including beam, truss, and shell elements. Utilizing both deterministic and probabilistic techniques, the results obtained from the models are evaluated and compared, indicating that simplistic two-dimensional models which neglect the effects of longitudinally and transversely eccentric loading and possible internal horizontal stresses induced by stringer/deck interaction are not conservative.

1. Introduction

Most bridge evaluation studies have been directed towards single bridge component checks. This traditional evaluation methodology thereby neglects bridge system effects and often leads to inappropriate decisions. The ability is at hand to measure these system effects given the numerous finite element modeling options available to the structural engineer. However, many bridge evaluators employ the simplest models and assume that these are conservative. There are two problems with this approach: 1) gross conservatism may unjustifiably close or post a bridge, and 2) the assumption in actuality may be unsafe.

2. Objective and Scope

It is the objective of this research to determine the effects of the variation in finite element modeling on the reliability evaluation of steel through-truss bridges. The results are used to formulate conclusions and recommendations in regard to the safety evaluation of these and other types of existing bridges. This is accomplished by: 1) studying loading criteria, 2)

A. S. Nowak (ed.), Bridge Evaluation, Repair and Rehabilization, 307–321.

presenting system safety evaluation methodology in a deterministic and probabilistic framework, 3) investigating finite element modeling possibilities, 4) investigating modeling results of a representative through-truss bridge, and 5) formulating recommendations with regard to reliability evaluation of bridges.

2.1 BRIDGE LOADING AND LOAD CASES

The actions to be considered in this study include dead and AASHTO live and impact loads. The live loads to be considered are both HS20-44 passing trucks and lane loadings [8,12].

In order to determine the positions of the trucks or lane loads for the maximization of stresses in each structural component, the influence lines for each of the elements must be generated. The live loads must then be superimposed on these influence lines to determine their governing position for each element of the structure. The influence lines are also utilized to compute the loaded portion of the span to produce the maximum stress in each element. The impact factor is a function of this length.

2.2 MODELING

This study investigates and compares the results of the following truss bridge models under all necessary loading conditions:

1. Two-dimensional, all truss elements (steel);
2. Two-dimensional, beam and truss elements (steel);
3. Three-dimensional, mixed beam and truss elements (steel); and
4. Three-dimensional, mixed beam and truss elements, including the floor system (steel, reinforced concrete).

These models will henceforth be referred to as Model 1, Model 2, Model 3, and Model 4, respectively. They are ordered with respect to their degree of complexity. Model 1 is commonly used to currently evaluate truss elements, Model 2 measures the effects of representing the chords as continuous beam elements, three-dimensional effects are investigated with Model 3, and Model 4 measures the influence of a horizontal diaphragm and composite action of the deck on the system.

2.3 INTERPRETATION OF RESULTS

The total normal stress output obtained from computer finite element analysis is used to perform both deterministic and probabilistic system safety evaluation of the structure. For those elements where this stress varies across the section (beam elements), the maximum normal stress is used.

2.3.1 *Interpretation of Deterministic Results.* Presentation of deterministic results is made in the form of plots of the load increment factor l versus the individual element stress ratios. The stress ratio is defined here as the ratio of the maximum total normal stress in an element to its allowable stress. The allowable stress in primarily bending and/or tension members is assumed to be equal to the yield stress, and the allowable stress in primarily compression members is assumed to be equal to the buckling strength of the member. The load increment factor l is the multiplier of the AASHTO truck or lane load plus the impact load applied to the structure. Thus, an increment factor of zero corresponds to a stress ratio due to dead load alone. From such plots the critical elements of the structure are identified.

2.3.2 *Interpretation of Probabilistic Results.* Two methods of system reliability evaluation are employed herein to determine the probability of failure of a structural system:simple bounds [2,3,13] and a more direct approach which shall now be discussed.

Studying the theory behind parallel and series systems used in the simple bounds approach, it is discovered that expedient application of the equations can only be conducted on systems composed of ductile truss elements. For beam-columns the relationship between loads and load effects is not linear and accurate computation of the reliability index b for these elements becomes quite complex. Moreover, at various stages of loading, both in the elastic and plastic range, the element reliability evaluation must consider distinctions between stresses caused by moments and those caused by axial forces. Hence, when evaluating the results obtained via Models 2, 3, and 4, which utilize beam elements, a more simplistic approach used in bridge reliability will be employed. For these models, the system reliability index is determined from the total load applied and the capacity (collapse load) of the entire system. The variables in the system reliability index computation are therefore the mean and variance of the applied load and the resistance (failure load).

The method used to compute system reliability by simple bounds will henceforth be referred to as "Method I" and the approach just described as "Method II". Method I requires several phases to determine the system reliability index: 1) compute element reliability indices, 2) compute element probabilities of failure, 3) compute system probability of failure bounds, and 4) compute system reliability index bounds. Method II requires only one step: computation of the system reliability index base d on the applied load and the failure load. The reader may wish to refer to Trautner [15] for a more comprehensive description of Method II.

For comparison, both methods will be used to compute the system reliability of the bridge based on Model 1's analyses. Only Method II will be used for reliability evaluation of the remaining models. Although Method II is somewhat crude, it may serve as a basis for comparison of the system reliability obtained utilizing finite element models.

Mean values are a product of a bias factor and the nominal value. The bias factor is the ratio of the mean to nominal values. For dead load, Nowak proposes a bias factor between 1.03 and 1.05 and a coefficient of variation between .04 and .08 [11] . The values used herein are a bias factor of 1.03 and a coefficient of variation of .08. For live load, Nowak suggests a bias factor of 1.34 and a coefficient of variation of .21. These values seem conservative. Values of 1.2 and .2 are therefore used for the bias factor and the coefficient of variation, respectively. The impact factor is assumed to have the same bias factor as the live load but a higher coefficient of variation of .30. The coefficient of variation of the structure's resistance is assumed to be equal to .12.

The above measures of variability are used to compute b based on total loads (Method II). However, the computation of the element reliability indices by Method I employs the element stresses and resistances. The variability (bias and coefficient of variation) used for the stresses are the same as that used for the loads with one exception. To account for the inexactness of the finite element method in stress computations, the variance of the load effect is increased by including an additional coefficient of variation of .05.

The mean tensile strength is that mean strength obtained from mill test data and the mean buckling strength is calculated using the mean yield strength. The coefficients of variation for tensile and compressive resistances are assumed to be 5% and 1 0%, respectively.

2.4 REPRESENTATIVE BRIDGE DESCRIPTION

2.4.1 *Physical Description of Representative Bridge.* From an existing bridge data base provided by the Colorado Department of Highways the bridge chosen for representative analysis was Colorado State bridge H-18-A, located on Colorado State Highway 24. This bridge was designed in 1933 and constructed in 1935. The 150 foot long superstructure is composed of two 8-panel Pratt trusses, spaced at 32' - 10". They are connected by overhead sway bracing above and by the floor beams below. The floor system consists of a reinforced concrete deck, stringers and floor beams. The floor beams are eccentrically pinned into the posts at 25" above the line of action of the lower chord. All truss connections are composed of riveted gusset plates. The mean yield strength of the structural steel calculated from mill test data documents of Colorado State Bridge H-18-A was 37.4 ksi. For a more detailed description of the bridge refer to Trautner [15].

2.4.2 *Loading Applied to Representative Bridge.* The overall dead load of the representative structure was calculated as 750.4 kips. Eleven live load cases resulted as per section 2.1. These load cases are presented in Figures 1 and 2. Impact factors for each load case are indicated on the figures.

The wheel loads are treated as point loads and distributed to the beam elements assuming a simple span between deck support locations (for 3D models) and then to the lower chord nodal points for 2D analysis. The lane loads are similarly distributed.

3. Results

3.1 TWO-DIMENSIONAL, ALL TRUSS ELEMENTS MODEL (MODEL 1)

Model 1, shown in Figure 3, is composed entirely of 2-noded truss elements. The boundary conditions at the extremities of the lower chord simulate a pin and a roller. The finite element model consists of 25 nodes, 50 elements and 75 degrees of free dom. Investigated are loading cases 1 through 10 assuming both brittle and ductile material behavior.

The load paths for the two types of material behaviors are plotted in three separate element groups: chords, diagonals and posts. These results may be found in the appendix of Trautner [15]. Figure 4 shows two representative load paths. Some general characteristics may be noted from the load path curves. For some elements the load paths are of constant slope with the exception of members whose dead load and live load stresses are of opposite sign. These members exhibit a change in slope as the stress ratio moves from the tension to compression range (or vice versa). Other elements exhibit shifts in slope (ductile) or kinks (brittle) in their load path curves when another element fails. A summary of the critical items obtained from the load path and deflection curves for Model 1 are presented in Table 1. Probabilistic results using Methods I and II are presented in Table 2. The simple bounds of Method I are calculated by dividing the system into two sub-systems: one parallel system consisting of the diagonals of the central two panels and the other system consisting of the remaining elements in series. These two sub-systems are then combined in series to compute the total system's probability of failure bounds.

Several interesting items are discovered from Table 2. The more accepted method of system reliability computation, simple bounds (Method I), indicates that the degree of correlation between system components is not a critical factor. It is also apparent that Method II's results are more conservative. What is most impressive, however, is that when using Method II the range of b over all load cases is relatively wide. Method I's b range is from 4.73 to 5.13 while Method II's is from 2.19 to 4.50. Note that the live + impact increment factor at failure 1 ranges from 3.12 to 5.68, also a relatively wide range.

3.2 TWO-DIMENSIONAL, MIXED BEAM AND TRUSS ELEMENTS MODEL (MODEL 2)

The second finite element, two-dimensional representation of the bridge investigated uses the same element numbering scheme as Model 1. Hence, Figure 3 may also be used as reference for this model. The top and bottom chords were modeled as continuous Euler-Bernoulli beams. The posts were also modeled as Euler-Bernoulli beams and the bending degrees of freedom were released at the end nodes, representing a pin connection into the chords. The model consists of 67 nodes, 50 elements and 120 degrees of freedom. Investigated are load cases 1 through 10. Significant results are summarized in Table 3.

3.3 THREE-DIMENSIONAL, MIXED BEAM AND TRUSS ELEMENTS MODEL (MODEL 3)

The finite element mesh of Model 3 is presented in Figure 5. Element and node numbers are omitted to avoid congestion of this Figure. For presentation of the results the element numbering scheme is similar to that used for Models 1 and 2. To include the response of both trusses, the least loaded truss shall be referred to as the "East Truss" and elements are labeled with an additional "E" (i.e., 1E, 2E, etc.). Floor beams are identified in reference to their associated lower chord node (i.e., B0 through B8). The load distribution differs in that the dead loads are distributed symmetrically throughout the structure, as opposed to the lower chord nodal points exclusively as is the case with Models 1 and 2. Also, the live and impact loads are placed on the floor beam nodes. The floor beams intersect the trusses 25.5" above the lower chord nodal points. Therefore beam reactions occur above the placement of loads on the two-dimensional models. The resultant model consists of 485 nodes, 427 elements and 1936 degrees of freedom. Investigated are load cases 1 through 11 assuming both brittle and ductile material behavior. Results are summarized in Table 4.

3.4 THREE-DIMENSIONAL, MIXED BEAM AND TRUSS ELEMENTS, INCLUDING FLOOR SYSTEM MODEL (MODEL 4)

Model 4 is the final and most detailed model studied. It is identical to Model 3 with the exception of the additional modeling of the 8" thick reinforced concrete deck and stringers. The stringers are modeled as Euler-Bernoulli, I-sections having dimensions which correspond to the designed sections. From material sampling data the mean concrete maximum compressive strength was determined to be 2.75 ksi and the mean yield strength of the reinforcing bars was 55.0 ksi. The resultant finite element modeling which approximates the moment curvature relationship of this described deck is 4" thick, 4 - noded quadrilateral shell elements with six degrees of freedom per node and material properties consisting of a modulus of elasticity of 64 00 ksi and a yield strength of 6 ksi.

The designed slab is not directly connected to the floor beams. The stringers are pinned into the floor beams with the slab continuous above the floor beams. The top flange of the stringers are imbedded in the deck. To simulate this stringer/deck connection while keeping the number of nodes at a minimum, the following configuration is used: the stringers and shell elements (slab) share nodes. However, intersection of the horizontal and vertical axes of the stringer elements with their i - j line of action is set at the center of the top flange. This is a reasonable approach for approximating the concrete deck and stringer interaction.

The resultant model consists of 1213 nodes, 997 elements and 4960 degrees of freedom. Only Load Case No. 11 is chosen for evaluation with this model. This decision is based on two factors: 1) Model 4 is an extension of Model 3 in which Load Case No. 11 is the load case which governs failure; and 2) Model 4 computing time for plastic material behavior is excessive due to its complexity. Hence, the computing effort needed to achieve plasticity results for all load cases is not warranted at this time. Results are summarized in Table 5.

3.5 MODEL COMPARISONS

3.5.1 *Comparison Between Models 1 and 2.* Results associated with Models 1 and 2 are very similar for brittle material behavior under passing truck load cases 2 through 9. The collapse live + impact increment factors are approximately the same, with Model 2 exhibiting slightly higher reliability for most load cases. The element failure sequence generally follows the same pattern for all load cases. As the live + impact increment factor increases, one of the "redundant" diagonals in the central two bays fails and then either another redundant diagonal fails, causing an instability in Model 1 and an over stressing of the top and bottom chords of Model 2 (load cases 3 through 8), or one of the first interior posts fails (elements 30 and 31) to cause final system collapse (load cases 2 and 9). In regards to the lane load cases (1 and 10) the reliability obtained with Model 1 is significantly higher than that of Model 2 for brittle material behavior. Critical members causing system collapse are unique for each model. Using Model 1, failure occurs as a result of the first internal posts reaching their capacity. With Model 2, the end posts dictate system collapse.

Assuming ductile material behavior, Model 2 exhibits higher reliability than that of Model 1 for all load cases. The explanation for this is quite simple. When a redundant member fails in a brittle manner and is no longer functional in the system, bending stresses are introduced in the chords. This results in an over stressing of these elements. Therefore, it is not significant if the chords are modeled as truss or beam elements. In the case of truss elements, an instability results; wit h beam elements, over stressing results. However, if the element remains functional in the system and sustains its yield load, the bending stresses in the chords do not become excessive immediately and additional loading may be sustained. If the chords are modeled as truss elements, bending stresses are undefined and a singularity in the global stiffness matrix results.

3.5.2 *Comparison Between Models 2 and 3*. Model 3 yields significantly lower reliability for brittle and ductile material behavior than the two-dimensional models. The reason for this is that additional non-redundant members were added to the system which have lower reserve capacity than that of the trusses, the floor beams. Also longitudinally eccentric live loads coupled with eccentric post/floor beam connections create additional compressive stresses in the portion of post members below t he connection. The increase in the axial stresses of the second interior posts (L2U2, L6U6) is significant enough to govern the collapse of the structure for several load cases.

3.5.3 *Comparison Between Models 3 and 4*. By incorporating the deck and stringers into Model 3 and assuming fixity between them, a significant difference in the distribution of stresses in the structure is seen. First, the distribution of loads is based on the rigidity of the deck and the stress distribution between the deck and stringers. The stresses are, therefore, reduced in the floor beams. Second, internal horizontal forces are introduced by the slab, causing longitudinal reaction components from the floor beams into the posts and contributing to the failure of the interior posts. This is illustrated in Figure 6, where the deflected shape of one truss under Load Case No. 11 for Models 1, 2, 3 and 4 is presented. For brittle material behavior, Model 4's results indicate the least capacity of this structure under Load Case No. 11. Even with the addition of the deck, the bridge becomes unstable and collapse is unavoidable after the brittle failure of a post. However, for ductile material behavior, the failure load exceeds that of Model 3 by almost 30%.

4. Conclusions

It was the purpose of this study to justify or discredit the need for complicated finite element modeling of bridges. If there were little difference between the results of the models studied, the conclusion would be brief: a simple, two-dimension al representation of trusses, and simple span evaluation of beam elements in a floor system, is the most suitable and economical procedure to use when evaluating the reliability of steel bridges. However, the discrepancies in the results of the four models were significant enough that a simple answer is not possible. Many interesting and pertinent items were learned from this study and shall be discussed.

The effects of computer modeling were studied herein on a particular structure. The structure was quite typical of through-truss bridges in it's general element topology and load carrying system. Therefore, several items uncovered in the study of this structure can be applied to most short to medium span through-truss bridges and, where applicable, to girder bridges. The concluding remarks for comparison between models 1, 2 and 3 are very clear: a) the two-dimensional model with continuous beam elements in the top and bottom chords results in more system reserve than that of the all-truss element model, for ductile material behavior only; and b) it could be unsafe to neglect the effects of transversely and longitudinally eccentric live loads coupled with any eccentric post/floor beam connections when evaluating existing through-truss bridges.

The modeling of chords is dependent upon the type of material behavior. For brittle material behavior, little added system reserve will be measured by modeling the top and bottom chords of the trusses as continuous beams. For ductile material behavior, however, a significant increase in the system reserve will be discovered. Hence, the material properties of the existing structure are significant when choosing the bridge's finite element model.

Model 4's results indicate that incorporation of the concrete deck in system evaluation may reveal a decrease in the bridge safety particularly for those structures with eccentric post/floor beam connections. Also, the greater the composite action between the stringers and floor beams, the greater the development of internal horizontal stresses. There is a beneficial result from composite action of the stringers and slab in that a greater distribution of loading among the floor beams is realized. However, it is emphasized that a careful inspection must be undergone to determine the degree of fixity between the deck and stringers. This is difficult to determine without field testing. Hence, it is recommended not to rely on total fixity of the stringers. This can result in an underestimation of floor beam stresses, which may be a weak link in the system. Upgrading of the floor beams is recommended when simple analysis indicates their stresses are beyond yield.

5. Acknowledgements

Support of this work by the National Science Foundation under Grant No. ECE-8609894, with Dr. John Scalzi as Program Director, to the University of Colorado at Boulder, is gratefully acknowledged.

6. References

1. Ang, A. H-S. and Cornell, C.A. (1974) "Reliability bases of structural safety and design," Journal of the Structural Division, ASCE, Vol. 100, No. ST9, 1755-1769.
2. Ang, A. H-S. and Tang, W.H. (1984) Probability Concepts in Engineering Planning and Design, Vol. II, John Wiley and Sons, New York.
3. Cornell, C. A., (1967) "Bounds on the reliability of structural systems," Journal of The Structural Division, ASCE, Vol. 93, No. ST1, 171-200.
4. Frangopol, D.M. (1988) "Probabilistic evaluation of existing highway bridges including component interaction," in P.L. Bourdeau (ed.), Proc. Symp. on Reliability-Based Design in Civil Engineering, Lausanne, Switzerland, Vol. 2, 37-49.
5. Frangopol, D.M., Nakib, R., and Fu, G. (1988) "Bridge reliability evaluation using 3-D analysis and damage scenerios," in P.D. Spanos (ed.), Probabilistic Methods in Civil Engineering, ASCE, New York, 177-180.
6. Frangopol, D.M., Goble, G.G., Trautner, J.J., and Scholfield, M. (1987) "Redundancy evaluation of existing bridges," in L. Tall (ed.), Bridge and Transmission Line Structures, ASCE, New York, 1-14.
7. James, R.W., Zimmerman, R.A., Loper, J.H. (1988) "Effects of repeated heavy loads on highway bridges," Texas Transportation Institute, Report No. FH-WA/TX-87/462-1F.
8. Manual for Maintenance Inspection of Bridges (1983), AASHTO, Washington, D.C.
9. Moses, F. (1989) "New directions and research needs in system reliability research," in D.M. Frangopol (ed.), New Directions in Structural System Reliability, University of Colorado, Boulder, 7-16.
10. Moses, F. and Verma, D. (1987) "Load capacity evaluation of existing bridges," NCHRP Report No. 301, Washington, D.C.
11. Nowak, A.S. (1987) "Probabilistic basis for bridge evaluation criteria," in A.S. Nowak and E. Absi (eds.), Bridge Evaluation, Repair and Rehabilitation, University of Michigan, Ann Arbor, 311-321.
12. Standard Specification for Highway Bridges (1983), AASHTO, 13 ed., Washington, D.C.
13. Thoft-Christensen, P. and Baker, M.J. (1982) Structural Reliability Theory and Its Applications, Springer-Verlag, New York.
14. Trautner, J.J. and Frangopol, D.M. (1989) "Computer modeling and reliability evaluation of steel through truss bridges," Structural Safety, Elsevier, Amsterdam (in press).
15. Trautner, J. J. (1989) Computer Modeling and Reliability Evaluation of Truss Bridges, Thesis presented to the University of Colorado, Boulder, in partial fulfillment of the requirements for degree of Doctor of Philosophy.

Table 1 Model 1 Deterministic Results

Load Case	λ at Failure		Element Failure/Yield Sequence		Deflection at Failure (in.)	
	Brittle	Ductile	Brittle	Ductile	Brittle	Ductile
1	5.68	5.68	31	31	3.699	3.699
2	4.89	4.89	29,30,31	29,30,31	2.456	2.411
3	4.54	4.88	29,30,25,26	29,30,17	3.323	3.436
4	3.71	4.32	30,26	30,29,2	3.475	3.817
5	3.56	3.40	30,28,26	30,29,27,28	3.634	3.387
6	3.12	3.70	27,23	27,28,23,2	3.211	3.509
7	3.44	4.07	27,23	27,28,23,2	3.212	3.553
8	5.05	5.38	28,24	28,27,18	3.356	3.480
9	4.66	4.66	28,32	28,27,32	2.143	2.097
10	5.68	5.68	32	32	3.699	3.699

Table 2 Model 1 Probabilistic Results.

Load Case	SYSTEM RELIABILITY INDEX, β (λ = 1)		
	Method I	Method II	
	Bounds (Ductile)	Brittle	Ductile
1	4.84,494	4.50	4.50
2	4.93,4.95	3.44	3.44
3	4.90,5.09	3.25	3.45
4	4.94,4.95	2.67	3.12
5	4.74,4.89	2.57	2.43
6	4.73,4.84	2.19	2.69
7	4.73,4.80	2.42	2.91
8	4.91,5.13	3.53	3.72
9	5.05,5.06	3.30	3.30
10	4.84,4.95	4.50	4.50

Table 3 Model 2 Result.

Load Case	λ at Failure		Element Failure /Yield Sequence		Deflection at Failure (in.)		β (λ = 1)	
	Brittle	Ductile	Brittle (Failure)	Ductile (Yield)	Brittle	Ductile	Brittle	Ductile
1	5.33	5.92	9	9,1,2,31,5,30,7	3.54	4.75	4.32	4.62
2	4.93	5.31	29,31	29,30,31,1,2,9,5	2.46	2.82	3.47	3.68
3	4.42	5.24	29,30	29,30,9,11,13,1,2	3.25	5.96	3.17	3.66
4	3.77	4.53	30,26	30,29,7,25,26	3.50	4.06	2.72	3.26
5	3.76	4.61	30,28,26	30,29,7,25,26,8	3.75	4.60	2.74	3.33
6	3.27	4.37	27,23	28,23,24,8,7,13,15,16,5	3.30	5.43	2.32	3.17
7	3.58	4.69	27,28,23,24	27,28,23,24,8,13,15,16	3.20	4.92	2.53	3.32
8	5.20	5.67	28,24	28,27,6,18,10,12,3,4,14	3.12	3.20	3.62	3.86
9	4.70	5.01	28,32	28,27,32	2.15	2.19	3.32	3.51
10	5.33	5.92	10	10,3,4,32,6,27,8	3.54	4.75	4.32	4.62

Table 4 Model 3 Results

Load Case	λ at Failure		Element Failure/Yield Sequence		Deflection at Failure (in.)		β (λ = 1)	
	Brittle	Ductil e	Brittle (Failure)	Ductile (Yield)	Brittle	Ductile	Brittle	Ductil e
1	3.36	3.54	B1	B1	2.87	3.15	2.63	2.79
2	3.20	3.52	29,B1	29,30,B1	2.35	2.54	1.82	2.07
3	3.17	3.30	29,30E,B2	29,30,B2, 30E,35	3.08	3.12	1.81	1.91
4	2.93	3.62	30,30E,37	30,29,30E, 37,B3	3.45	4.40	1.63	2.17
5	2.69	3.03	30,28,30E, 37	30,29,30E, 27,37,B4	3.62	3.85	1.43	1.73
6	2.75	3.56	27,27E,37	27,28,37E, 37,B4,28E	3.56	4.70	1.48	2.15
7	2.73	3.12	27,27E,37	27,28,27E, 37,B5	3.33	3.66	1.43	1.75
8	2.78	3.20	28,37	28,27,37,B6	2.67	2.99	1.47	1.82
9	3.05	3.54	28,B7	28,B7	2.12	2.82	1.70	2.08
10	3.36	3.54	B7	B7	2.87	3.15	2.63	2.79
11	2.59	3.02	30,30E,B4	30,30E,B4	3.48	4.27	1.30	1.67

Table 5 Model 4 Results, Load Case No. 11

Parameter	Material Behavior	
	Brittle	Ductile
λ at failure	2.13	3.88
Element failure/yield sequence	30&30E, 39&39E	30&30E,39&39E,37& 37E,B4,29&29E
Deflection at failure (in.)	2.81	4.18
β, λ=1	.85	2.32

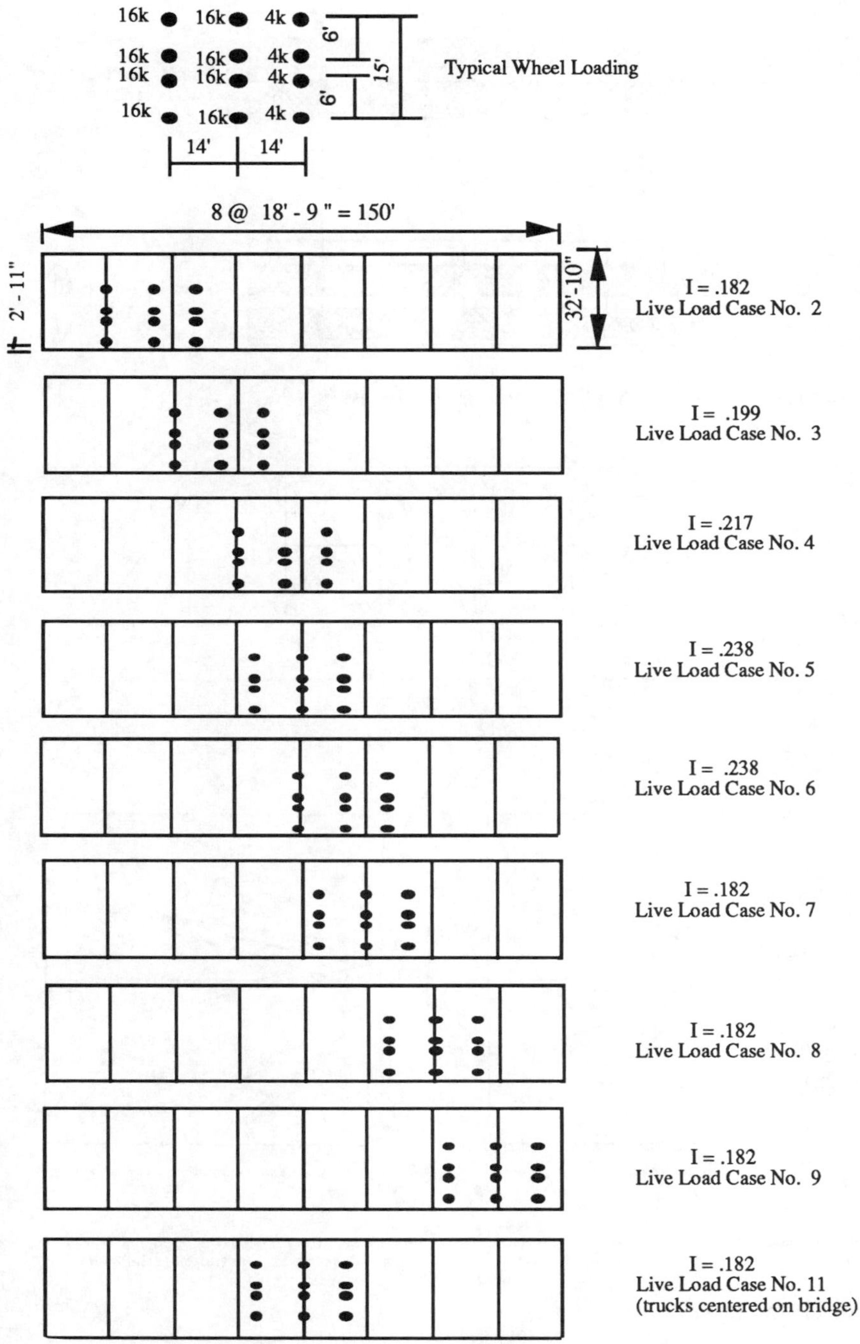

Fig. 1 AASHTO Passing Truck Loads Applied to Representative Bridge.

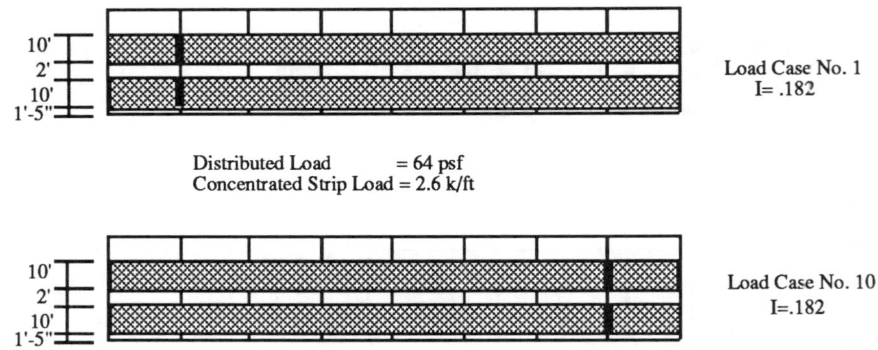

Fig. 2 AASHTO Lane Loads Applied to Representative Bridge.

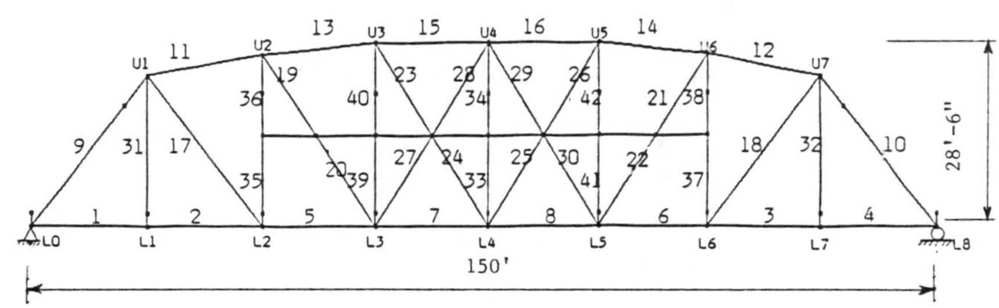

Fig. 3 Model 1.

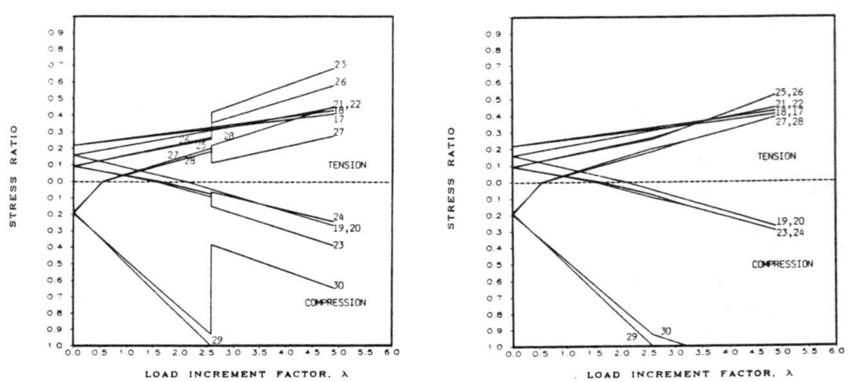

Stress Results for Model 1, Diagonal Elements
Load Case No. 2, Brittle Material Behavior

Stress Results for Model 1, Diagonal Elements
Load Case No. 2, Ductile Material Behavior

Fig. 4 Load Path Curves.

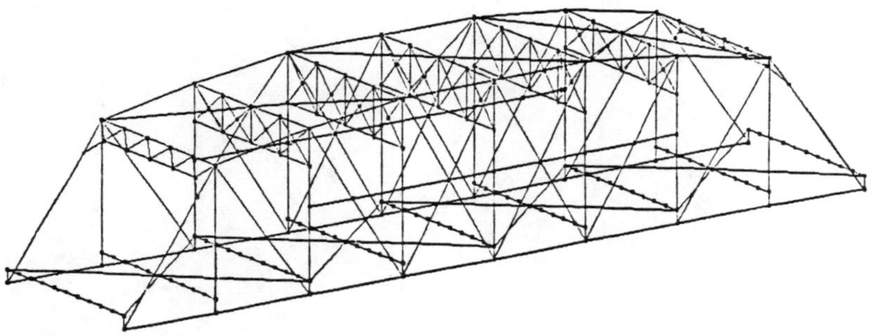

Fig. 5 Model 3.

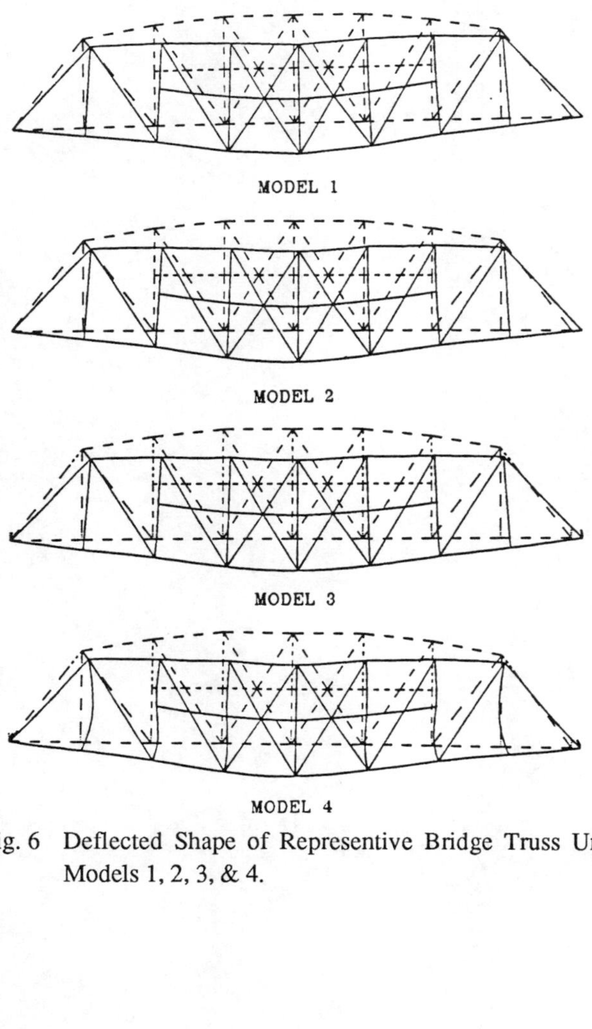

Fig. 6 Deflected Shape of Representive Bridge Truss Under Load Case No. 11 Utilizing Models 1, 2, 3, & 4.

PROBABILITY BASED LOAD MODELS OF HIGHWAY TRAFFIC FOR ULS AND SLS ANALYSES

ROLF BEZ and MANFRED A. HIRT
ICOM - Steel Structures
Swiss Federal Institute of Technology (EPFL)
1015 Lausanne, SWITZERLAND

ABSTRACT. This paper presents a probability based concept for the modeling of highway traffic loads which includes the effects of dead load, live load and the resistance of the structure. It consists of the following parts:

- the principles of traffic simulation,
- the concept of the probabilistic analysis of ultimate limit state (ULS),
- the load models obtained for ULS,
- the particular problems for the analysis of serviceability limit state (SLS).

The results of this study show that it is possible to justify, on a probabilistic basis, load models for highway traffic which assure a uniform degree of safety for different structures and which are independent of type of material, type of cross section and other structural characteristics such as span length and deck width.

1. Introduction

This paper summarizes the doctoral thesis of Bez (1989), which proposes a probability based concept for the modeling of highway traffic loads. The goals of this research were to:

- clearly distinguish between the ultimate limit state (ULS) and the serviceability limit state (SLS),
- define the design criteria for each of these limit states,
- establish the concept which defines the load levels for ULS and SLS,
- assess the influence of the type of material and cross section,
- propose load models for codes which are easy to use in practice.

The fatigue limit state is not treated, because it requires a slightly different approach. This is shown in studies by Fothergill et al. (1973), Moses and Garson (1974), Baar and Bruls (1975), Jacquemoud (1980) or Moses and Ghosn (1985). The work reported here is also limited to the study of a particular section of a structure, i.e. the mid-span section of a continuous girder, essentially in order to set a clear basis for the proposed methodology.

A. S. Nowak (ed.), Bridge Evaluation, Repair and Rehabilization, 323–337.

The different steps of this study are illustrated in Figure 1, namely:

- establishment of the concept which defines the load levels for ULS and SLS,
- probabilistic analyses of ULS and SLS in order to establish the numerical values of the corresponding traffic loads effects,
- proposition of load models which are consistent with the probabilistic analyses.

A traffic simulation is done in order to simulate traffic jams. It takes into account the characteristics of traffic and those of the structure studied. The latter also directly influence the probabilistic analyses, such as the appropriate material resistance.

This paper gives some details about the following parts:

- the procedure of the traffic simulation,
- concept, assumptions and results of the probabilistic analysis of ULS,
- the load model proposed for ULS,
- the particular problems for the analysis of SLS.

More information about these and other points are given by Bez (1989).

2. Traffic Simulation

2.1 PROCEDURE

The purpose of traffic simulation is to establish the statistical characteristics of the highway traffic loads effects, in order to introduce them into the probabilistic analyses. The first step of the traffic simulation consists in choosing the characteristics of the structure which will support the traffic loads, namely the static system, the influence line, the type of cross section and the deck width which is directly correlated with the number of traffic lanes. On the basis of measured traffic and its statistical characteristics, namely:

- geometry of the vehicles (length, axle spacing),
- gross weight and its distribution to the axles,
- distance between moving or stopped vehicles,
- composition of the traffic (percentage of cars, trucks, tractor trailers and semi trailers),

it is possible to simulate a traffic jam by a random generation of vehicles which are placed on the chosen influence line. This procedure is repeated over and over again (NSIM times), in order to simulate the number of traffic jams which could occur during the service life of the bridge. This results in obtaining the extreme value of the traffic loads effects. The simulation of this number of traffic jams is repeated a number of times (NREP times), in order to establish the statistical characteristics (mean and standard deviation) of the extreme value of the traffic loads effects. The statistical distribution obtained corresponds to a Gumbel distribution, which is a distribution of upper bound values.

This traffic simulation is done with the help of a specially developed computer program called SIMBOU (1989), where the number of simulation of jams (NSIM) was taken equal to 50,000 (corresponds to about 4 jam situations per day during a service life of 50 years, taking into account a number of 250 working days per year), and the number of repetitions of the procedure (NREP) equal to 50.

2.2 TRAFFIC CHARACTERISTICS

The traffic characteristics used for the traffic simulation are established by field measurements of stopped and fluid traffic in Switzerland between 1976 and 1988. Figure 2 shows, for example, the truck gross weight distribution measured in 1988 on a weigh-in-motion station located at Mattstetten (north of Berne, on the N1 highway Berne - Zurich).

In addition, the following characteristics are adopted for the traffic simulation with SIMBOU (1989) (see Bez (1989)):

- only stopped traffic is considered to yielding about the same results as slowly moving traffic (this has the advantage that it is not necessary to take into account the dynamic effects, which is a research area by itself),
- the distances between vehicles is fixed at 1.0 meter, which is a conservative value for stopped traffic,
- the percentage of trucks and cars is fixed as follows, according to the measurements done in Switzerland: 75 % of cars, 10 % of trucks, 10 % of tractor trailers and 5 % of semi trailers,
- one and two directions of circulation are considered.

2.3 RESULTS

Traffic simulations are done considering the following structural characteristics (see Figure 3):

- continuous girder with equal spans of L = 20.0, 32.75, 50.0, 70.0 and 100.0 m,
- two-girder composite steel-concrete and concrete box girder cross sections,
- deck width of b = 4.5, 8.0, 11.5 and 15.0 m (which corresponds respectively to one, two, three and four traffic lanes),
- bending moment at mid-span.

The traffic simulations using the above defined structural characteristics have permitted obtaining, for each case, the mean value and the standard deviation of the mid-span moment M_L in the main girder due to the highway traffic. Figure 4 shows the results obtained for the mean value of M_L.

3. ULS Probabilistic Analysis

3.1 INTRODUCTION

The purpose of the ULS probabilistic analysis is to establish the design values of the different variables. The design situation where the highway traffic represents the predominant load giving the maximum mid-span moment of a continuous girder of equal spans (see Figure 3) is chosen. This design situation is chosen because it is often used for the preliminary design of bridges. In addition, it takes into account all the major load and resistance variables. The traffic loads are considered on one span only, because this is the situation where the maximum mid-span moment is obtained.

3.2 BASIC CONCEPT OF THE PROBABILISTIC ANALYSIS

The theory by Hasofer and Lind (1974), which is a first order second moment analysis, is used for the probabilistic analysis of the ULS. The theory will not be given here, since it can be found in detail in several references (see for example Thoft-Christensen and Baker (1982) or Ang and Tang (1984)). This theory permits the analysis of a limit state function Z which can be expressed as follows:

$$Z = g(X_1, X_2, ..., X_m, X_{m+1}, ..., X_n) \quad (1)$$

$g(...)$: function of load and resistance variables,
$X_1, ..., X_m$: resistance variables,
$X_{m+1}, ..., X_n$: load variables.

This limit state function represents a curved limit state area. The most probable point of damage yields the design values x_i of the different variables X_i involved. The design values are expressed as a function of the safety index β.

3.3 LIMIT STATE FUNCTIONS

The limit state functions describing a mid-span cross section subjected to bending are given here for the two cross sections studied, i.e. the open cross section of a two-girder composite bridge and the concrete box girder cross section. The limit state functions involve all the load and resistance variables necessary to describe the ULS:

Open cross section:

$$M_D + M_L = F_y S \quad (2)$$

Box girder:

$$M_D + M_S + M_L = 0.95 A_r d F_{yr} \quad (3)$$

M_D : bending moment due to dead load (steel and concrete, asphalt layer and parapet),
M_S : secondary bending moment due to prestressing,
M_L : bending moment due to live load, i.e. the highway traffic loads (obtained by traffic simulation, see section 2),
F_y : yield stress of the main girder steel,
F_{yr} : yield stress of the reinforcing steel,
S : elastic section modulus of the composite cross section,
A_r : idealized area of the reinforcing steel of the concrete box girder cross section,
d : static height of the cross section.

The variables involved in these limit state functions are considered with the statistical characteristics given in Table 1. These characteristics were proposed by Vrouwenvelder and Siemes (1987). All other parameters are considered as deterministic values.

Some references (see Vrouwenvelder and Siemes (1987) or Schneider (1988)) introduce additional variables, so-called model variables, on the load and the resistance sides of the above limit state functions in order to take into account the inaccuracy of the load and resistance models used. These variables have a direct influence on the value of the safety index β, and can change the results of the probabilistic analysis. Although it would be necessary to consider such model variables in order to be coherent with the general concept of the probabilistic analysis, it could not be done in this study, because their statistical characteristics are not yet sufficiently known.

3.4 SAFETY INDEX

The target value of the safety index β, proposed in the literature for ULS, varies between about 3.5 and 5.0. The use of these values is not possible for the probabilistic analysis of this study, specifically because no model variables are used in the limit state functions. The safety index value is therefore established on the basis of existing bridges. The range of the actual β-values of real structures permits obtaining the same level of safety as for existing structures.

The evaluation of the safety index value was based on two bridges, shown in Figure 5. The first one is an open composite steel-concrete cross section, and the second one was a project for a concrete box girder cross section.

The probabilistic analysis of these bridges, typical for Switzerland, gave β values of 9.45 and 10.8 for the open and the box girder cross sections, respectively. These relatively high values, in comparison with those given in the literature, are partially due to the fact that no model variables are considered in the limit state functions (see section 3.3). Another reason is that the design of those bridges has been made using loads and safety factors which were not established on a probabilistic basis. However, this does not exclude the use of the probabilistic analysis, because the safety index β is a target value calibrated on existing structures, which cannot be correlated in this case with any failure probability. Hence, a target value of $\beta = 9.5$ is used for the ULS probabilistic analysis.

3.5 RESULTS

The ULS probabilistic analysis is done with the FORM (1987). The design values of the different variables involved in the limit state functions studied (see equations 2 and 3 and Table 1), as well as the nominal values of the dimensional variables (i.e. respectively S and A_r for the open and the box girder cross sections) are computed. The results for S and A_r, illustrated in Figure 6, are very important because they will be used further for the modeling of the highway traffic loads. It is indeed necessary to propose a load model which approaches as closely as possible the nominal values of S and A_r, obtained with the probabilistic analysis.

4. ULS Load Model

4.1 CODE DESIGN CONCEPT

The traffic load models cannot be established independently, but they must be incorporated into the general design concept comprising the whole set of code regulations and rules. In Switzerland, this concerns essentially the codes SIA 160 (1989), SIA 161 (1989) and SIA 162 (1989), where the following partial safety factors are already defined:

- resistance side : $\gamma_R = 1.1$ for steel or composite structures,
  $\gamma_R = 1.2$ for concrete structures,
- load side : $\gamma_D = 1.3$ for dead load,
  $\gamma_L = 1.5$ for the predominant live load.

The limit state function given by equations 2 and 3 for the open and the box girder cross sections, respectively, can be rewritten as follows, using the partial safety factor design concept of the SIA codes:

Open cross section:

$$\gamma_D M_{D,nom} + \gamma_L\, M_{L,nom} = \frac{F_{y',nom} S_{nom}}{\gamma_R} \tag{4}$$

Box girder:

$$\gamma_D M_{D,nom} + M_s + \gamma_L + M_{L,nom} = \frac{0.95\, A_{r,nom}\, d\, F_{yr,nom}}{\gamma_R} \tag{5}$$

In these equations, the index "nom" refers to the nominal values given in the code.

If we now assume that the design value $M_{L,d}$ of the bending moment due to the traffic loads is given by:

$$M_{L,d} = \gamma_L\, M_{L,nom} \tag{6}$$

it becomes possible to establish the design values $M_{L,d}$ (see Figure 7) which are consistent both with the probabilistic analysis (nominal values of the dimensional variables S and A_r) and the design code concept (partial safety factors).

As an additional step, appropriate load models need to be established in such a way that the resulting bending moments will be as close as possible to the design values $M_{L,d}$.

4.2 LOAD MODEL

The load model which approximates as closely as possible the design values $M_{L,d}$ is given in Figure 8a. It is composed of the following parts:

1) a lane load (design value 9 kN/m^2, without impact factor), 3 m wide, representing a line of heavy trucks,

2) four concentrated loads, representing a tandem axle (design value 100 kN plus dynamic impact factor Φ), superimposed on the above lane load within the 3 m width,
3) a lane load (design value 6 kN/m^2, without impact factor), to be applied to the remaining deck area, representing the load due to a mixture of trucks and cars.

The load model given in Figure 8a approaches the results of the probabilistic analysis very well, and also has the advantages of being independent of the type of material, type of cross section and other structural characteristics such as span length and deck width. Note, that a constant dynamic impact factor of $\Phi = 1.8$ is used for the concentrated loads only.

The same procedure was applied to the Eurocodes on the basis of a heavy German highway traffic. This shows the influence of the maximum legal truck gross weight, which is 280 kN in Switzerland and 400 kN Germany as well as in many other West-European countries. The resulting load model is shown in Figure 8b. The same load model may be used except that a second tandem axle has to be added. This is necessary in order to preserve the same safety index as was obtained for existing Swiss structures.

5. Particular Problems for SLS Analysis

5.1 BASIC PRINCIPLE

The SLS is definitely different from the ULS, because the SLS describes a traffic situation which concerns the serviceability of the structure and not its carrying capacity. In order to be consistent with this fundamental difference, it is very important to consider this limit state on its own. Basically, the SLS check is a comparison between the effect due to service loads and a limiting value. This can be expressed as follows:

$$w(Q_{ser}) \leq w_{lim} \tag{7}$$

$w(Q_{ser})$: effect (deformation, vibration, crack opening, ...) due to the service load Q_{ser},

w_{lim} : limiting value.

5.2 SAFETY INDEX

In the literature, values of the safety index β for SLS between 1.0 and about 3.5 are proposed. The use of the safety index for the SLS is, however, not easy because there exists almost no information about the statistical parameters of equation 7. Even though information exist concerning traffic loads and their service loads Q_{ser}, it is indeed very difficult to find adequate statistical information for the limiting values w_{lim}. The limiting values often depend on people's personal, arbitrary and subjective opinion. In addition, no basic reference is available which could define the value of the safety index β to use. It is therefore difficult to justify a value of β for the SLS.

This explains why the safety index β was not used for the SLS probabilistic analysis.

5.3 PROPOSED PROCEDURE

In order to realistically assess the SLS, it is very important that the correlation between the serviceability requirement, the check criteria and the service load of highway traffic is well established. This connection is illustrated in Table 2, where it is shown that two types of traffic should be considered for the SLS check: a moving traffic with a short loading interval (i.e. a high probability of occurrence or a short return period, for example one day) or a stopped traffic with a long loading interval (i.e. a low probability of occurrence or a long return period, for example ten years).

A load model for the serviceability check must then be established in accordance with Table 2, in order to perform a realistic SLS analysis. Some proposals respecting the above basic principle are given by Bez (1989). They are not detailed here, because they depend strongly on the limiting values given in the codes.

6. Conclusions

This study shows that it is necessary to consider separate ULS and SLS in order to be consistent with the fundamentally different nature of both limit states. The following conclusions, considering an open composite steel-concrete and a concrete box girder cross section have been obtained by this study:

6.1 ULTIMATE LIMIT STATE

1. It is very important to consider simultaneously the different variables describing dead load, live load and the resistance of the structure for a realistic modeling of the highway traffic loads. This is possible with a first order second moment analysis and a safety index β.
2. The safety index β, established by an analysis of existing bridges, has a value of about 9.5. This value can be used for calibration and as a target value. However, it is not a safety indicator, since no model variables were used in the ULS probabilistic analysis.
3. The analysis has shown that it is possible to justify, on a probabilistic basis, load models for highway traffic which assure a uniform degree of safety for different structures and which are independent of type of material, type of cross section and other structural characteristics such as span length and deck width.

6.2 SERVICEABILITY LIMIT STATE

1. It is very important that the correlation between the serviceability requirement, the check criteria and the service load of the highway traffic is well established. This leads to considering, for example, two different highway traffics for SLS: a moving traffic with a short loading interval and a stopped traffic with a long loading interval.

2. A load model for the SLS check can be established according to the above principle. It is, however, not discussed here, because it depends strongly on the limiting values given by different codes.

6.3 FUTURE WORK

1. This study needs to be extended to other cross sections, to the entire structural system, to other types of highway bridges and to influence lines of other structural systems, such as cable-stayed bridges.
2. The statistical characteristics of the model variables, which are very important in order to obtain a realistic value of the safety index β, must be better known. Much research is needed in this area.
3. An important effort must be made to better justify the limiting values used for SLS. This would permit a better correlation between the check criteria and the service highway traffic load models.
4. This study may be extended to the evaluation of existing bridges.

7. Acknowledgments

The authors would like to thank the Swiss Federal Highway Administration for its interest in this study, and who provided the statistical traffic data. Thanks are also due to the Swiss Engineers and Architects Association, especially the commission of the loading code SIA 160 and the members of the working group "Highway Traffic Loads" for their financial support and for the many helpful contacts and discussions. Appreciation is also expressed for the preparation of the manuscript to the staff of ATLSS at Lehigh University (Bethlehem, U.S.A.) where the first author was a visiting research engineer between January and June 1990.

8. References

Ang, A.H.S. and Tang, W.H. (1984) Probability concepts in engineering planning design, Vol II : Decision, risk and reliability. New York, John Wiley & Sons.

Bez, R. (1989) Modélisation des charges dues au trafic routier. Lausanne, Swiss Federal Institute of Technology (EPFL), doctoral thesis no 793.

Baar, S. and Bruls, A. (1975) Action d'un trafic routier dense. Liége, Station d'essai et de recherche de la construction métallique (Rapport Sm. 103).

FORM (1987) First Order Reliability Method computer program. Muenchen, RCP GmbH.

Fothergill, J.W. et al. (1973) Prediction of long-term stress range. Washington D.C., Federal highway administration (Report FHWA-RD-73-42).

Hasofer, A.M. and Lind, N.C. (1974) Exact and invariant second-moment code format. Proceedings ASCE, Journal of the engineering mechanics division, New York, vol. 100, pp. 111-121.

Jacquemoud, J. (1980) Analyse du comportement á la fatigue des ponts-routes. Lausanne, Swiss Federal Institute of Technology (EPFL), doctoral thesis no 793.

Moses, F. and Garson, R. (1974) New procedures for fatigue design of highway bridge girders. Washington, D.C., Transportation research board, pp. 58-67 (TRR 507).

Moses, F. and Ghosn, M. (1985) A comprehensive study of bridge loads and reliability. Cleveland, Case Western Reserve University (Report FHWA/OH-85/ 005).

Schneider, J. (1988) Sicherheit und Zuverlaessigkeit von Tragwerken. Zurich, Swiss Federal Institute of Technology (ETHZ), IBK.

SIA 160 (1989) Actions on structures. Zurich, Swiss Engineers and Architects Association (also available in French and German).

SIA 161 (1989) Steel structures. Zurich, Swiss Engineers and Architects Association (code under revision).

SIA 162 (1989) Concrete structures. Zurich, Swiss Engineers and Architects Association (also available in French and German).

SIMBOU (1989) SIMulation BOUchon computer program (written by Bez, R.). Lausanne, Swiss Federal Institute of Technology (EPFL), ICOM - Steel structures.

Thoft-Christensen, P. and Baker, M.J. (1982) Structural reliability theory and its application. Berlin, Springer Verlag.

Vrouwenvelder, A.C.W.M. and Siemes, A.J.M. (1987) Probabilistic calibration procedure for the derivation of partial safety factors for the Netherlands building code. Heron, Delft, University of Technology, Stevin-Laboratory of the Faculty of Civil Engineering, vol. 32, no 4, pp. 9-29.

TABLE 1. Statistical characteristics of the load and resistance variables.

VARIABLE		MEAN VALUE	STANDARD DEVIATION	TYPE
Load	M_D	1.05 * nominal value	0.07 * mean value	Normal
	M_L	obtained with traffic simulation		Gumbel
Resistance	F_y	355 N/mm^2	22.4 N/mm^2	Log-normal
	F_{yr}	460 N/mm^2	0.10 . mean value	Log-normal
	S	1.00 . nominal value	0.04 . mean value	Normal
	A_r	1.03 . nominal value	0	Constant

TABLE 2. Connection between serviceability requirements and service loads of the highway traffic loads.

SERVICEABILITY REQUIREMENT	GOALS	CHECK CRITERIA	SERVICE LOAD OF THE HIGHWAY TRAFFIC
Function of the structure	Sufficient stiffness	Deformation Vibration	Moving traffic with a short loading interval
Durability	Limitation of water seepage	Crack opening	Stopped traffic with a long loading interval
Appearance	Good aspect Camber	Crack opening Deformation	Highway traffic may be ignored

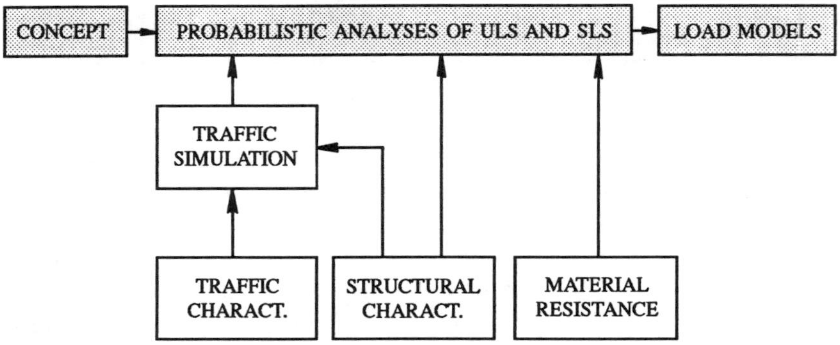

Fig. 1 Schematic Representation of the Main Steps of the Study.

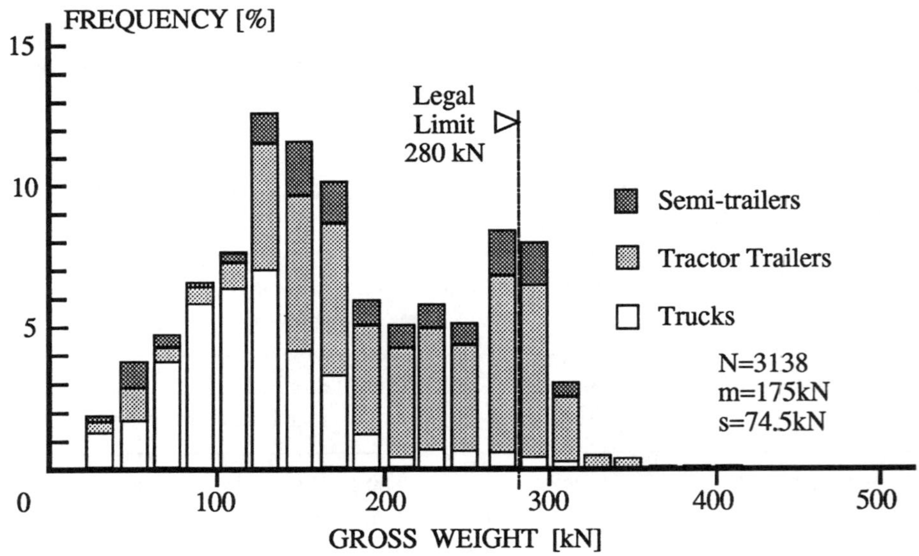

Fig. 2 Histogram of Truck Gross Weight Measured at Mattstetten (Switzerland) in 1988.

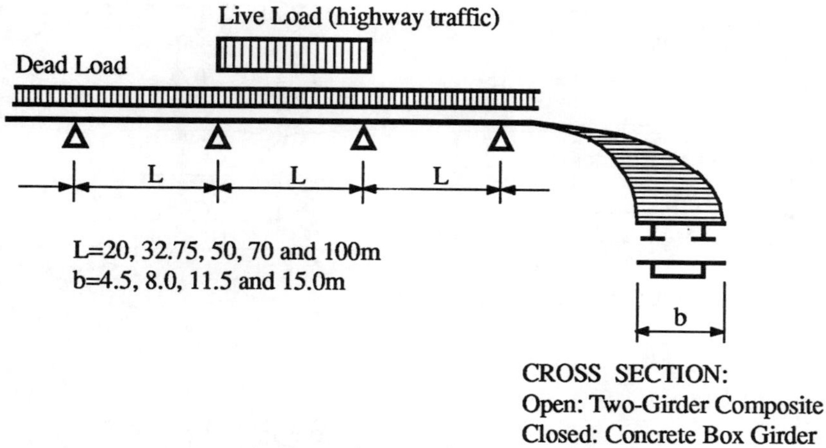

Fig. 3 Structural Characteristics and Loads Considered for the Maximum Mid-Span moment.

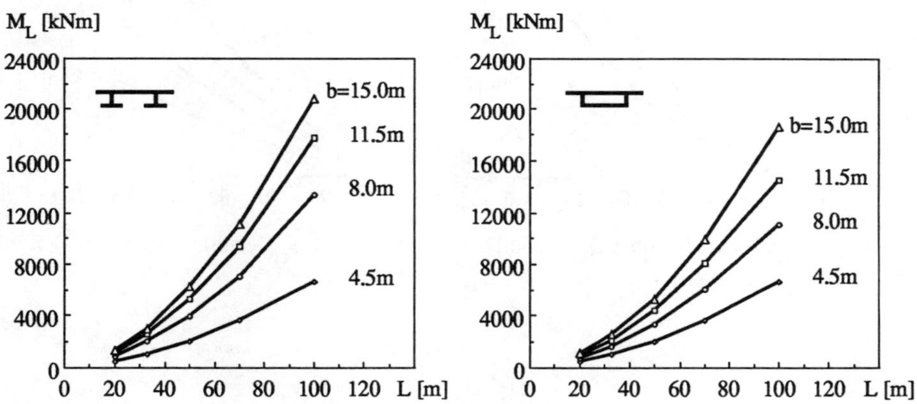

Fig. 4 Mean Value of the Mid-Span Moment M_L Due to the Traffic Obtained with the Traffic Simulation for the Main Girder of the Open Cross Section and for the Box Girder.

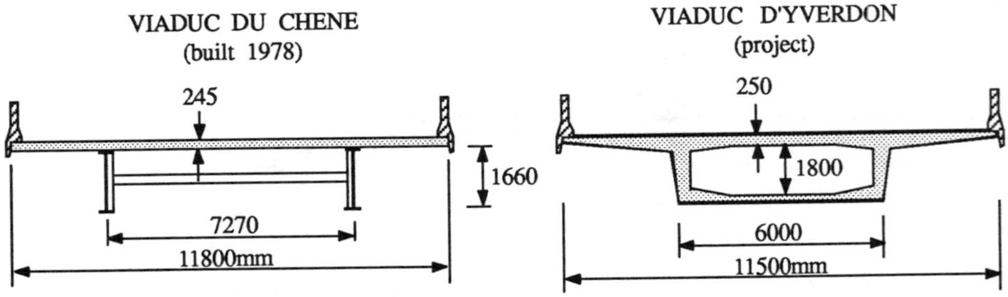

Fig. 5 Characteristics of the Two Bridges Examined.

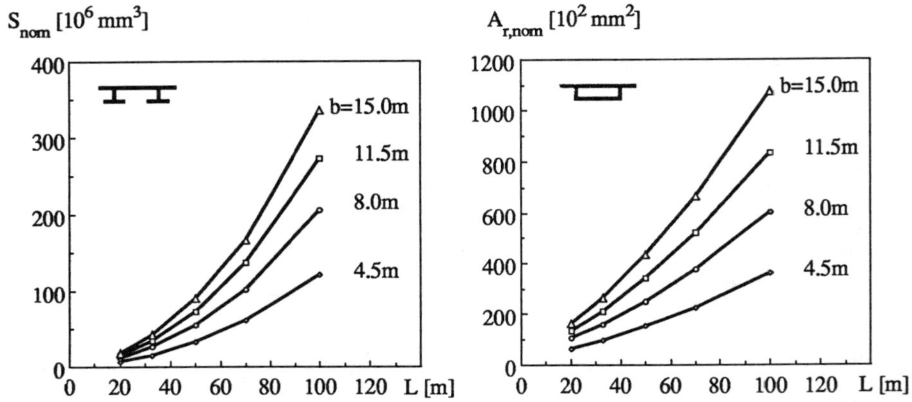

Fig. 6 Nominal Values of the Dimensional Variables S and A_r Resulting from the ULS Probabilistic Analysis.

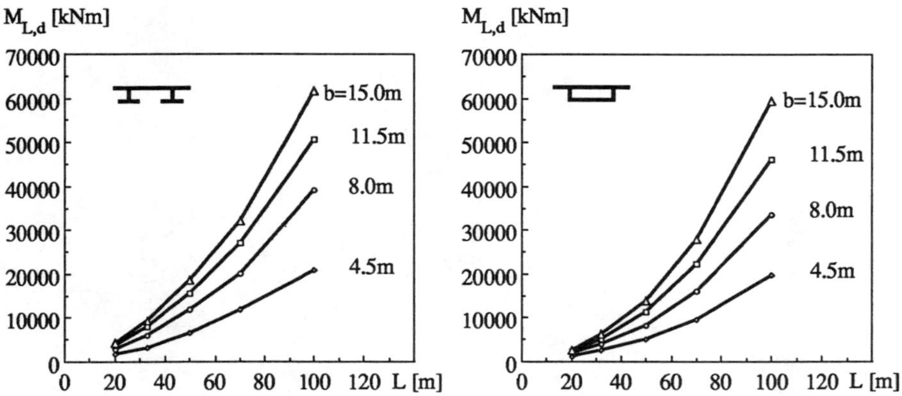

Fig. 7 Design Values $M_{L,d}$ of the Bending Moment Due to the Traffic Loads Taking into Account the Results of the Probabilistic Analysis and the Code Design Concept.

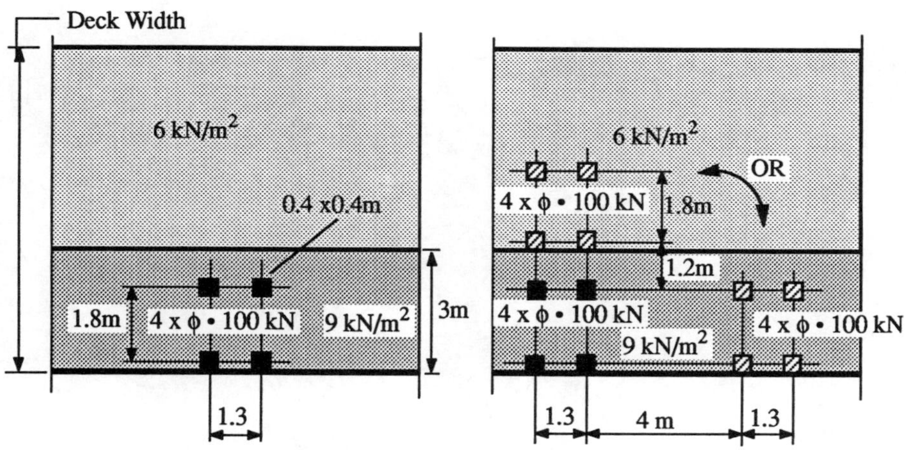

a) Based on the Swiss Traffic and the SIA Codes.

b) Based on the German Traffic and the Eurocodes.

Fig. 8 Proposed Load Models (the Values Shown are Design Values; in the Codes, Representative (or Nominal) Values are Given Corresponding to the Design Values Divided by γ_L).

DISTORTION-INDUCED FATIGUE CRACKING OF A NON-COMPOSITE STEEL GIRDER-FLOORBEAM-STRINGER BRIDGE

J. M. KULICKI, D. R. MERTZ, R. E. MURPHY & S. R. ESHENAUR
Modjeski and Masters
P.O. Box 2345
Harrisburg, PA 17105 USA

ABSTRACT. Recently, a non-composite steel girder-floorbeam-stringer bridge in Pennsylvania has experienced previously unanticipated fatigue cracking. The basically horizontal cracking has occurred in floorbeam webs at girder-floorbeam connections. The distress is apparent only at or adjacent to the supports (in this case, the abutments and the piers). The cracking is attributed to differential displacements between the girders and the reinforced concrete deck-stringer system which is transmitted through the girder-floorbeam connection. The solution of the floorbeam cracking problem began with strain measurements under normal random traffic. These strain measurements form the basis for comparison for both trial retrofits in the field and for the computer-based Finite Element Method (FEM) analysis. Herein, the distortion-induced fatigue cracking problem is described, the results of the FEM investigation are discussed, and the proposed retrofitting alternatives are reviewed.

1. Introduction

Twin bridges carry Interstate 83 across the Susquehanna River at the southern end of Harrisburg, Pennsylvania. Although these bridges were designed and constructed more than 20 years apart, the southbound bridge in 1956 and the northbound bridge in 1979; the structures are conceptually very similar. The similarities include: dual haunched steel girders continuous over 21 spans with two sets of hangers, a floor system consisting of steel stringers, floorbeams and outrigger cantilever brackets, and a concrete deck which is not directly attached to the girders. A typical bridge cross section is shown in Figure 1. The primary difference between the two structures is that the earlier bridge is of riveted construction while the newer bridge is welded.

Cracks were discovered in the floorbeam webs of the northbound structure adjacent to the girder-floorbeam connection. The location and orientation of these cracks suggest that they are the result of out-of-plane distortion. The floor system includes outrigger cantilever brackets which are attached to the top flange of the floorbeam by a tie plate which is not attached to the top flange of the girder. The distortion is apparently caused by differential displacement between the deck which is rigidly attached to the stringers and consequently the floorbeam top

A. S. Nowak (ed.), Bridge Evaluation, Repair and Rehabilization, 339–348.

flange and the top flange of the girder. The majority of this distortion is concentrated in the floorbeam web between the flange termination and the girder-floorbeam connection.

Distortion-induced fatigue cracking can be retrofitted in two different manners. The problematic detail can be made:

1) stiffer to preclude the distortion, or
2) more flexible to accommodate the distortion.

In general, the stiffness approach is more reliable, yet more costly. Sometimes in providing a more flexible detail, the distortion itself is increased, resulting in little or no improvement [1].

The Pennsylvania Department of Transportation retained Modjeski and Masters to analyze the as-built bridge using the Finite Element Method (FEM). The as-built bridge model was modified representing several retrofits which incorporate the principles of the two approaches described above. Based upon the FEM analysis, one or two more promising trial retrofits will be chosen for field testing. After the trial retrofits are evaluated, one retrofit will be selected for application on the entire bridge.

2. Crack Description

Cracks developed in the floorbeam web, as shown in Figure 2, at two distinct locations adjacent to the floorbeam-to-girder connection:

1) at the termination of the floorbeam flange-to-web weld where the flange is terminated short of the girder web to facilitate connection,
2) along the upper edge of the transverse connection plate which is welded to the girder web.

The cracking is apparent only at or adjacent to the girder supports (in this case, the abutments and the piers).

Cracks extend normal to principal stresses. The basically horizontal orientation of the cracking in a global stress field with horizontally oriented principal stresses suggests that the cracking is the result of distortion-induced local stresses.

The termination of the floorbeam flange is a large change in out-of-plane stiffness, magnified by the stress concentration of the weld toe.

The upper edge of the transverse connection plate also represents a large change in out-of-plane stiffness when the floorbeam web, moving out of plane, impinges on the connection plate.

3. Background

Similar cracking has been observed in two other bridges which have comparable floor systems; the Pennsylvania Turnpike's Allegheny River Bridge near Pittsburgh, Pennsylvania, and the Lehigh Canal and River Bridges along U.S. Route 22 near Allentown, Pennsylvania.

The more documented of these examples is the Lehigh Canal Bridge[2]. Several fatigue cracks were discovered in the tie plates connecting the floorbeams to outrigger cantilever brackets. The tie plates were tack-welded to the girders during fabrication. All of the cracks propagated from the tack welds at the edge of the tie plates.

Strain measurements on the Lehigh Canal Bridge showed that the cracking was the result of differential displacement between the deck-stringer system and the non-composite girders. The cracking was observed at locations along the bridge where the largest changes in girder rotation or slope occur. The time-histories (i.e., plots of strain with time due to the passage of live load) of measured horizontal bending strains in the tie plate corresponded with influence lines for girder slope. The incompatible displacement between the deck-stringer system and the girder is the difference between the expansion or contraction of the deck-stringer system with live load and the rotation of the girder times the distance between the centroid of girder and that of the deck-stringer system. This displacement was concentrated in the tie plate which represented a flexible interface between the deck-stringer system and the girder.

In the case of the I-83 bridge, the tie plate is only attached to the girder flange at the center floorbeam of each span. At the other floorbeam locations, the tie plate is not attached to the girder flange. Any differential displacement between the deck-stringer system and the girders is transmitted through the floorbeam web adjacent to the girder, where the floorbeam flanges have been removed to facilitate connection to the girder, which represents a relatively flexible interface. Thus, cracks developed in the floorbeam web.

4. Field Experimentation

The National Science Foundation's Engineering Research Center for Advanced Technology for Large Structural Systems at Lehigh University, under the direction of Dr. John W. Fisher, field-measured strains and distortions of the northbound bridge during July 1988. Strain and clip gages were placed on four floorbeams (two interior floorbeams, one pier floorbeam, and one floorbeam adjacent to a hanger) and their accompanying outrigger cantilever brackets. At the gaged floorbeam locations, the tie plates are not connected to the girder top flanges.

Strain measurements on the web in the coped regions of the floorbeam and bracket verified the hypothesis of significant out-of-plane distortions. The stress range excursions, corresponding to the larger measured strains, exceeded 10 kips per square inch, the constant amplitude fatigue threshold (Category C) hypothesised as being applicable to distortion-induced fatigue details. Thus, with enough stress range excursions, cracking is expected in these coped regions.

The strain gages were protected and left in place on the bridge with the intention of monitoring trial retrofits.

5. Analysis

The basis for the FEM analysis of the I-83 bridge is the zooming substructure technique. This technique was employed rather than modelling the bridge with one single discretization containing areas of very fine mesh relative to other areas of the model, possibly producing an ill-conditioned structural stiffness matrix. In this procedure, subsequent, separate analyses are performed on smaller and smaller "regional" discretizations or substructures encompassing the localized area under consideration. Each substructure is loaded along its external boundaries by displacements obtained from the immediately previous analysis. Thusly, the procedure "zooms" into the localized area of investigation.

A global discretization was generated which was representative of the entire bridge. This model initially was a five-span portion of the bridge recognizing symmetry about the centerline of the bridge. Because this model included the girders, the floorbeams, the stringers and the deck, in excess of 13500 degrees-of-freedom (DOF) resulted. To develop a more practical model, the five-span model was divided in two by recognizing symmetry about the midspan of the third span. This resulted in a two and a half-span model with 6500 DOF, as shown in Figure 3.

The substructure discretization included one floorbeam with girders, stringers and deck spanning one stringer span on each side of the floorbeam. Symmetry about the centerline of the bridge was again recognized producing a model with 2650 DOF, as ahown in Figure 4.

The FEM analysis involved two distinct phases. First, the as-built bridge (i.e., the condition which proved problematic in terms of fatigue cracking) was modelled. The results from this initial solution were compared with the field-measured strains to assure the integrity of the FEM models. Then, various retrofitting schemes were tried by modifying the original FEM models to incorporate the features of these schemes. With satisfactory retrofitting schemes chosen, trial retrofits will be applied to the bridge.

6. Proposed Retrofits

Four proposed retrofits have been modelled at both the global and substructure discretization level.

6.1. RETROFIT 1

In this model, a rigid attachment is made between the deck and the top flange of the girder for the entire length of the bridge. This scheme enforces composite behavior between the deck-stringer system and the girders. Although this retrofit ensures the elimination of the relative distortion between the deck-stringer system and the top flange of the girder, the implementation of the actual field retrofit is extremely difficult unless it is made during deck replacement.

6.2. RETROFIT 2

Retrofit 2 consists of attaching every other floorbeam tie plate to the top flange of the girder. This retrifit also attempts to reduce the distortion by increasing the stiffness of the distortion-sensitive detail. It is imperative to examine any adverse side effects which may be caused by superimposing an additional bending moment on the tie plate.

6.3. RETROFIT 3

Retrofit 3 provides increased flexibility in the girder-floorbeam connection by increasing the gap between the top of the floorbeam web and the first point of connectivity with the girder. To investigate the effectiveness of increasing the flexibility of the detail, web gaps of 6, 15 and 24 inches were evaluated.

6.4. RETROFIT 4

Finally, an X-bracing system in the plane of the top flanges of the floorbeam and girder was added to the models as Retrofit 4. The bracing intersects the girder flange at the midpoint between floorbeams and intersects the floorbeam flanges at the stringers on each side of the girder.

7. Analytical Results

The results of the FEM analysis of the as-built bridge, under an assumed live load, were of the same general magnitude as the field-measured random-traffic data. This was taken as verification of the FEM model.

Table 1 shows a comparison of the results of the analyses of the proposed-retrofit models, normalized to the as-built bridge model results. This table forms the basis for the following discussion of the retrofits.

7.1. RETROFIT 1

Retrofit 1 (providing a rigid attachment between the deck and the girder top flange) produces a drastic reduction in the pier floorbeam top flange moment and the maximum pier floorbeam web bending moment in the coped area. Since Retrofit 1 produced the best results, it was advanced to the next stage of trial retrofits.

7.2. RETROFIT 2

Retrofit 2 (attaching every other floorbeam tie plate to the girder top flange) produces only a modest decrease in the maximum pier floorbeam web bending moment while the pier floorbeam top flange moment is greatly increased. Retrofit 2 has been eliminated from further consideration because of its poor results and the distinct possibility of moving the distortion-induced fatigue problem from the floorbeam web to the tie plate.

7.3. RETROFIT 3

Retrofit 3 (providing an increased gap between the top of the floorbeam web and the first point of connectivity with the girder) had three subsets with web gaps of 6, 15, and 24 inches. The 24-inch web gap was the only one which produced satisfactory results. Retrofit 3 with a web gap of 24 inches was advanced to the next stage of trial retrofits since it produced the second best results of all the retrofits and its cost should be much less than Retrofit 1.

7.4. RETROFIT 4

Retrofit 4 (providing an X-bracing system in the plane of the top flanges of the floorbeam and girder) also produced a satisfactory results. Although the results for Retrofit 4 are similar to the results of Retrofit 3 with a web gap of 24 inches, its projected cost and potential for negative side effects are greater than those of Retrofit 3 with a web gap of 24 inches. Therefore, Retrofit 4 has been eliminated from further consideration.

8. Future Agenda

Detail modifications incorporating Retrofit 1 and Retrofit 3 with a 24-inch web gap will be designed and detailed. These trial retrofits can be included in another construction project scheduled for 1991 on this bridge. The retrofits will be monitored by collecting data from strain and clip gages.

A recommendation will be made as to which retrofit should be applied to the entire bridge. The possibility of developing a simplified method for evaluating retrofits for bridges with similar girder-floorbeam connection problems is to be investigated.

9. Summary

The FEM analysis of the I-83 bridge across the Susquehanna River at Harrisburg,Pennsylvania, has provided a means of evaluating several distortion-induced fatigue crack retrofits for the girder-floorbeam connection detail. Retrofit 1 and Retrofit 3 with a 24 inch web gap have been selected for trial retrofits through this evaluation. Retrofit 1 employs the approach of stiffening the detail to restrict the distortion by making a rigid attachment between the deck-stringer system and the girder top flange. Retrofit 3 with a 24 inch web gap increases the length over which the distortion is taking place by disconnecting the top portion of girder-floorbeam connection.

The remaining portion of this project will provide an opportunity for further comparison of the FEM analysis with the field measurements from the trial retrofits.

10. References

1. Mertz, Dennis R. (1984) "Displacement-Induced Fatigue Cracking in Steel Bridges," Ph.D. Dissertation, Lehigh University.
2. Fisher, John W., Yen, Ben T. and Daniels, J. Hartley (1977) "Fatigue Damage in the Lehigh Canal Bridge from Displacement-Induced Secondary Stresses," Transportation Research Record 607, 56-62.

Table 1 Retrofit Comparison

	AS-BUILT STRUCT	RETROFITS					
		1	2	3			4
				6"	15"	24"	
PIER FLBM TOP FLG MOMENT	1.00	0.04	2.58	0.43	0.30	0.17	0.32
MAX FLBM WEB BEND. MOMENT	1.00	0.01	0.63	0.75	0.38	0.16	0.15

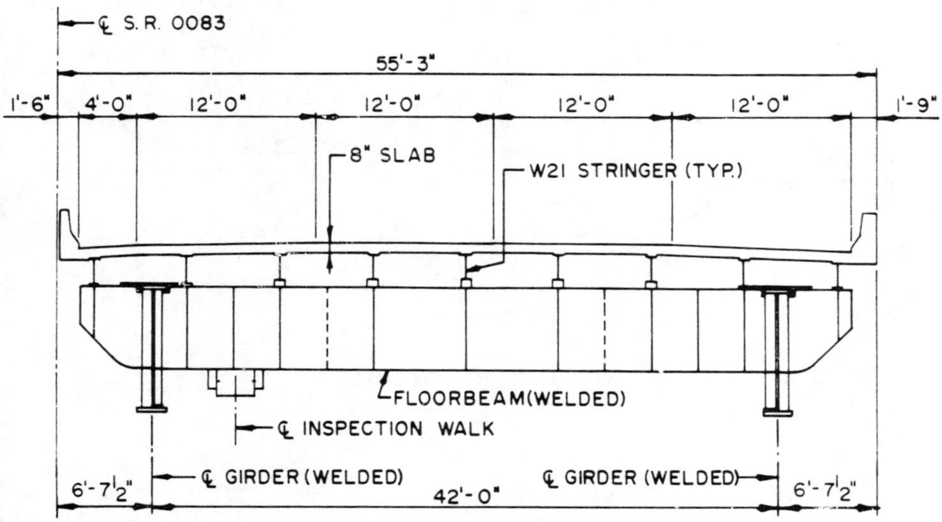

Fig. 1 Typical Bridge Cross Section.

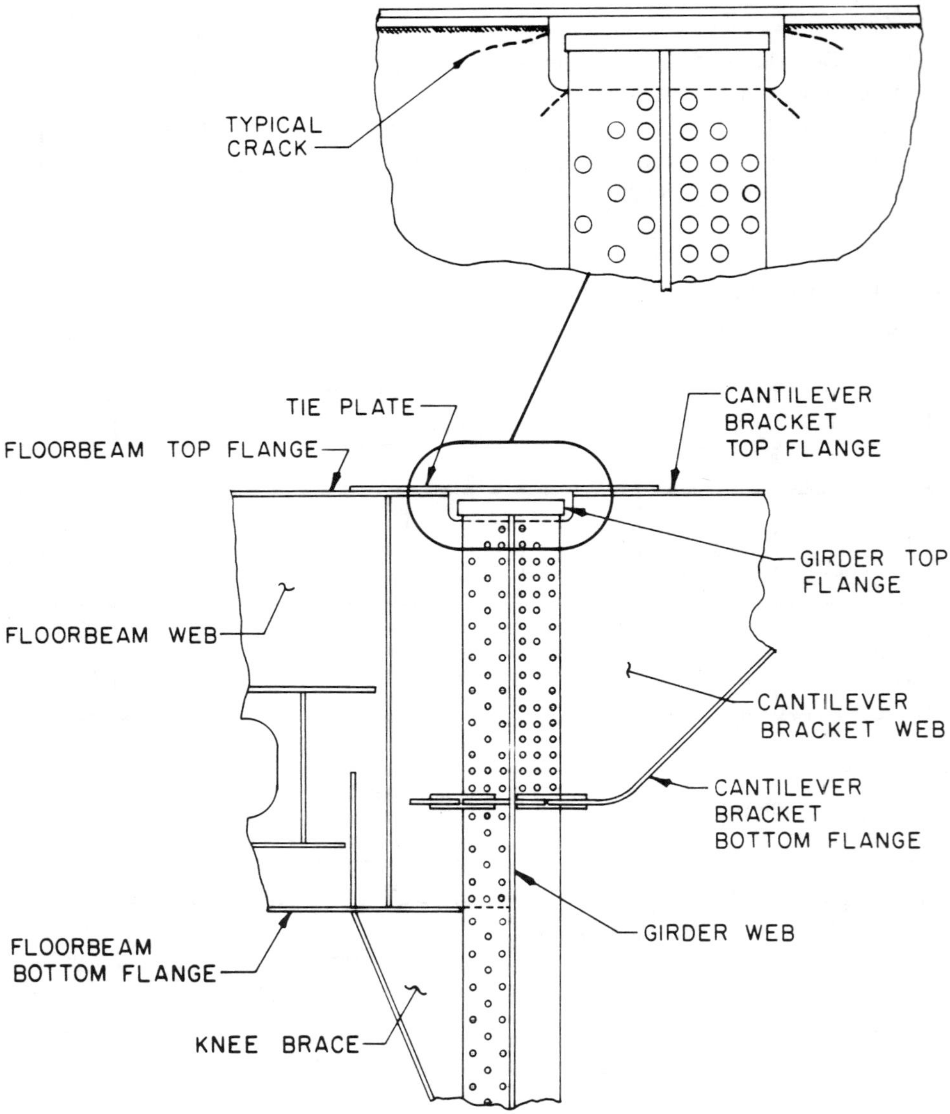

Fig. 2 Crack Locations Near Girder-Floorbeam Connection.

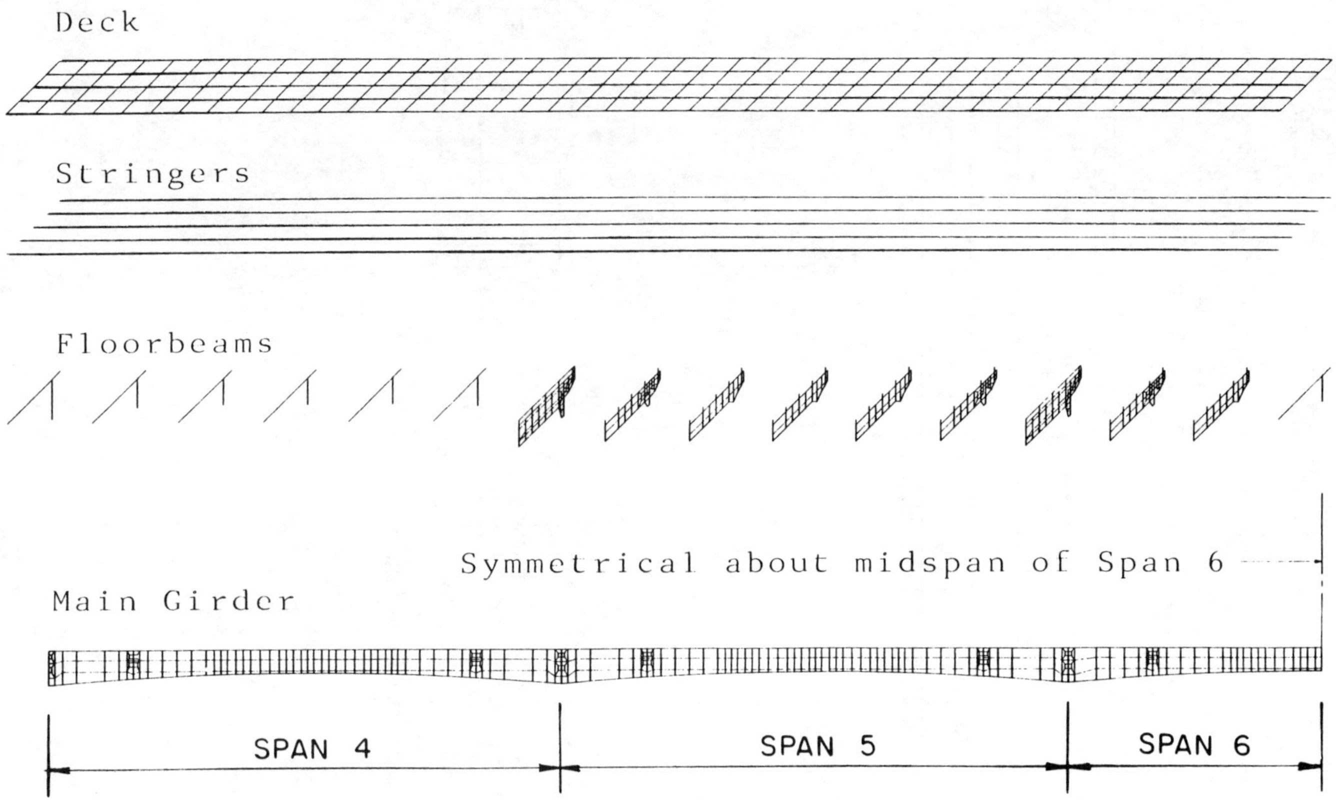

Fig. 3 Global Discretization.

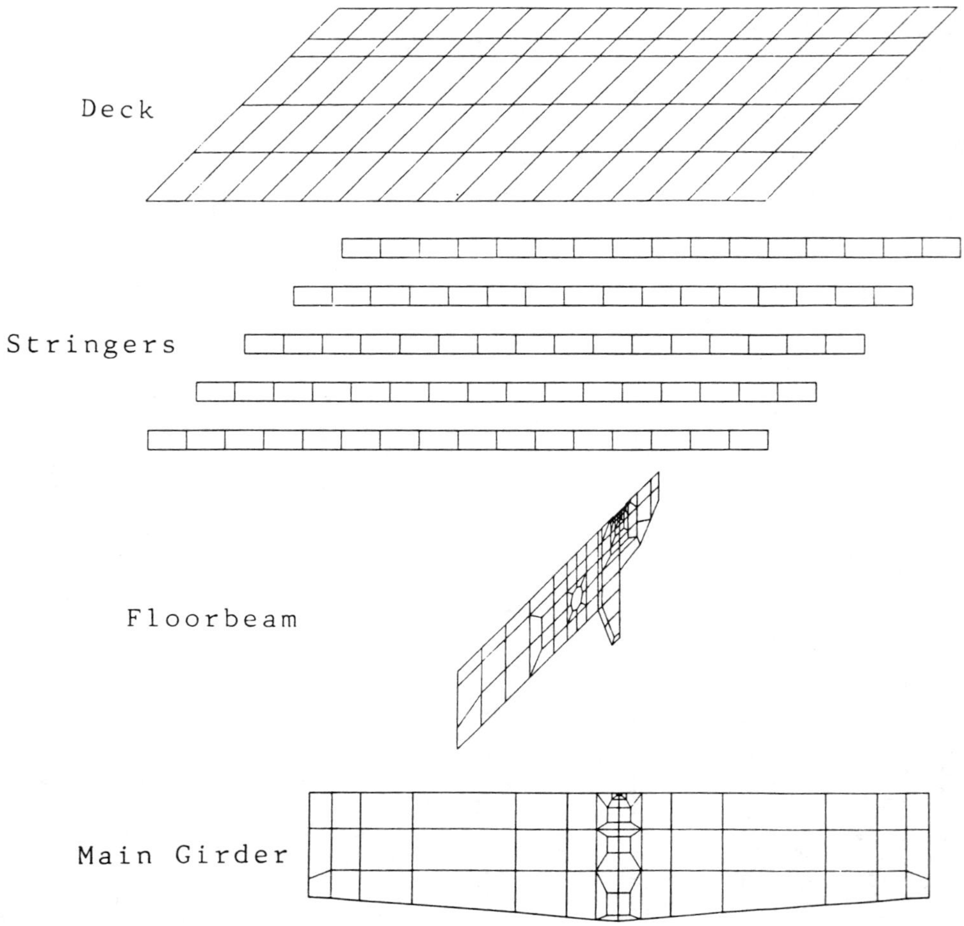

Fig. 4 Substructure Discretization.

NON LINEAR INTERACTION OF LONGITUDINAL AND TRANSVERSE LOAD EFFECTS IN CONCRETE BOX GIRDER BRIDGES

G. FANTI
Sintecna
Via Piffetti 16
Turin, ITALY

G. GAMBERINI and G. MANCINI
Dept. of Structural Enrg.
University of Cagliari
Piazza d'Armi
Cagliari, ITALY

ABSTRACT. A mechanically non linear procedure for the evaluation of the interaction between longitudinal and transverse resisting mechanisms in folded plated structures is proposed.

Both the analysis procedure and the considerations concerning the properties of the materials and reliability are developed in accordance with the recommendations of the CEB-FIP Model Code 1990 Predraft.

The proposed procedure is applied to a prestressed concrete bridge subjected to concentrated live loads applied in different positions.

A comparison is made so as to evidence the higher level of safety attained by means of a non linear analysis compared to a linear one.

The effect of the folded plate action is outlined, also, as determined by means of a linear and non linear procedure applied to a frame of unit length assumed to be independent of the whole structure.

1. Introduction

Concrete box-girder bridges essentially amount to folded plate structures, in which the loads are carried three-dimensionally.

The evaluation of the internal forces and moments in the different plates forming the box section is generally performed in the linear field by means of the following calculation procedures: direct application of the theory of elasticity [1], beam on elastic foundation analogy (BEF analogy) [2], folded plate theory [3], finite strip method [4,5], finite element method.

In current design practice, the second approach (BEF analogy) is frequently used because it can be easily handled and it offers a sufficient degree of accuracy [6].

However, a correct evaluation of the ultimate safety of the bridge can be achieved only by means of a nonlinear analysis, performed step-by-step up to the attainment of the ultimate limit state of the structure.

The advantage of the non linear approach is that it emphasizes the very effective interaction between longitudinal and transverse resisting mechanisms.

A. S. Nowak (ed.), Bridge Evaluation, Repair and Rehabilization, 349–363.

In this paper, the authors illustrate an updated version of the BEF analogy procedure, extended to the mechanically non linear field. The analysis was performed in accordance the criteria suggested by '90 CEB-FIP Model Code Pre-draft concerning the reliability and the stress-strain properties of the materials.

2. BEF Analogy Approach

As is known, the BEF analogy procedure entails the following steps for the verification of a box section beam.

2.1 FIRST STEP

For each load arrangement, the first step of the analysis in the longitudinal direction is performed as for a one-dimensional beam-element, assuming the cross-section to have a rigid shape; the output of this analysis is the longitudinal internal forces and moments (S_L).

2.2 SUBSEQUENT STEPS

In the transverse direction, the presence of nonsymmetrical loads entails a strict correlation between longitudinal and transverse effects (folded plate action), the evaluation of which can be completed in several steps (Figure 1):

2.2.1 *Analysis of a unit length transverse frame*, subjected to all the acting loads (dead weight, g_1, permanent loads, g_2, variable loads, q) and to the internal effects, applied along the edges, of the corresponding internal forces and moments as determined in the longitudinal analysis (Figure 1); at this stage, the frame is assumed to be unrestrained. The output of this step consists of a first set of transverse internal forces and moments (S_T).

2.2.2 *Analysis of the antisymmetrical part of the acting loads*, by restraining the frame through the adoption of two very stiff internal diagonals (figure 2). The purpose of this step is an evaluation of the restraining diagonal forces (N/2).

The components of such fictitious forces, N/2, along the plates are carried by the webs and flanges, not only lengthwise (through a flexural mechanism) but also crosswise (the entire box section behaving as a closed frame).

2.2.3 *Evaluation of the subdivision of the diagonal forces* into longitudinally carried (N_L) and transversely carried (N_T) parts. The objective is achieved by considering the box beam as a fictitious beam on an elastic foundation described by the equation:

$$EI_i \frac{d^4y}{dx^d} + Ky(x) = q(x)$$

in which:

I_i stands for an ideal moment of inertia of the box section, taking into account compatibility conditions along the connection between the different plates (flexural mecha nism);

K corresponds to an ideal Winkler modulus, representative of the resistance of the box section against diagonal forces (frame mechanism);

q(x) is the component of the diagonal forces in a web or plate.

2.2.4 *Application of the longitudinally carried part of the diagonal forces* (N_L) to the free-shape frame (see Figure 3) and evaluation of the resulting internal forces and moments (S_{NL})

2.2.5 *Superposition of the internal forces and moments* evaluated at paragraphs 2.2.1 (S_T) and 2.2.4 (SNI), providing the elastic solution to the problem in the transverse direction.

2.2.6 *Calculation of the variation in longitudinal stresses longitudinal stresses in the plates* due to the longitudinally carried part of the diagonal forces (N_L).

3. Extension of the BEF Analogy to the Non-linear Field

A realistic evaluation of the behavior of the R.C. elements making up the box beam, either in the longitudinal or the transverse direction, can be achieved by adopting the nonlinear moment rotation diagrams (Figure 4) plotted in accordance with the '90 CEB-FIP Model Code provisions [7].

The significant nonlinearity of the M -ϑ relationship (owing to the intervention of cracking, the yielding of the reinforcement and final plastic response) entails the impossibility of analyzing the problem in closed form; the only viable alternative is to follow an incremental step-by-step procedure. In this manner a possible flow-chart of the procedure is worked out; subsequently, the same method is applied again to evaluate the ultimate safety of a prestressed box girder bridge.

3.1 DRAWING A MESH OF THE STRUCTURE

First of all, a mesh of the structure has to be drawn in the transverse direction (the box frame is represented by a longitudinal beam element); the length of the mesh elements should be 0.5 to 2 times the height of their section, which corresponds to the normal extension of the plastic zones.

The first step is to select a transverse beam element (reference element) and to define the load arrangement under which this element is critical. In agreement with recent proposals concerning reliability provisions in non-linear analysis [8], the following mechanical properties are ascribed along the structure:

- the behavior of the reference element is described by means of factorized values of the mechanical strength (γ_c = 1.5 for concrete and γ_s = 1.15 for steel);
- in all the other elements mean strength values are adopted.

3.2 PRELIMINARY ELASTIC ANALYSIS

A preliminary elastic analysis in the presence of service loads is carried out as follows:

- the reference longitudinal beam element behaving as an unrestrained frame is analyzed in the presence of the whole set of actions (see para. 2.2.1)
- the other longitudinal beam elements to which the external load is applied are analyzed in the same conditions in the presence of unitary actions (actual internal forces and moments calculated by proportionality);
- all the longitudinal elements (including the critical one) behaving as restrained frames with very stiff diagonals (para. 2.2.2) are analyzed in the presence of the antisymmetrical part of the load. In this case too, proportionality can be applied. The basic equation of the BEF approach is then substituted with a numerical analysis of a longitudinal beam supported on elastic restraints; the BEF model includes all the longitudinal beam elements, each of them connected to the ground via an elastic restraint (Figure 5) representing the frame mechanism effect (see para 2.2.3).

The acting load on the elastically restrained beam coincides with the diagonal force components along the web of the section, as evaluated in section 2.

The output of this analysis, i.e., the displacements of the elastic restraints, is then used to calculate the longitudinal distribution, along the beam, of the diagonal forces resulting from the folded plate action.

The last step of this preliminary linear analysis is now the evaluation, in all the cross box frames, of the transverse internal forces and moments produced by the effects of all the acting loads, including the diagonal forces (para. 2.2.5).

3.3 STEP-BY-STEP PROCEDURE

First of all the "controlling event", defining the size of a step, is identified when one of the following "conditions" is attained in any longitudinal or transverse beam element (the governing elements):

- cracking bending moment;
- reinforcement yielding bending moment;
- limit plastic rotation or ultimate resisting bending moment.

At each step, the distribution of local rigidities is updated, by taking into account the level of the internal forces and moments and the corresponding rotations. Moreover, an element by element comparison is made between acting and resisting internal forces and moments. After that, the minimum load amplification factor necessary to reach the next event is evaluated. This procedure is then repeated up to the attainment of the ultimate limit state.

If the "event" defining the step occurs in the transverse direction, two different procedures are adopted depending on whether the governing longitudinal element is loaded directly by the antisymmetrical load or is not loaded. In the first case, the diagonal forces are updated by referring to the rigidities corresponding to the actual level of the internal forces and moments; the parameter of the plastic restraint of the longitudinal element is also updated for the calculation of the beam on elastic restraints. In the other case, the load condition of the beam on elastic restraints remains unchanged and the updating only concerns the elastic restraint corresponding to the longitudinal beam element which governs the step.

If the "event" occurs in the longitudinal direction, the updating is limited to longitudinal flexural rigidities.

The next step consists of calculating the new distribution of internal forces and moments within the transverse frames (on the basis of the updated values of the rigidities). This gives the value of the local load amplification factor which determines the next "event". The size of the next step is defined by the value of the minimum load amplification factor in the different elements.

The iteration can be halted when a limit plastic rotation or a limit ultimate bending moment is reached in an element, unless the structure is transformed into a pseudomechanism at a lower load level.

4. Application to a P.C. Box-Girder Bridge

The non linear procedure described above is now applied to a prestressed girder bridge, spanning over 40 meters, whose cross-section is shown in Figure 6. The bridge is taken to be simply supported and restrained by rigid frames at the support sections.

The constitutive materials have the following properties;

- concrete: f_c = 37.5 MPa; f_{ck} = 33.2 MPa; f_{cd} = 22.1 MPa
- steel: f_{ym} = 541 MPa; f_{yk} = 500 MPa; f_{yd} = 43.5 MPa

The acting loads are:

- dead weight of the box beam;
- paving: 25 MPa;
- new jersey: 8 + 8 KN/m;
- live load applied to the cantilever region of the section, representing a truck with the characteristics given in Figure 7 (total weight 600 KN); this load is amplified by the dynamic coefficient ϕ = 1.4.

Distribution conditions of the live load on the cantilever region have been determined by means of a FEM analysis, as shown in Figure 8.

Two different load conditions are taken into consideration, corresponding to the live load acting at 1/2 or 1/4.

The mesh used for the cross sections of the beam is shown in Figure 9. Table 1 lists the geometrical parameters of the different transverse beam elements. Table 2 lists the values of the "event" resisting bending moment and Table 3 gives the corresponding rotation value in an element with a length equal to its height.

The mesh adopted in the longitudinal direction is not the same for the different load cases; the meshes relating to the different positions of the live load on the beam are illustrated in Figures 10 and 11.

4.1 SYMMETRICAL LOAD CONDITION

In this case of a symmetrical load condition (live load in 1/2), the ultimate limit state of the structure is reached in section 9 after 90 steps (starting from the serviceability level), in the initial node of beam element 36 (node 11). A description of the non linearities appearing during the evolution of the analysis is given in Table 4. In Figures 12, 13, 14 and 15, the evolution of the bending moment vs. live load amplification factor, γ_F, is illustrated for a number of characteristic sections located in the proximity of the critical region. The redistribution of internal forces and moments from the web to the top flange is very effective, in particular after the yielding of the reinforcement in the critical region.

As concerns the safety coefficient, γ_F, it should be underscored that its value is found to be 1.717 from linear analysis and 2.139 from the non linear analysis, which means that an increase of 24.6% in load carrying capacity is determined on the basis of the latter.

4.2 NON SYMMETRICAL LOAD CONDITION

In the case of a non symmetrical load condition (live load in 1/4), the ultimate limit state of the structure is reached after 94 steps, at the same point as in the previous case (initial node of element 36 belonging to longitudinal beam element 8). A description of the non linearities appearing during the evolution of the analysis is given in Table 5.

Figures 16, 17, 18 and 19 describe the evolution of bending moment vs. γ_F, with reference to the more characteristic sections lying near the region in which the ultimate limit state is reached (the same sections as were used in the previous case).

It is of interest to notice that the diagrams virtually coincide with the one corresponding to the symmetrical load condition, which means that the longitudinal beam element which includes the localized defect (γ_c = 1.5 and γ_s = 1.15) and is subjected to the maximum value of the live load, receives about the same help from the remaining part of the beam as in the previous case. This phenomenon is associated with the high level of longitudinal prestressing introduced for the cantilever construction of the bridge. Because of high prestressing, the influence of longitudinal in the presence of a single truck is very small.

As for the safety coefficient, γ_F, its value is found to be 1.722 from linear analysis and 2.127 from non linear analysis; hence, in this case, the adoption of a non linear approach leads to an increase in the calculated value of the load carrying capacity of 23.5%.

4.3 SIMPLIFIED APPROACH

A simplified approach has been defined by taking into account the longitudinal beam element carrying the highest proportion of the load and neglecting the folded plate effect. This hypothesis is equivalent to the application of all loads to the reference frame supposed to be unrestrained.

Under these circumstances, γ_F takes on the following values:

- linear analysis: $\gamma_F = 2.084$

- non linear analysis: $\gamma_F = 2.286$

The ratio of the two values is 1.083, which means that the increase in load carrying capacity is reduced to 8.3%, due to the fact that in this case the redistribution can work only in the transverse direction.

Moreover, in this case, the ULS is reached in the initial node of beam element 36, but, with a value of 2.225, the yielding bending moment is reached in the final node of beam element 31 too.

5. Conclusions

The Non linear procedure for the evaluation of the actual bearing capacity of box girder bridges can be likened to an extension of the well known non linear procedure used for one-dimensional elements.

Obviously, it would also be theoretically possible to extend to the non linear field the more highly sophisticated calculation procedures listed in the introduction. But such an attempt would require a redefinition of the two-dimensional properties of two-dimensional R.C. elements, a field in which the available literature is still uncertain.

6. References

[1] Kristek, V. (1979) "Theory of Box Girder Bridges," John Wiley and Sons, Chichester, new York, Brisbane, Toronto.

[2] Dabrowski, R. (1965) "Der Schubverformungseinfluss auf fie Wölbkrafttorsion der Kastenträger mit Verfombarem Biegesteifen Profil', Der Bauingenieur, 11/65.

[3] Goldberg, I.E., Leve, H.L., (1975) "Theory of prismatic folded plate structures," IVBH, Vol. 17.

[4] Cheung, Y.K. (1976) "Finite Strip Method in Structural Analysis," Pergamon Press, Oxford, New York, Toronto, Sydney, Paris, Frankfurt.

[5] Chung, Y.K. (1971) "Analysis of Box Girder Bridges by the Finite Strip Method," ACI Publications, SP-26, Vol. 1.

[6] Schlaich, J., Scheef, H. (1982) "Concrete Box Girder Bridges," Structural Engineering Document 1e, IABSE, ETH Hönggerberg,, Ch-8093, Zurich, Switzerland.

[7] CEB-FIP, "Model Code 1990 Predraft" (1988), CEB Bulletin 190a-190b, Dubrovnik, September.

[8] Levi, F., Mancini, G., and Napoli, P. (1983) "The Problem of Model Uncertainty in the Structural Analysis of Concrete Structures," Atti dell'Accademia delle Scienze di Torino, Volume 122, Fasc. 3-4, May-August.

Table 1 Geometrical Parameters of the Different Transverse Sections

Section Type	Width (m)	Height (m)	A_1 (cm^2)	A_2 (cm^2)	Arrangement of Reinforcement
1	1.00	0.68	10.05	10.05	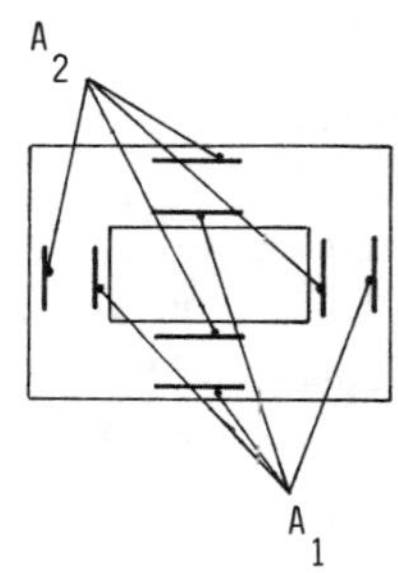
2	1.00	0.29	10.05	10.05	
3	1.00	0.22	10.05	5.65	
4	1.00	0.18	5.65	10.05	
5	1.00	0.61	22.62	22.62	
6	1.00	0.40	22.62	22.62	
7	1.00	0.90	22.62	22.62	
8	1.00	0.83	23.56	5.65	
9	1.00	0.48	23.56	5.65	
10	1.00	0.29	23.56	15.71	
11	1.00	0.20	15.71	15.71	
12	1.00	0.40	22.62	22.62	

Table 2 "Event" Resisting Bending Moment in the Different Sections.

Section Type	$M_{cr}<0$ (KN·m)	$M_y<0$ (KN·m)	$M_u<0$ (KN·m)	$M_{cr}>0$ (KN·m)	$M_y>0$ (KN·m)	$M_u>0$ (KN·m)
1	269.4	269.4	269.5	272.9	468.8	500.5
2	58.4	135.8	139.4	58.4	135.6	139.2
3	34.3	129.2	140.9	33.9	101.1	110.4
4	23.2	44.1	48.8	23.6	76.6	81.9
5	227.9	424.6	433.8	233.8	763.5	917.1
6	111.4	424.8	435.8	111.4	424.8	435.6
7	461.5	867.7	971.0	462.1	915.8	1030.9
8	398.3	398.3	398.7	408.8	1010.3	1079.4
9	145.1	145.1	145.2	150.0	532.6	549.6
10	59.2	173.8	210.7	61.5	326.3	379.8
11	30.3	133.3	137.5	30.3	133.3	137.6
12	79.9	338.5	348.4	79.9	338.5	348.6

Table 3 Rotation Value Corresponding to One "Event" ($\varphi = 1000\ h/r$)

Section Type	$\varphi_{cr} < 0$	$\varphi_y < 0$	$\varphi_\mu < 0$	$\varphi_{cr} > 0$	$\varphi_y > 0$	$\varphi_\mu > 0$
1	0.20	0.20	26.70	0.21	3.24	29.74
2	0.22	3.46	14.90	0.22	3.46	14.90
3	0.23	3.93	11.60	0.23	3.82	11.89
4	0.24	3.68	10.08	0.24	3.97	10.65
5	0.20	5.40	28.30	0.21	3.52	25.70
6	0.22	3.53	19.17	0.22	3.53	19.17
7	0.20	4.94	35.44	0.20	4.71	37.60
8	0.20	0.20	32.56	0.20	3.26	35.63
9	0.21	0.21	18.89	0.22	3.50	22.01
10	0.22	3.64	14.94	0.23	4.09	12.20
11	0.23	4.04	11.25	0.23	4.04	10.95
12	0.17	2.92	18.57	0.17	2.92	18.42

Table 4 Map of Non Linear Behaviour at the Last Load Step.

ELEMENT No.	CRACKED ELEMENTS (i, f = initial, final section)	YIELDED ELEMENTS	BROKEN ELEMENTS
Ø	36i/19f	-	-
1	36i/19f	-	-
2	36i/19f	-	-
3	19f	-	-
4	19i/18f/19f	-	-
5	19i/18f/19f	-	-
6	19i/36i/37i/38i/18f/19f/31f/32f/36f/37f	-	-
7	19i/31i/36i/37i/38i/18f/19f/30f/31f/32f/36f/37f	-	-
8	19i/24i/31i/36i/37i/38i/39i/18f/19f/30f/31f/32f /33f/36f/37f/38f	-	-
9	3i/4i/5i/6i/18i/19i/24i/25i/30i/31i/32i/33i/36i 37i/38i/3f/4f/5f/12f/13f/17f/18f/19f/24f/29f/ 30f/31f/32f/33f/36f/37f	36i	36i

Table 5 Map of Non Linear Behaviour at the Last Load Step.

ELEMENT No.	CRACKED ELEMENTS (i, f = initial, final section)	YIELDED ELEMENTS	BROKEN ELEMENTS
1	36i/19f	-	-
2	19f	-	-
3	19f	-	-
4	19i/18f/19f	-	-
5	19i/36i/37i/38i/18f/19f/31f/32f/36f/37f	-	-
6	19i/31i/36i/37i/38i/18f/19f/30f/31f/32f/36f/37f	-	-
7	19i/24i/31i/36i/37i/38i/39i/18f/19f/30f/31f/32f 36f/37f/38f	-	-
8	3i/4i/5i/6i/18i/19i/24i/25i/30i/31i/32i/33i/36i 37i/38i/3f/4f/5f/11f/12f/13f/17f/18f/19f/24f/ 29f/30f/31f/32f/33f/36f/37f	36i	36i
9	19i/24i/31i/36i/37i/38i/39i/18f/19f/30f/31f/ 32f/33f/36f/37f/38f	-	-
10	19i/31i/36i/37i/38i/18f/19f/30f/31f/32f/36f/37f	-	-
11	19i/36i/37i/38i/18f/19f/31f/32f/36f/37f	-	-
12	19i/18f/19f	-	-
13	19i/18f/19f	-	-
14	19f	-	-
15	36i/19f	-	-
16	36i/19f	-	-
17	36i/19f	-	-
18	36i/19f	-	-
19	36i/19f	-	-

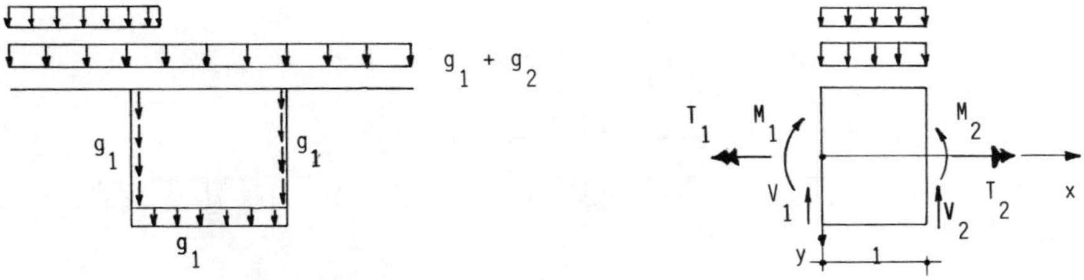

Fig. 1 Acting Loads on the Unrestrained Frame.

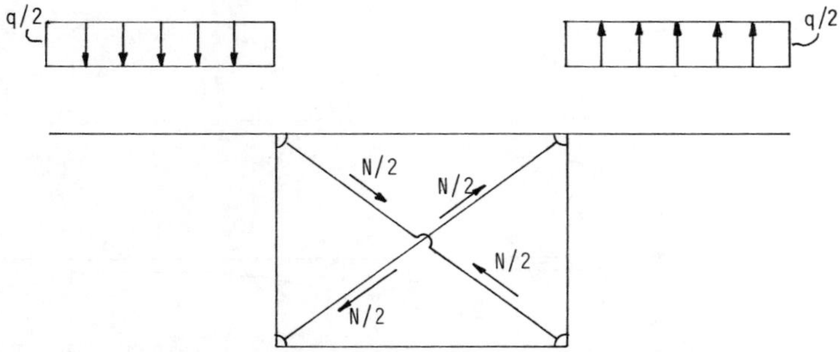

Fig. 2 Antisymmetrical Load on the Rigid Shape Frame.

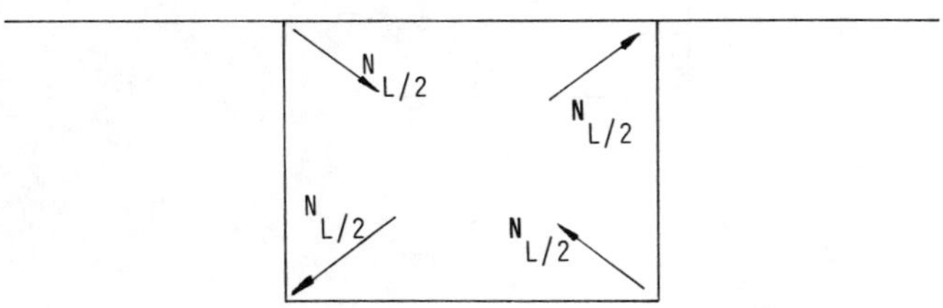

Fig. 3 Longitudinally Carried Diagonal Forces on the Free-Shape Frame.

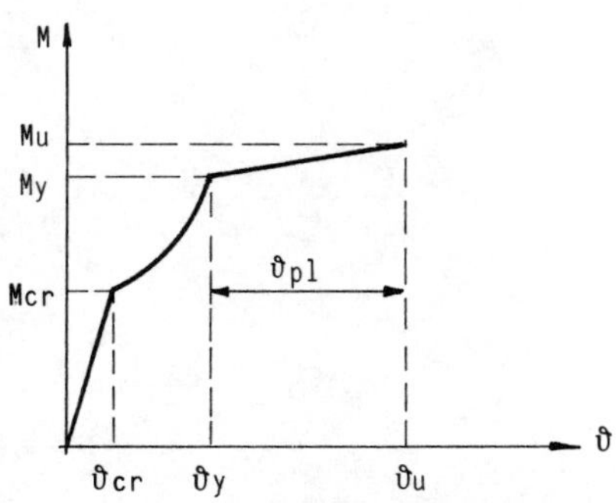

Fig. 4 Moment-Rotation Diagram for a Reinforced Concrete Section.

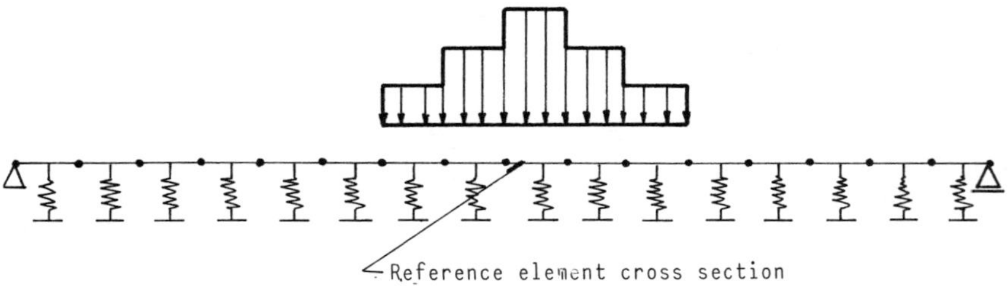

Fig. 5 Beam on Elastic Restraints.

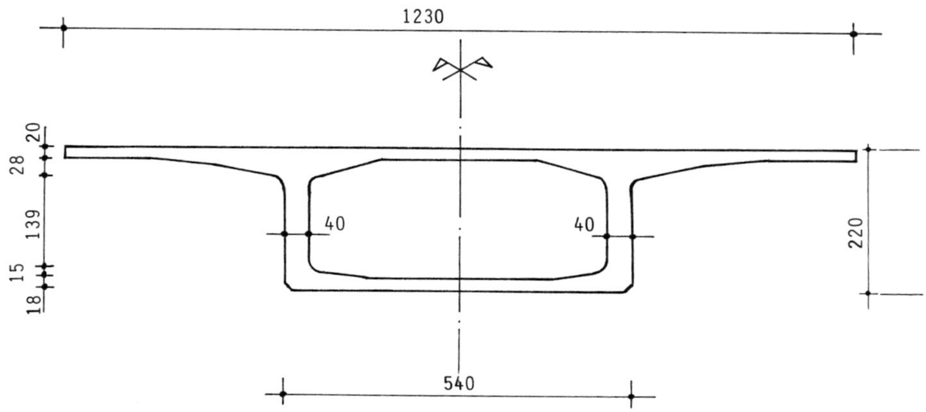

Fig. 6 P.C. Bridge Cross Section.

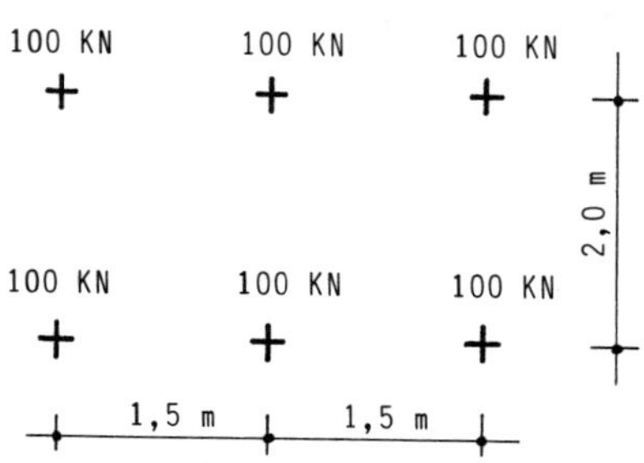

Fig. 7 Live Load Applied on the Bridge.

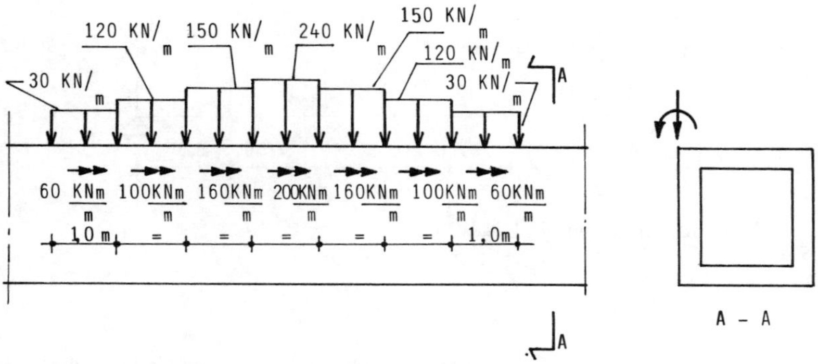

Fig. 8 Live Load Distribution on the Longitudinal Edge.

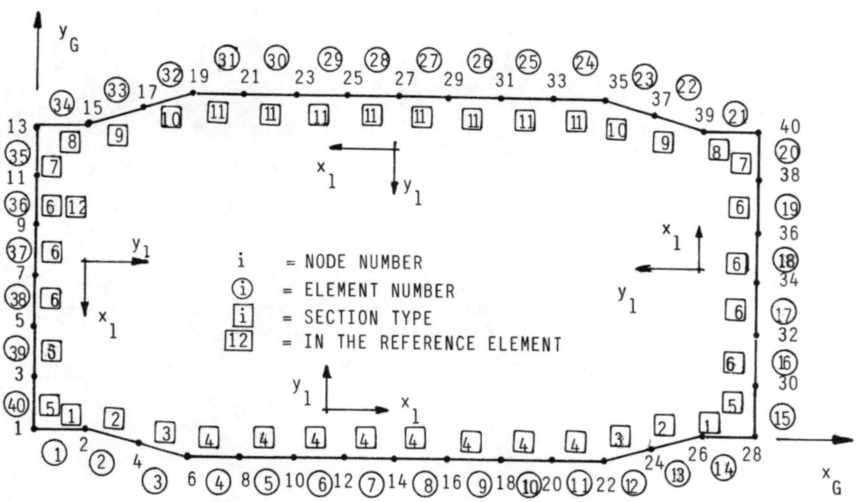

Fig. 9 Cross Section Mesh.

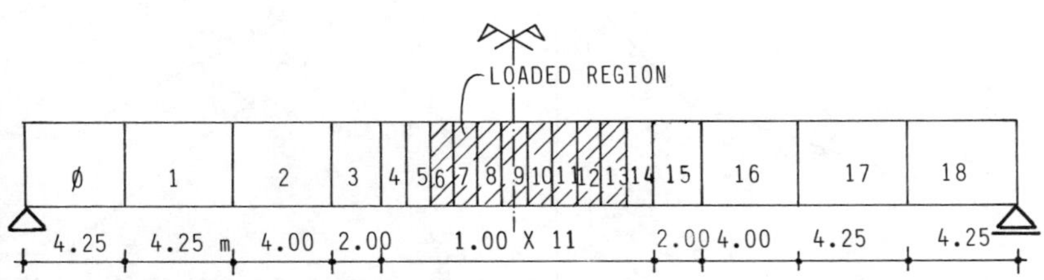

Fig. 10 Longitudinal Mesh for Live Load in L/2.

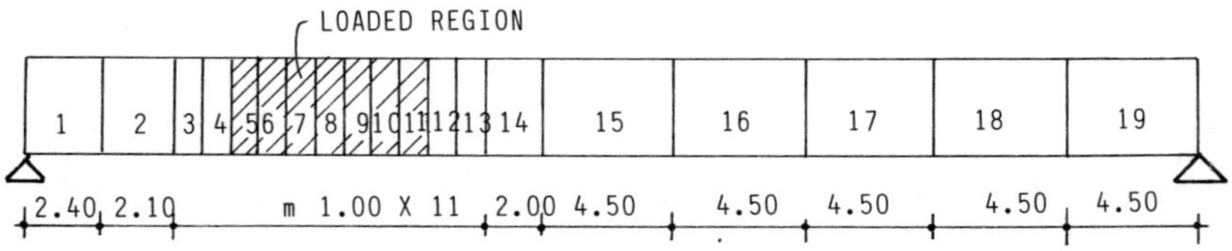

Fig. 11 Longitudinal Mesh for Live Load in L/4.

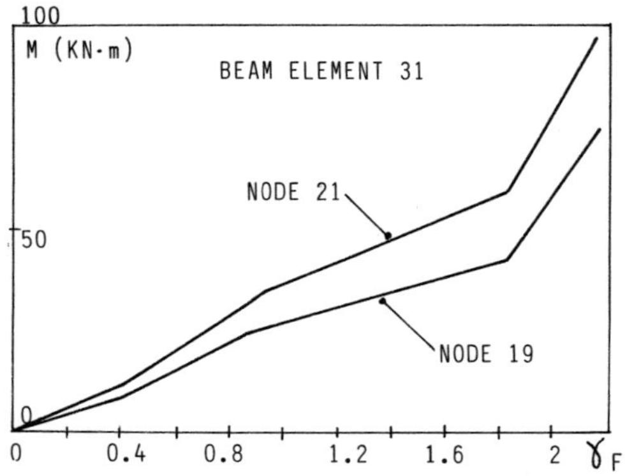

Fig.12 Bending Moment Evoluation Versus γ_F.

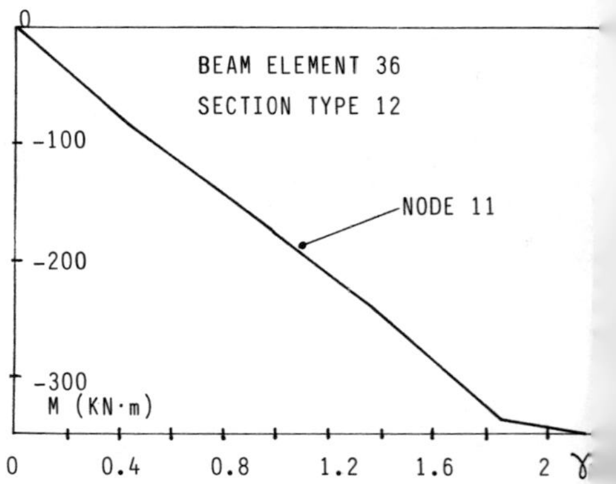

Fig. 13 Bending Moment Evolution Versus γ_F.

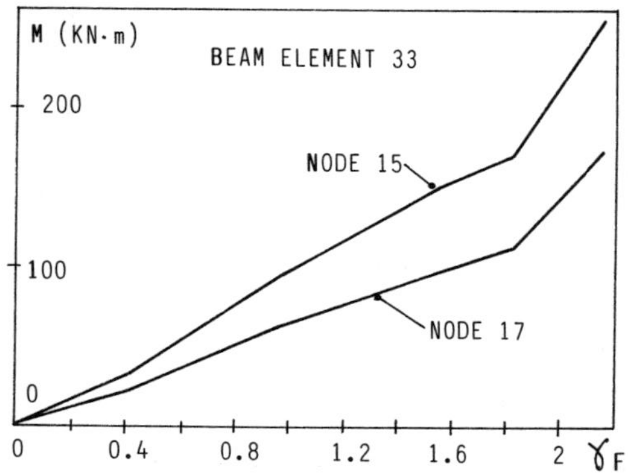

Fig.14 Bending Moment Evolution Versus γ_F.

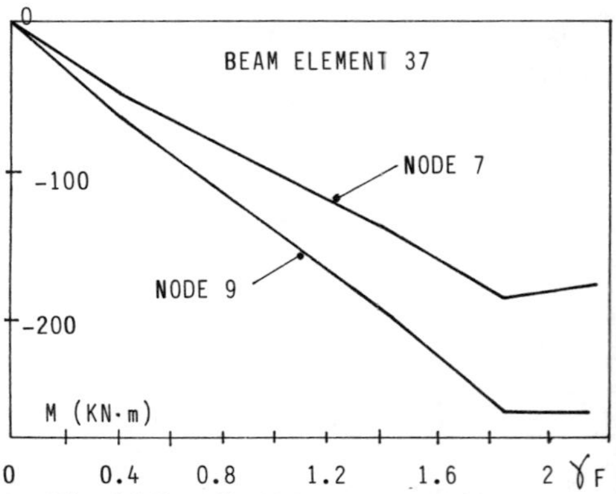

Fig. 15 Bending Moment Evolution Versus γ_F.

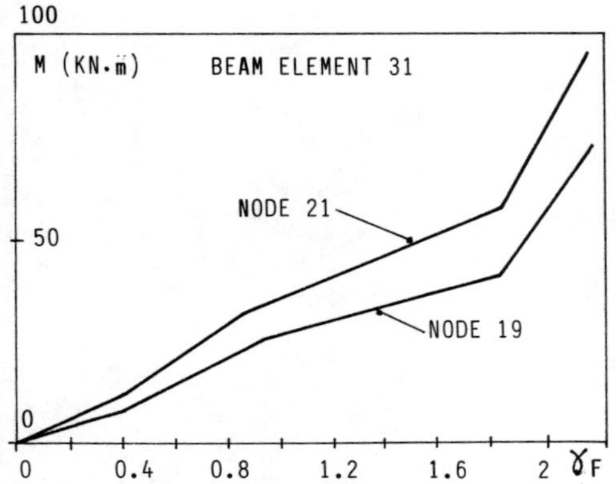

Fig.16 Bending Moment Evolution Versus γ_F.

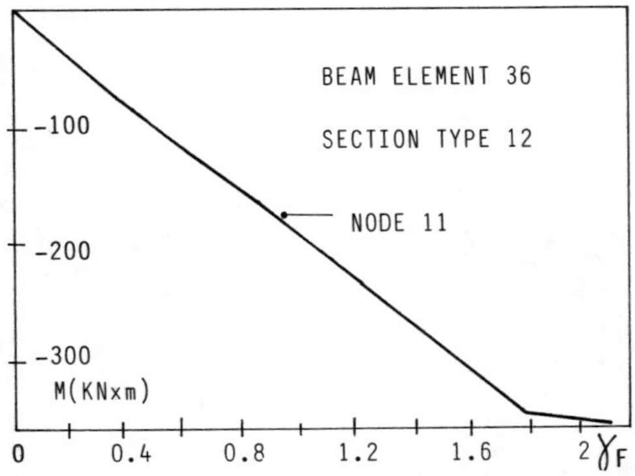

Fig. 17 Bending Moment Evolution Versus γ_F.

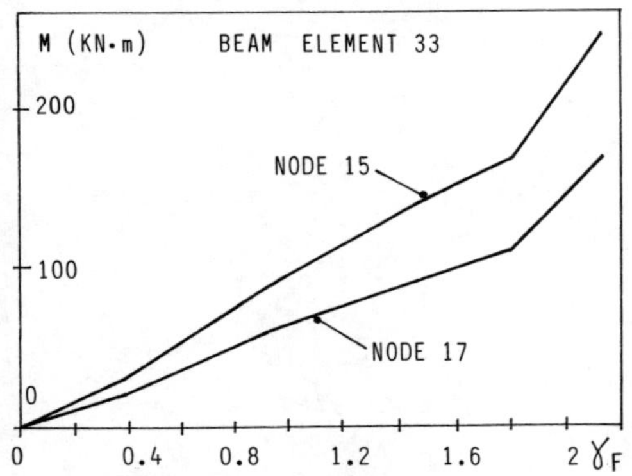

Fig.18 Bending Moment Evolution Versus γ_F.

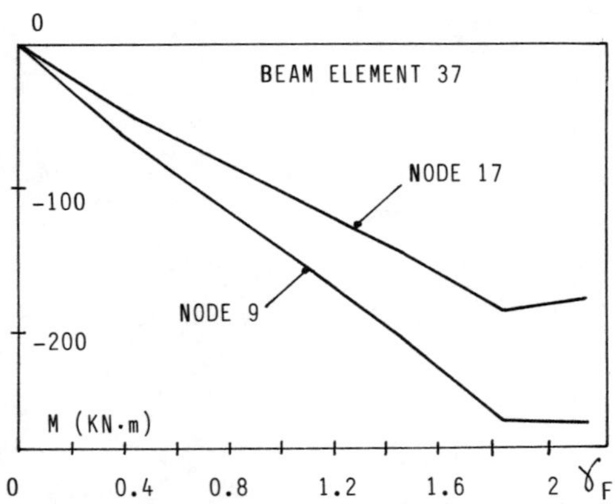

Fig. 19 Bending Moment Evolution Versus γ_F.

IV - Evaluation and Tests

EVALUATION OF BRIDGES - EUROPEAN PERSPECTIVE

G. P. TILLY
Transport and Road Research Laboratory
Crowthorne
Berkshire, UNITED KINGDOM

ABSTRACT. Concrete bridges have been in use for some 90 years but are now beginning to present problems for assessing corrosion of the reinforcing and prestressing steel. Corrosion is being experienced in most European countries due to the use of deicing salt during cold weather. Another factor is that design codes have changed over the years and many concrete bridges are having to be reassessed to determine their load carrying capacity under present loading.

This paper is concerned with evaluation of bridges in relation to corrosion and load carrying capacity.

1. Introduction

In Europe there is a wide range of types of highway bridge representing the different ages of construction and different materials. Strengths vary from one country to another according to loading requirements and it is only recently that a start has been made on the harmonisation of loading standards.

Corrosion of reinforcing and prestressing steel is being experienced in most European countries and there has been an increasing awareness of the susceptibility of post tensioning tendons. Until recently The Netherlands have been an exception in that corrosion has not been reported and this was ascribed to using good quality concrete and great care in quality control during construction. However the position is changing and it is now recognized that chlorides are penetrating the concrete.

There are many historic bridges that are still carrying unrestricted traffic apparently without undue trouble. A large proportion of these are masonry arches mostly over 100 years old and some that are much older. There are also numerous cast iron and wrought iron bridges having special features and classified as being of historic interest. Many would fail a normal assessment but actually have adequate reserves of strength or can be strengthened fairly easily. In consequence special assessments have had to be carried out to take account of materials and structural sections not designed into modern bridges.

In the time since the late 1940s lightweight bridges having orthotropic steel decks have been developed for long span crossings. These present special considerations for assessing the fatigue strengths of welded connections subject to wheel loadings. Collaborative research to

A. S. Nowak (ed.), Bridge Evaluation, Repair and Rehabilization, 367–377.

determine traffic loading, stress ranges and classification of the connections has been supported by the European Coal and Steel Community. The programme has involved research establishments from France, West Germany, Italy, Netherlands, Belgium and Britain.

Following earlier collapses of several steel bow girders much research was done to improve the evaluation of compressive stability. This has been mainly completed and attention has shifted to concrete bridges.

2. Detection of Corrosion

Visual inspection locates deterioration only when it has reached a fairly advanced stage and it is necessary to use instrumentation, preferably non destructive, to obtain an earlier warning.

2.1 REINFORCING STEEL

The methods commonly available for detecting corrosion of the buried steel can give ambiguous if not misleading results and have to be handled with care. It is therefore best to use several different methods so that the resulting data can be analysed to obtain a more informed assessment of corrosion (1), for example measurements of the thickness of concrete cover, depth of carbonation, electrode potential, chloride content and resistivity. Carbonation develops from the surface and has been found to have reached the depth of steel after 50 years or so in the UK but seemingly has not been reported in the USA. Interpretation of electrode potentials, chlorides and resistivity requires special care.

2.1.1 *Electrode Potential.* It has been known for a number of years that the electrode potential of steel in concrete is an indicator of corrosion activity and such measurements have become a popular method of detecting corrosion albeit there has been a lot of debate about precisely how the measurements should be carried out. The reference electrode generally used is saturated copper/copper sulphate because it is cheap, rugged and reliable. Developments of the method have enabled it to be automated so that a map of electrode potentials can be obtained for the concrete surface being investigated (2), see Figure 1. Values of electrode potential (E) can be interpreted using the following guide:

E greater than -0.20 volt, probability of corrosion is less that 5 per cent
E between -0.20 and -0.35 volt, probability uncertain
E less than -0.35 volt, probability greater than 95 per cent.

2.1.2 *Chloride Content.* Chloride content has attracted a lot of attention and, like electrode potentials, requires care in its interpretation. The critical concentration of chloride for corrosion to occur is ill defined but it has been suggested that for a probability of less than 10 per cent, the chloride concentration should be less than 0.2 per cent chloride ion by weight of cement. Criteria suggested by BRE for assessing chloride data in buildings are broadly in line with data obtained from investigations of bridge decks, and may be used as a guide:

0 to 0.4 per cent chloride, low risk corrosion
0.4 to 1.0 per cent chloride, moderate risk
greater than 1.0 per cent chloride, high risk.

Measurement of chloride content is expensive and there has been quite a lot of activity towards developing techniques that are both reliable and cheap. To this end, methods of determining chloride concentrations in insitu concrete were surveyed (3). It was concluded that there are several promising techniques:

- the Kansas vacuum sampling drill
- the Gran potentionatic silver salt titration
- x-ray fluorescence spectrometry or ion chromatrography

2.1.3 *Resistivity*. The corrosion process involves the passage of an ionic current between the electrodes and through the concrete. Suggested relationships between resistivity and rate of corrosion are:

Resistivity less them 5 KOhm cm, high rate of corrosion
Resistivity between 5 and 12 KOhm cm, intermediate rate
Resistivity greater than 12 KOhm cm, low rate

2.2 PRESTRESSING STEEL

Detection of corrosion in prestressing steel presents different problems to reinforcement bars. Until recently it has been considered that the bridges most at risk are post tensioned and there is little likelihood of corrosion in pretensioned steel. However it has become evident that corrosion can occur in some types of prestressed concrete although it is still considered unlikely in conventional factory made precast pretensioned beams.

In post tensioned bridges the locations most at risk are joints in segmental construction and voids in ducts, see Figure 2. The collapse of Ynys y Gwas bridge in 1985 has served as a strong pointer to the likelihood of corrosion at segmental joints even when the ducts are fully grouted, Woodward (4). With regard to the efficacy of grouting, Woodward (5) surveyed twelve post tensioned concrete bridges and found voids in the ducts of ten of them and in just over half of the ducts examined. Exposed wires were usually covered by a protective film of cement paste but this cannot be relied on to prevent corrosion if chlorides get into the void.

Numerous techniques for detecting voids in ducts and fractures in the prestressing steel have been investigated. The techniques include hole drilling, radiography and radiometry, magnetic methods and ultrasonics. Although the other methods have shown some degree of promise, hole drilling, as well as being cheap, is the most direct and practical method of inspecting ducts in post tensioned concrete. The procedure is to drill 25mm diameter holes into the duct and inspect it using an endoscope. Photographs can by taken to provide a permanent record. If a series of holes is drilled and a vacuum is applied at one end, the pressure drop at the other gives an indication of the continuity of any voids present. An approximate measure of void volume can be obtained by measuring the volume of air required to refill the void when the vacuum is released.

Applying a small pressure to the duct and measuring the rate of leakage gives a measure of the ease with which moisture, atmospheric gases and other deleterious materials can enter the duct.

Although hole drilling provides useful information it is not appropriate for routine inspection and should really be used for structures where there is reason to believe that there are defects in the ducts. Neither hole drilling, nor any other method, will detect broken wires unless the fractures occur within the field of view of the endoscope.

3. Structural Assessment

3.1. REMNANT MECHANICAL STRENGTH

In cases when it is known that the steel reinforcement is corroded it may be necessary to carry out mechanical tests to determine the remnant strength. Static strength, as defined by the ultimate tensile strength, has been found to be broadly proportional to the net cross sectional area of the corroded section. The "as received" strength of reinforcement of steel is usually some 50 N/mm^2 above the characteristic strength so that there is a margin of reserve to be tapped. Reinforcing steel is ductile and surprisingly forgiving of corrosion pits so that it is unlikely to be embrittled. If it is considered that there is any likelihood that brittle fracture could occur, it is possible to carry out impact tests on lengths cut from the bar and impacted in air. The toughness of 'as received' reinforcement bars has been calculated for different temperatures, in concrete and in air, by Armstrong et al. (6).

Fatigue strength appears to be little affected by general corrosion but for local corrosion it is reduced by more than would be calculated from the loss on cross section area. It has been suggested (7) that the remnant fatigue strength , as defined by cycles to failure at a given range of stress, can be estimated from the reduced area:

level I corrosion (up to 25 percent loss in area) the reduction factor is 1.35;

level II corrosion (more than 25 per cent loss in area) the reduction factor is 1.70.

These losses are very high but can, and do, occur locally in the reinforcement.

Prestressing strand, like reinforcement bars, is surprisingly tolerant of corrosion. The fatigue strength is reduced in much the same way and in tests on corroded strand removed from structures that were 16 to 21 years old it was found that the lower bound performance curve was below the "as received" curve by a factor of 1.35 on stress range. The samples of strands tested had between 5 to 15 per cent reduction in area and very badly corroded sections having more than one wire broken, were excluded. It is not possible to characterise the degree of corrosion in the same way as suggested for reinforcement bars.

There have been no reported cases of fatigue failures of pre-stressing strand in service but in a curved concrete bridge in Germany several fractures were discovered at the mechanical sleeve joints used to connect tendons in the vicinity of construction joints (8). There had been warning signs of damage because 2mm wide cracks had developed at the surface of the concrete. It follows that connections of this type need to be examined when a structural assessment is being carried out. In Germany 583 post tensioned bridges built since 1977 were inspected and it was found that 160 had cracks up to 0.2 mm wide and 227 had cracks exceeding 0.2 mm.

Guidance on fatigue design of concrete structures is given in CEB Bulletin 189 containing the first pre draft of the Model Code to be published in 1990.

3.2 LOAD TESTING

3.2.1 *Field Testing.* Thompson (9) reviewed the literature on 252 load tests conducted on bridges up to 1981. Of these 35 concerned research work in support of improved design.

In load testing of bridges it is commonly found that structural response is somewhat less than calculated and it is sometimes quite difficult to produce enough strain to be able to obtain measurable data. An example of this arose when the Quinton bridges were being strengthened with bonded steel plating and it was decided to measure the response before and after plating. DTp's heavy (HB) axle was used to obtain longitudinal influence lines but it turned out that measured strains were too low to make meaningful comparisons. However, behaviour of crack widths changed so that it was possible to demonstrate a beneficial effect due to the plating.

In cases where it is required to demonstrate that a given traffic load can be carried with safety, it is necessary to apply test loads to some value incorporating an appropriate factor of safety. Although such tests can give assurance, there remains a degree of doubt as to the actual ultimate load. This is exemplified by an experience in Russia in 1986 when an 80 m span of the Keguns bridge across the Daugava River in Latvia collapsed during a routine load test causing the deaths of 10 people and injuries to 14 others (10). The bridge was a 40 year old steel arch structure and was being loaded by 14 gravel filled trucks according statutory regulations calling for strength tests every five years. It is believed that in the time since the previous test the dead weight had been increased by resurfacing so that the safety factor on live load was reduced.

Another example of a load test that led to collapse of the structure is the Clevedon Pier in the UK. In this failure there were no injuries and the test could be said to be successful in that it settled any question of the pier being safe for public use.

Load tests to collapse are currently being carried out by Page (11) as part of the TRRL research programme on bridge assessments, see Figure 3. These tests have been on redundant masonry arches and are shortly to be extended to reinforced concrete structures. Line loading has been applied at the position on the span calculated to give the weakest response, usually at 1/4 span. The field work has been supported by structural analyses using mechanisms and finite elements. It has been possible to learn a lot from the eight tests to date, not just for the types of arch being tested but for arches in general:

- Different types of structural action have been identified and analysed 2).
- Different collapse modes have been obtained.
- It has been found that collapse can take place after few warning signs and cracking does not necessarily occur before the final increment of load.

As part of the TRRL programme on earth structures Temporal et al. (13) have tested five culverts, four were loaded with the HB axle and the fifth was taken to collapse, see Figure 4. Important points that emerged were:

- Stiffness of the backfill material was greater than expected so that the presence of load effects was underestimated.
- A relatively thin flexible pavement could produce a significant reduction in bending moments.
- In the test to failure there was a lot of deformation giving warning of the impending collapse.

Data obtained from these tests have been used to check current methods of analysis and for development of design rules for culverts.

Load tests have been used successfully to uprate old bridges that fail to satisfy modern assessment criteria. Falconer (14) tested two bridges after assessments had indicated that loads should be restricted to well below values routinely being imposed by existing traffic. The first bridge was a 3.35 m span simply supported and composed of a reinforced concrete deck having inadequate transverse reinforcement. Due to flexural inadequacy of the deck in respect to single wheel, a restriction of 3 tonnes was found to be required. Load testing was applied using a concentrated load applied to an area of 300 sq mm. Gauges were fixed to the underside of the deck to enable the response to be monitored. Loading of up to 15 tonnes was successfully applied and stresses were shown to be very low. In consequence it was possible to raise the restriction to 25 tonnes. The second bridge was a 6.47 m span simply supported deck composed of rectangular fabricated steel troughing with weak concrete infill. An assessment had shown that loads should be restricted to 7.5 tonnes. Load testing was carried out using two test vehicles of 30 and 32 tonnes gross weight respectively. Gauges were fitted beneath the deck to enable the response to be monitored and as a safety precaution because the ends of the troughs exhibited signs of corrosion. These loads were applied successfully and from subsequent analysis it was possible to suggest that the restriction be raised to 25 tonnes. The two tests had resulted in more realistic levels of weight restriction and saved the costs of immediate replacement of the bridges. Their value was exemplified by the fact that they cost between 4 and 7 per cent of the replacement costs of the bridge.

3.2.2 *Laboratory Testing.* In order to investigate the degree of conservatism in Design Codes, and to determine whether there are secondary structural actions that give extra strength, two half-scale bridges have been tested at TRRL. The first was a prestressed beam and in situ slab having a 9 m simply supported span. The scale was selected so that proprietary materials could be used with the smallest dimensions to ensure reality. The bridge was designed to comply with the design code BS 5400. Combinations of serviceability and ultimate load were applied to determine behavior up to collapse. Failure occurred at a loading of 3.17 times Design Ultimate and this was very close to the value of 3 times Design Ultimate implied by the design partial safety factors. The mode of collapse was unexpected and is believed to be unique. It was considered that the presence of diaphragms may have reduced the ability for transverse distribution to take place so that the collapse load was lower than it would otherwise have been, Withey (15).

Analyses of the test using elastic grillage and finite element calculations, mechanism calculations and non-linear finite element calculations, were carried out by Crisfield et al. (16, 17). Both grillage and finite element methods gave good agreement with experimental results in the linear loading stages. A combination of linear finite element analyses and unfactored code calculations produced values of shear and tendon failure that were lower than the experimental results. Introduction of elastoplastic shear and tension properties into a non linear finite element model led to reasonable agreement both for the load deflection response and maximum load. When only flexural effects were considered, the calculated strength was too high.

The second half scale bridge was composite having steel I-beams and an insitu concrete deck. Tests are being completed at the time of writing.

4. Assessment Using Codes

Many aspects of design codes are unsuitable for assessment purposes and if used without due allowance could result in perfectly adequate bridges being classed as under strength. The difference between design and assessment is that design codes deal with maximum levels of uncertainty whereas much of this can be resolved for existing bridges. Partial safety factors have to take into account variabilities in materials and construction for new designs but these can be measured directly when assessing existing bridges. For example strength of reinforcing steel is commonly some 50 N/m^2 above the characteristic value. Concrete is not only stronger that the characteristic design value but older types continue to gain strength for years after placement. Design codes tend to use simplified methods of analysis having additional conservatism which is appropriate for design but not for assessment. A further factor is that design codes may prohibit the use of undesirable features or simply ignore outdated construction techniques which may nevertheless be present in existing bridges. For example masonry arches are rarely constructed nowadays albeit they are an efficient and durable form of construction.

In order to cope with the assessment of existing bridges, required by the increasing vehicle loading brought on by harmonization of loads, an assessment version of the concrete design code BS 5400 part 4 is being prepared by the Department of Transport. Each clause is being reviewed and where necessary adapted for assessment purposes.

The situation with regard to the assessment of steel structures is less pressing because structural steel is a more consistent and closely controlled material than concrete. Furthermore the substantial volume of research into methods on non-linear analysis of steel box girders in the 1970s has resulted in more sophisticated design and greater reliability.

5. Concluding Remarks

Assessment of highway bridges presents new problems some of which cannot be tackled by traditional methods but require adaptation of techniques from disciplines outside engineering.

In the past few years, research has enabled much progress to be made in establishing ways to detect corrosion. Techniques such as potential mapping, chloride sampling and measurements of resistivity can now be carried out on a fairly routine basis. It remains to "ruggedise" these methods so that they are cheaper and can be used more widely. There is a need for non-destructive methods of detecting local losses in section in reinforcing bars and broken wires in prestressing strand.

Remnant mechanical strength of reinforcing and prestressing steel can be estimated from the minimum corroded cross section. In cases where more precise data are required, special tests can be carried out.

Load testing has an important role to play in quantifying reserves of strength that are not taken into account by conventional analysis. However, it is best used as a research tool rather than for routine assessment. Tests to collapse enable structural actions to be identified and aide the development of improved methods of analysis.

To aid the assessment procedures in the UK an assessment version of the concrete design code BS 5400 part 4 is being prepared.

6. Acknowledgements

The work described in this paper forms part of the programme of the Transport and Road Research Laboratory and The paper is published by permission of the Director.

7. References

1. Vassie, P.R. (1980) "A survey of site tests for the assessment of corrosion in reinforced concrete," TRRL Report LR953, Crowthorne.
2. Winnett, M.A. (1985) "Microprocessor controlled multiple half cell measurements," TRRL Application Guide AG3, Crowthorne.
3. Figg, J.W. (1986) "Methods to determine chloride concentrations in insitu concrete," TRRL Contract Report CR32, Crowthorne.
4. Woodward, R.J. and Williams, F.W. (1988) "Collapse of Ynys-y-Gwas bridge, West Glamorgan," Proc. Instn. Civ. Engrs., Part 1, 84, Aug., pp. 635-669.
5. Woodward R.J. (1980) "Conditions within concrete ducts in post-tensioned prestressed concrete bridges," TRRL Report LR980, Crowthorne.

6. Armstrong, M., et al. (1988) "Toughness of reinforcing bars and its influence on the impact resistance of reinforced concrete beams," TRRL Contract Report CR85, Dept. of Transport, Crowthorne.
7. Tilly, G.P. (1988) "Durability of concrete bridges," Highways and Transportation Journal, Vol. 35, No. 2, pp. 10-19, February.
8. Sensible, F. (1987) "Coupling joints of prestressing tendons in continuous post-tensioned concrete bridges," Transportation Research Record, 1044, Washington, DC.
9. Thompson, D.M. (1981) "Loading tests on high bridges: a review," TRRL Report LR1032, Department of Transport, Crowthorne.
10. Anon (1986) "Latvian bridge fails test; fall kills 10, injures 14," pp. 13, Engineering News Record 23, October.
11. Page, J. (1988) "Load tests to collapse on two arch bridges at Torksey and Shinafoot," TRRL Research Report RR159, Department of Transport, Crowthorne.
12. Crisfield, M.A. (1987) "Numerical methods for the non linear analysis of bridges," Civil-Comp 87, Civil Comp Press Edinburgh, Vol. 2, pp. 67-74, (also to be published in Computers and Science).
13. Temporal, J. and Johnson, P.E. (1988) "Loading tests on two long span corrugated steel culverts," TRRL Research Report RR141, Department of Transport, Crowthorne.
14. Falconer, R.E. (1988) "Full-scale load testing of in-service bridges. Two case studies," unpublished paper to the Concrete Society.
15. Withey, K.H. (1990) "Tests on a half-scale prestressed beam and slab bridge deck," TRRL Research Report RR198, Department of Transport, Crowthorne.
16. Wills, J. and Crisfield, N.A. (1989) "No-tension finite element and mechanism analyses of a half-scale beam-and-slab bridge deck," TRRL Research Report RR217, Department of Transport, Crowthorne.
17. Wills, J., Crisfield, M.A. and Cheung, J.T. (1989) "Numerical analysis of a half-scale prestressed beam-and-slab bridge deck," TRRL Research Report RR218, Department of Transport, Crowthorne.

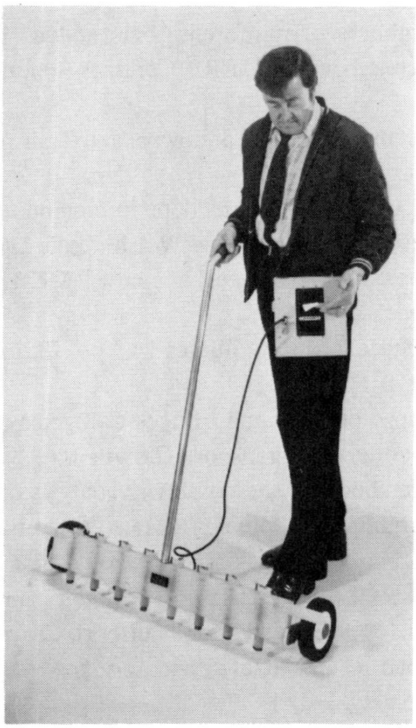

Fig. 1 Microprocessor Controlled Half Cell Monitoring System.

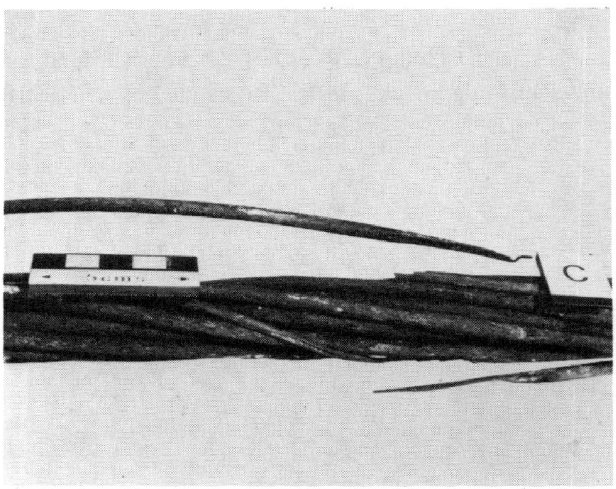

Fig. 2 Corroded 19 - Wire Prestressing Strand Removed from Post Tensioned Bride After after 21 years.

Fig. 3 Prestwood Bridge Immediately Before Collapse.

Fig. 4 Failure of Culvert at 48.8 Tonne Axle Load and 0.36 m Depth of Cover to the Crown.

UTILIZATION OF SERVICE LOADS IN BRIDGE EVALUATION

BAIDAR BAKHT
Ministry of Transportation of Ontario
Downsview, Ontario
CANADA M3M 1J8

LESLIE G. JAEGER
Technical University of Nova Scotia, Halifax, N.S.
CANADA B3J 2X4

ABSTRACT. A design or evaluation procedure has to make conservative allowances for the possibility, that a structural component may have strength that is well below the nominal. A proof test on a component, by reducing the uncertainty regarding the existence of very low strengths, can be helpful in eliminating some of these conservative allowances and thus making available some extra nominal strength for applied loads. A technique is presented by which proof loads can be used with advantage to upgrade the load carrying capacities of existing bridges. The in-service performance of a bridge is itself a kind of proof testing. A probabilistic approach is presented using which the maximum in-service proof loads can be estimated. Once the in-service proof loads are established, they can be incorporated into the analytical design procedure in the same way as are the proof loads of formal tests. The techniques presented are general in nature. However, their application requires some charts and numbers which depend not only upon local code provisions, but also upon the variability of the strengths of the components and of their load effects.

1. Introduction

A well-conceived proof test which is conducted under known test loads, can be helpful in the analytical evaluation of the load carrying capacity of an existing bridge in two ways. Firstly, it identifies correctly the manner in which a bridge distributes applied loads amongst its various components. Secondly, and sometimes more significantly, the proof test confirms that the tested bridge has strength in excess of the strength corresponding to the test load. Because of the increased confidence in its strength, the calculated safety margin for the tested bridge can be shown to be wider than the margin if the test were not conducted and a definite lower-bound of its strength were not thereby established. In practical terms, the increased reliability of the strength of a tested bridge can be translated into a higher nominal strength and, consequently, into heavier loads that can be permitted safely on the bridge. The in-service performance of a bridge is itself a kind of proof testing, albeit without the first advantage mentioned above of a formal test. Techniques are available using which it is possible to determine the maximum loads that a bridge carrying a given volume of traffic, is virtually certain to have sustained

A. S. Nowak (ed.), Bridge Evaluation, Repair and Rehabilitization, 379–389.

during its life. By treating these maximum loads as proof loads, advantage can be taken of the in-service performance of a bridge in upgrading its analytically-obtained nominal strength.

The technique of taking account of proof loads in the analytical evaluation of bridge capacities is discussed in this paper along with the method of determining the maximum service loads. Limitations on the length of the paper make it impossible to give extensive details of the proposed method. However, it is shown conceptually that the inclusion of "proof loads" derived from service loads can be very helpful in upgrading the analytical assessment of the load carrying capacity of an existing bridge.

2. Proof Testing

Proof testing is a standard means of establishing that a structure possesses adequate strength to continue to carry the loads that it is supposed to carry. The minimum test load that must be applied to a structure for it to be considered to have adequate strength, is sometimes referred to as "the target proof load."

2.1. CODE PROVISIONS FOR TARGET PROOF LOAD

Codes for structures other than bridges specify the target proof load to be such that it induces the load effect X_p which is given by

$$x_p = \alpha_L L_n + \alpha_D D_n \tag{1}$$

where D_n and L_n are the load effects due to nominal, or design, dead and live loads and α_D and α_L are the respective factors that should be applied to them to obtain the proof load. The implication of the second term on the right-hand side of Eq. (1) is that the load effect due to dead load carried by the structure is equal to D_n. If α_D is specified to be 1.0, the superimposed test load should account for only the factored nominal load.

There is considerable diversity in the values of α_D and α_L specified by the various codes. For example in the references that form part of this paper, [1] requires α_D and α_L to be 1.19 and 1.45, respectively; [2] specifies them to be 1.5 and 2.0, respectively; and [5] has α_D=1.00 and α_L=1.25.

The specification of a value of α_D greater than 1.0 may have emerged from the realization that the variability of the capacity of a structure to sustain live loads is strongly influenced by its dead to live load ratio. As pointed out by Fujino and Lind (1977), some of the code provisions for the magnitude of proof load may have required much wider safety margins for structures evaluated by testing than for structures evaluated analytically. The procedure proposed in this paper for obtaining the target proof load, does take account of the dead to live load ratio of structures, but it nevertheless requires αD to be effectively equal to 1.0.

It is noted that bridge codes usually do not specify the magnitude of the test load that must be applied to a bridge for it to be deemed to have sufficient capacity to carry unrestricted traffic.

The Ontario code (1983) does permit testing as a means of evaluating the load carrying capacity of bridges; however, even this code does not specify the target levels for proof loads.

2.2. EFFECT OF PROOF TESTING ON FAILURE PROBABILITY

In the context of the discussion of the topic at hand, the reliability of a bridge is being dealt with for only one bridge component at a time, thus disregarding the beneficial "systems effect" whereby the reliability of the structure exceeds the reliability of its individual components.

It is well known that the resistance R of a component and the load effect S to which it is subjected to, are not deterministic quantities; they are variable and may have different frequency distribution patterns. For example, as shown in Fig. 1, both R and S may be normally distributed. The overlapping area of the R and S distributions, shown shaded in Fig. 1, is a qualitative, although not a quantitative, measure of the probability that S may exceed R, indicating failure.

A quantity g defined by g = R-S, provides a quantitative measure of the probability of failure. From a frequency distribution diagram of this quantity shown in Fig. 2, it can be seen that a certain portion of the diagram falls into the negative region. It is this portion of the frequency diagram of g that quantifies the probability of failure. As shown in Fig. 2, the negative values of g lie a distance b x s_g to the left of m_g, where s_g and m_g are respectively the standard deviation and mean of g. The quantity b is the quantitative measure of the probability of R exceeding S; it is usually referred to as the safety index. For normal distributions of R and S, the safety index can be shown to be given by

$$\beta = \frac{\mu_R - \mu_S}{(\sigma_R^2 + \sigma_S^2)^{1/2}} \tag{2}$$

The overlapping area of the R and S distributions which is shown shaded in Fig. 1 for a specific example, is a qualitative measure of the influence of the lower tail of the strength distribution on the probability of failure. If the loads are reduce d, which has the effect of moving the S distribution to the left, the overlapping area is diminished thus indicating that the probability of failure is reduced.

Another way of reducing the overlapping area of the R and S distributions, and hence the probability of failure, is by proof-testing the component and thus ensuring that it has a certain minimum strength. The effect of proof-testing is to truncate the strength distribution by eliminating the lower tail of the R diagram curve and thus reducing the overlapping area as illustrated in Fig. 3. By comparing this area with the corresponding overlapping area of Fig. 1 which corresponds to the pretest distribution of strength, it can be seen that proof testing reduces the probability of failure.

2.3. CALCULATION OF TARGET PROOF LOAD

The subject of the reliability of proof-tested structures has been dealt with fairly extensively in the recent technical literature. The work of Fujino and Lind (1977), however, provides a method which can be conveniently applied to the determination of the target proof load for bridges by using readily available data on the variation of strengths and load effects in highway bridges. For the sake of brevity, only the steps of calculation required for the determination of the target proof load are given below. Details of the formulation of the method are given by Bakht and Jaeger (1990). It is emphasized that the proposed method can be applied only in conjunction with the analytical evaluation which is based either on the Limit States Design principles such as in [7], or on the Load and Resistance Factor Design approach.

2.3.1. *Step 1.* By analytical evaluation, determine the reduction factor RF from

$$RF = \frac{\phi R_n - \alpha_D D_n}{\alpha_L L_n (1+I)} \tag{3}$$

where f = the resistance factor; R_n = the nominal strength; α_D = the dead load factor, being different from α_D used in Eq. (1); D_n = the nominal dead load effect; α_L = live load factor, being different from α_L used in Eq. (1); L_n = the nominal live load effect; and I = the impact factor. If RF is greater than 1.0, the bridge has sufficient strength to sustain unrestricted traffic, and the steps which follow are redundant.

2.3.2. *Step 2.* From charts such as the one reproduced in Fig. 4 for reinforced concrete components, find the value of pre-test safety index $\bar{\beta}$ corresponding to the ratio h of nominal live to dead loads. It is noted that these charts may be different for different jurisdictions as they depend upon the variability of both the loads and resistances. The implication of the pretest safety index is that if the substandard bridge is permitted to carry unrestricted traffic, its safety index will drop from the target value β to $\bar{\beta}$. It may be noted that the typical value of β is 3.5.

2.3.3. *Step 3.* Using a chart such as the one reproduced in Fig. 5, obtain the value of the non-dimensionalized proof load index k corresponding to the type of the component and the value of $\bar{\beta}$ calculated above. A separate chart is needed for each different value of β. The chart presented in Fig. 5 is for $\beta = 3.5$.

2.3.4. *Step 4.* Find the value of the load effect X_p induced by the target proof load, from

$$X_p = \lambda R_n (1 + kV_R) \tag{4}$$

where λ = the ratio of mean to nominal resistance and V_R is the coefficient of variation of resistance.

These parameters also do not have the same values for all situations because of being dependent on code specifications and the degree of quality control. As an example, it is noted that in North America, λ is typically 1.05 and 1.09 for prestressed concrete and reinforced concrete components, respectively, while V_R for these components is typically 0.065 and 0.11, respectively. X_p has the same units as the resistance. For example, if the resistance is expressed as the flexural strength, then X_p is the moment induced by the proof load.

3. Quantification of Maximum Service Loads

As mentioned earlier, the in-service performance of an existing bridge carrying a known volume of traffic is itself a kind of a proof test. However, unlike a formal test, the in-service performance of a bridge cannot lead to a deterministic value of the proof test. The determination of the maximum loads that a bridge is likely to have sustained in the past requires a probabilistic analysis which is discussed in this section of the paper.

3.1. FACTORS AFFECTING DESIGN

The design capacity of a highway bridge to carry normal traffic safely is based upon the implicit consideration of many factors including the following

a. Legally permitted vehicle weights.
b. Accidental and deliberate exceedance of legally permitted vehicle weights.
c. Transverse vehicle position.
d. Simultaneous presence of more than one vehicle in the transverse direction.
e. Simultaneous presence of more than one vehicle in a lane.

All these factors, except (a), are probabilistic in nature. For example, it is not known beforehand what fraction of the total vehicle population will exceed legally permitted weights, nor what the probability is that such overweight vehicles will t ravel in extremely eccentric transverse positions.

It is usual to include the most unfavourable of the above factors in design or evaluation calculations. Calculations based on estimating the probability of individual occurrence of these unfavourable factors rapidly lead to the conclusion that certain combinations of them are extremely unlikely to occur even once in the lifetime of a bridge. It follows, therefore, that the maximum loads that a bridge is likely to have sustained during its lifetime are not the same as those specified for design and evaluation, and that these maximum loads should be determined on a probabilistic basis.

3.2. BASIC ASSUMPTIONS

It is assumed that the maximum weights of those vehicles which are critical for a given bridge are normally distributed with a mean of m and a standard It is assumed that the maximum weights of those vehicles which are critical for a given bridge are normally distributed with a mean of m and a standard deviation s; that the bridge under consideration has at least two lanes, and the volume of traffic is known and has been the same either during the lifetime of the bridge, or for a given number of years.

3.3. EXAMPLE OF MULTIPLE PRESENCE IN TWO ADJACENT LANES

The process of determining the maximum loads that a bridge is likely to have sustained during its lifetime can be explained readily with the help of an example in which the simultaneous presence of vehicles in two adjacent lanes is studied.

Consider a two-lane bridge with a span of 90 m that carries a low density of heavy trucks, estimated at 300 per day. Let it be supposed that the bridge is 20 years old and that it has been carrying the same traffic since its construction. The statistical distribution of the weights of these heavy trucks is assumed to be normal. An estimate is sought of the probability that two trucks at least 60% heavier than the mean will be simultaneously situated on the middle third of the bridge, one in each lane, at least once during the 20-year life of the bridge.

It is first necessary to establish a time interval within which any given traffic may be on the central 30 m of the bridge. Taking a slow vehicle speed of 6 m/sec, a time interval of 5 sec is postulated. From the properties of normal distribution, it is found that 0.00135 of the total vehicle population has weights 60% or more above the mean. Thus, the expected number of heavy vehicles per day that cross the bridge and are this heavy is 300 x 0.00135 = 0.405 vehicles per day. In one day the re are 1,728 intervals of 5 sec each. Hence, the probability that a vehicle 60% or more heavier than the mean will be on the middle third of the span during any single 5-sec interval is 0.405 x (1/17,280) = 2.34×10^{-5}. The probability, P, that two such vehicles will be present during any given 5-sec interval is the square of this quantity: i.e. P = 5.47×10^{-10}.

The combination of a low probability that such an event will happen in a given 5-sec interval together with the large number of intervals that represent the lifetime of a bridge is ideally suited to the use of the Poisson distribution of probability of occurrence. With the customary notation, the probability that the stated occurrences will happen in any given 5-sec interval is denoted by P and the total number of time intervals by N. The product NP is the Poisson parameter α_1, so that NP = α_1

If the probability of m occurrences is defined as P(m), the following equations are obtained by using Poisson distribution theory: $P(0) = \exp(-\alpha_1)$; $P(1) = \alpha_1 \exp(-\alpha_1)$; $P(2) = (\alpha_1^2/2!) \exp(-\alpha_1)$. During the 20-year life of the bridge the total number of 5-sec intervals (N) is equal to 1.26×10^8. Hence α_1 is found to be 0.069, and it is found that P(0) = 0.933 and P(1) = 0.063. It is concluded that on the basis of the assumptions made it is unlikely (only a 6.7 percent chance) that the two vehicles, each at least 60 percent heavier than the mean, will be on the middle third of the bridge simultaneously. However, the situation changes swiftly if the simultaneous presence of somewhat lighter vehicles is investigated. For example, for vehicles 40 percent heavier than the mean, α_1 is found to equal 19.65 and P(0) to be approximately 10^{-8}. Thus it is virtually certain that there will be two vehicles, each at least 40 percent heavier than the mean, simultaneously present on the middle third of the bridge at least once during its lifetime.

Using the above technique, the details of which are given in (4), the weights of the two heaviest vehicles can be found that are virtually certain to have been simultaneously present on the bridge. These maximum weights can be regarded as the proof load for the bridge.

The weight of each of these in-service proof load vehicles is assumed to be r standard deviations above the mean, i.e. = $\mu + r\sigma$. The design vehicle usually represents vehicles whose weights are typically, as for example in (7), 3.5 standard deviations above the mean, i.e., its weight = $\mu + 3.5\,\sigma$. The proof load can now be related to the design vehicle through a factor F_1.

$$F_1 = (\mu + r\sigma)/(\mu + 3.5\,\sigma) \tag{5}$$

The factor F_1 is similar to the multipresence modification factor specified in some codes with the difference that the former relates to the situation that is most likely to have taken place and the latter to the situation which is remotely probable in the future.

The values of F_1 calculated by the above procedures are plotted in Fig. 6 against the number of vehicles per day per lane for different time intervals and for two specific life spans. It is comforting to note that this factor is not very sensitive t o either the life of the bridge or to the time interval assumed. From charts such as those presented in Fig. 6, it is possible to get a safe, i.e. low, estimate of F_1.

3.4. TRANSVERSE POSITIONS OF VEHICLES

The in-service proof loads obtained by the procedure discussed above correspond to the normal travelling positions of the vehicles which are the centres of their respective marked traffic lanes. These positions are different from the code-specified transverse positions of the design loading which correspond to the maximum load effects on a given transverse section of the bridge. It is necessary, therefore, to modify the proof load to account for these differences in the transverse positions of the vehicles. This can be achieved through a factor F_2 which is defined as

$$F_2 = S_t/S_c \tag{6}$$

where S_t and S_c are the maximum load effects at a given transverse section obtained by load distribution analysis corresponding to the design loading in the code-specified transverse positions and in the usual travelling positions of the vehicles, respectively. It will usually be found necessary to develop different values of F_2 for different responses.

4. Outline of Proposed Feature

From the above discussion, a preferred procedure emerges naturally for taking account of the in-service performance of a bridge, in the analytical evaluation of its load-carrying capacity. The steps involved in this procedure are outlined below.

a. Calculation of F_1. Based upon the life of the bridge and the representative traffic volume over its life, determine the value of F_1 with the help of charts such as that produced in Fig. 6.
b. Calculation of F_2. Perform a load distribution analysis of the bridge under consideration to obtain S_t and S_c respectively, and hence calculate F_2 from Eq. (6).
c. Calculation of In-service Proof Load. Obtain the in-service proof load L_p from

$$L_p = F_1 \times F_2 \times W \qquad (7)$$

where W is the weight of the specified design loading.
d. Calculation of the Rating Factor RF. Perform the analytical evaluation of the bridge to obtain the rating factor RF for the weakest component from Eq. (3) using the code-specified values for f, α_D, α_L and I. As mentioned earlier, no further calculation is required if RF is greater than 1.0.
e. Calculation of Pretest Safety Index $\bar{\beta}$. Using charts such as those presented in Fig. 4 obtain the value of the pre-test safety index $\bar{\beta}$ for the component under consideration corresponding to the value of RF calculated above.
f. Calculation of Target Proof Load. Using charts such as those presented in Fig. 5, obtain the value of k corresponding to the value of $\bar{\beta}$ obtained above. Thereafter, calculate X_p from Eq. (4) by using the values of λ and V_R which are relevant to the code being used and the type of construction of the component. If the load which induces the load effect X_p in the component under consideration is smaller than, or equal to, L_p, then it can be stated with confidence that, contrary to conclusions reached by the analytical evaluation, the bridge has adequate strength to carry unrestricted traffic.

4.1. A WORD OF WARNING

It is emphasized that the implication of using L_p as the target proof load is that the strength of the bridge has not deteriorated over the period for which the likelihood of the multiple presence of vehicles is sought. If the condition of a bridge has deteriorated significantly, then this period should be reduced accordingly.

4.2. SUBSTANDARD BRIDGES

There may be some bridges for which even L_p is not large enough to upgrade them to full capacity. For such substandard bridges, Bakht and Jaeger (1990) provide a procedure using which advantage can be taken of the proof load to upgrade the rating factor RF beyond the value obtained by only the analytical evaluation; this procedure can also be used in conjunction with the in-service proof load L_p.

5. Conclusions

It has been shown that the in-service performance of a bridge over a number of years is itself a kind of proof testing, which can be quantified by a safe estimate of maximum loads that a bridge is virtually certain to have sustained during its lifetime. The effect of a proof load is to establish a lower-bound of the strength of the bridge and thus effectively curtail the lower tail of the frequency distribution of its strength. As a consequence of the curtailment of the lower tail of strength distributions, the analytically-assessed capacity of the bridge can be upgraded by using the procedures outlined.

6. Acknowledgements

The authors wish to acknowledge the Ministry of Transportation of Ontario and the Natural Sciences and Engineering Research Council of Canada for their support of the research which led to this paper.

7. References

1. American Concrete Institute. (1984) ACI 318M-83, Building Code Requirements for Reinforced Concrete, Detroit, Michigan, U.S.A.
2. American Institute of Steel Construction. (1977) Cold-Formed Steel Design Manual Washington, D.C., U.S.A.
3. Bakht, B. and Jaeger, L.G. (1990) "Proof Load Testing," In preparation.
4. Bakht, B. and Jaeger, L.G. (1984) "A rational procedure for overweight permits," Transportation Research Record 950, Vol. 1, Transportation Research Board, Washington, D.C., USA, pp. 59-70.
5. British Standards Institutions. (1972) CP 110, Part 1: 1972, The Structural Use of Concrete, London, England.
6. Fujino, Y. and Lind, N.C. (1977) "Proof load factor and reliability," ASCE Journal of the Structural Division, 102(ST4), pp. 853-870.
7. Ontario Highway Bridge Design Code, Second Edition. (1983) Ministry of Transportation of Ontario, Downsview, Ontario, Canada.

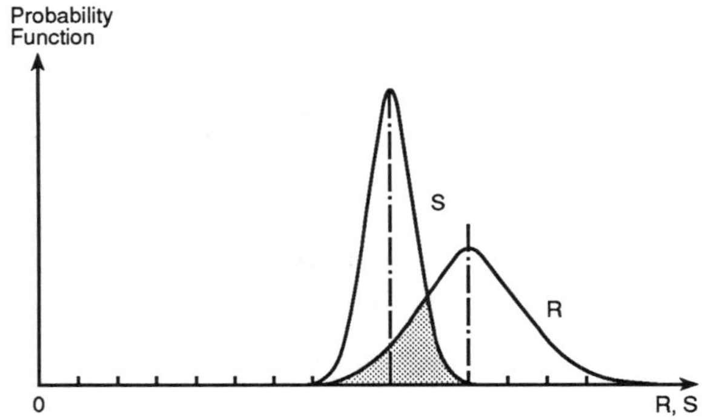

Fig. 1 R and S Plotted on the Same Chart.

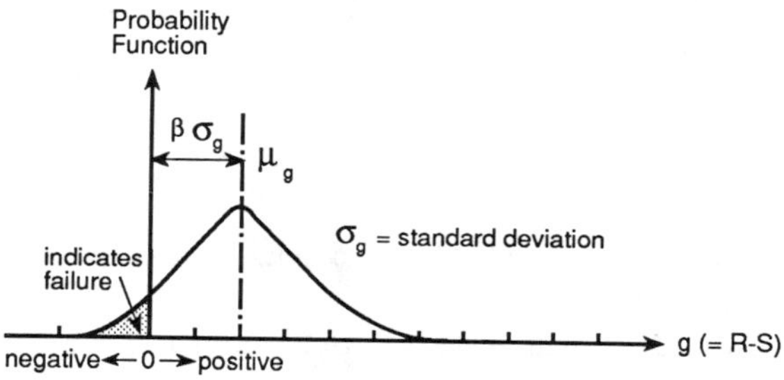

Fig. 2 Distribution of g.

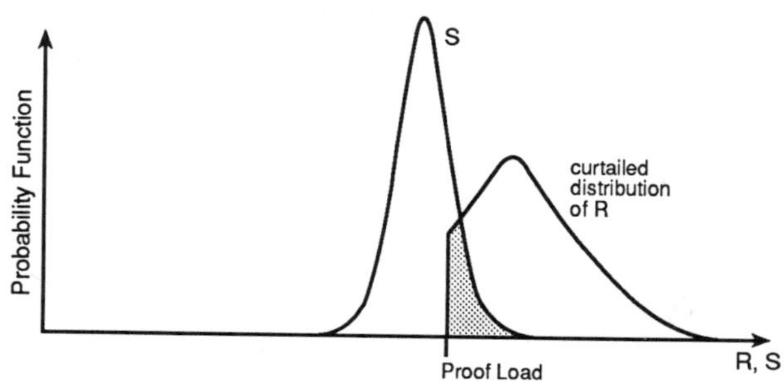

Fig. 3 Distribution of R and S After Proof Testing.

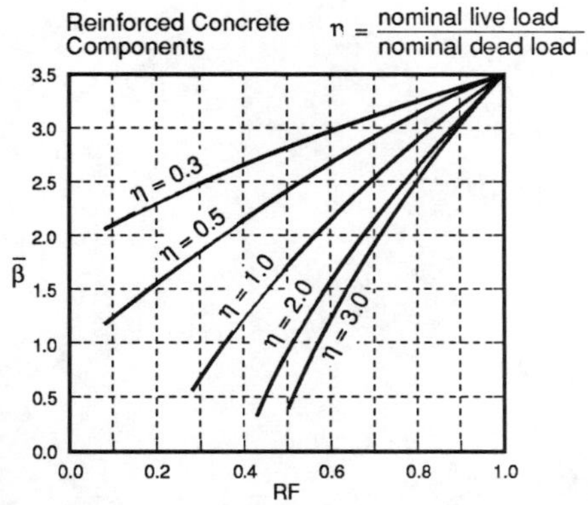

Fig. 4 Relationship Between the Rating Factor and Safety Index.

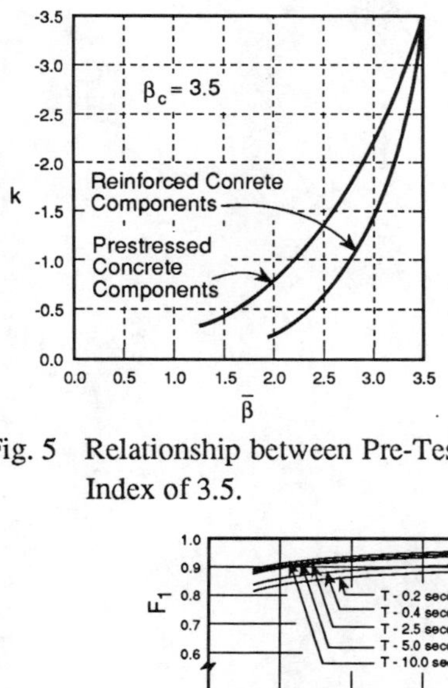

Fig. 5 Relationship between Pre-Test Safety Index β and k corresponding to a Target Safety Index of 3.5.

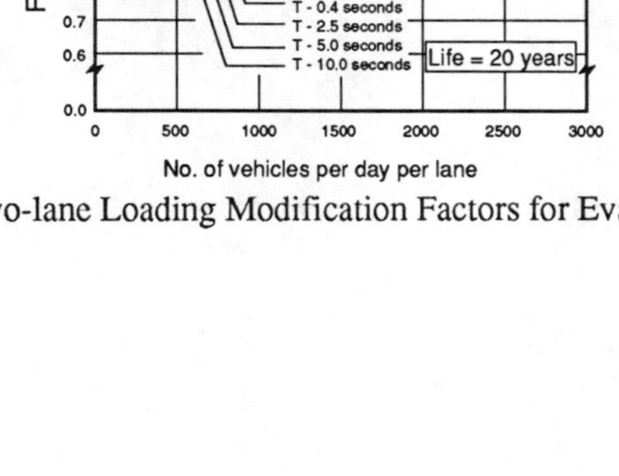

Fig. 6 Two-lane Loading Modification Factors for Evaluation of Safe Overweights.

PRESTRESSED CONCRETE GIRDERS AFTER 20 YEARS IN SERVICE

S. A. OLSON, C. W. FRENCH, and R. T. LEON
University of Minnesota
Civil and Mineral Engineering Department
122 Civil and Mineral Engineering Building
500 Pillsbury Drive S.E.
Minneapolis, MN 55455-02220 USA

ABSTRACT. An experimental investigation of four prestressed bridge girders was conducted at the University of Minnesota. The girders were standard AASHTO Type III, cast in July 1967, and subsequently placed in an interstate overpass. The girders became available for testing after nearly twenty years of service, when they were removed in 1986. The principal objectives of the investigation were fourfold: (1) to determine the prestress losses incurred over the life of the girders and compare them with those given by code-prescribed equations; (2) to determine the current material properties using nondestructive techniques (NDT) and compare the results with core samples; (3) to examine the performance of the girders under various fatigue loading levels; and (4) to determine the amount of impact damage the girders could sustain before repair is required, and to evaluate the fatigue performance of strand splice and external post-tensioning repair schemes.

1. Introduction

Four twenty-year-old prestressed bridge girders, located in an interstate overpass in the Minneapolis metropolitan area, became available for testing when the bridge was dismantled as a result of road realignment work. The girders were cast in July 1967, placed in the overpass shortly afterwards, and then removed in 1986. Fig. 1 shows the layout of the bridge, cross section of girders, and the cross section of the test specimen. The girders were AASHTO Type III, 45 in. (1140 mm) deep, 64 ft. 8 in. (19.7 m) long, prestressed with thirty 250 ksi (1720 MPa) stress-relieved seven-wire strands. Twenty-two of the strands had a straight profile within the bottom flange of the girders. The remaining eight strands were draped from two hold-down points located 5 ft. (1720 mm) either side of centerline of the span. A 6 in. (150 mm) thick, 64 in. (1630 mm) wide deck slab was added to each girder in the lab. The new deck was reinforced in a manner similar to that of the deck in the field. Each girder was tested independently on a simply supported span.

The objectives of the tests were:

A. S. Nowak (ed.), Bridge Evaluation, Repair and Rehabilization, 391–403.

(1) To determine the effective prestress of each of the specimens and compare these values with those given by current code-prescribed equations. Little is known about actual prestress left in pretensioned girders after years of service, primarily because quantification of thermal and time effects (creep and shrinkage of the concrete, strand relaxation, cyclic load and fatigue) is difficult. Code provided formulas for prestress losses have not been adequately verified. The amount of effective prestress in an element affects its behavior under loading. The larger the effective prestress, the larger the load needed to crack the cross section and to subsequently reopen the cracked cross section. In addition, if cracks open under cyclic loading, the strand stress range is a function of the effective prestress.

(2) To determine current material properties using nondestructive techniques (NDT) and compare the results with core samples. A major problem for bridge rating is the characterization of material properties; NDT techniques are ideal for this application, but few comparisons are available on the actual material properties obtained from destructive testing versus those obtained from NDT tests.

(3) To examine the performance of the girders under various fatigue loading levels. Originally prestressed concrete bridges were designed with no tension allowed to appear in the cross section under service loads. Later, this restraint was relaxed and $3\sqrt{f'c}$ psi ($0.25\sqrt{f'c}$, MPa) was allowed under service loads. Presently, the AASHTO code allows $6\sqrt{f'c}$ psi ($0.5\sqrt{f'c}$, MPa) nominal tension.

(4) To determine the amount of impact damage that such a girder can sustain before replacement girder can sustain before replacement or repair is required. Prestressed concrete bridges are sometimes damaged by overheight vehicles traveling beneath the bridge. Little information is available to the bridge engineer to decide whether to leave the section unrepaired, to repair, or to replace the damaged girder [1,2]. To address the repair issue, one of the girders was subjected to incremental damage and subsequently cycled to investigate the amount of damage sustainable before repair/replacement becomes necessary. Two additional girders were subjected to simulated impact damage and subsequently repaired and fatigued to assess the performance of two repair schemes: an internal strand splice and an external post-tensioning method.

2. Test program

2.1 PRESTRESS LOSSES

The fabrication reports of the girders indicated that the strands were tensioned to within three percent of the design tension of 0.70 fpu. This value for the initial tensioning was used as the baseline from which the magnitude of the prestress losses was determined. The girders were uncracked when they arrived in the laboratory. They had been designed for no allowable tension under service load conditions. The girders were loaded to cracking and then reloaded above the crack opening load several times to determine the load at which the cracks reopened. Figure 2 shows the initial cracking load deflection plot of G1, the first girder tested, and a subsequent reloading. Figure 2 also shows the output of an linear voltage differential transformer (LVDT) placed across a crack on the bottom flange of G1. The output of the LVDT is plotted versus live load. The plots indicate that the response of the girder softened and the LVDT's displacement rate increased at approximately 85 kips (378 kN). This load was used as the decompression load. Knowing the decompression live load, dead load, and section properties the effective prestress can be determined.

In all, four techniques were used to estimate prestress losses: (1) formation of first crack; (2) reopening of cracks; (3) discontinuities in the load-deflection curves; and (4) direct instrumentation and cutting of the strands. In situ prestress levels given by methods (1)-(3) were very close to one another; they indicated a remaining prestress of 130 ksi (895 MPa) in the strands, or about 74 percent of the original prestressing. The total prestress losses calculated were approximately 45 ksi (310 MPa), very close to the lump sum prestress losses predicted by current AASHTO specifications. Method (4) gave somewhat lower values (64%) which can be explained by slight misalignment of the strain gages and differences in prestress from strand to strand.

2.2 METHODS USED TO MEASURE MATERIAL PROPERTIES

Nondestructive tests used to determine current material properties included: (1) rebound hammer, (2) Windsor probe, (3) ultrasonic pulse velocity, and (4) break-off tests. The results were subsequently compared with 2 and 4 in. (50 and 100 mm) core samples removed from the girder.

To correlate the core tests with the nondestructive test methods, correction factors were considered to account for nonstandard diameter (2 and 4 in. (50 and 100 mm) cores rather than 6 in. (150 mm) standard diameter) and aggregate size to core diameter ratio, vertical height location of core within the girder, and length-to-diameter (L/d) aspect ratio. Of these, the correction factor for the L/d aspect ratio had the greatest effect on the results. The correction factors determined from the literature, ASTM C42 and Kesler factors [3,4], led to unconservatively high strength results for the smaller diameter (2 in. or 50 mm) cores. This might be attributed to the small ratio of core diameter to aggregate size (2.13). Using regression analyses to correct the core strengths yielded an average compressive strength of 8400 psi (58 MPa) for G1. The results represent a 22 percent strength gain over twenty years in comparison with the reported 28-day standard concrete cylinder test strengths. These values are consistent with those reported elsewhere [5].

Table 1 gives an indication of the concrete uniformity of the girders. All of the tests with the exception of the break-off tests gave coefficients of variation which were within the reported "within test variations" determined from tests on identical concrete samples [6-10]. The large coefficient of variation (16.2%) and very low values indicated by the manufacturer's correlation (51.7% of the core strengths) indicated by the break-off test were attributed to the large aggregate size in the girders (1 in. or 25 mm) and damage to the cores during the drilling process.

As a consequence of the concrete uniformity across the girder, it was not possible to obtain meaningful results from regression analyses or evaluations of correlation curves which required core strength calibration. The uniform characteristic of the concrete yielded only one average compression strength with which to compare the NDT tests. To conduct a study of the applicability of correlation curves it is desirable to have a range of concrete strengths and corresponding NDT results.

2.3 CRACKING, FATIGUE AND ULTIMATE LOADING OF GIRDERS

A number of different load tests were applied to the specimens. Initially several static tests were performed on each girder to crack the cross section and to estimate the load at which cracks reopened (to investigate prestress losses). Following the initial static tests, fatigue loading was applied to each specimen. At intervals during the cyclic loading, static tests were performed to determine the amount of stiffness degradation incurred with fatigue loading. Following cyclic loading, an ultimate strength test was performed on each girder.

The control girder (G1) was subjected to a fatigue loading consisting of 400,000 cycles at the decompression load; 1,000,000 cycles at $3\sqrt{f'c}$ ($0.25\sqrt{f'c}$, MPa); 1,300,000 cycles at $6\sqrt{f'c}$ ($0.50\sqrt{f'c}$, MPa); and 60,000 cycles at $12\sqrt{f'c}$ ($1.00\sqrt{f'c}$, MPa). Figure 3 shows the load-deflection curves obtained at the end of each of these fatigue loadings. Only slight strength and stiffness deterioration were evident at the end of this very severe load history.

Figure 4 shows the load-deflection curves for the damaged specimen G2, which was cycled at $6\sqrt{f'c}$ after progressive damage was induced. A large loss of strength and stiffness was noted between the initial tests on the undamaged girder and those performed after the concrete was removed and the first two strands cut. The centerline deflection, for an equivalent load, increased from 0.97 in. (24.6 mm) to 1.35 in. (34.3 mm) between the end of these two steps. The deflections continued to increase, albeit more slowly, as more damaged was introduced. During the final fatigue run the deflections did not stabilize.

Figure 5 shows a comparison of the tests to ultimate from the first three girders. Both G1 and G2 exhibited very large ductilities, with the centerline deflection exceeding L/30, where L is the span length. The difference in strengths between these two tests is attributable to the strands cut. Girder 3, which was repaired using internal splices performed well in terms of strength, but did poorly with respect to ductility because strands began to fracture early due to the fatigue damage.

2.4 IMPACT DAMAGE AND REPAIRS

To simulate the damage of an overheight vehicle striking a prestressed bridge girder from below, the following procedure was employed. Beginning with the second girder (G2), the concrete over approximately a 4 ft. (1220 mm) region of the bottom flange near centerline was removed exposing the four strands most susceptible to being severed by impact damage. This consisted of three strands in the bottom layer and one in the second layer. Once these strands were instrumented, stress ranges for the exposed strands were obtained under fatigue loading.

After a short fatigue loading sequence with the strands intact, two of the exposed strands were severed. Once again the beam was subjected to fatigue loading, and data on stress ranges was collected. At this point, further damage was simulated by severing more strands in the damage control specimen G2 and fatigue loading was continued. Damaged specimens G3 and G4 were subjected to the same initial impact damage simulation (two strands severed) and load history, and then repaired before further fatigue loading.

There are currently no criteria in the design rating specifications to determine whether a damaged girder needs to be replaced. For the purposes of this test it was decided that the performance of the damaged or repaired girders was deemed acceptable if the test specimen was able to sustain 500,000 fatigue load cycles at 0, 3, and 6√f'c psi (0, 0.25, and 0.5√f'c MPa) without undergoing continuously increasing deformations. The behavior was then assessed with static tests to investigate the stiffness deterioration. If the performance was satisfactory, the load level or damage level was increased.

2.4.1 *Damage Control Specimen Results* The impact damage control specimen, G2, demonstrated susceptibility to fatigue failure with fairly moderate strand damage (6.7% severed). The fatigue loading level was held constant for the specimen as increasing amounts of damage were induced by cutting strands. All fatigue loading was at a level which would produce a 6√f'c psi (0.5√f'c MPa) nominal tension in the concrete. The f'c utilized to compute the nominal tension was compute the nominal tension was the 5000 psi (35 MPa) design strength.

After the first two strands were severed, the stress ranges in the remaining two strands increased from an average of 22 ksi (152 Mpa) to 37.5 ksi (260 MPa). Approximately 300,000 cycles into the fatigue loading with two strands severed, the third instrumented strand broke. The stress range in the remaining exposed strand was 42 ksi (290 MPa). After 500,000 cycles had been reached, the fourth exposed strand was severed. A short fatigue loading was applied to the specimen at this damage level. Behavior of the element did not stabilize under fatigue loading. Fatigue loading of the specimen was discontinued and the ultimate load test performed.

2.4.2 *Internal Strand Splice Results* The third test girder, G3, was used to study the performance of an internal strand splice to repair two severed prestressing strands. (It was discovered at this time that all of the strands within this girder were placed 1 in. (25 mm) below the location specified in the plans). The strand splice consisted of two standard 1/2 in. (13 mm) strand chucks retensioning the strand via a turnbuckle as shown in Figure 6. The turnbuckle included an internal load cell which in conjunction with the strain gages on the severed strand was used to measure the force reintroduced into the strand. Stress ranges of the strands were measured more extensively with this specimen as compared to G2. Once the four susceptible strands had been exposed, strain gages were applied to the strands. Stress ranges were measured when cycling at the 0, 3, and 6√f'c psi (0, 0.25, and 0.5√f'c MPa) levels with the design f'c. Two strands were then cut one at a time, and the stress ranges at the different load levels were determined.

The first turnbuckle was then tensioned and stress ranges were measured again. The turnbuckle was subsequently retensioned to account for some initial slip between the strand and the chucks. The second turnbuckle was then tensioned to repair the remaining severed strand. Again stress ranges were collected. A preload equal to the decompression load was applied to the specimen and concrete was replaced around the turnbuckles and strands. This resulted in the concrete patch being in compression when the preload was removed and reducing the stress range in the strands under cyclic loading.

Loading began with 500,000 cycles of decompression. After satisfactory performance through this loading history, stress ranges were collected and the fatigue loading level was increased to $3\sqrt{f'c}$ psi ($0.25\sqrt{f'c}$ MPa) nominal tension. Once again the specimen survived 500,000 cycles of loading. Stress ranges were collected and the loading increased to $6\sqrt{f'c}$ psi ($0.5\sqrt{f'c}$ MPa) nominal tension. After 282,000 cycles of loading at this level a strand failed. The fatigue loading was stopped and the specimen was tested statically to failure.

2.4.3 *External Post-Tensioning Repair* The final test specimen, G4, will undergo a test program similar to that of G3. The external post-tensioning scheme to be incorporated with this specimen is shown in Figure 7. It comprises casting corbels on either side of the damage region and tensioning a high strength rod between the corbels to restore the prestress.

In this test series, the specimen will be cracked, concrete removed, strands instrumented, and stress ranges measured. Then the strands will be severed, stress ranges recorded, the repair conducted, and finally repaired stress ranges will be measured. Fatigue loading will then begin with $0\sqrt{f'c}$ and continue through $6\sqrt{f'c}$ psi ($0.5\sqrt{f'c}$ MPa). Similar to G3, following fatigue loading, the specimen will be loaded to failure statically.

3. Summary and Conclusions

(1) The experimentally determined prestress losses of the composite section were estimated to be on the order of 45 ksi (310 MPa). This value was within the range of the detailed prestress loss estimates (40 to 60 ksi (275 to 410 MPa)) given by the AASHTO code [11], the PCI handbook [12], Lin and Burns [13], and Collins and Mitchell's [14].

(2) The compressive strength of the girders increased approximately 22 percent over the 28 day strength reported in the fabrication records. The girders also exhibited uniformity of strength throughout the section.

(3) Results of the tests conducted on the girders indicate that the girders had experienced minimal deterioration while in service. The performance of the girders under fatigue loading was excellent as evinced loading was excellent as evinced by the stable behavior of the intermittent static tests. The undamaged specimen carried nearly three million load cycles before cycling was stopped and the girder loaded to failure.

(4) At failure, each of the girders exhibited a ductile failure with a centerline deflection exceeding 20 in. (510 mm) at failure. The ultimate strength of the undamaged specimen

(after 20 years of service and extensive fatigue loading applied in the lab) was larger than the Mp derived from standard strength equations.

(5) The girders appeared quite sensitive to relatively minor amounts of strand damage. The fatigue behavior of the damage control specimen, G2, deteriorated rapidly with 13 percent of the strands severed.

(6) The internal strand splice as a strand repair method worked well. The repaired specimen required 1.3 million fatigue load cycles before a strand failure occurred.

4. Acknowledgments

The researchers gratefully acknowledge the generous support of the Minnesota Department of Transportation in financing this project. The opinions expressed in this paper are those of the authors and do not necessarily represent those of the sponsoring agency.

5. References

1. Shanafelt, G.O., and Horn, W.B. (1980) "Damage Evaluation and Repair Methods for Prestressed Concrete Bridge Members", National Cooperative Highway Research Program Report 226, Washington.
2. Shanafelt, G.O., and Horn, W.B. (1985) "Guidelines for Evaluation and Repair of Prestressed Concrete Bridge Members", National Cooperative Highway Research Program Report 280, Washington.
3. American Society for Testing and Materials, ASTM C42, "Standard Method of Obtaining and Testing Drilled Cores and Sawed Beams of Concrete."
4. Kesler, C.E. (1957) "Effect of Length to Diameter Ratio on Compressive Strength," ASTM Bulletin, No. 221, April, pp. 68-73.
5. Washa, G.W. and Wendt, K.F. (1975) "Fifty Year Properties of Concrete," Journal of the American Concrete Institute, Vol. 72, January, pp. 20.
6. Bloem, D.L. (1968) "Concrete Strength in Structures," Journal of the American Concrete Institute, Vol. 65, March, pp. 176-187.
7. Swamy, R.N. and Al-Hamed, A.H. (1984) "Evaluation of Small Diameter Core Tests to Determine In Situ Strength of Concrete," In Situ/Nondestructive Testing of Concrete, V.M. Malhotra, ed. ACI Publication SP-82, pp. 4411-440.
8. Greene, Gordon W. (1954) "Test Hammer Provides New Method of Evaluating Hardened Concrete," ACI Journal, Vol. 51, Nov., pp. 249-256.
9. American Society for Testing and Materials, ASTM C597, "Standard Test Method for Pulse Velocity Through Concrete."
10. Malhotra, V.M. (1976) Testing Hardened Concrete: Nondestructive Methods (ACI Monograph No. 9), American Concrete Insititute.

11. The American Association of State Highway and Transportation Officials (1983) "Standard Specifications for Highway Bridges", Thirteenth Edition, Washington.
12. Prestressed Concrete Institute (1985) "PCI Design Handbook", Third Edition, Chicago.
13. Lin, T.Y., and Burns, N.H. (1981) "Design of Prestressed Concrete Structures", John Wiley & Sons, New York.
14. Collins, M.P., and Mitchell, D. (1987) "Prestressed Concrete Basics", Canadian Prestressed Concrete Institute, Ottawa, Ontario, Canada.

Table 1 Concrete Uniformity

Test	COV		Within test COV	
4 in. cores	5.8%	(6)[1]	6.0%	(3)[1]
2 in. cores	7.4%	(15)	11.6%	(8)
Rebound hammer	3.3%	(10)	7.4%	(18)
Pulse velocity	1.2%	(10)	2.0%	
Windsor probe	3.9%	(9)	3.4-7.7%	

[1]Numbers in parentheses indicate the number of samples.

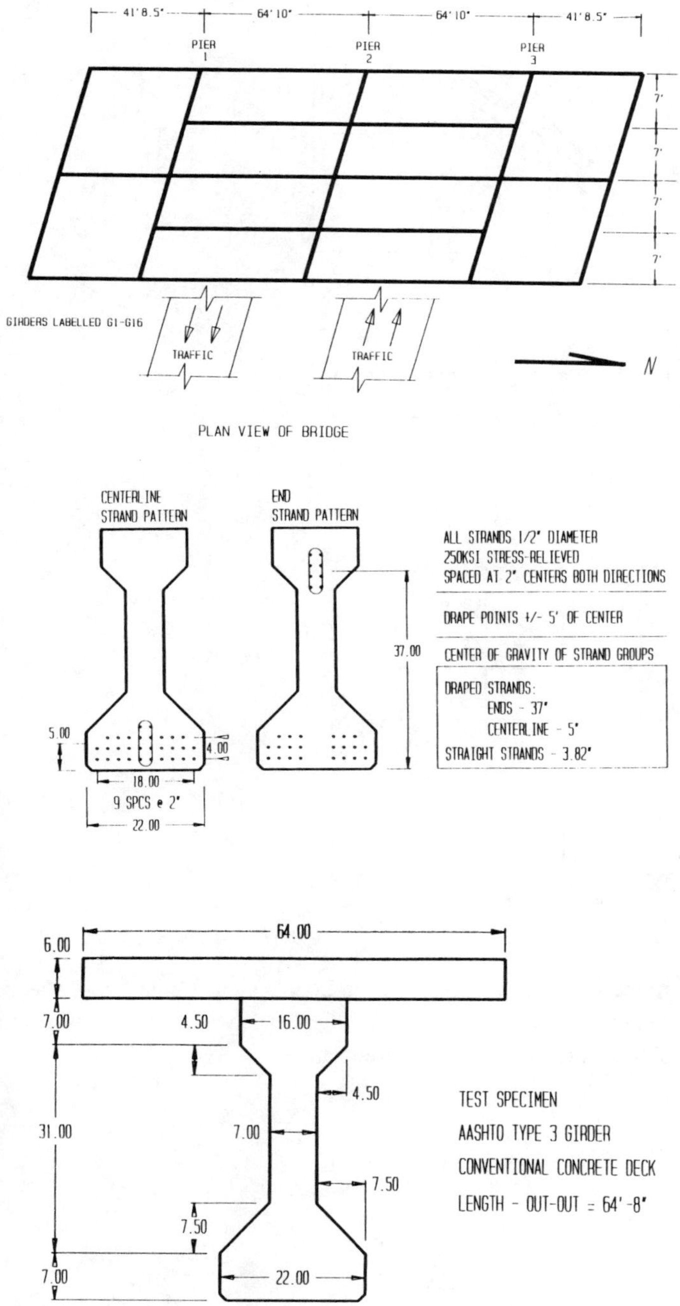

Fig. 1 Bridge Layout and Cross Sections.

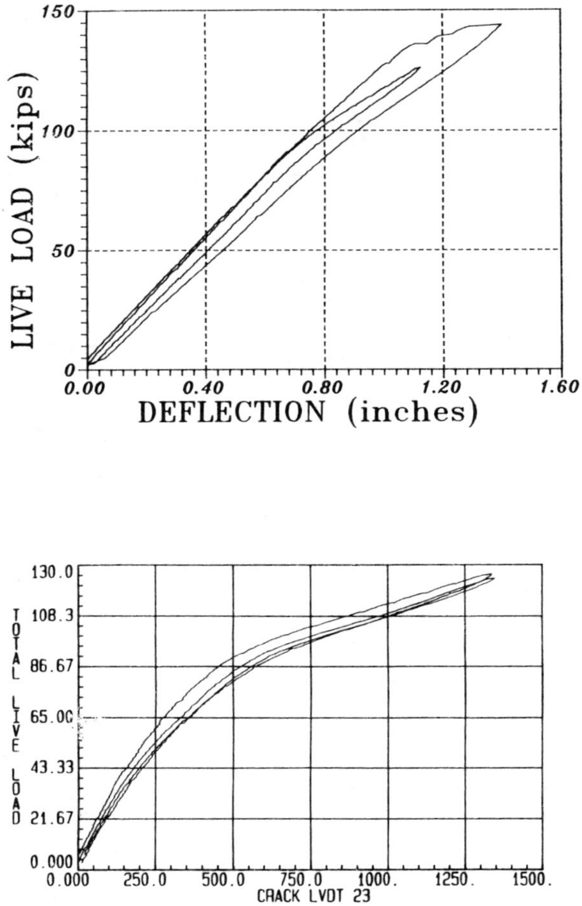

Fig. 2 Plots Used to Determine Prestress Losses; (a) Live Load Versus Centerline Deflection During Initial Cracking and Subsequent Reloading, (b) Live Load Versus Bottom Flange Crack LVDT Monitored During Crack Reopening.

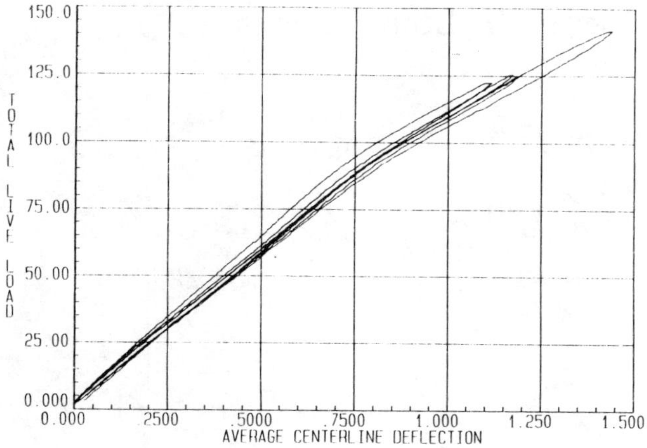

Fig. 3 Tests Conducted Between Increases in Fatigue Load for G1.

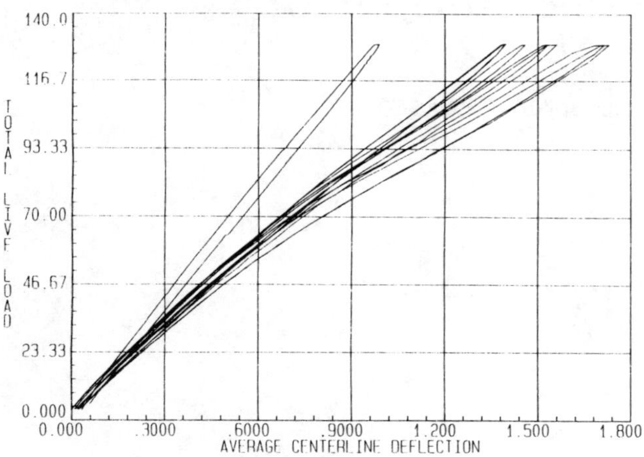

Fig. 4 Tests Conducted Between Incremental Damage Levels for G2.

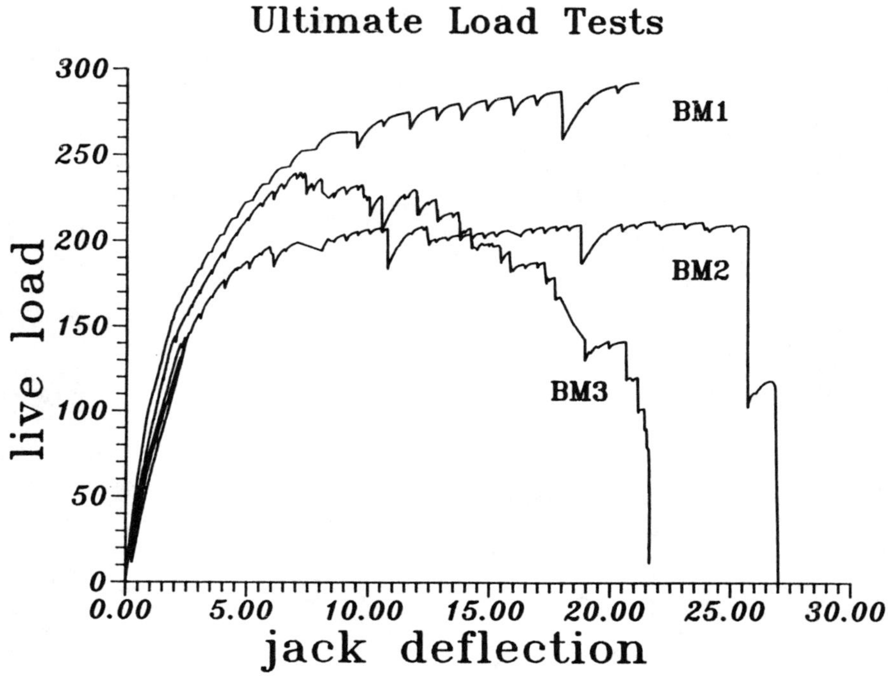

Fig. 5 Static Tests to Ultimate for G1, G2, and G3.

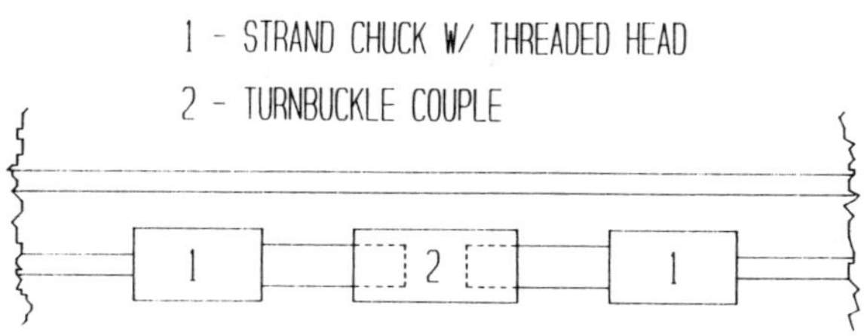

Fig. 6 Internal Strain Splice Repair Technique.

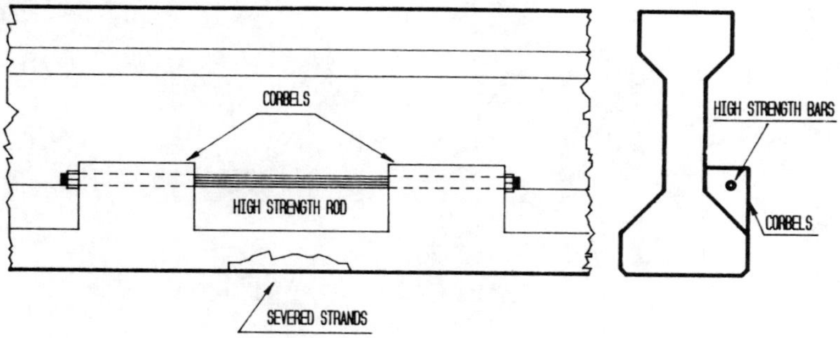

Fig. 7 External Post-Tensioning Repair Technique.

DYNAMIC TESTING OF BRIDGES USING TRAFFIC-INDUCED VIBRATION

JUAN R. CASAS and ANGEL C. APARICIO
Technical University of Catalunya
Civil Engineering Department
Jordi Girona Salgado-31
08034 Barcelona, SPAIN

ABSTRACT. In this paper the increasing necessity of rehabilitation or replacement of bridges is stated. The actual inspection methods in bridges are quite limited, due mainly to a visual examination basis, new techniques of assessment and inspection in bridges must be developed. The structural dynamic response as a non destructive tool in assessing integrity of actual bridges is proposed. The theoretical and experimental background of the proposed method of inspection is presented. The theoretical problem is reduced to a parameter identification procedure (inverse problem), i.e., to estimate the actual dynamic p arameters of the bridge (frequencies, damping and mode shapes), starting from the structural dynamic response and to compare them with the parameters estimated in the test performed when the bridge was open to traffic. Therefore, the success of the method is only accomplished if a good estimation of the dynamic parameters is achieved. The traditional estimation is strongly limited when ambient traffic vibrations in bridges recorded by accelerometers are examined. In this paper, a new method is presented for the dynamic parameter estimation which avoids these limitations. This estimation is applied to experimental vibrations recorded in two dynamic tests performed on real viaducts in Barcelona (Spain). The results obtained are compared with those provided by the application of traditional methods whenever possible or by theoretical models. The conclusions of this experimental research shows the effectiveness of the proposed method of estimation when applied properly.

1. Introduction

In the last several years the oncoming problem of deteriorating infrastructure has been a subject of interest to different National and International Administrations responsible for the evaluation, rehabilitation and replacement of bridges [1]. An adequate policy concerning both maintenance and rehabilitation can only be achieved with important investments. Nowdays, the economic resources available are still insufficient. As a consequence it is of maximum interest to establish a priority schedule and guidelines to be gradually developed as a new budget is available. Therefore the inspection and evaluation of existing bridges in order to assess the actual resistance and solicitation parameters are very important when a decision concerning rehabilitation or replacement has to be made.

A. S. Nowak (ed.), Bridge Evaluation, Repair and Rehabilization, 405–420.

Actual inspection methods in bridges are mainly based on visual examination and consequently important limitations are involved. Besides this, many theoretical and experimental works have shown the possibility of using the dynamic response to assess the actual behaviour of a structure (2,3...10]. The use of a non destructive technique is largely appreciated when the structure is open to traffic. Therefore the analysis of the dynamic response of a bridge seems to be a useful non-destructive tool in the evaluation of existing bridges. Moreover, using the traffic-induced dynamic response of a bridge measured by acceleration transducers, many important goals in the inspection procedure can be achieved when compared with static tests:

a) Dynamic testing of bridges is easier because less auxiliary equipment in the experimental set-up is needed.
b) The bridge must not be closed to traffic during tests. This possibility is of great interest to the administration responsible for the bridge.
c) Use of reference measuring points outside the bridge (often very difficult or even not possible) is not necessary.

2. Theoretical Background

The diagnostic procedure can be theoretically stated as follows:

1) Dynamic response of the bridge depends on the structural performance and can be expressed in terms of the dynamic parameters of the structure.
2) Variation in the dynamic parameters of the bridge is related to its actual integrity.

So the evaluation of existing bridges starting from dynamic response is formulated from a theoretical point of view as a system identification problem (inverse problem): "We know an experimental response of the system (bridge). We assume a theoretical model for the system. The parameters of the model should be found (parameter estimation) in such a way that responses from the model and experiment are as similar as possible."

Once the experimental dynamic response of the bridge is recorded, the first step is the choice of the theoretical model to be used in the identification procedure. There are several dynamic models available [11]. The question is: which is the most appropriate? The answer is related to two requirements in the practical application of the method.

1) The model must be applied easily and with high efficiency. It should be useful in the parameter estimation problem, both from a theoretical and an experimental point of view.
2) The model (its parameters) must allow the identification of incorrect performance of the bridge in order to accomplish the inspection objective.

   <u>First step</u>: Estimation of modal parameters (frequencies, damping and mode shapes) starting from an experimental dynamic response. The modal and frequency response models can be used.

   <u>Second step</u>: Estimation of spatial parameters (inertia, mass, rigidity, etc...) from modal parameters.

The estimation of modal parameters in structural engineering has been a subject of investigation and important results during recent years [2]. Consequently there are many conventional and standard techniques available in the parameter estimation of bridges (mainly using servo-hydraulic or servomechanic actuators for the excitation and displacement transducers in the response measurement). The use of traffic induced vibrations recorded by acceleration transducers introduces some important variations when compared with a conventional procedure. It requires estimation techniques different from those used in the conventional estimation (mainly Fast Fourier Transform (FFT)) as we will see in the next paragraphs.

The vibration caused by traffic crossing the bridge is a random process with a wide band frequency range. Therefore, using Frequency Response Functions, F.R.F. (or transfer functions) seems to be the most adequate method [11,12]. Minimizing the interpolation error:

$$\epsilon = \sum_{j=1}^{N_p} \sum_{l=1}^{N_j} (H_j^E(\omega_l) - H_j^T(\omega_l))^2 \tag{1}$$

N_p = Number of experimental frequency response functions derived from data
N_j = Number of frequency points to fit.
H^E = Experimental frequency response function (derived from experimental vibration)
H^T = Theoretical frequency response function (derived from structural model and related with modal parameters)

The experimental F.R.F. of a random process between two points i,j is defined as:

$$H_{ij}^E(\omega) = \frac{S_{ij}^{xf}(\omega)}{S_{jj}^{ff}(\omega)} \tag{2}$$

$S_{ij}^{xf}(\omega)$ =cross-spectral density between response in i and excitation in j

$S_{jj}^{ff}(\omega)$ =auto-spectral density of excitation in j

The spectral density is usually obtained by averaging the FFT estimates [13]. However, in the application of this method some implicit hypotheses are assumed:

1) In the dynamic test, excitation and response are simultaneously measured and recorded
2) The vibrational process involved is stationary.

But some problems arise when the process is due to the traffic crossing the bridge:

1) Measurement of dynamic excitation is very difficult or not even possible.
2) The dynamic response is clearly a non-stationary process [2].

We can avoid 1) by using the Frequency Response Pseudofunction (FRP):

$$|\hat{H}_{ij}^E(\omega)|^2 = \frac{S_{ii}^{x}(\omega)}{S_{jj}^{x}(\omega)} \tag{3}$$

$S^{x}_{\ddot{i}\ddot{i}}(\omega)$=auto-spectral density of response in i

$S^{x}_{\ddot{j}\ddot{j}}(\omega)$=auto-spectral density of response in j

We can avoid 2) by:

1) Eliminating the deterministic part of the response by using digital pass-band filtering [2].
2) Using of short-time records from the total record length [2].

But in this way a new problem arises because the length of the record is too small to properly estimate the spectral density of the process by means of Fast Fourier Transform with sufficient frequency resolution and accuracy [13]. The solution to this problem is the use of parametric estimation techniques in the evaluation of the spectral density; in particular the autoregressive model [14].

3. Experimental Results

This section deals with the main results obtained in the evaluation of the actual dynamic parameters of two bridges. It starts from the signals recorded in dynamic tests performed with both single and multiple vehicles.

3.1 ALFONSO X BRIDGE

The main objectives of the tests were:

1) To compare the dynamic parameters estimated using conventional and proposed methods.
2) To compare results derived from displacement and acceleration transducers in order to obtain some conclusions available for the design of future tests.

3.1.1 *Description of the Structure.* Alfonso X viaduct was a seven span (250 meters long) continuous prestressed box girder bridge (Figure 1) which was located at Alfonso X square in B arcelona (Spain). The cross-section (Figure 1) was a constant 1.60 m in depth and the deck width was 12.49 meters increasing gradually from pier 5. The bridge was partially located on a 300 meter-radius curve, so it could be assumed straight in the analysis.

Piers and abutments were reinforced concrete, and the deck was supported on each pier and abutment by elastomeric bearing pads. Construction of the bridge was completed in November 1974 and it was demolished in May 1985 due to urbanization requirements. Before its demolition, several static and dynamic tests were performed. The static tests included a load test, loading the bridge up to failure to evaluate its ultimate strength capacity [15].

3.1.2 *Test Procedures.* Before the test execution the bridge was cut in span four (see Figure 1), so the part tested was formed by spans 4,5,6,7. The bridge was dynamically excited by (Figure 2):

a) passages of a single, fully loaded two-axle truck. The gross weight of the vehicle was 136 kN and the axle spacing was 3.8 meters. Passage were performed with (WTH) and without (WOU) the standardized RILEM obstacle.
b) loading the bridge in the middle of the sixth span with a concrete cube (gross weight about 120 kN) and quick-releasing the load to simulate the free vibration of the bridge with an imposed initial displacement.

The test vehicle was driven at constant speed, whenever possible along the longitudinal bridge axis and always in the same direction (from pier 4 to abutment 7). The initial speed was 10 km/h and was increased after every passage in steps of 10 km/h up to the maximum allowable speed of 40 km/h.

3.1.3 *Data Acquisition and Recording.* In order to obtain the longitudinal bending and torsional vibrations, eight displacement transducers and two acceleration transducers were located at several points on the bridge (mid-point and quarter-point of the spans), both at the middle and the ends of the bottom deck (Figure 2).

Each transducer was completed with a signal amplifier and signal-conditioning. The experimental set-up was controlled by a micro-computer. The signals were digitized in an analog-digital converter before recording. Thus, 9 channels were simultaneously measured, digitized and recorded in the computer during the test. After every test the experimental data was stored in flexible diskettes for later data processing and analysis. In order to control the experimental results during the test procedure, one acceleration transducer, which has been previously located at the central section of the sixth span (channel 10), was attached to a real-time spectrum analyzer. The use of both displacement and acceleration transducers in the same measuring point was adopted to accomplish objective 2.

3.1.4 *Data Processing and Results.* Dynamic parameters derived from different vehicle passages and from several transducers are summarized in Table 1. The part of the total record used in the analysis is represented by the relative position of the vehicle on the bridge (arrow). The a pproximate record length used is 4 seconds. Starting from modal amplitudes in the central section of the spans, the experimental mode shape is derived using the proposed method and is compared with that derived from the theoretical model and damped free vibrations of the bridge (Table 2). D ynamic parameters derived from the whole tests are listed in Table 3 and compared with the theoretical model and an application of extended Prony's method [2] to quick-releasing tests.

3.1.5 *Conclusions.* Some important conclusions were available from the tests:

1) Frequency and mode shape results show the effectiveness of the proposed method when short-length records ($\simeq$ 4 sc) from a forced vibration by a vehicle crossing the bridge must be analyzed.
2) Resolution and accuracy in the dynamic parameters estimated are always better than results derived from FFT analysis (relative error = 8% in frequency and 145% in mode shape).

3) Results derived from displacement or acceleration transducers are quite similar.
4) Damping is the most difficult parameter to estimate.

3.2 DIAGONAL VIADUCT

Based on the conclusions derived from the Alfonso X bridge, a new dynamic test was designed and carried out for the following reasons:

1) New verification of the proposed estimation method.
2) In the previous test the correct behaviour of accelerometers located inside the box was checked. In this test the performance of the accelerometer directly located on the pavement is investigated. In this way the dynamic test is even more simplified.
3) In order to reduce the number of accelerometers used in tests, the possibility of mode shape estimation using only two accelerometers and the traffic-induced vibration would be largely appreciated. A part of the test is designed in order to confirm this possibility. The tests were carried out in May 1988 and are fully described in [16].

3.2.1 *Description of the Structure.* The viaduct is a 3 span (total length = 127 m) continuous prestressed box girder bridge crossing the main south exit highway in Barcelona (Figure 3). The typical cross-section is a constant 1.90 m in depth and the deck width is 10.95 m. The bridge is skewed (58.157^{g} (grads)) and clamped to bending and torsion in piers and abutments by means of two elastomeric bearings.

3.2.2 *Test Procedures.* The vibration of the bridge was forced by:

a) Passages of 1 fully-loaded two-axle truck (gross weight = 140 kN. Axle spacing = 3.8 m). Initial speed was 10 km/h and was increased in 10km/h steps up to the maximum speed of 80 km/h.
b) Passages of the same truck over the RILEM obstacle placed in the central section of the center span.
c) Passages of two full-loaded two-axle trucks (140 kN each) with different relative positions and velocities.

The vehicles were driven from North to South (Figure 3).

3.2.3 *Data Acquisition and Recording.* Instrumentation consisting of four displacement transducers and four accelerometers was located as shown in Figure 3. The use of displacement transducers was required because the impact coefficients should also be evaluated. Three accelerometers were located on the box and one (no. 7) was directly attached to the pavement. In this way it would be easier to move the accelerometer in the test where it had to be placed in different cross-sections to evaluate mode shape. The data acquisition system was the same as used in Alfonso X bridge. In this test accelerometers 6 or 8 were connected to the spectrum-analyzer system.

3.2.4 *Data Processing and Results.* Table 4 is devoted to the results related with the mode shape estimation using two accelerometers (fixed and mobile) and traffic-induced vibrations. The modal amplitudes corresponding to a fixed (6) and mobile (7) accelerometer are presented for every test, X_7 being the distance from accelerometer 7 to the North abutment. Figure 4 is drawn with these results and those derived from a theoretical model as well. In Table 5 we can see the dynamic parameters relative to the first and second vibration modes, for both one or two trucks crossing the bridge (traffic simulation), obtained with proposed method (FRP) and compared with conventional techniques.

3.2.5 *Conclusions*

1) Estimation of mode shape with 2 accelerometers plus traffic is possible with the proposed method.
2) Results concerning frequency and mode shape are fully satisfactory (2.5% error in frequency) and are always better than using Fast Fourier Transform estimates.

4. Conclusions

The effectiveness of the proposed estimation method when applied to traffic vibrations recorded with accelerometers is successfully verified. Moreover, with this method we can obtain the dynamic parameters with a minimum of instrumentation (2 accelerometers). The accuracy is always greater than accuracy achieved with a conventional estimation method. The instrumentation and auxiliary equipment needed to carry out the test are dramatically reduced when compared with a conventional dynamic test.

5. Acknowledgements

This project has been possible thanks to the economic support of the Spanish Ministry of Public Works (Road Department, Bridges and Structural Service).

6. References

[1] Federal Highway Administration (1986) "Highway bridge replacement and rehabilitation program," 7th Annual Report to Congress, U.S. Department of Transportation, Washington, DC.

[2] Casas, Juan R. (1989) "Study of dynamic response as a method of control and surveillance of bridges" (in Spanish), Ph.D. Thesis, Technical University of Catalunya Publications, Barcelona.

[3] Javor, T. (1986) "General Recommendation for vibration test in long term observation of concrete bridge structures," Materiaux et Constructions 19, No. 113, pp. 401-411.

[4] Ward, H.S. (1984) "Traffic generated vibrations and bridge integrity," Journal of Structural Engineering (ASCE) 110, N. 10, pp. 2487-2499.

[5] Stifkens, J. and Demars, P. (1981) "Surveillance des ouvrages d'art par le suivi de leur comportement dynamique," in International Conference on gestion des ouvrages d'art, Proceedings Anciens ENPC, pp. 243-246, Paris.

[6] Davis, A.G. (1987) "Application of dynamic test methods to the evaluation of bridge structures," Proceedings of the First US-European Workshop on Bridge Evaluation, Repair and Rehabilitation, pp. 284-295, Andrzej S. Nowak and Elie Absi Editors, the University of Michigan, Ann Arbor.

[7] Buckland, P.G., Hooley, R., Morgenstern, B.D., and Rainer, J.H. (1979) "Suspension bridge vibrations: computed and measured," Journal of the Structural Division (ASCE) 105, pp. 859-874.

[8] Douglas, B.M., and Reid, W.H. (1982) "Dynamic tests and system identification of bridges," Journal of the Structural Division (ASCE), 108, pp. 2295-2312.

[9] Shimada, S., Tanabe, T.,and Kato, M. (1982) "Safety inspection system of existing structures," IABSE Workshop: Informatics in Structural Engineering, pp. 119-133, Bergamo.

[10] Kato, M., and Shimada, S. (1986) "Vibration of PC bridge during failure process," Journal of Structural Engineering (ASCE) 112, N. 7, pp. 1692-1703.

[11] Ewins, D.J. (1986) "Modal Testing: Theory and Practice," Research Studies Press Ltd., Letchworth.

[12] Newland, D.E. (1983) "Vibraciones aleatorias y análisis espectral," Editorial A.C., Madrid.

[13] Bendat, J.S., and Piersol, A.G. (1980) "Engineering applications of correlation and spectral analysis," John Wiley and Sons, New York.

[14] Kay, S.M., and Marple, S.L. (1981) "Spectrum analysis. A modern perspective," Proceedings of the IEEE, Vol. 69, No. 11, pp. 1380-1419.

[15] Aparicio, A.C., and Casas, J.R. (1986) "Results and conclusions deriving from the tests carried out on the Alfonso X flyover on the first Ringroad in Barcelona: A source of relevant data," Bridge Department Report, Technical University of Catalunya.

[16] Casas, J.R., and Aparicio, A.C. (1989) "A study on the experimental estimation of bridge dynamic parameters using non conventional methods. Results and Conclusions," (in Spanish), Interior Report, Spanish Ministry of Public Works, Bridge and Structures Department, Madrid.

Table 1 Dynamic Parameters on Different Tests in Alfonso X Bridge.

		Transducer			
TEST	DYN. PARAMETERS	9	3	1	8
WOU 10 (b)	Frequency (Hz)	3.00	3.00	2.98	3.00
	Damping (%)	1.55	1.69	1.15	1.20
	Modal Amplitude	4.40	5.18	6.07	7.35
WTH 10 (1)	Frequency (Hz)	3.02	2.96	2.99	3.00
	Damping (%)	1.92	2.02	2.27	2.23
	Modal Amplitude	6.90	7.30	12.25	13.40
WTH 10 (2)	Frequency (Hz)	3.02	2.94	2.97	2.99
	Damping (%)	2.65	2.26	2.07	2.82
	Modal Amplitude	7.47	7.37	10.45	14.12
WOU 20 (a)	Frequency (Hz)	3.15	3.12	3.07	2.90
	Damping (%)	2.01	1.80	1.04	2.11
	Modal Amplitude	7.54	7.20	7.86	10.14
WOU 20 (b)	Frequency (Hz)	3.41	2.95	2.96	3.13
	Damping (%)	7.40(*)	6.34(*)	1.54	1.66
	Modal Amplitude	13.77	11.40	6.78	9.00
WOU 20 (c)	Frequency (Hz)	2.77	2.82	2.74	2.81
	Damping (%)	2.21	2.66	1.00	2.14
	Modal Amplitude	9.60	11.04	11.90	15.90
WTH 20 (1a)	Frequency (Hz)	3.16	3.21	3.00	2.88
	Damping (%)	2.26	2.41	1.56	2.42
	Modal Amplitude	6.50	7.40	7.91	10.17
WTH 20 (1b)	Frequency (Hz)	3.20	3.07	3.05	3.14
	Damping (%)	2.60	2.34	2.43	2.53
	Modal Amplitude	5.10	4.40	6.16	7.85
WTH 20 (2)	Frequency (Hz)	3.09	3.09	3.00	2.98
	Damping (%)	2.47	1.30	1.07	1.19
	Modal Amplitude	7.82	4.62	6.48	7.80

(*)–invalid records

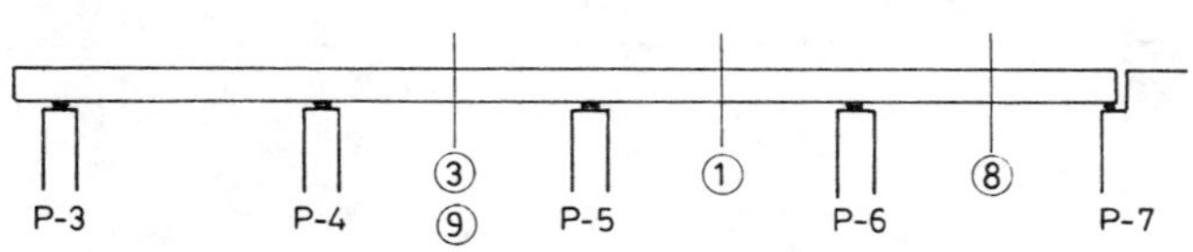

Table 2 Mode Shape N. 1 Deduced from Forced and Free Vibrations and Compared with Theoretical Models.

	ϕ_1/ϕ_3	ϕ_8/ϕ_3	ϕ_9/ϕ_3	ϕ_8/ϕ_1	ϕ_1/ϕ_9	ϕ_8/ϕ_9
Model	1.59	1.65	1.00	1.03	1.59	1.65
Quick-Releasing Test (Free vibration)	1.43	1.23	1.10	0.86	1.30	1.12
Truck (Free vibration)	1.52	1.60	1.04	1.05	1.46	1.54
WOU 10 (b)	1.17	1.42	0.85	1.21	1.38	1.67
WTH 10 (1)	1.68	1.84	0.94	1.09	1.79	1.96
WTH 10 (2)	1.41	1.91	1.01	1.35	1.40	1.90
WOU 20 (a)	1.09	1.41	1.04	1.30	1.05	1.36
WTH 20 (b)	Records	3,9 not	valid	1.32	Records 3	,9 not valid
WTH 20 (c)	1.08	1.44	0.87	1.34	1.24	1.66
WTH 20 (1a)	1.07	1.37	0.88	1.29	1.22	1.56
WTH 20 (1b)	1.40	1.78	1.16	1.27	1.21	1.53
WTH 20 (2)	1.40	1.68	--	1.20	Record 9	not valid

Table 3 Dynamic Parameters Corresponding to First Vibrations Mode with Proposed Method and Comparation with Method and Free-Vibration Techniques.

TEST	f (Hz)	ξ (%)	ϕ_3	ϕ_1	ϕ_8	ϕ_9/ϕ_3
MODEL	3.05	--	1.00	1.59	1.65	1.00
QUICK-REL. (Ex. Prony's Met.)	2.90 ($\sigma \simeq 0.$)	1.50 ($\sigma = 0.4$)	1.00	1.37	1.18	1.10
TRUCK (Ex. Prony's Met.)	2.95 ($\sigma \simeq 0.$)	1.08 ($\sigma = 0.13$)	1.00	1.49	1.57	1.04
TRUCK (FRP Method)	3.00 ($\sigma = 0.11$)	1.96 ($\sigma = 0.54$)	1.00	1.31 ($\sigma = 0.22$)	1.63 ($\sigma = 0.21$)	0.96 ($\sigma = 0.11$)

Table 4 Modal Amplitudes in Fixed and Mobile Accelerometers for Different Positions of Mobile Accelerometer

TEST	X_7 $(m.)$	ϕ_6	ϕ_7	ϕ_7/ϕ_6
DIAC1	7.8	0.245	0.098	0.4
DIAC2	15.6	0.187	0.215	1.15
DIAC3	23.4	0.184	0.232	1.26
DIAC4	31.2	0.148	0.141	0.95
DIAC5	45.14	0.114	0.070	0.61
DIAC6	51.28	0.164	0.221	1.35
DIAC7	57.42	0.073	0.132	1.81
DIAC8	63.56	0.039	0.074	1.90
DIAC9	69.70	0.249	0.480	1.93
DIAC10	75.84	0.195	0.279	1.43
DIAC11	81.98	0.11	0.070	0.64

Table 5 Dynamic Parameters (Vibration modes 1 and 2) from Proposed Method and Cnoven-tional Techniques

				Vibration	Mode	1	
TEST	$f\ (Hz)$	ξ (%)	ϕ_6	ϕ_7	ϕ_8	ϕ_2	ϕ_5/ϕ_6
Model	2.25	--	1.0	2.03	1.09	1.0	1.0
1 Truck (Free vib.)	2.25 ($\simeq$ 0.)	1.48 (0.31)	-- --	-- --	-- --	-- --	-- --
1 Truck (FFT)	2.22 (0.1)	-- --	-- --	-- --	-- --	-- --	-- --
1 Truck (FRP)	2.25 (0.005)	1.45 --	1.0 --	2.07 (0.12)	0.98 (0.1)	1.36 (0.07)	1.36 (0.07)
Traffic (FRP)	2.33 (0.06)	2.80 (1.)	1.0 --	2.36 (0.28)	1.21 (0.2)	-- --	1.29 (0.11)
				Vibration	Mode	2	
TEST	$f\ (Hz)$	ξ (%)	ϕ_6	ϕ_7	ϕ_8	ϕ_2	ϕ_5/ϕ_6
Model	3.53	--	1.0	0.00	1.16	1.0	1.0
1 Truck (Free vib.)	--	--	--	--	--	--	--
1 Truck (FFT)	3.10 (0.35)	-- --	-- --	-- --	-- --	-- --	-- --
1 Truck (FRP)	3.47 (0.014)	1.73 --	1.0 --	0.00 --	0.94 (0.13)	1.29 (0.11)	1.31 (0.08)
Traffic (FRP)	3.57 (0.08)	3.00 (1.3)	1.0 --	0.00 --	1.42 (0.35)	-- --	1.44 (0.23)

(*) standard deviation into brackets

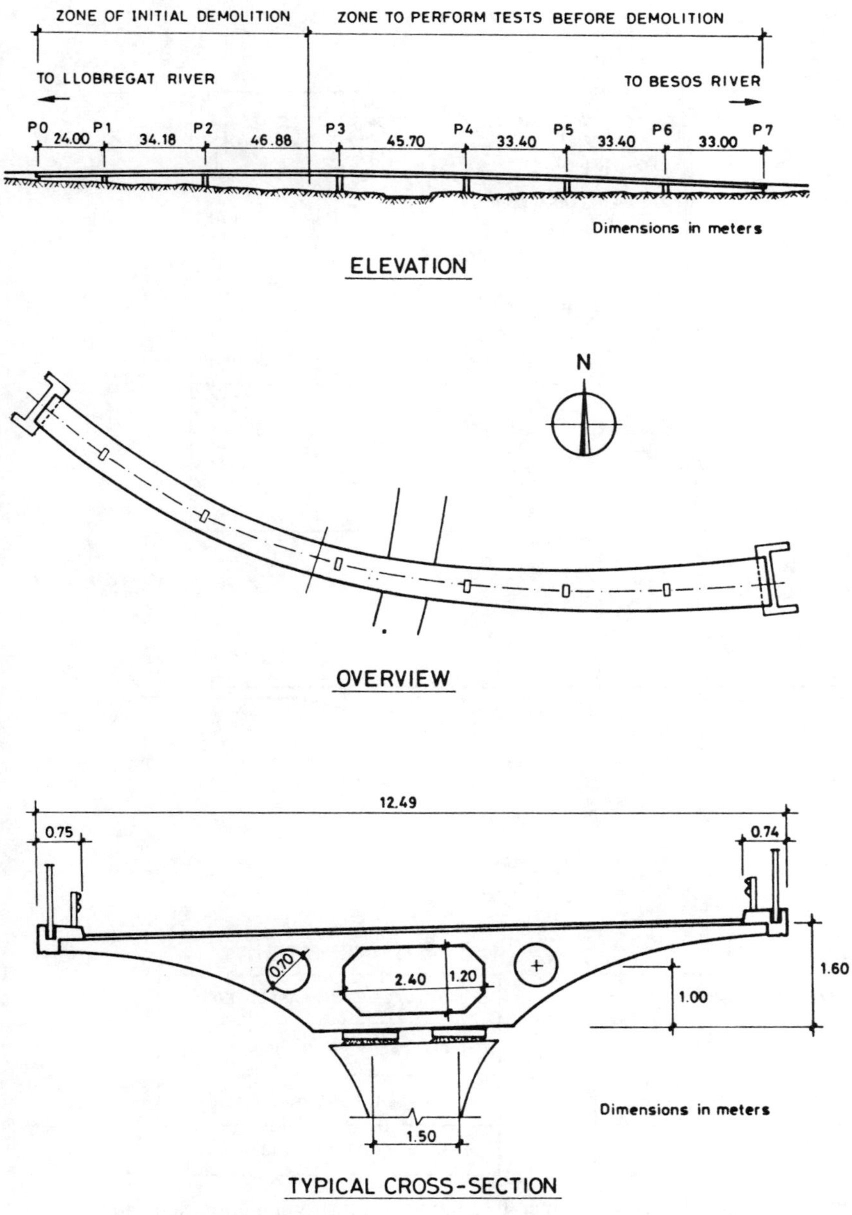

Fig. 1 Description of Alfonso X Bridge.

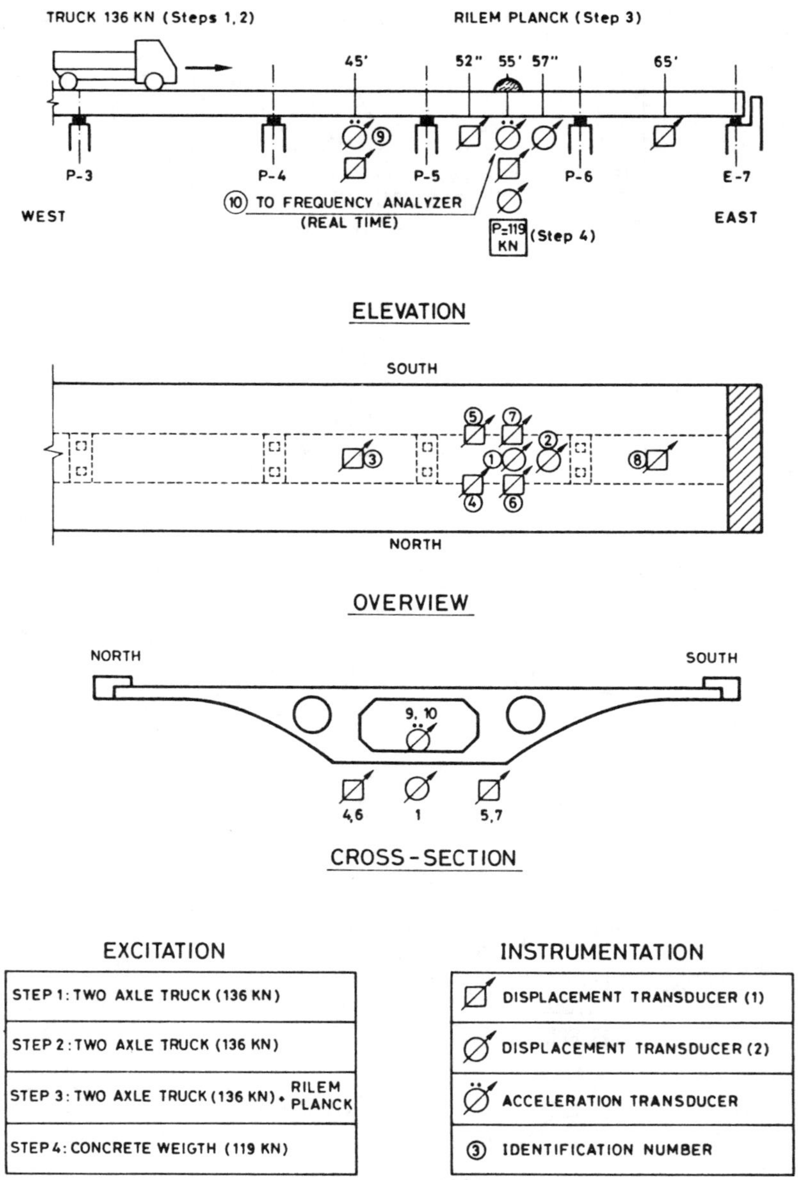

Fig. 2 Dynamic Tests om Alfonso X Bridge. Excitation and Instrumentation Schemem.

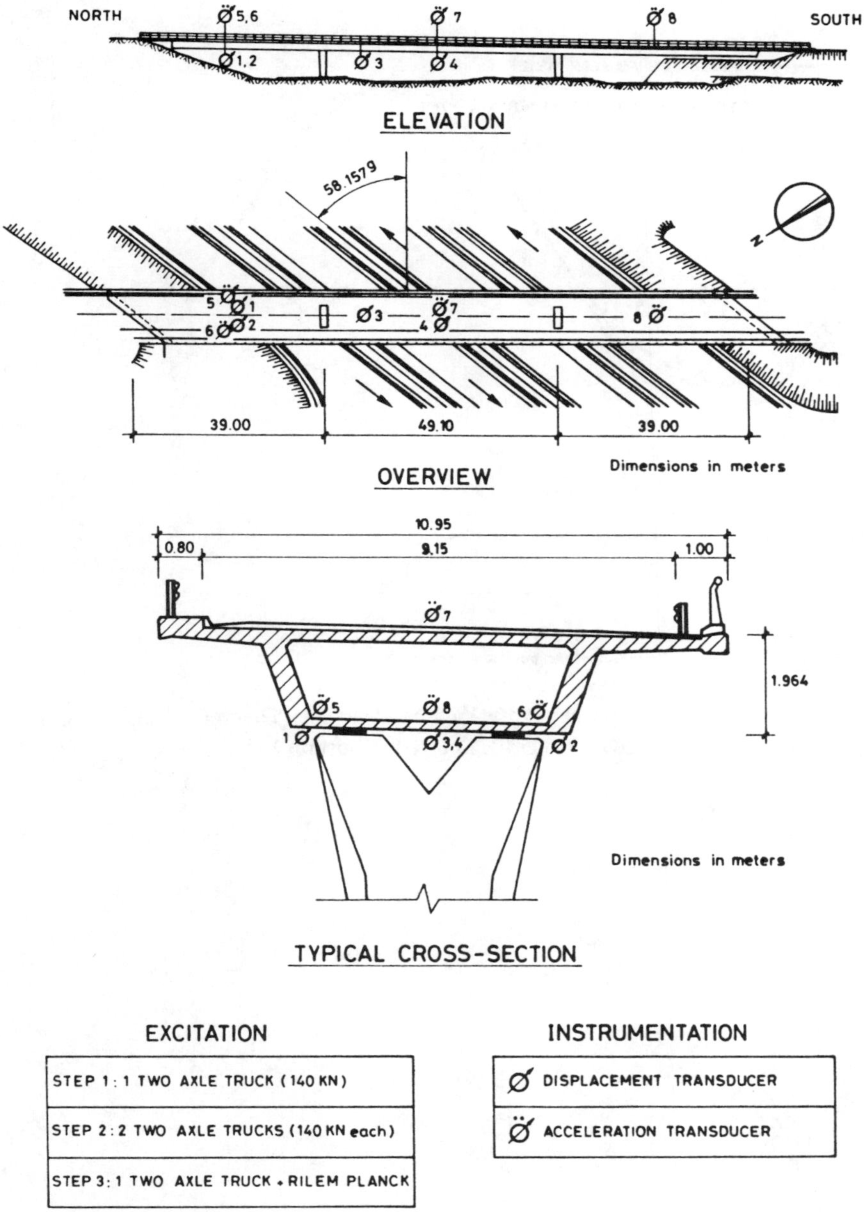

Fig. 3 Description of Diagnoal Viaduct. Exctation and Instrumentaiton Scheme.

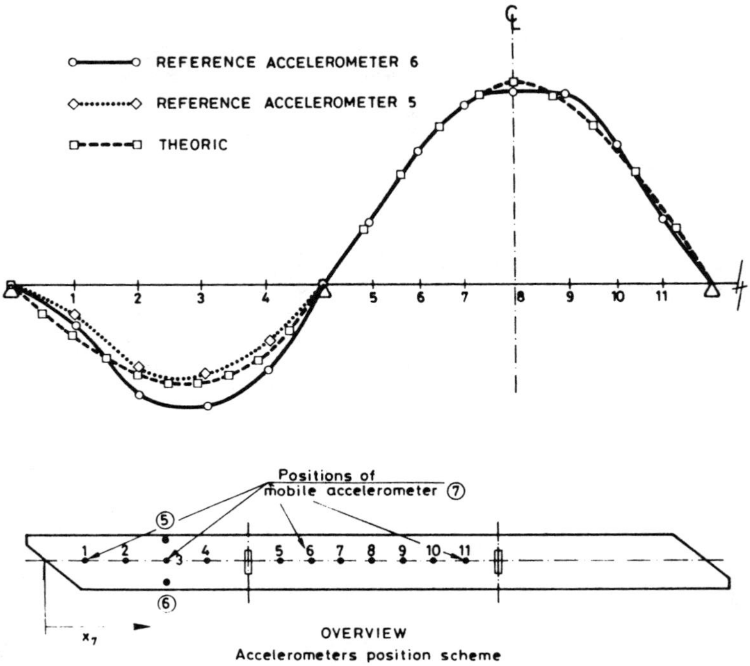

Fig. 4 Vibration Mode Sahpe N. 1 of Diagnoal Viaduct (Theoretical and with Proposed Method) Mobile Accelerometer (7) Located in Postion 1 to 11.

MEASURED STRESSES IN THE DECK SEGMENTS OF A CABLE-STAYED BRIDGE

F. W. BARTON, T.T. BABER
and R. D. RAMSEY
Dept. of Civil Engineering
University of Virginia
Charlottesville, VA 22903 USA

W. T. MCKEEL, JR.
Va. Trans. Research Council
P. O. Box 3817
University Station
Charlottesville, VA 22903 USA

ABSTRACT. This paper describes the results of a research program in which the field response of a cable-stayed bridge near Richmond, Virginia was experimentally determined using an extensive array of mechanical and electrical resistance strain gages. Of particular interest was the magnitude and distribution of in-plane stresses in the deck segments introduced by the stay cables and the post-tensioning strands. The transverse distribution of stress in the deck and lower flange of each segment was evaluated and found to compare favorably with that predicted by the loading. Results from this study provide valuable information regarding overall response of a cable-stayed bridge during construction.

1. Introduction

For a number of years, European designers have made use of both cable-stayed bridges and segmentally prestressed concrete box girder bridges as efficient solutions to certain bridge design requirements. But these two innovations in bridge design and construction have only recently been introduced and effectively utilized in the United States. Both cable-stayed and segmentally constructed bridges incorporate the efficient use of high strength materials. Moreover, the use of cable stays and cantilever construction in conjunction with segmentally post-tensioned box girder elements provide competitive designs for spans up to 1000 feet and longer.

Although such design concepts are attractive alternatives in a number of situations, much about their behavior in service environments is still unknown. A number of features make the analysis and design of cable-stayed and segmentally-erected post-tensioned bridges quite complex. Such bridges are typically continuous over several spans; the numerous stages of erection together with the associated spectrum of loading patterns contribute to the complexity; and the inherent indeterminacy of a multi-span bridge with multiple supporting stays and the nonlinear behavior of the stays themselves make prediction of the response of such bridges extremely difficult.

Considerable effort has been devoted to the development of suitable analysis and design procedures for the analysis of cable-stayed, segmentally-erected, post-tensioned bridges, but

A. S. Nowak (ed.), Bridge Evaluation, Repair and Rehabilization, 421–433.

only limited information is available on the actual field behavior of such structures. While laboratory and field studies have been conducted on several segmentally erected prestressed bridges, few of these studies completed to date have involved cable-stayed segmental bridges. And even less information is available on the behavior of such bridges during construction. Additional field measurements on the behavior and response of critical elements of such bridges during the construction and the in-service phases of these complex structures should materially aid in the development of improved procedures for future designs of these bridges.

Beginning in the summer of 1986 and continuing until the present time, a research team from the University of Virginia and the Virginia Transportation Research Council has undertaken an ambitious program of field instrumentation and response measurements on a number of key elements in a segmental, cable-stayed, box girder bridge now nearing completion in Virginia. This paper describes the results from one of the major phases in this extensive research program which was concerned with the instrumentation of the deck segments in this cable-stayed bridge and the subsequent analysis of the collected data.

The structure which was the subject of this investigation is a segmentally erected, precast, post-tensioned, cable-stayed box girder bridge which carries Interstate 295 over the James River near Richmond, Virginia. The bridge consists of thirty-one individual spans including the approach spans; these approach spans are each 150-ft. long precast box girder segments with external post-tensioning continuous over six spans and constructed by the span-by-span method. That portion of the structure which is the focus of this particular investigation consists of the central seven-span continuous section of the bridge which includes the 630-ft. main span over the river and the three approach spans on either side. An elevation sketch of this portion of the bridge is shown in Fig. 1.

As indicated in Fig. 1, the middle five spans of the bridge, including the main span, are supported by a total of 26 cable stays arranged in a single plane harp configuration and emanating from two pylons, one on either side of the river. The forces from the cable stays are transferred to the twin box-girders through a precast delta-frame assembly located between the girder segments at each stay location as shown in the cross section sketch of Fig. 2. The main span over the river was constructed as two cantilevers extending from the piers located adjacent to the pylons and made continuous by a midspan closure pour. Post-tensioning consisted of two parts, temporary tensioning bars used during construction, and permanent strand post-tensioning located inside the box segments after the erection was completed. The construction of the bridge is essentially complete and is expected to open to traffic in the fall of 1990.

2. Instrumentation and Data Acquisition

Instrumentation in the deck segments of the I-295 bridge was comprised of two major components, an extensive array of mechanical strain gages placed in the segments of the main span and the two adjacent approach spans, and an equally extensive array of electrical resistance strain gages installed in three deck segments of the main span.

2.1. MECHANICAL STRAIN GAGES

The program of instrumentation using mechanical strain gages was initiated at the request of the Federal Highway Administration to provide information relating to the major components of the bridge during construction. The response data obtained from the mechanical gages was intended to supplement data from the electrical resistance strain gages which were intended to be the primary source of strain data. This supplementary instrumentation was designed to provide additional information on construction stresses on the bridge, to measure and evaluate what effect, if any, shear lag played in the load distribution from the cables, and to assist in estimation of the bending moments and any torsional effects introduced in the deck segments.

A total of twenty deck segments on the north side of the bridge were instrumented with the mechanical strain gages. These segments represented ten longitudinal locations along the deck with two instrumented segments, one for each lane of traffic, at each location. Of these ten sections, six sections were located in the main span cantilever, three sections were located in the first approach span, and one section in the next adjacent approach span. In each of the twenty deck segments, eight mechanical gages were installed as shown in Fig. 3, five gage points located transversely along the top surface of the deck, and three gage points located across the lower flange of the segment inside the box. The gage points consisted of a pair of brass plugs, each with a small hole drilled to accept the points of the mechanical gage. The gage points were imbedded in the concrete after casting using nonshrink grout. The gage length on all of the gages was approximately 250 mm.

All of the mechanical strain data was measured and recorded manually. Since efforts were made to obtain strain readings at all significant stages of construction, data collection required an almost continuous presence on the bridge. This in turn required utilization of a variety of personnel which made consistent data recording difficult. The manually recorded strain data was subsequently entered in a personal computer in spreadsheet format to facilitate data analysis and provide printed and graphical output.

2.2. ELECTRICAL RESISTANCE STRAIN GAGES

This instrumentation was intended to provide information on primary deformation response of the main span, which was expected to include bending within the vertical plane of the bridge and torsional twist of the sections. The electrical resistance strain gages were placed in six different deck segments of the main span corresponding to three different longitudinal locations along the roadway. The foil strain gages were mounted on four-foot lengths of reinforcing bar and these bars were then imbedded in the selected deck segments prior to casting.

Segments 33, 48 and 62 in span 16 were instrumented with gages installed either as single gages or in a rosette configuration. The gaged reinforcing bars were placed in the reinforcing cages of the deck segments during the period between cage construction and placement of the cage in the precasting forms. The gaged bars were tied to one of the reinforcing bars in the cage and the lead wires were then run along the cage to a block-out in the wall of the box. After casting and placement of the segment, the lead wires were retrieved and run directly to a data acquisition system located within one of the box segments in the vicinity of the instrumentation.

The data acquisition system was a unit manufactured by the John Fluke Company. This system uses a Helios main controller to communicate with a number of individual remote scanning units each of which is located near an instrumented section. The data is then transmitted from the scanning units to the main unit in digital form and from there to a personal computer where it is recorded and stored on a hard disk. This data is periodically retrieved and transferred to a computer in the laboratory in spreadsheet format for subsequently analysis and manipulation. The personal computer in the field and the main controller are located in a specially designed and fabricated cabinet which is provided with complete environmental controls.

3. Results and Data Analysis

3.1. INTRODUCTION

A number of factors and events during the data collection phase may have had some effect on the response of the deck and should be noted prior to presenting and discussing results. The mechanical strain gage data was affected to some degree by the quality of the mechanical gage itself and by the personnel taking the readings. The pace of construction required that the gages be read almost continuously which required the utilization of some state inspection personnel on the job site not adequately trained in the use of the mechanical gage. The construction and erection schedule of the contractor also introduced some unexpected difficulties. There were frequent instances when the research team learned of a particular construction activity for which strain readings would have been desirable too late to make arrangements for data recording.

An attempt was made to obtain strain readings from all gages immediately before and after a particular major construction activity to determine its effect on deck response. Proper evaluation and interpretation of the strain readings obtained were complicated, however, by the variety of activities that occurred almost simultaneously during construction. This made it difficult to isolate the exact causes of changes in strain that were observed in readings taken only a few hours apart. For example, during the study, every effort was made to obtain strain readings before and after the tensioning of each stay cable. However, during the time required to tension a stay, partial post-tensioning of the deck may have occurred, major construction equipment may have been shifted, and other changes could have occurred which may have had an effect on the recorded strain data.

Although strain data was recorded almost daily, it was impossible, because of construction constraints, to obtain consistent readings corresponding to zero strain on each of the deck segments instrumented. Accordingly, the strain data presented and discussed in this section are in terms of strain increments or changes over a certain period rather than absolute strains. Also, the data are presented in terms of microinches per inch; however, if it is assumed that stresses in the deck segments, particularly those in the deck and lower flange regions of the box segments, are uniaxial in nature, a measure of the magnitude of the stresses can be obtained by multiplying the strain values given by the modulus of the concrete which was calculated to be between 4.0 and 4.46 million psi.

The longitudinal strains as determined by the gages installed in the box segments allow a reasonably complete determination of the longitudinal strain/stress distribution throughout the deck and lower flange portions of these elements as well as changes in strain that occur as a result of the various construction activities. In particular, the longitudinal strains along the top and bottom sections of the box girder permit a realistic estimate to be made of the axial compression developed in each section and the bending moments developed in these sections. These longitudinal strains provide an unusually clear picture of the distribution of stresses in the deck segments as a result of post-tensioning the spans or by the tensioning of the stay cables. In the case of stay tensioning, which produces a concentrated in-plane axial compression as well as gross bending of the section, the strains measured thus far permit the evaluation of the shear lag that can be expected in these segments when unusually large in-plane forces are introduced in these massive precast and post-tensioned elements.

To better understand the results to be presented, it will be helpful to understand the relative locations of the various segments and the stay cables in the main span. The deck segments are designated by a number which identifies the span, the segment number within the span, and whether the segment is part of the north- or south-bound lanes. Thus, for example, segment 16-N5 is the fifth segment in the north bound lane of span 16 which is the main span. Also in the main span, stay number 1 is connected to segment 4 of the main span, stay 2 to segment 6, etc., and finally stay 13 to segment 28.

3.2. RESULTS

The collection of strain data presented first will correspond to the effect of stay tensioning on the longitudinal strains and general behavior of typical box segments in the main span. The initial stay tensioning introduced large longitudinal compressive loads in the deck segments. For example, the initial tensioning of a stay introduced a force in the stay on the order of 1.4 to 1.8 million pounds or roughly 90,000 psi depending on the particular stay. This force would correspond to an axial compressive force of approximately 1.2 to 1.6 million pounds transmitted to the deck.

The effects of initial stay tensioning on the longitudinal strains in several deck segments are presented in the next several figures. Figures 4 - 6 depict the incremental strain experienced in segments N1, S1, and N3 of the main span due to the tensioning of two different stays. Segments N1 and S1 are immediately adjacent to the pier segment at the pier on the north side and likely reflect the effects of stay tensioning on a girder segment near a support. As suggested by the data in these three figures, stay tensioning had greater effect on deck segments further from the stay. Since the compressive strains measured in the segments are the result of both axial loads and bending, this distance effect might be explained in part by the increase in bending moment with distance between stay and segment.

Similar strain increment data due to the tensioning of a number of stays, as measured in segments N15 and S15 of the main span, are shown in Figs. 7 and 8. The changes in strain at this location near quarter span are somewhat more pronounced than was observed for segments near the pier and the changes in strain are less uniform. Also, in both of these segments, there is a definite suggestion of shear lag as indicated by larger compressive increments toward the interior of the segment. The symmetric distribution of the longitudinal forces introduced by stay tensioning is indicated by a plot of changes in strain in segments N15 and S15 of the main span due to the tensioning of stay 8 as shown in Fig. 9. In Fig. 10 are presented plots of strain increments in a number of different segments due to tensioning of a single stay. The data suggest that, although there is considerable variation in the distribution of the strain increments, the levels of strain are of the magnitude that would be expected given the level of stay forces and temporary post tensioning introduced.

In the case of a few deck segments, it was possible to establish a reliable value of zero strain and consequently monitor absolute strain values during the construction process. An indication of the level of absolute strain in segment S3 of the main span during a variety of construction activities is presented in Fig. 11. In this case, the maximum values of strain are, for all gages, less than 300 microin./in., or on the order of approximately 1000 psi. As indicated by the data, there was also an indication of tensile strain near the exterior edge of the segments, an occurrence which was verified visually in many instances by the observance of small openings in the joints between segments. The distribution of the strain in this segment also would suggest the presence of some shear lag effect due to the longitudinal and bending loads.

Another feature of interest was observed in strain data measured in segment S2 of span 15, an approach span immediately adjacent to the main span on the south side. The gages in this segment were installed prior to erection and thus it was possible to measure changes in strain due to certain erection procedures. In Fig. 12 are plotted values of strain increment caused by the post-tensioning of span 15 after erection of all segments in that span. The location of the post tensioning jacks is indicated in the figure and the data clearly show the variation and distribution of strain experienced by the segment. As indicated by the data, the post-tensioning produced compression in the region of the jacks, as would be expected, but also produced tensile strains toward the segment edges. These measurements were recorded prior to the closure pour and hence the segment edges are effectively free.

3.3. DERIVED RESULTS

Thus far, most of the data has been presented in the form of strains taken directly from the measured data. Additional information can be obtained by appropriate calculations using the measured strains. For example, in those segments where it was possible to obtain reliable strain readings in both the deck portion and the lower flange of a segment, and making use of the known location of the neutral axis, it was possible to extract information relating to both the pure compressive load as well as the bending moment as defined by the strain readings.

In segment S32, near the south pier and pylon, the effect of stay tensioning was again evaluated by determining the change in strain at each gage point. An example of the analysis performed will be illustrated by investigating the change in strain as a result of tensioning stay 4. The tensioning force in stay 4 was approximately 1.8 million pounds. Strains were converted to stresses, the location of neutral axis determined, and the stress distribution used to identify the resultant compressive thrust and bending moment. The average compressive stress determined from the strain readings was approximately 150 psi corresponding to an axial compressive load of approximately 1.9 million pounds. This compares favorably with the value of 1.6 million pounds which is the horizontal component of the tensile force in stay 4.

In a similar fashion, the bending moment in the section, using averaged stresses in the top and bottom elements of the segment, was calculated to be approximately 80 million ft-kips. The value of bending moment determined using the stay force and distance between stay and segment was calculated to be approximately 70 million ft-kips, a value which again compares favorably with the measured value. It should be emphasized, however, that not all of the measured and calculated values of axial load and moment compared as well. The number of different activities affecting the longitudinal strains frequently made it difficult to isolate each of the effects.

4. Conclusions

Analysis of data is still underway but some conclusions regarding the measured response data is possible. The method of instrumentation, including both the mechanical and electrical resistance strain gages, seems to provide an effective means of obtaining measured response data. Evaluation of the strain data indicated incremental strain levels on the order of 200 microin/in. or approximately 800-900 psi due to stay tensioning. The corresponding level of average compressive stress was determined to be approximately 150 psi for the same loading. Some evidence of shear lag was observed, but it was not pronounced. Finally, no evidence of unusually large strain increments was observed during the entire construction process.

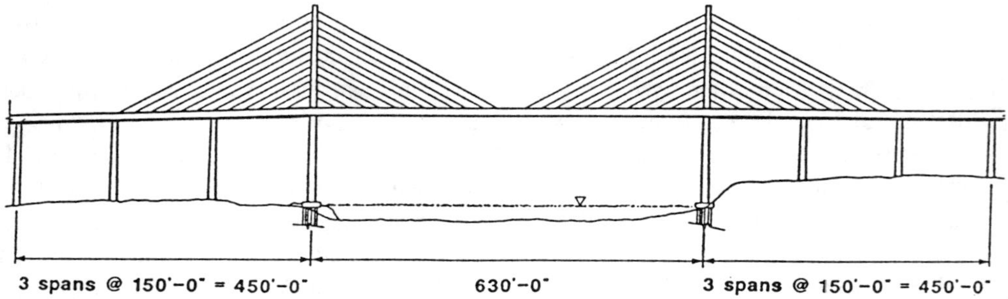

Fig. 1 Evaluation of I-295 Bridge.

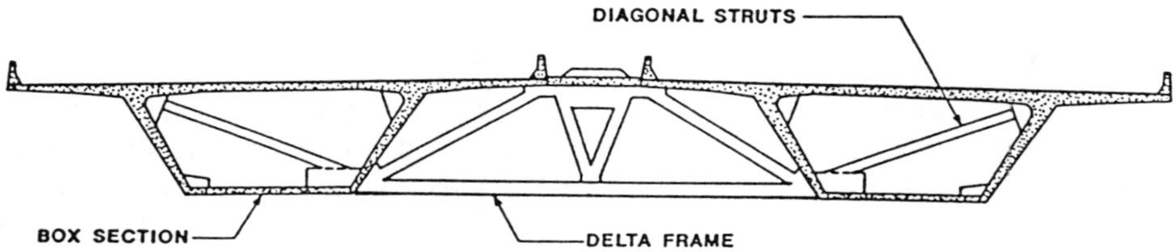

Fig. 2 Bridge Cross Section.

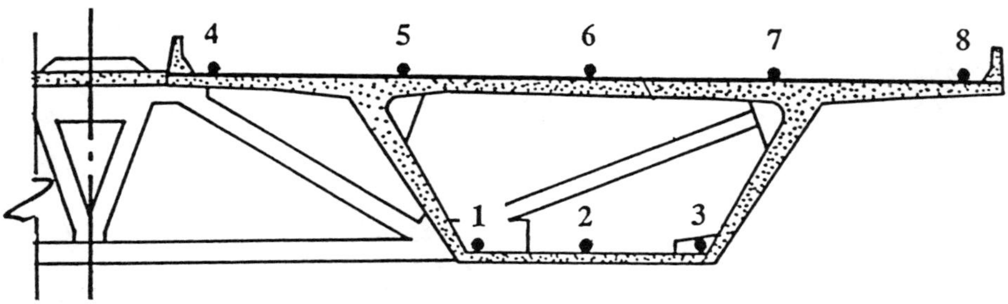

Fig. 3 Typical Gage Locations.

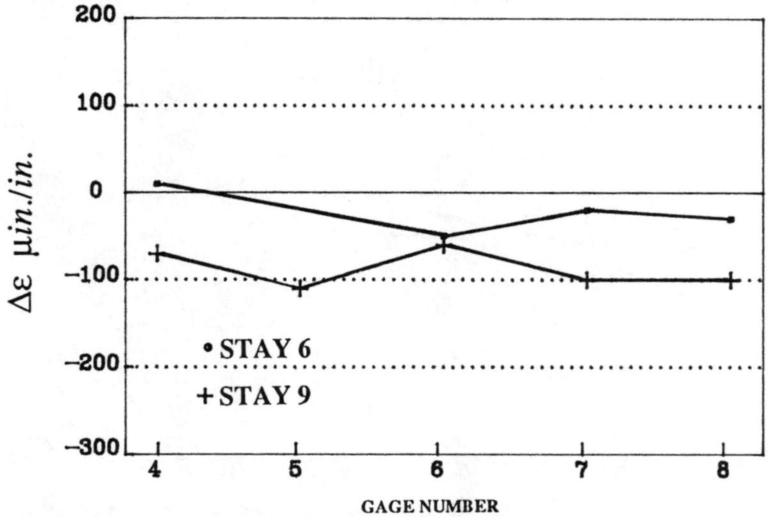

Fig. 4 Strain Increments in Segment 16-N1.

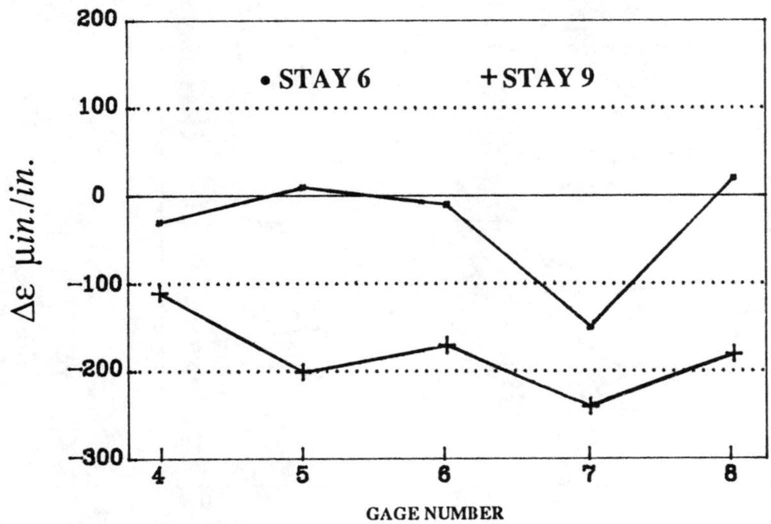

Fig. 5 Strain Increments in Segment 16-S1.

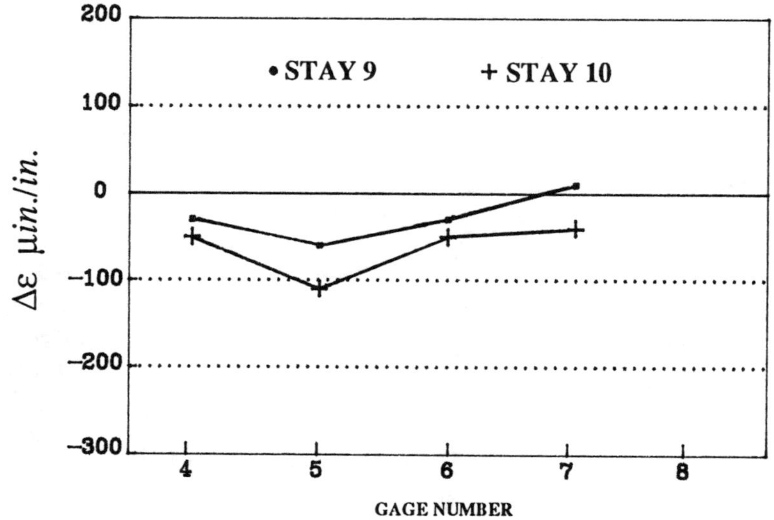

Fig. 6 Strain Increments in Segment 16-N3.

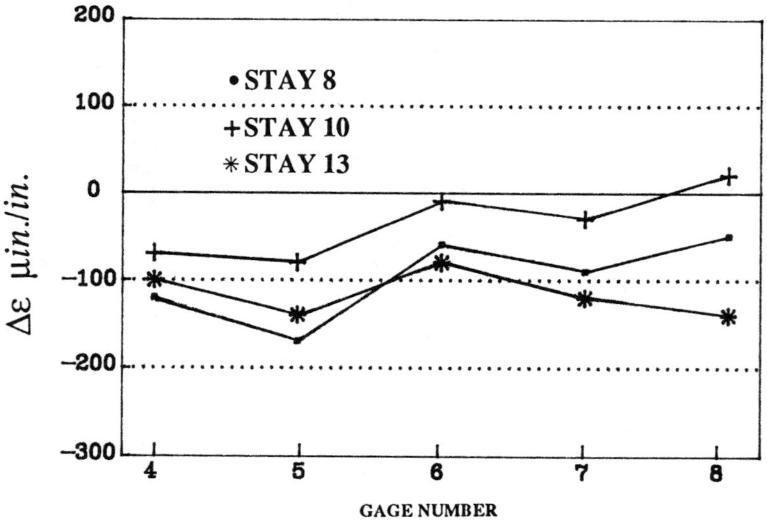

Fig. 7 Strain Increments in Segment 16-N15.

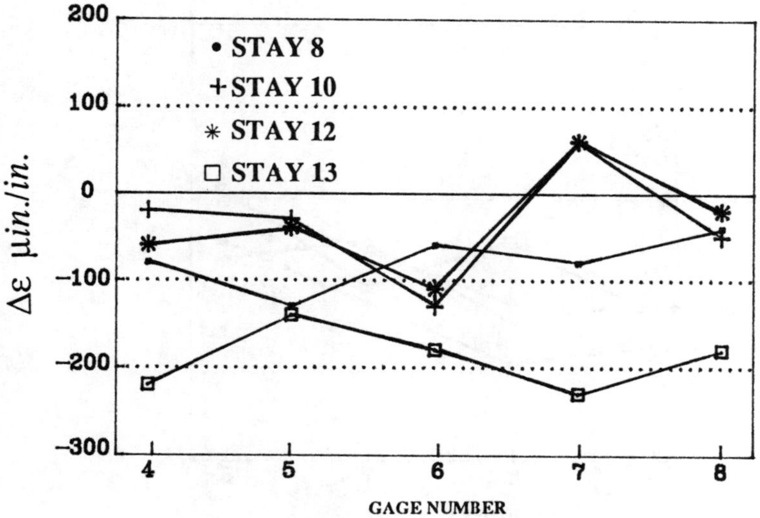

Fig. 8 Strain Increments in Segment 16-S15.

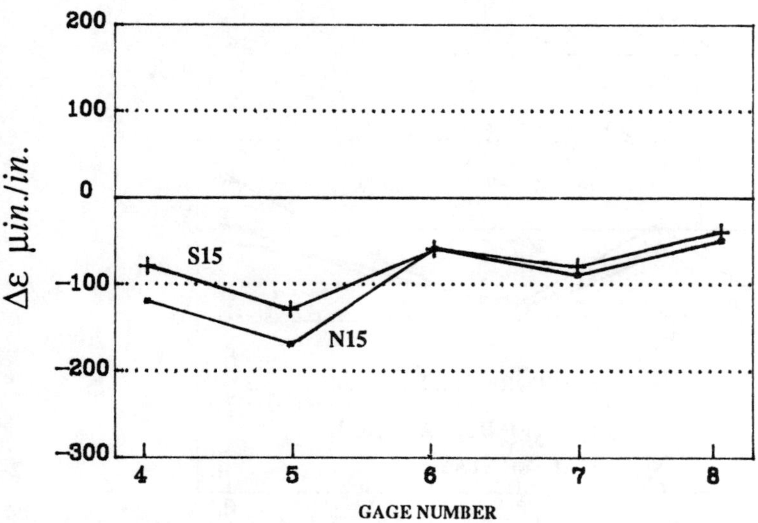

Fig. 9 Strain Increments in Symmetric Segments.

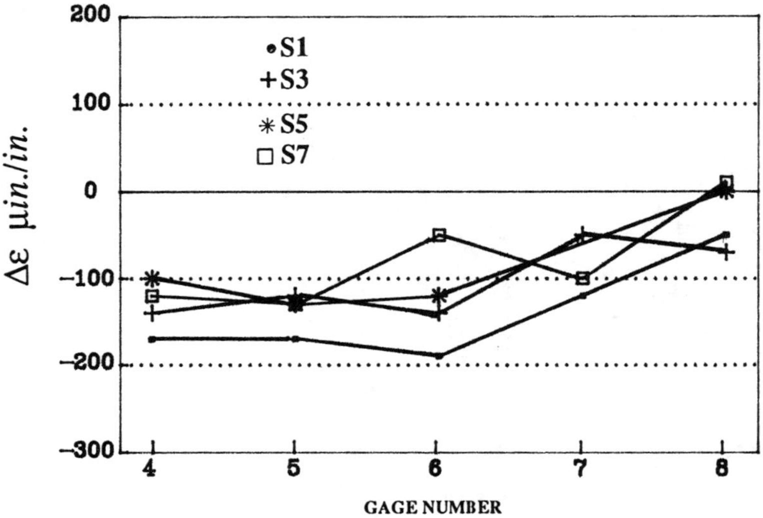

Fig. 10 Strain Increments in Segments 16-S1, S3, S5, S7.

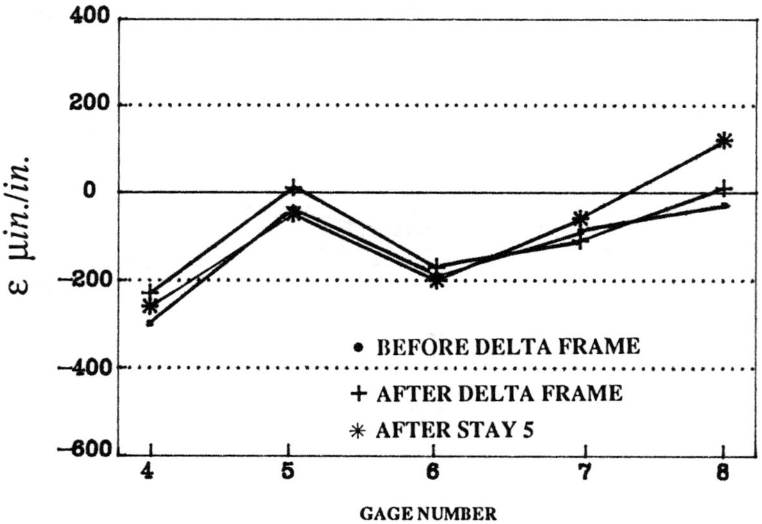

Fig. 11 Absolute Strain in Segment 16-S3.

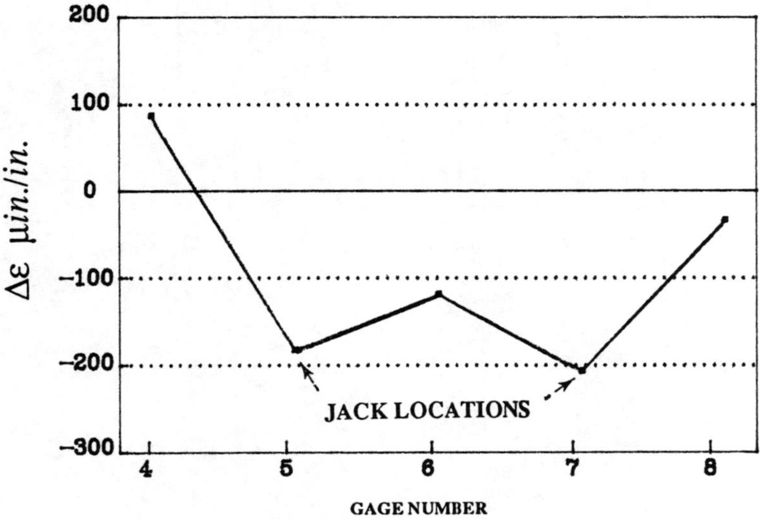

Fig. 12 Strain Increment in Segment 15-S2.

EVALUATION OF AN OLD PRESTRESSED CONCRETE BRIDGE

T. TASSIOS and K. ALIGIZAKI
National Technology University, Athens
Laboratory of Reinforced Concrete
Patission 42, 10682 Athens
GREECE

ABSTRACT. A 30 year old, 140 m long prestressed concrete bridge was inspected and evaluated. Several in-situ and in-lab test methods were used to determine strength, carbonation depth, chloride content, static and dynamic loading characteristics. The main findings are: very good quality of concrete, local destruction of some tendons by corrosion (due to a localized action of drainage waters), some bearings were compressed. A discussion of the global behavior of the bridge is included; the beneficial role of transversal prestressing is reiterated.

1. General Information

A prestressed concrete bridge crossing a river near the city of Arta (Greece) was built in 1959. It consists of three free supported spans of 46 m each; five prestressed beams, 2.13 m high, form the deck (Figures 1 and 2).

Based on the inadequate file of the bridge, each post-tensioned beam includes 15 tendons with 12 strands of diameter 7 each. A standard cube strength of 45 N/mm^2 was considered in the design. Loads corresponded to AASHTO-H20/44.

Two abnormal interventions have taken place on the bridge:

- Heavy lighting poles were fixed only on the protruding, 1.15 m wide, curbs of the bridges.
- A 600 mm diameter water pipeline was arbitrary fixed on the thin web of the upstream side-beams, by means of heavy duty fixing dowels.

A rather violent replacement of damaged asphalt pavement was carried out four years ago; the asphalt pavement was taken out (together with fragments of surfacing concrete of the deck) by means of a mechanical fork.

A. S. Nowak (ed.), Bridge Evaluation, Repair and Rehabilization, 435–443.

2. Visual Inspection

The following alterations or damages were observed:

2.1 SUBSTRUCTURE

- Neoprene bearings were covered by earth deposits; after uncovering, they were found to be in very good condition except for those of the north abutment which seem to be abnormally compressed.
- The plain concrete of abutments and piers shows no damage except for some local erosion and color alterations in some areas of the piers.

2.2 BEAMS

- The concrete seems generally to be in good condition, although some poor finishing and several honeycombs were observed, mainly near the bottom of some side-beams and cross-beams (Figure 3). Local color alterations, vegetation growth and surface salt deposits were also observed near the locations of water drains. Nonetheless, local spalling of concrete and some longitudinal cracking was observed at the locations of intensive corrosion of tendons near drain areas, as described later.

  Furthermore, local disruption of concrete was observed at the areas where lighting poles were fixed on curbs.

  Similarly, heavy damages were produced at the areas where the 600 mm diameter pipeline was fixed on the web of the upstream beams; fixing dowels are pulled out near the south abutment causing serious spalling of concrete to 80 mm in depth (Figure 4).
- Steel defects were observed as follows:
  - Due to inadequate cover, the location of stirrups was easily identified in several places, without, however, any serious corrosion.
  - Some tendons near the inner face of the side-beams are totally corroded (Figure 5) at the narrow areas where drainage pipes were discharging deck waters, occasionally containing deicing salts. Thus two to four tendons on some side-beams are locally destroyed; concrete cover (10 to 15 mm) was clearly insufficient. It is worth noting that in no other areas is any corrosive effect observed; even where tendons were exposed in heavily honeycombed areas, only slight surface rust is observed on ducts (Figure 6).
  - There are no indications of any damage in the anchorage zones of tendons.

2.3 BRIDGE DECK

Inspection of the underside of the deck slab identified locally deteriorated concrete, wet areas and efflorescence. No cracks, corrosion spalling, exposed steel, rust stains or other defects were detected. During rain falls, water does leak from some concrete hinges of the first span.

2.4 ACCESSORIES

The drainage system seems to have serious damage. Observed from underneath the drainage pipes are totally corroded and disrupted; this has resulted in water leaching the internal surface of the side-beams. Accumulation of debris and asphalt disposal on the deck drainage areas seems to have acted favourably because they have stopped water leakage.

The guard rails are in perfect condition.

The bituminous overlay which was replaced 4 years ago, suffers from serious damage mainly on the first span. Damages observed are localized subsidence followed by cracking along the road axis; wheel truck cracking is also observed. Cracking of the asphalt is observed at the joint areas of the slab (above the piers).

3. Experimental Investigation

In-situ and laboratory measurements were carried out in order to evaluate the condition of concrete and the global structural behavior of the bridge.

3.1 CONCRETE STRENGTH

Cores 100 mm by 170 mm were taken from the slab; their compressive strengths ranged between 47 and 50N/mm^2.

Rebound hammer tests indicated that concrete strength ranged from 40 to 50 N/mm^2 on beams, and 35 to 40 N/mm^2 on piers. Since the carbonation depth was very small (see subsection 3.2), rebound test results may be considered as rather reliable.

Pulse velocity measurements on beams resulted in compressive strengths on the order of 35 to 45 N/mm^2.

Tensile strength of concrete measured by means of splitting tests on 100 mm by 15 mm disks cored out from the slab of the bridge, was found equal to 2.3 N/mm^2 (ranged between 1.8 and 2.0 N/mm^2).

3.2 CARBONATION DEPTH

The carbonation depth of concrete was measured in-situ by spraying phenolphthalein solution on freshly chipped concrete surfaces. Despite the age of the bridge, the carbonation depth was very small (3 to 15 mm), even in the drainage areas where water leaching was expected to have accelerated the carbonation process.

3.3 CONCRETE COVER

The thickness of concrete cover was systematically measured. In the most critical lower part of the beams, cover ranges from 10 to 30 mm at the bottom, and 30 to 35 mm laterally. Most frequently, prestressed tendons were almost in contact with the horizontal leg of stirrups.

3.4 CHLORIDE CONTENT

The chloride content of concrete was measured at various depths in the deck slab. Even though the bridge is situated far from the sea coast (appr. 13 km away), it was judged useful to measure the chloride content of the concrete, in order to find out whether the occasional use of deicing salts during the last 15 years had contaminated the concrete. Cores taken from the slab were cut to slices of 15 mm in thickness. The slices corresponding to 0-15 mm and 30-45 mm depths were pulverized and examined in the laboratory by means of wet chemical analysis (chloride electrode model 941798, Russell). The average total chloride content was found to be 0.135% by weight of cement (ranging from 0.121 to 0.161% b.w.c) which is clearly below the maximum acceptable content for prestressed concrete.

Since no appreciable differentiation of chloride content was observed with increasing depth from the surface, it may be said that deicing salts have not contaminated the concrete.

3.5 COMPOSITION AND POROSITY OF CONCRETE

From the cores taken, it was observed that the grading and the distribution of aggregates were satisfactory. The concrete seems to have only a few small voids, indicating that compaction was sufficient.

An indication about the porosity of concrete can be obtained from the measurements of water absorption by immersion under vacuum according to a modified RILEM method developed in the Laboratory of the Association of Greek Cement Industry. Pieces of core slices which remained after the splitting test were used for absorption testing. The mean porosity measured for three cores was lower than 4%.

3.6 STATIC LOAD TEST

Two static load tests were carried out on the first and the second span of the bridge, using heavy vehicles arranged in the middle of the span. The weight of the vehicles was not accurately known but it should range between 130 kN and 185 kN. Concrete decompression strains were recorded at several points on the bottom of the beams at several cross sections. The measured strains ranged from 17×10^{-8} to 37.5×10^{-8} for the first span, and 17×10^{-8} to 33×10^{-8} for the second span.

The analytically expected strains for the loads actually applied range from 22.2×10^{-8} to 27.3×10^{-8}, corresponding to a modulus of elasticity, E_c, of 35,570 N/mm^2 (estimated from the formula: $E_c = 10^4\,(fcm)^{1/3}$ for a compressive strength of 45 N/mm^2). Despite the large scatter in the observed data, this range of values is taken as a confirmation of a globally sound structural behavior. More detailed tests are planned.

3.7 DYNAMIC TESTING

The response of the bridge was monitored under normal traffic circulation at peak hours. Seismometers were placed on the deck, and the displacements, velocity and acceleration of vibrations were measured. The measurements were repeated on the first and second span of the bridge.

The evaluation of these measurements is in progress.

4. Discussion

a) The most striking result of this investigation was the local corrosion of some tendons of side-beams where deck drainage waters were flowing in narrow areas; clearly insufficient concrete covers (15 mm) have accentuated the phenomenon. Once more, a couple of well established durability rules are confirmed: Drainage waters should be channeled out of the structure without any contact with concrete surfaces. Besides, sufficient concrete covers should be secured, especially on prestressed tendons. However, it is worth while to remind here that even when the tendons' cover was insufficient (e.g. less than 20 mm), nothing had happened; it was only by the synergy of the local effects of drainage deck water that damage was produced in areas of insufficient cover.

b) The satisfactory overall behaviour of the bridge, however, should be attributed to both the good quality of the concrete and the presence of transversal prestressing which (in conjunction with an adequate number of cross-beams) ensures a better distribution of action effects across the deck. Besides, transversal prestressing of the slab offers a better function of the longitudinal joint between prefabricated beams and in-situ poured slab-strips. Thus, transversal prestressing in this particular case has contributed to the implementation of favourable alternative scenaria for post-normal behaviour conditions.

c) Local damage to the asphalt pavement has to be separately examined and evaluated. Possible local flexibility of the supporting slab, combined with additional vibration effects caused by "compressed" bearings (especially in the first north span), may explain the related pavement damages. Eventual local damage induced to the slab during the peeling off of the asphalt pavement, may also explain some of these defects.

d) Defective bridge management has resulted in several abnormal interventions effectuated on the bridge, such as inappropriate fixing of lighting poles and of large diameter pipe-lines, as well as "wild" peeling-off of the asphalt pavement. Besides, inspection and maintenance were clearly inadequate.

Here again, well established bridge management rules would have prevented almost any one of the actual defects in the bridge.

e) Detailed computations, actually carried out, will lead to the final assessment of the bearing capacity of the bridge in terms of the actual European Standards for bridge loads. Accordingly, appropriate methods of repair intervention will be decided, including external unbonded tendons.

5. Acknowledgements

The authors wish to acknowledge the financing of this investigation by the Ministry of Public Works and the personal collaboration of the directors G. Bassas and D. Tsoutis. Thanks are due to Assistant Professor G. Batis for carrying out chemical measurements.

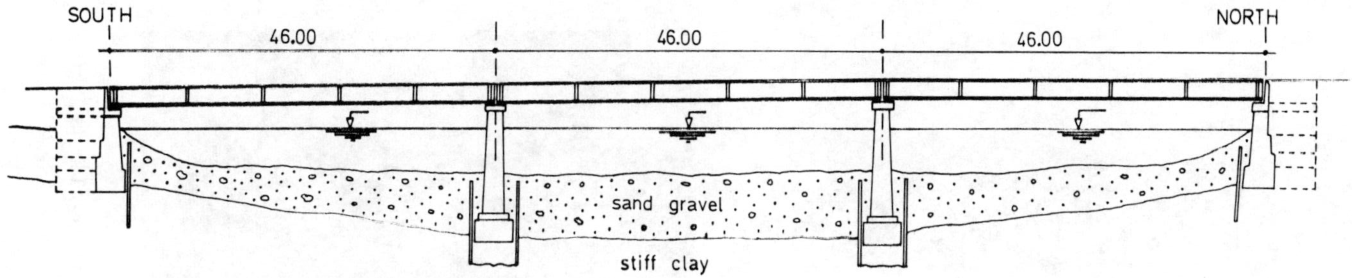

Fig. 1 Elevation of the Bridge.

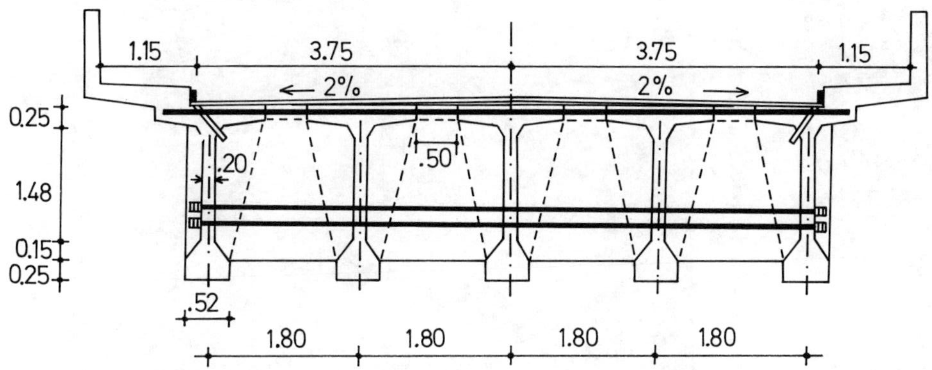

Fig. 2 Cross Section of Bridge Substructure.

Fig. 3 Honeycombs Near the Bottom of Side Beams (Bottom View).

Fig. 4. Spalling of Concrete of Upstream Side-Beams' Web Due to Pull-out of Fixed Dowels.

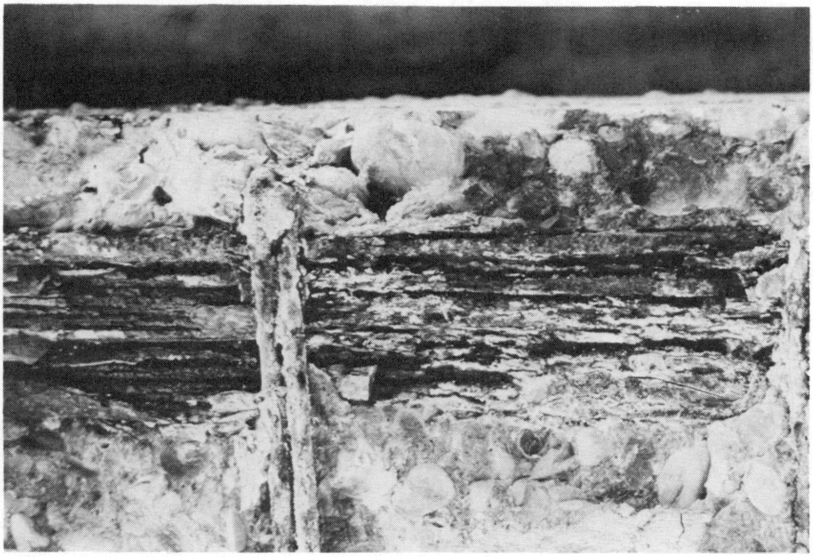

Fig. 5 Total Corrosion of Tendons in the Inner Face of Some Side-Beams Near Drain Areas (Bottom View).

Fig. 6 Slight Surface Rust on ducts in Heavily Honeycombed Areas.

INSTRUMENTATION OF A CABLE STAYED BRIDGE AT SKAGWAY, ALASKA

J. L. HULSEY, D.K. DELANEY and R.B. BUNCH
University of Alaska
Fairbanks, AK USA

R.W. BRIGGS
Alaska Dept. of Transportation
Juneau, AK USA

ABSTRACT. A 300 ft (91.4 m) cable stayed bridge near Skagway, Alaska carries in addition to normal traffic 160,000 lb (71.576 kg) ore trucks. It has a laminated timber deck supported by transverse floor beams spanning two stiffened ASTM A588 steel box girders. Two inclined cables stay the girders near midlength to a single tower. The abutments and tower were rock anchored to the canyon walls.

Information to address questions due to heavy loads and support degree of fixity was provided by instrumenting and field testing the bridge for static loads. Static loads were: a) a snooper truck; and b) ore trucks. Strains and environmental conditions were monitored.

Experimental results and a partial comparison with a two-dimensional analysis are presented. The maximum experimental strains were 224 micro-strain in the girder and 134 micro-strain in the tower support for the ore truck tests. Ore truck loads did not vary significantly and like loading produced excellent repeatability.

1. Introduction

The behavior of any structure subjected to load is dependent upon the type of support conditions. Consider a bridge with prestressed rock anchor supports and pose the question: Are the supports pinned, fixed, or other? This type of information is not available by mathematical means.

A 300 ft (91.44 m) long cable stayed fracture critical bridge near Skagway, Alaska carries approximately fourty 160,000 lb (72.576 kg) ore trucks per day. Uncertainties associated with the degree of fixity provided by the rock anchor supports and the effect of climatic conditions on the state of stress due to the ore truck traffic provided the basis for conducting an experimental study of this structure.

It is the objective of this paper to present the instrumentation techniques and experimental static load field test results used to evaluate the fixity of the rock anchor supports. This paper also provides a partial two-dimensional comparison of girder strains and deflections for a selected load condition.

A. S. Nowak (ed.), Bridge Evaluation, Repair and Rehabilization, 445–462.

1.1 BRIDGE GEOMETRY

The Captain William Cooper Moore Creek Bridge, is a two lane bridge located on the highway between Skagway, Alaska and Carcross, Canada. The length is divided into four spans of 30-123-127-20 ft (9.14-36.16-38.71-6.10 m). The deck consists of 7 in. (17.78 cm) thick glue laminated timber planks supported by floor beams spaced 12 ft (3.66 m) on center. These beams are connected to two longitudinal ASTM A588 stiffened box girders. Each box girder is supported by: a) end bents; b) the tower; c) two inclined cables stayed near midlength; and d) a brace near the upstation end (Fig. 1).

The girder support provided by the tower is an unusual configuration consisting of an elastomeric bearing pad and an adjacent curved steel plate. The girder stay provided by the cables are galvanized structural strands which extend through the girder and are anchored at the underneath side of the bottom flange.

The fore and back stays are supported by a tapered ASTM A588 steel frame H-shaped tower. The cross-section is box-shaped. The back stays terminate downstation at tripod supports affixed to the hillside by prestressed rock anchors.

2. Previous Experimental Studies

2.1. MODEL STUDIES

Kayser and Binkhorst (1975) used strain gages on a 1:50 scale model to verify erection stresses calculated by the finite element method for the main girder and diaphragm of a 9 span 1055 m bridge located on Highway 75 over the Waal river near Ewjik in the Netherlands.

The construction sequence for the fan type double plane multi-cable Meiko-Nishi Bridge in Japan was investigated through the use of model studies in the laboratory by Koizumi, et. al (1985). A 1:55 scale model of an A-shaped tower was tested in a wind tunnel and a 1:2.5 partial model of six main girder panels and anchorages were load tested.

Troitsky and Lazar (1971) investigated experimental laboratory techniques with small scale structural models as possible tools for the analysis and design of cable stayed bridges. Two models (A and B) were used in the study. The prototype had six lanes of traffic supported by two box girders. The structure was a double tower single plane cable stayed bridge with cables in a harp configuration. Strain gages and dial gages were used to obtain influence lines for model A and nonlinear behavior for model B.

2.2. FULL SCALE TESTS

Results in English of static tests for cable stayed bridges were not evident in the literature.

The Tjorn Bridge in Sweden was subjected to dynamic tests one week prior to opening the bridge to traffic (Ohleson (1986)). This structure is a cable stayed steel girder bridge with approximately a 366 m free span. Vertical acceleration measurements were made at 14 points located inside the main box girder at the outer stiffeners of the bottom flange for three types of tests. These were dynamic load; free decay; and forced vibration using a 2000 ton exciter.

3. Methodology

The study was conducted by field testing the bridge for two types of static loads (Hulsey et al. (1990)) and comparing the experimental results with a 2-D linear finite element analysis (Hulsey and Delaney (1990)). A comparison of both 2-D and 3-D finite element analysis with experimental results are in progress and will be presented at a later time.

3.1. INSTRUMENTATION

Instrumentation consisted of weldable strain gages, extensiometer, transit and level, velocity seismoprobe, thermistors, thermocouple, anemometer, and pryanometer.

Because of a limited budget, rosettes were not used. An extensiometer with an LVDT and dial gage was designed to examine the feasibility of monitoring cable strains. Girder deflections were obtained by recording elevations with a transit/level.

Thermistors were attached to the exterior face of the right girder to measure the temperature profile. Other measurements included instrument temperatures, deck velocity to insure measurements were taken under static conditions, air temperature, wind speed, and horizontal solar radiation. Twenty-six 5 conductor Belden cables were used to connect the instruments to the monitoring equipment located in a van used as a portable laboratory.

A summary of the instruments used in the study were:

- Twenty-five full bridge 350 ohm HITEC weldable strain gages;
- One 10 in. (25.4 cm) clamp-on custom extensiometer manufactured from aluminum with teflon bearings and LVDT/dial gage;
- One Wild T-16 Theodolite transit and a Zeiss N-i2 automatic level;
- A Bently-Nevada velocity seismoprobe with a visual needle read-out used to reach static conditions;
- Eight YSI 44030 3000 ohm thermistors mounted on magnetic fixtures for girder temperatures;
- One Type-T thermocouple for ambient air temperature:
- One Met-One 3 cup anemometer for wind speed; and
- One Star pyranometer.

Figure 2 illustrates the gage number scheme, location and installed cable number-to-gage number arrangement.

3.2 MONITORING EQUIPMENT

The data acquisition system was assembled with two Hewlett Packard 3421A data loggers controlled by an HP-71 handheld computer. Peripheral devices included a printer for hard copy and a 3.5" 640 K disk drive for the program and data storage (Fig. 3). A 12 volt DC battery was used for power to eliminate the need for a portable generator and to avoid sources of electric and acoustic noise.

A menu driven program was written to automate the collection, linearization, printing and storage of data. Data storage and hard copy were provided by a HP9114A disk drive and a HP2225B ink jet printer, respectively. Both units are powered by rechargeable batteries and can receive a trickle charge from the data logger.

3.3 LABORATORY PREPARATION AND CALIBRATION

3.3.1. *Cables/Connectors.* Belden 8723 22 gage, individually shielded, 2 pair cable suited for both static and dynamic measurements were used for extension cables. The extension cables were cut to length, numbered with waterproof tape and soldered to Lemo waterproof connectors. The loop resistance of each pair of wires was measured and recorded under controlled temperature conditions to account for desensitization effect.

After the connectors were installed on the cables and sensors, the waterproof connectors were tested in a 24 hour soak to confirm O-ring integrity after assembly.

3.3.2. *Extensiometer.* The extensiometer was manufactured from aluminum with teflon bearings in the University machine shop. It was found in the field that aluminum oxide formed on the teflon, causing friction. Thus, the instrument did not provide sufficiently accurate results and these data were abandoned (Hulsey et al. (1990)).

3.3.3. *Thermistors.* Eight YSI 44030 thermistors were mounted to two hardwood fixtures with magnetic clamps for mounting to the exterior side of the steel box girder to measure the girder temperature profile.

The assembly was prepared by soldering the 8723 cable to a miniature printed circuit board and mechanically fastening the thermistor's leads to the board. The assembly was coated with two coats of Dow 3140 RTV, a non-corrosive silicone rubber, encapsulated in heat shrink tubing and covered with two coats of Scotchguard(TM).

The eight thermistors were checked in an ice bath for accuracy. All were within the manufacturer's specifications.

3.4. FIELD INSTALLATION

All weldable gages were welded in place after rust and scale were removed by electric grinders. Girder gages were mounted inside the box on the ceiling and floor. Gages at the base of the pylon and left abutment column were mounted on the exterior face in the direction of the principal axis.

The extensiometer was installed with integral thumbscrew clamps to facilitate removal at the close of each day. The two thermistor assemblies were placed on the right girder by the pylon.

A three cup anemometer was mounted on the left side south end of the bridge. The air temperature thermocouple and pyranometer were located on the right side approximately 150 ft (45.72 m) south of the end of the bridge.

Extension cables were then connected to the gages and these cables were connected to the data logger near the downstation end.

3.5. FIELD TEST PROCEDURE

Each test was designed to provide static results for a known truck load at a given position on the bridge. This was accomplished by referencing the: a) type of truck; b) position across the deck; c) location of the front axle along the girder; and d) direction of movement (Fig. 4).

The structure was loaded either in a lane or on the bridge center line. The terms left and right were used to reference the truck position across the roadway when looking upstation.

Each test series was given a file name and the file name described the type of test, lane position, direction of the truck and type of truck used for the test (Hulsey et al. (1990)).

A typical static test series involved instrument initialization, truck position placement and use of the velocity probe to insure vibration had dampened; then time and voltages for each strain gage, LVDT, thermistor, anemometer and pryanometer were recorded.

4. Experimental Results

4.1. SNOOPER TRUCK

The snooper truck was a standard flat-bed truck with a man basket connected to a maneuverable folding hydraulic boom mounted on the bed (Fig. 5a).

Data were recorded for a total of 106 load positions on the first day of testing. There were 53 load positions with the truck in the left lane, 28 with the truck in the right lane and 25 with the truck on the centerline. Girder elevations were recorded for 9 of the 106 load positions. Table 1 summarizes these static load tests.

4.1.1. *Strains*. The results for the girders, pylons, and upstation column were examined for each load case and the maximum strains for each were summarized. A summary of the maximum strain results are provided in Tables 2 and 3.

4.1.2. *Girder Deflections*. Maximum girder deflections were produced by the truck front axle at position 9. Maximums occurred at position 8 for eccentric loads (truck in the left lane) and at position 12 for symmetric loads (truck on bridge center line). Eccentric load deflections were 0.84 in. (2.13 cm) and symmetric load deflections were 0.48 in. (1.22 cm).

4.1.3. *Other Findings.* The experimental results clearly showed that when the bridge was subjected to a symmetrical load it performed symmetrically with respect to its longitudinal center line.

4.2 B-TRAINS

On the day following the snooper truck tests, B-trains were tested. The B-Trains were semi tractors towing two lowboy trailers containing a single dual axle and two sets of tandems. The B-Trains carried ore in four pots mounted on the lowboy trailers (Fig. 5b). The loaded gross weight of these trucks varied between 155 k (70.31 kg) to 158 k (71.67 kg). Axle weights for the B-Trains used in the tests were obtained from weight tickets.

Twelve static tests were conducted with the B-trains: six in the left lane; three on the bridge center line; and three in the right lane. Table 4 summarizes the static load tests that were carried out for the B-Trains.

4.2.1. *Strains.* Like the snooper truck tests, strains were recorded for both girders, both pylon bases, and the axial strain in upstation left column at the bridge abutment. The summary of maximum strain results are presented in Tables 5 and 6. Table 7 provides maximum experimental girder strains for the B-Train tests.

4.2.2. *Girder Deflections.* The truck front axle at position 9 produced maximum girder deflections at positon 12 on the girder. The maximum deflection was found to be 2.76 in. (7.01 cm) for eccentric loads (left lane) and 2.16 in. (5.49 cm) for symmetric loads (on the bridge center line).

4.2.3. *Other Findings.* For like loading conditions, excellent repeatability in recorded values was observed, thus endorsing the instrumentation technique and testing procedure. B-Train axle weights did not vary significantly from truck to truck.

5. Finite Element Comparison

Parametric studies using a two-dimensional finite element model were initiated to investigate the degree of fixity at the supports for the load cases with the trucks on the bridge center line. Preliminary results for symmetric/asymmetric loads indicate that the 2-D model is satisfactory for static loads on the bridge center line, and a 3-D model should be used for other cases. A partial comparison between calculated and experimental girder strains and deflections for one B-Train truck symmetric load condition are presented in Figures 6 and 7 for consideration. A complete 2-D model comparison is available (Hulsey and Delaney (1990)).

6. Summary and Conclusions

The experimental results indicated that: a) over a two day period, the structure experienced some changes in strain due to the environmental exposure; b) wheel loads on the loaded ore trucks did not vary significantly; and c) like loading conditions produced excellent repeatability. Preliminary finite element results show that a 2-D linear model gives good results for symmetrically applied loads but a 3-D model is needed to account for asymmetrical loads. The experimental maximum strains due to ore trucks were 224 micro-strain in the girders and 134 micro-strain in the tower support. The experimental maximum top of girder strain with an ore truck on the bridge center line was 169 micro-strain in the left girder and a corresponding 164 in the right. The maximum bottom of girder strains were 171 micro-strain in the right and a corresponding 158 in the left.

Based on the study, it was determined that the prestressed rock anchor supports acted as fixed supports for this structure; all others were pinned.

7. References

Hulsey, J.L., Delaney, D.K., Briggs, R.W. (1990), "Cable Stayed Bridge Static Tests, An Experimental Study for the Captain William Cooper Moore Bridge," FHWA report AK-RD-90-08 (part 1), Alaska Department of Transportation, in preparation.

Hulsey, J.L., Delaney, David K. (1990), "Cable Stayed Static Linear Analysis for the Captain William Cooper Moore Bridge," FHWA report AK-RD-90-08 (Part 2), Alaska Department of Transportation, in preparation.

Kayser, T.H., Binkorst, J. (1975), "Computer Calculations of a Complex Steel Bridge Verified by Model Investigations (Applied at the Steel Bridge Spanning the River Waal Near Ewijk)," Ministry of Transport, Netherlands, Rijkswaterstaat The Hague, Netherlands, No. 23.

Koizumi, M., et. al (1985), "Construction of Meiko-Nishi Bridge," Nippon Kokan Technical Report, Report No. 45, pp. 22-12.

Ohleson, Sven (1981), "Dynamic Properties of the Tjoern Bridge Experimental Program," Chalmers University of Technology, Sweden Institute of Transportation, Onsternik, Staaloch Traebvggnad, Report No. 81:3, 36 pages.

Trotski, M.S. (1988), Cable Stayed Bridges, An Approach To Modern Bridge Design, Van Nostrand Reinhold Company, New York.

Trotsky, M.S., Lazer, B.E. (1971), "Model Analysis and Design of Cable Stayed Bridges," Inst. of Civil Engrs., Proc., London, England, Proc. Vol 48, Paper 17375, pp. 439-464.

TABLE 1. Snooper Truck Test Series
(Left & Right refer to looking upstation)

Test	Front Axle Pos.	Truck Locations	Truck Movement	Measured Deflection Locations
SLDOS01	0-24	Center of lt lane	downstation	none
SLDOS02	0-24	Center of lt lane	downstation	none
SLRUS03	24-0	Center of rt lane	upstation	none
SCDOS04	0-24	Center of Bridge	downstation	14,9,4
SCDOS05	4,9,14	Center of rt lane	upstation	4,9,14
	14,9,4	Center of lt lane	downstation	14,9,4

TABLE 2. Maximum Strains Produced by the Snooper Truck

Load Condition	Type of Member	Tension (strain, micro-in/in)	Compression (strain, micro-in/in)
Symmetrical:	Along the girders	53	52
	Pylon base	15	32
	Upstation column (left side)	5	6
Asymmetrical:	Along the girders	93	97
	Pylon base	42	24
	Upstation column (left side)	4	10

TABLE 3. Maximum Strains for Snooper Truck versus Lane Location

Truck Location	Item	Strain (micro-in/in)
Left lane		
	Girder (98')	88
	Pylon base (left pylon)	42
	Left column upstation end	10
Center line		
	Girder (98')	53
	Pylon base (left pylon)	32
	Left column upstation end	6
Right lane		
	Girder (98')	97
	Pylon base (right pylon)	27
	Left column upstation end	7

TABLE 4. B-Train Tests
(Left & Right refer to looking upstation)

Test	Front Axle Pos.	Truck Locations	Truck Movement	Measured Deflection Locations
SLDOB06	4,9,14	Left lane wheel path	downstation	none
SLDOB08	4,9,14	Left lane wheel path	downstation	14,9,4
SCDOB10	4.9,14	Center of Bridge	downstation	14,9,4
SRDOB11	4,9,14	Center of Rt lane	downstation	none

TABLE 5. Maximum Strains Produced by B-Trains

Load Condition	Type of Member	strain(micro-in/in) Tension	Compression
Symmetrical:	Along the girders	163	171
	Pylon base	40	99
	Upstation column (left side)	22	
Asymmetrical:	Along the girders	205	224
	Pylon base	54	134
	Upstation column (left side)	26	

TABLE 6. Maximum Strains for B-Train versus Lane Location

Truck Location	Truck wt. (kips)	Item	Strain (micro-in/in)
Left lane	157.379		
		Girder (98')	224
		Pylon base (left)	134
		Left column	26
Center line	157.556		
		Girder (278')	171
		Pylon base (left)	99
		Left column	22
Right lane	157.093		
		Girder (98')	207
		Pylon base (right)	103
		Left column	17

TABLE 7. Strain Extremes for Symmetric and Asymmetric B-Train Truck Loads
(All Strains in Micro-Strain)

	Top of Girder				Bottom of Girder			
	Left Girder		Right Girder		Left Girder		Right Girder	
Gage Loc.(ft)	Max. Tens.	Max. Comp.	Max. Tens.	Max. Comp.	Max. Tens.	Max. Comp.	Max. Tens.	Max. Comp.
Symmetric								
33.25	154	-	152	-	-	157	-	158
98.00	-	169	-	164	146	67	163	74
146.00	-	69	-	72	34	31	33	31
206.00	14	144	15	165	116	17	111	15
278.00	117	-	127	-	-	158	-	171
Asymmetric								
33.25	204	4	191	4	-	208	-	198
98.00	-	224	-	207	196	82	205	85
146.00	-	94	-	93	51	38	47	35
206.00	17	189	20	204	151	22	138	18
278.00	157	-	160	-	-	211	-	215

Note: 1) Blank entries imply girder did not go into compression/tension.
2) Gage locations are measured from downstation end of bridge.

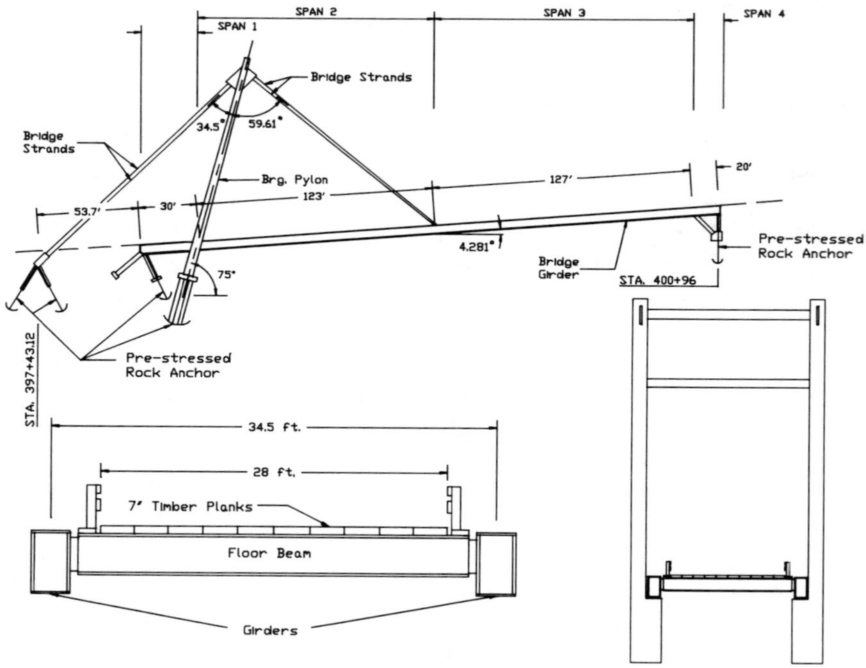

Fig. 1 Captain William Cooper Bridge Geometry.

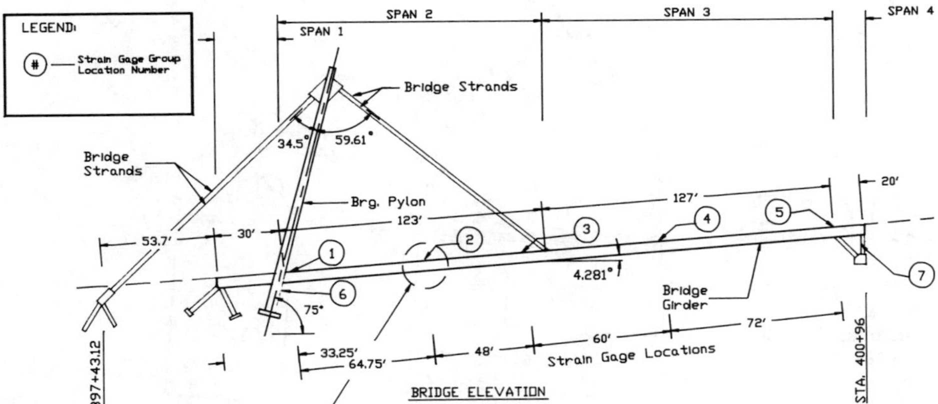

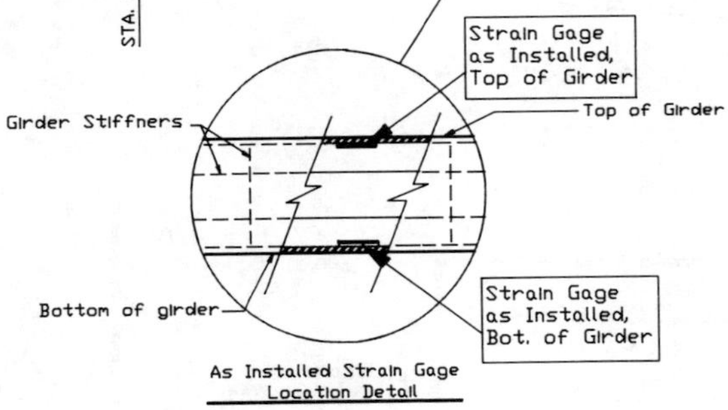

Strain Gage Group Location No.	(Gage No.) – (Cable No.)			
	Top of Girder		Bot. of Girder	
	Lt. Side	Rt. Side	Lt. Side	Rt. Side
1	(18)-(18)	(16)-(16)	(20)-(20)	(19)-(19)
2	(2)-(6)	(1)-(1)	(4)-(8)	(3)-(3)
3	(14)-(14)	(13)-(13)	(17)-(17)	(15)-(15)
4	(12)-(12)	(9)-(9)	(10)-(10)	(11)-(11)
5	(6)-(2)	(7)-(7)	(8)-(4)	(5)-(5)
	Down Station Side		Up Station Side	
6	(22)-(22)	(21)-(21)	(24)-(24)	(23)-(23)
7	(25)-(25)			

Fig. 2 Strain Gage Locations and Numbering Scheme.

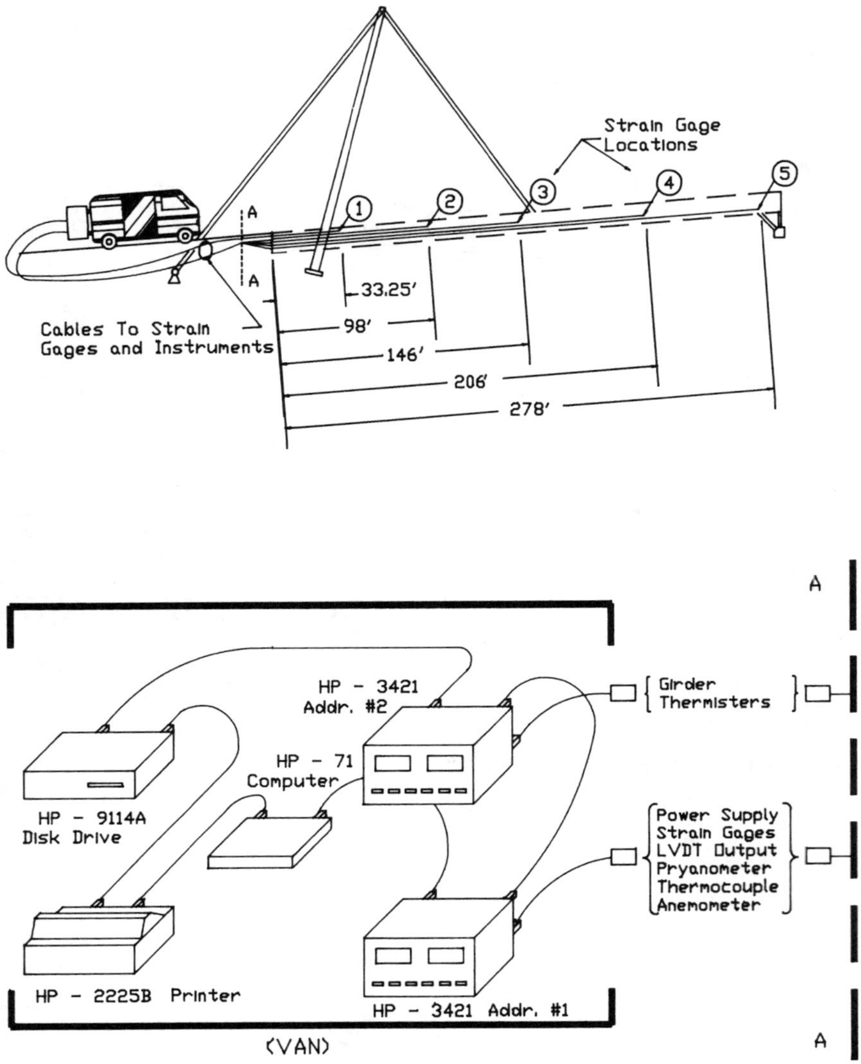

Fig. 3 Instrument Monitoring System.

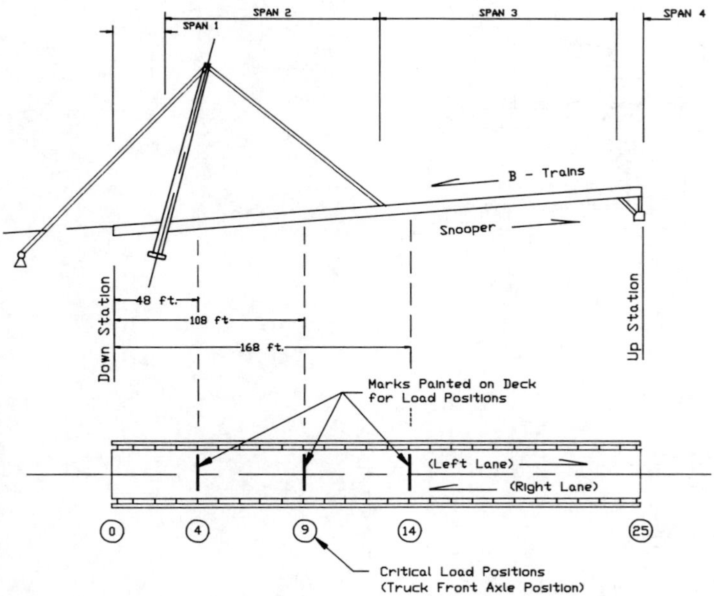

Fig. 4 Bridge Frame and Critical Load Positions.

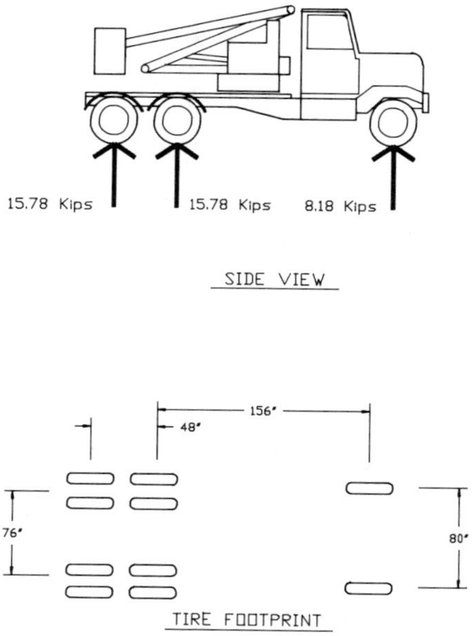

a) Snooper Truck

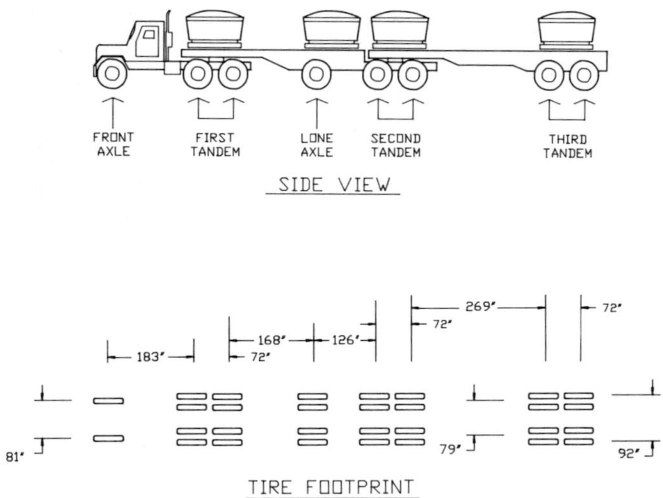

b) B-Train

Fig. 5 Truck Loads for the Static Tests.

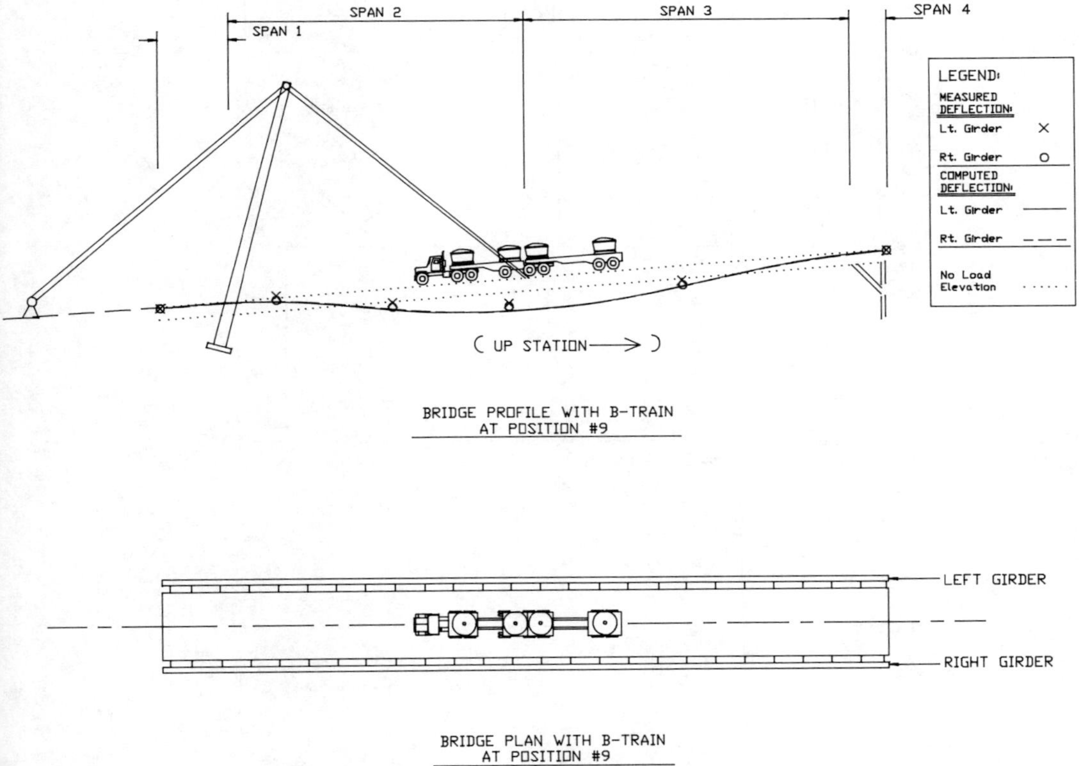

Fig. 6 Girder Deflections for B-Train at Position 8, Symmetrical Load.

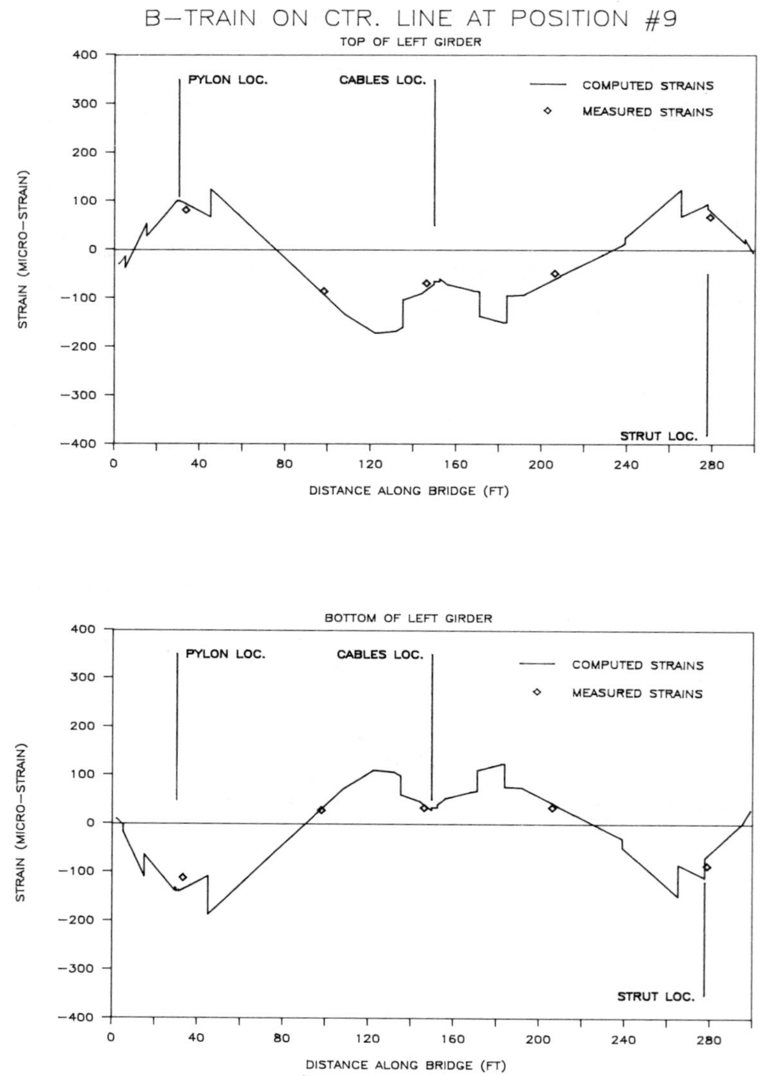

Fig. 7 Girder Strains for B-Train at Position 9, Symmetrical Load.

FULL SCALE TESTING OF STEEL RAILWAY BRIDGES

T. T. PIETRASZEK
Canadian National Railways
Technical Research Centre
3950 Hickmore Avenue
St. Laurent, P.Q. H4T 1K2, CANADA

ABSTRACT. Fundamental purposes of full scale bridge tests conducted by the Canadian National Railways (CNR) are bridge safety and serviceability. Results collected during these tests were of a broader sense and allowed the study of more general aspects of bridge behaviour under moving loads such as static and dynamic stresses and displacements, impact effects, vibrations and fatigue. This paper concentrates on major technical aspects of load, strain and displacement measurements, recording apparatus, signal processing techniques and evaluation of results. Samples of obtained results concerning three different bridge structures tested recently are presented and major conclusions discussed.

1. Introduction

Each full scale bridge test is a complex and expensive undertaking involving cooperation of various departments in a railway organization. This fact limits this type of bridge evaluation to a few cases per year which are deemed to be of extreme importance to the railway operations. Results collected from a test are to answer such questions as whether a bridge is to be strengthened, replaced or its behaviour is satisfactory under present or limited traffic conditions (i.e. imposed speed limits).

To answer these questions several parameters including strains in critical members, displacements in selected locations and imposed live loads by various railway equipment (locomotives, cars) are to be measured and evaluated. The paper presents major technical aspects of load, strain and displacement measurements which are employed by the CNR in bridge testing. Also samples of results collected during three recent bridge tests concerning different types of bridge structures are presented and final results discussed.

A. S. Nowak (ed.), Bridge Evaluation, Repair and Rehabilization, 463–478.

2. Load, Strain and Displacement Measurements

2.1. LOAD MEASUREMENTS

A full scale test means that a known load (closely resembling real live load) is to be moved on a bridge and sets of measurements are to be taken for various positions of a moving load. In railway bridges which are loaded by long trains a special test train is set up which has axle loads of various magnitudes predetermined before the test. For practical reasons the test train consists of one (or more) locomotives and a few loaded and empty cars in various combinations. The load effects measured may be therefore easily related to predetermined train loads (each train vehicle is weighed before the test on a special scale). In this case however real (dynamically applied) loads and load positions corresponding to a particular effect (i.e. peak stress or displacement) are not known. To overcome this shortcoming, rail shear circuits must be installed.

The purpose of a rail shear circuit is to measure vertical and horizontal wheel loads as the train moves across the bridge. Shear circuits are installed in a rail web (for vertical load measurement) or in rail base (for horizontal load measurement). They comprise multiple strain gage rosettes combined into full bridge circuits (Figure 1). After installation shear circuits are calibrated by applying a known static load to the rail head (special beams and jacks are used for this purpose). This method of measurement is quite precise in obtaining individual wheel loads applied to the bridge by a moving train but permits the measurement of those loads at the location of the shear circuits only (several axles may be present on a bridge at one time).

2.2. STRAIN MEASUREMENTS

The best available today and most reliable method of strain measurement is the one utilizing electrical resistance strain gages. This technique has been known for a few decades but still several difficulties are to be overcome each time the strain gages are applied in a real structure. Surface preparation, wiring, laying power cables, etc. in not necessarily good weather conditions are just few constraints to be mentioned. Selection of gage location and circuit configuration (a half bridge or a full bridge) has to be decided in each case to eliminate secondary stress effects (uniaxial, biaxial and 3-D state of strains and stresses may be present). Special configurations are used to measure axial only stresses in girder flanges or bracings (Figure 2). Different configurations must be used to determine stress distributions across a built-up section or to determine possible stress concentrations in a given location.

Since the application of strain gages is a time consuming and precision demanding process alternate methods were developed. Vibrating wire gages are used to measure axial strains under predominant static loads. Reusable transducers not bonded permanently to the members surface are also being developed. They are, however, less reliable than bonded strain gages.

2.3. DISPLACEMENT MEASUREMENTS

Since one of the serviceability criteria in bridge design is a limit of deflections (or displacements - in general) these are also measured and recorded during a full scale test. Measured deflections may also serve as an additional check for strain measurements and will indicate an overall behaviour of the structure with actual boundary restraints.

Linear variable differential transformers (LVDT) serve most often as displacement transducers. They are located usually at mid spans and at bearing points of girders to detect vertical deflections (displacements) at respective points. Lateral displacements at some locations are also measured to determine dynamic behaviour of the bridge span. In places where a fixed reference point for a displacement transducer is not available an alternate method of measuring accelerations has to be used. This method is not as precise as direct displacement measurement due to additional signal integration involved specially in low frequency ranges (quasi static loads and very slow motions).

2.4. SIGNAL RECORDING AND PROCESSING

Successful installation of strain gage circuits or displacement transducers on a bridge is only a part of a bridge test. Signals recording and processing becomes an even more demanding part of a test when more than a few strain signals are to be recorded and analyzed. A passage of a test train or a regular train produces a set of strain analog outputs which can be recorded directly in an analog form on a multichannel chart recorder. Analog signals can be stored for later reproduction in an analog or digital format. Analog storing requires use of a multichannel tape recorder while a digital storing requires use of specialized computer equipment with analog to digital convertors and dedicated software for this specific purpose. During recent test referred to in this paper both methods were used in a configuration as shown in Figure 3 below.

Two important limitations in the use of digital recordings are the sampling rate and the number of channels recorded which are usually predetermined by the available hardware. The digital system used in the described tests allowed recording of up to 48 channels with a standard sampling rate of 200 Hz; six channels were able to record signals at 1000 Hz sampling rate.

Signal processing for relatively short test and few channels only is possible by a manual method directly from chart records. With longer tests involving several train runs and several strain circuits an automated processing is a must. In this as?? either an analog signal (recorded on tape) is processed by a dedicated instrument giving out peak and cycle counts or digital records are processed by a specialized software to produce similar outputs. In a latter case, scaling, combining signals and preparing reports may be automated as well. An in-house developed software for HP-21MX computer was successfully used in bridge tests described here.

3. Examples of Bridges Tested

3.1. BRIDGE # 1: DECK PLATE GIRDER BRIDGE

This bridge, consisting of 4 identical simply supported spans (l=19.1m) per track was built in 1902/1903 according to Grand Trunk specifications of 1900. General plan and typical cross sections are shown in Figure 4.

Each span comprises two plate girders (depth h=1.69 m) at 2.13 m spacing. A system of crossbracing (2 end frames and 3 intermediate frames) and top and bottom laterals are made of single angles. Each girder rests on flat cast iron bearing block Two separate tests were carried out during 1987 and 1988 on one of the four spans of that bridge to determine its behaviour under high speed freight trains. Also a test train was used to obtain strain and displacement data under controlled loads an d train speeds. Twenty strain gage circuits (4 - in girder flanges at mid span, 4 - in top and bottom laterals, 8 - in cross braces and 4 - in rail shear circuits) were installed. Vertical and lateral displacements of one of the girders were measured by 6 LVDT transducers. All strains were measured with 5 microstrains resolution. Typical strain records for girder flanges and rail shear circuits are shown in Figure 5. Changes in vertical deflections, girder strains and axial forces in end laterals versus train speeds are shown in Figure 6 to Figure 8.

Stresses in girders increased about 4% and vertical deflections increased about 10% with speeds varying from 0 to 96 km/h. Lateral deflections and axial forces in bracings or laterals increased few folds with the same speed increases. Lateral deflections measured under high speed freight trains were somewhat smaller than those observed under test train conditions. That observation led to setting a speed limit for those freight trains at 105 km/h. Apart from immediately available results an extensive amount of data regarding strains in instrumented bridge members under various types of traffic was collected (a total number of 60 trains of various lengths and weights were recorded). All records will be used in future evaluations of remaining fatigue lives of selected bridge members.

3.2 BRIDGE #2: DECK PLATE GIRDER SPANS ON STEEL TRESTLES

This bridge built in 1914 consists of twelve simply supported spans (l=12.2m and l=18.3m) resting on steel trestles of variable height (max. h= 22.6m to top of rail). An extensive test was carried out on this bridge in 1988 to examine static and dynamic behaviours of one of the spans and the supporting trestle (Figure 9). Ten strain gage circuits were installed to monitor strains in two girder flanges, two top laterals, two braces at the bottom of the trestle and the trestle column itself. All strain gage circuits except in the column were configured as full bridges to eliminate flexural effects in those members. Column gages were installed as half bridges measuring strains at four corners of the column section to detect column bending (Figure 10). Three additional half bridge circuits were installed on both faces of the web of the girders pedestal. This configuration enabled determination of principal stresses at this location (45-degree rosette was formed by the three gage circuits). Several pedestals were found cracked previously near their bottom edges and were replaced by the identical detail substituting bolts for rivets. Lateral displacements (only these were of interest) were measured by acceleration transducers mounted at girders support line as shown in Figure 9. This method allowed for detecting major peaks of lateral motion of the trestle and girder. A test train was used to generate known loads (monitored also by two rail shear circuits) at controlled speeds. Several re cords of regular trains were also collected which were compared with the test train effects.

Speed effects on girders stresses and forces in trestle members (brace and column) are shown in Figure 11 and 12 respectively. A relatively small increase in girders stresses was observed (effective impact factor was less then 10%); increases in force levels in trestle members were more noticeable (force in brace L2 increased more than 50%). Maximum force (compression) in the column increased only slightly while minimum force (tensile) changed significantly and not linearly.

Strains recorded in the pedestals web indicated a presence of large tensile stress component approximately perpendicular to the direction of observed cracks. The peak value of this stress occurred when the load was on the span. This observation was confirmed by a detailed analysis of the pedestals web by a finite element method. Figure 13 shows a part of the analyzed model with principal stress contours obtained under one of the assumed boundary conditions. Stress concentrations and girder restraints at bearing points were found to be responsible for cracking of these pedestals.

It is worth noting that most of the work required for data processing after this test was done by the in-house developed software which was mentioned before.

3.3. BRIDGE # 3: THROUGH TRUSS BRIDGE.

This bridge built in 1903 in accordance to the Grand Trunk specifications of 1900 consists of a double truss girder (l=58.6m) and a system of floor beams (l=9.45m) and stringers (l=6.51m) connecting to the bottom chords. The purpose of the test which was conducted in 1989 was to determine stress distributions and stress ranges in a typical floor beam and four stringers. Also strains in two hangers connecting to a floor beam were measured. Fourteen full bridge circuits were installed to record strains in 4 stringers (B1 to B4), one floor beam (FB) and 2 hangers (H) - see Figure 14.

Only regular traffic was recorded (freights and passenger trains) - a test train was not used this time. All data were recorded on a 14 channel tape recorder and strain analysis (peaks and range cycles counts) was performed using a real time stress wave analyzer. Also a hard copy of all records were obtained from a chart recorder.

Stresses at bottom flanges of four stringers tested were not identical (maximum peaks were equal to 52%, 115%, 136% and 97% of the average stress for stringers B1, B2, B3 and B4 respectively). Similar differences in stringer stresses were discovered for maximum stress ranges, and fatigue life of each stringer varied considerably from the estimated one based on equal stress ranges for each beam. Fatigue life ratios with respect to the average life were equal to 3.6, 0.13, 0.05 and 0.27 for stringers B1 to B4 respectively. A frequency distribution curve of recorded stress ranges for approximately one day traffic is shown in Figure 15.

Another interesting observation made concerned hanger stresses (stresses were measured independently for each flange of the section). Strain values for inside flange of the hanger (i.e. flange facing the track) were very high compared to the opposite flange (inside flange indicated peak stresses reaching 55 MPa while outside flange showed stress of 2 to 5 MPa at the same time). This phenomenon had to be attributed to a presence of very high bending moment induced by a rigid connection of floor beam to hanger and a large ratio of moment of inertia of the floor beam to that of the hanger. This type of stress distribution in hanger section is not often considered in the design process.

4. Closing Remarks

A limited space allocated for this paper did not permit presentation of full details of the tests described in the previous chapter or to include more very interesting examples of other bridges tested across the CN system.

Bridges tested in recent years encompassed a variety of structures ranging from deck plate girders, through trusses to deck trusses and combinations of those basic types. Observations made during and after tests indicated that measured stresses in main structural members were usually lower than obtained from theoretical analyses. Such analyses however did not include true boundary conditions existing in the structure (rigid joints, immobilized bearings etc.) and interaction between main an d secondary members. Also the test loads were lower than the design loads specified by design codes (usual test train represented about 60% of the maximum axle load used in the design). All these reasons may have contributed to the observed (lower than expected) stress levels. Impact factors were also lower than those stipulated by design codes (measured impacts were only 10% to 15% of the static load).

Secondary elements and some deficient details of tested bridges seemed to be mostly affected by loads generated by todays traffic. Lateral forces arising from higher speed limits are now a limiting factor in meeting bridge serviceability criteria (limits on deflections of spans and support movements are frequently exceeded).

Evaluation of fatigue life of main and secondary members is playing a major role in bridge maintenance and replacement programs. Stress ranges caused by real loads and affected by progression of corrosion can only be determined by direct strain measurements covering adequate time span in a life of a bridge. Reliable traffic load spectra have to be determined for major traffic types and bridge members. Similar work has already been completed and results implemented in the development of UIC 71 bridge design code in Western Europe. Inadequate design details from the fatigue life point of view have still to be checked and eliminated from existing bridges if their lives are to be extended further.

Each full scale test required careful planning and preparation period during which all necessary information about the structure to be tested was gathered and assessed. Gage locations, instrumentation and other details such as accessibility of t he bridge site, power supply sources and other facilities had to be confirmed by earlier site inspection. Safety measures such as scaffolding, warning systems against incoming trains, track closures and speed limits were to be arranged for in advance. In spite of those constraints full scale bridge tests will still be indispensable for correct assessment of structural integrity and serviceability of these structures.

5. Acknowledgements

The author wishes to acknowledge numerous contributions from research and technical staff of CN Technical Research Centre in conducting all tests and assisting in various stages of preparation of this paper.

All tests described here were part of bridge evaluation and assessment programs organized by Engineering Department of CN Operations in cooperation with district engineers.

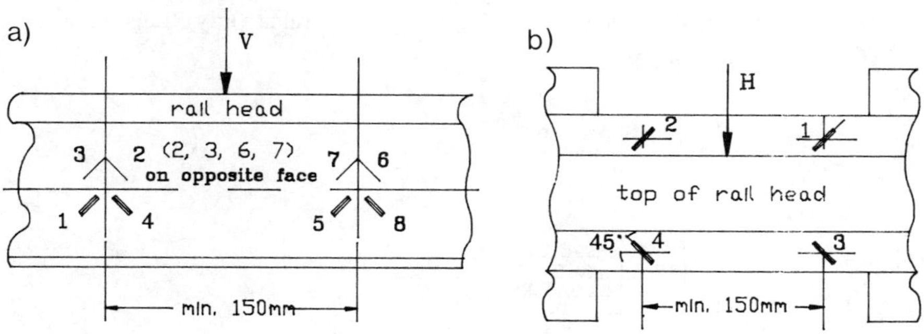

Fig. 1 Rail Shear Circuits: a) For Vertical Loads, b) For Horizontal Loads.

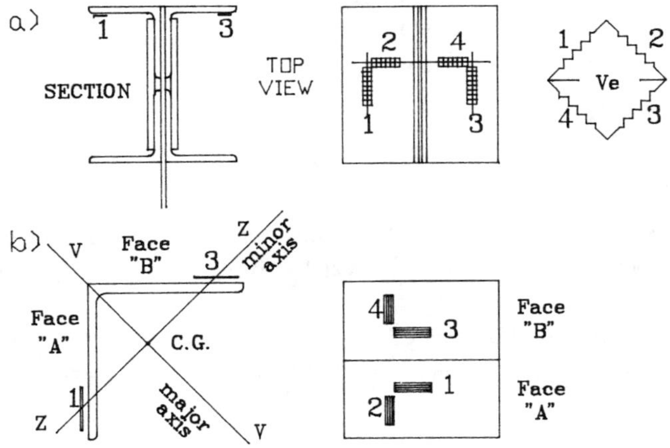

Fig. 2 Examples of Strain Gage Configurations: a) For Girder Flange (Axial Strains Only), b) For Bracing (Single Symmetrical Angle - Axial Strains Only). Legend: 1&3-Active Gages, 2&4-Dummy Gages.

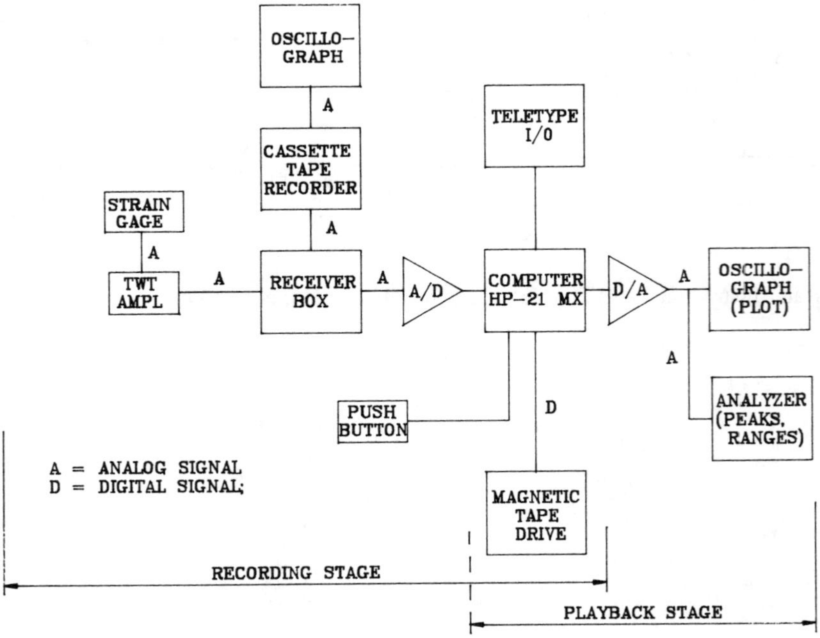

Fig. 3 Schematic Outline of Recording and Playback System for Strain Signals.

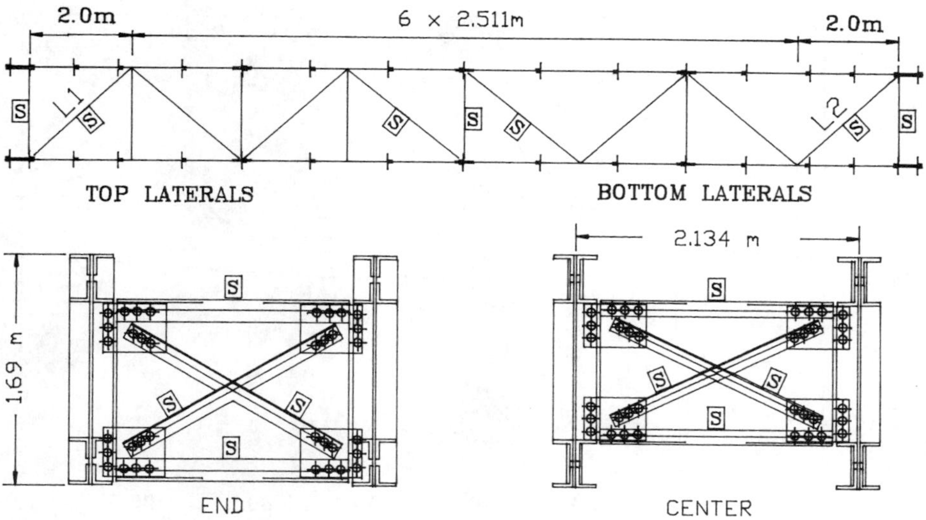

Fig. 4 Plan and Cross Sections of a Typical Span (Strain Gage Locations Shown by letters S).

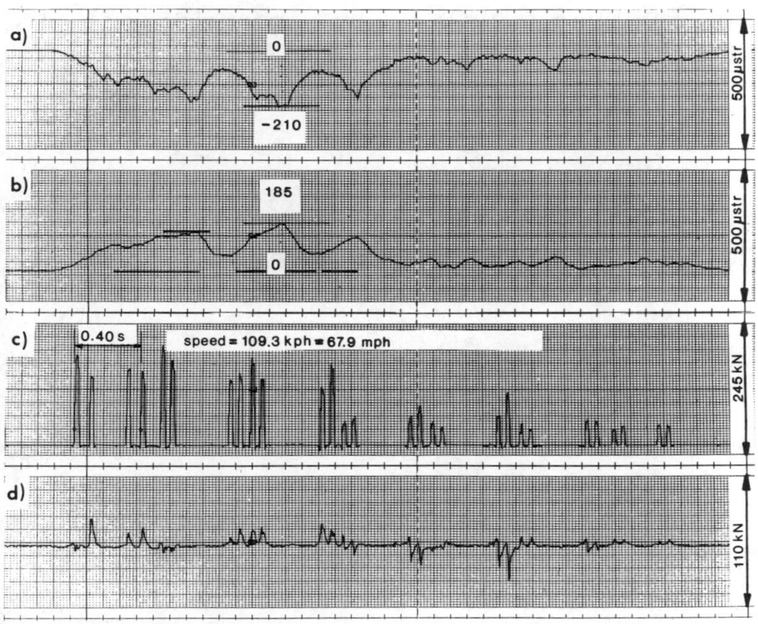

Fig. 5 Strains Recorded for a Test Train Run at 110 km/h: a & b Top and Bottom Girder Flanges, c & d Vertical and Lateral Rail Shear Circuits.

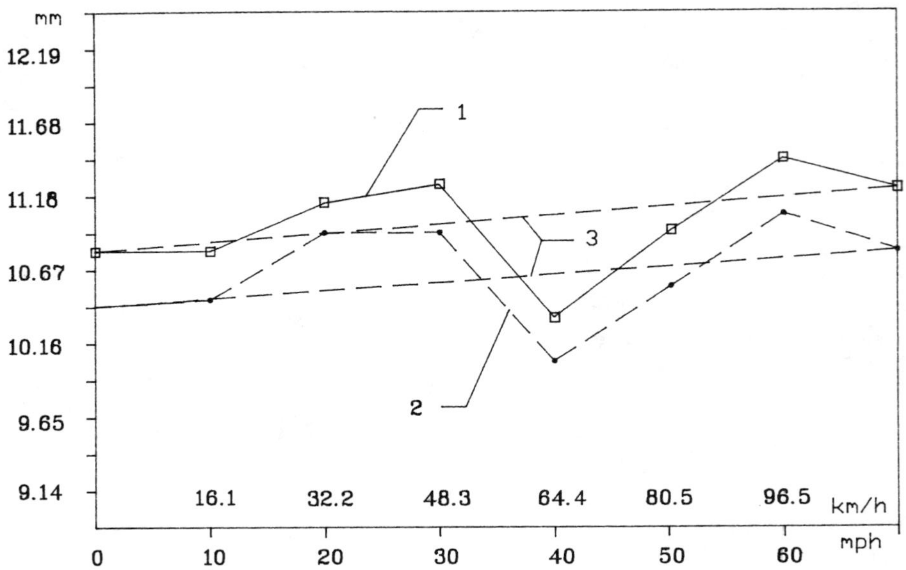

Fig. 6 Girder Deflections at Mid Span Versus Test Train Speed (Legend: 1 - Gross Deflection Including Bearings Deflections; 2 - Net Deflections; 3 - Averaged Between 0 and 70 mph).

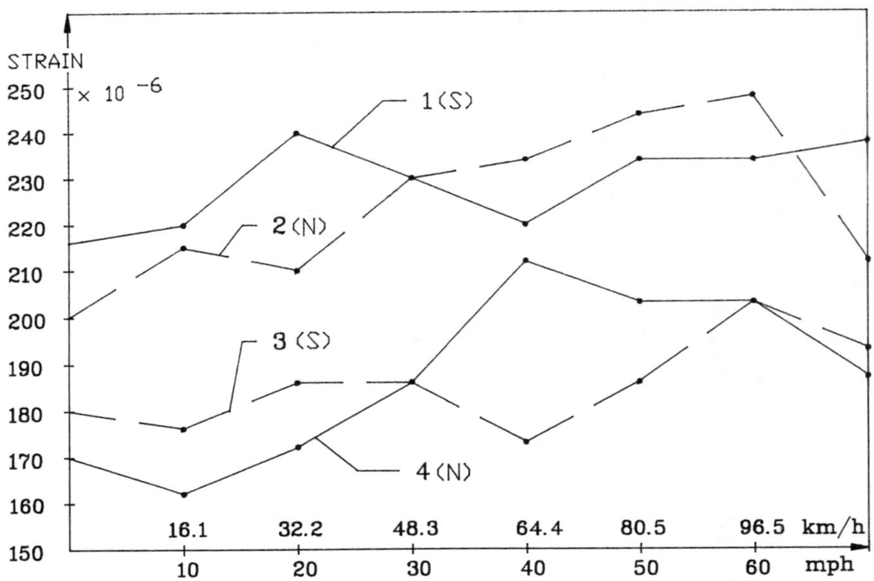

Fig. 7 Peak Girder Strains at Mid Span Versus Test Train Speed for Two Girders. Legend: 1 & 2 Top Flanges, 3 & 4 - Bottom Flanges; S - South Girder, N - North Girder).

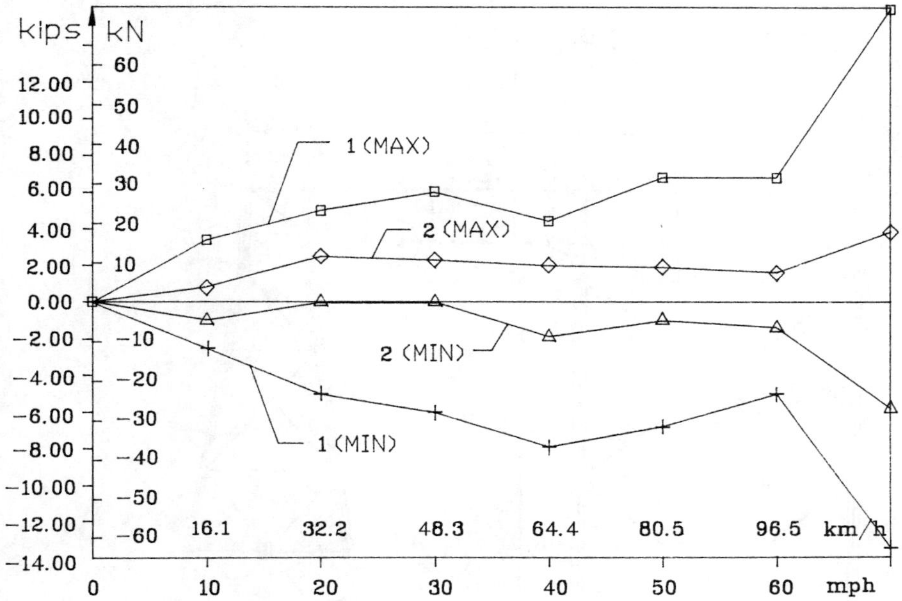

Fig. 8 Forces in End Laterals Versus Test Train Speed (Legend: 1 - Member L1, 2 - Member L2; see Fig. 4 for Locations). 3.2. Bridge # 2: Deck Plate Girder Spans on Steel Trestles.

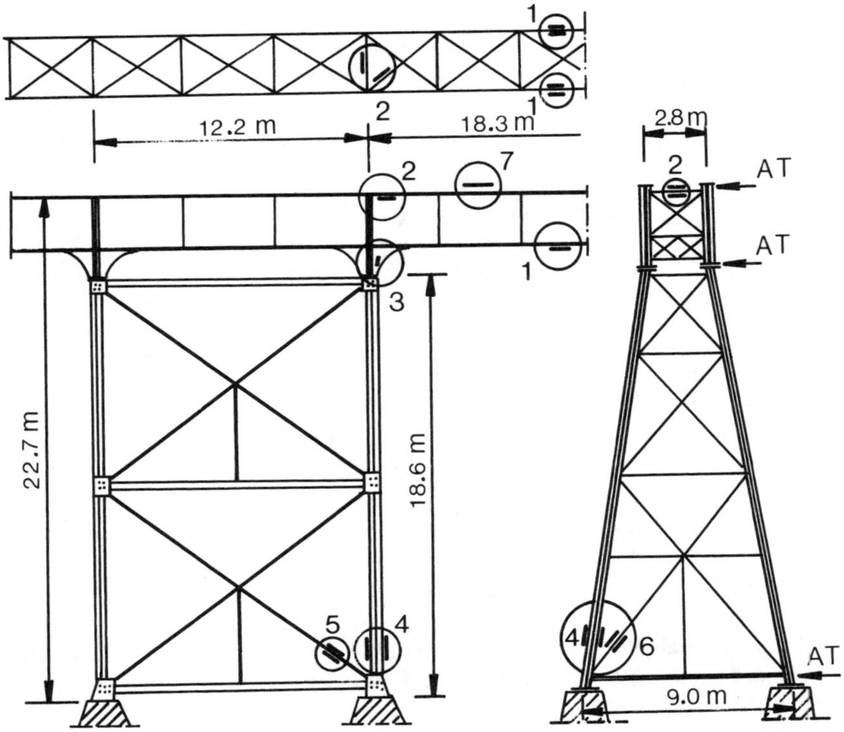

Fig. 9 Location of Test pPints for Bridge # 2: Strain Gage Circuits (1 to 7) and Acceleration Transducers (AT).

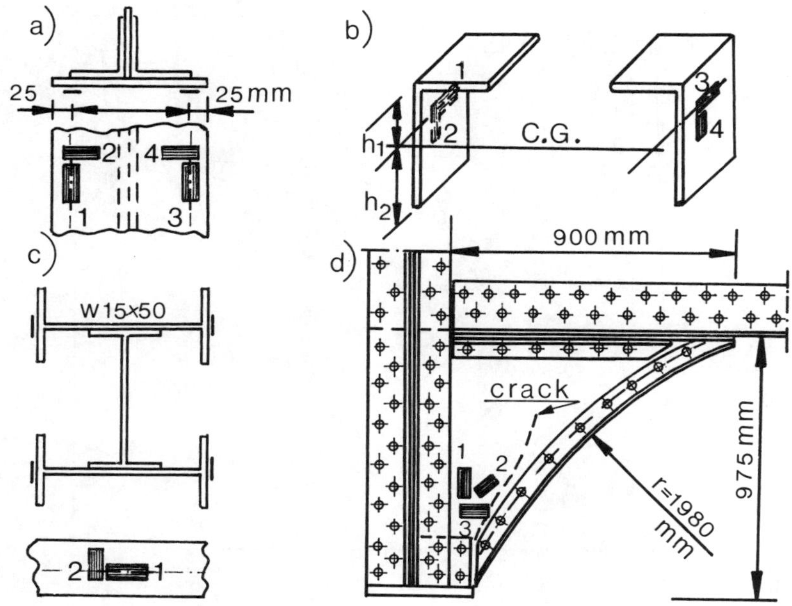

Fig. 10 Details of Strain Gage Configurations: a) For Girder Flanges, b) For Trestle Bracings (Unequal Angles), c) For Trestle Column, d) For Girders Pedestal.

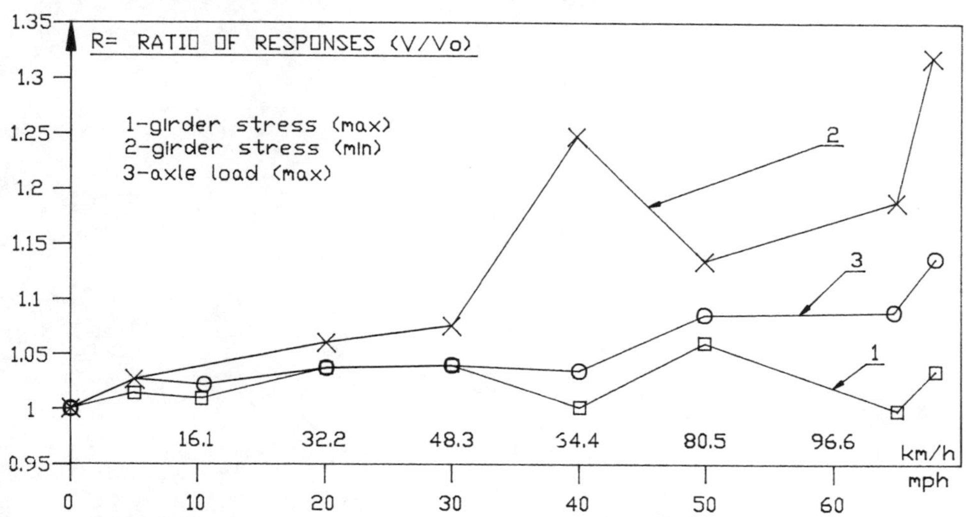

Fig. 11 Girder Stresses and Maximum Axle Loads Versus Test Train Speed (Legend: Vo - Response at Crawl Speed, V - Response at Actual Speed; Response Defined as Measured Stress or for Fig. 12.

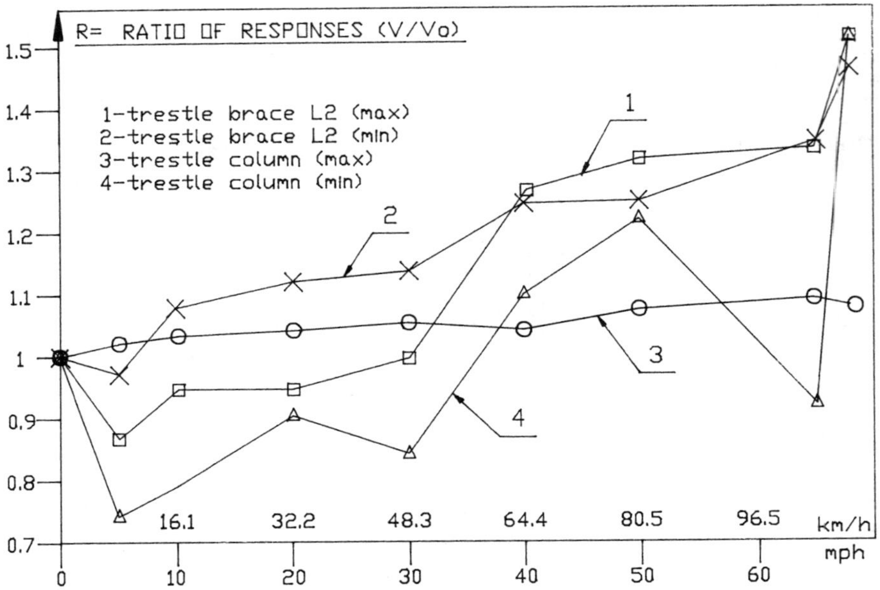

Fig. 12 Forces in Trestle Members Versus Test Train Speed (Brace L2 - Location 5 in Fig. 9; Legend as in Fig. 11).

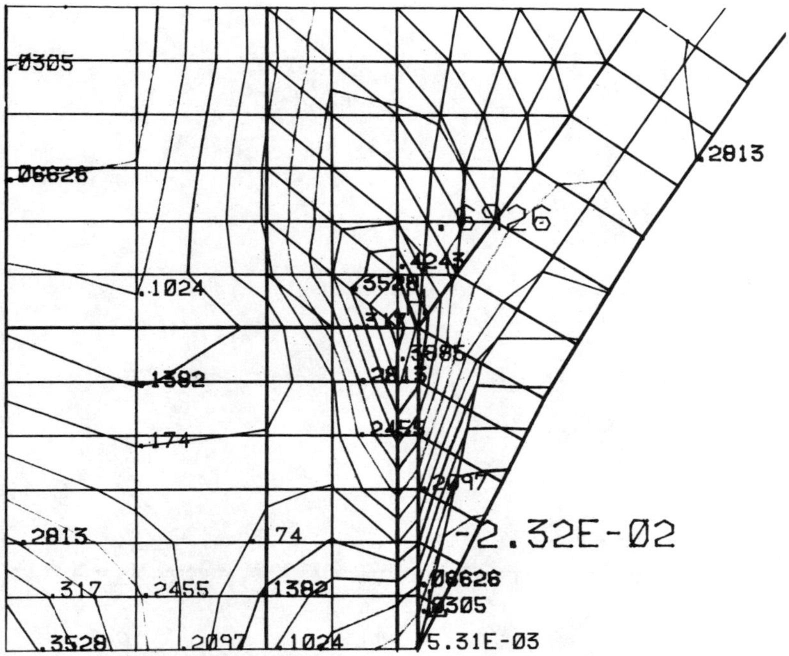

Fig. 13 Stress Concentrations Near the Pedestals Bottom Edge as Determined by 2-D Finite Element Model.

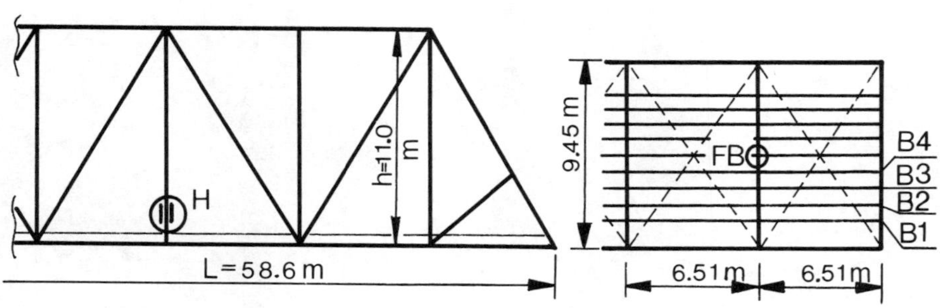

Fig. 14 Elevation and Floor Plan of Bridge # 3 (B1 to B4: Stringers, FB: Floor Beam, H: Hangers).

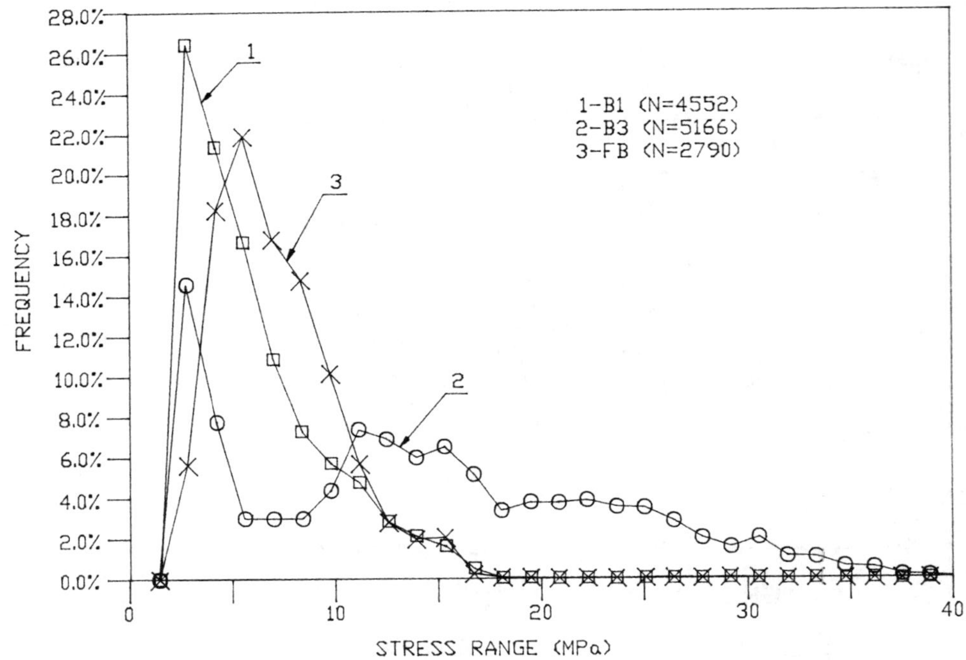

Fig. 15 Frequency Distribution Curve of Stress Ranges for Selected Stringers (B1 and B3) and Floor Beam (FB). Recorded Numbers of Stress Cycles are in Parenthesis (N).

STATIC STRENGTH TESTS OF STEEL PLATE STRENGTHENED CONCRETE BEAMS

S. H. AHMAD and S. A. HAMOUSH
Civil Engineering Department
North Carolina State University
Raleigh, NC 27695 USA

ABSTRACT. The behavior of damaged concrete beams strengthened by externally bonded steel plates is investigated experimentally. The variables tested were the thickness of the external steel plate, length and location of the interfacial crack, and the degree of surface preparation of the steel plate. In all five beams the thickness of adhesive was kept constant. The results indicate that (i) the behavior of the damaged open sandwich beam is similar to that of a singly reinforced concrete beam when no debonding occurs between the concrete and the adherent steel plate. (ii) the case with interfacial crack between the steel and the adhesive is more critical than that the case when the interfacial crack is between the adhesive and the concrete. (iii) the failure load and the mode of failure are dependent of the degree of the surface preparation of the steel plate. The results suggested that the interface separation (debonding) is the critical mode of failure of the open sandwich beam if a large interfacial crack between the steel plate and the adhesive exists. An analytical investigation to predict the interfacial debonding is summarized, and its results suggest that linear elastic fracture mechanics is suited for predicting the failure load for the case of debonding in open sandwich beams.

1. Introduction

Structurally deficient and functionally obsolete bridges need retrofitting measures to make them serviceable. Concrete members deteriorate at a rapid rate when exposed to loads in excess of service loads. At loading levels approaching the ultimate strength, the concrete members exhibit spalling and cracking of the concrete cover. Thus, the repetitive overloading of structural concrete member is not desirable. This phenomenon is more likely to occur in concrete members designed to perform in the elastic range for the service loads smaller than the current loads. In addition to the repetitive overloading problem, there are bridge structures which may not be able to carry the increased loads of vehicular traffic. These structures were designed for vehicle weights much less than those in present use and on load frequent rates that are only a small percentage of current values. These concrete members require strengthening to permit increased loads of vehicular traffic.

A. S. Nowak (ed.), Bridge Evaluation, Repair and Rehabilization, 479–492.

Strengthening deficient of critical members involves adding new material to the existing member or replacing part or all of the member. The concrete members can be strengthened by adding external reinforcement, by external post tensioning, or by a combination of these. Strengthening of reinforced concrete structure has become increasingly desirable in recent years [1]. One promising method of strengthening concrete beams is the use of an externally bonded steel plate, where the steel plate is attached to the underside of the concrete beam. The technique is relatively cheap, effective, and versatile, and can also be used for new construction as steel-concrete open sandwich beams for reducing the weight of the roadway [2]. The viability of this technique under short term conditions [3,4] and more recently under long term conditions [5] has been the focus of recent investigations. Since stronger and more durable types of adhesive are being developed, the need to study the beneficial effects of these types of adhesive for structural uses becomes urgent. There have been a few cases [6-10] where this technique has been used in practice.

In this paper, attention is focused on monotonic symmetrically loaded open sandwich beams. The open sandwich beam is a concrete beam with a bonded steel plate on the tension side. The parameters studied in this paper include (1) the thickness of the steel plate, (2) the critical location of the interface crack, and (3) the effect of the surface preparation of the steel plate. It is recognized that other aspects such as fatigue, creep, and temperature effects are important and these aspects need to be evaluated to ensure a proper use of this procedure for strengthening and for new construction applications.

2. Previous Work

Some work has already been reported [11-15] but much more detailed investigation is needed before the technique can become generally accepted.

Macdonald [11] conducted tests on four reinforced concrete beams having steel plates bonded to their tension flanges with structural epoxy resin. The tests results showed that all the beams failed by horizontal shear in the concrete adjacent to the steel plate. It was noted that the calculated normal shear stress at the interface between the concrete and the steel at failure (ignoring the thickness of adhesive) were less than the nominal shear strength of the concrete used.

Jones and Swamy [12] discussed the composite behavior of eight strengthened reinforced concrete beams. All the beams tested were simply supported over a span of 7.4 feet with equal point loads loading. The tests results indicated that the external plate increases the ultimate strength and the bending stiffness of the basic beams, and the plate separation occurs when the interface shear stress reach a critical value.

In England, Irwin [13] performed tests on two reinforced concrete beams in flexure, one of which had external steel plates epoxy-bonded to the tension flange. The test results indicated that (1) highly localized stresses in the concrete and adhesive layers may be produced; (2) progressive failure of the bond may begin at a crack bridged by the steel plate, when loading is increased; and (3) the bond strength of the epoxy resin may be affected by cyclic loading and long-term creep.

Solomon [14] conducted tests on seventeen sandwich beams, in which all the beams collapsed after slippage between the tensile steel plate and the concrete core. The surface of the plate showed that a thin layer of concrete stripped off the concrete core during slippage between the tensile steel plate and the concrete core. The surface of the bottom plate showed that a thin layer of concrete stripped off the concrete core during slipping. Beams in which the tensile plate yielded before slip occurred were considered to have failed in bending.

Six plain concrete beams with external steel plate reinforcement were tested by Cusens and Smith [15]. The failure was primarily due to breaking down of the adhesion of the concrete surface to the steel plate. Yielding of the steel plate took place followed by the crushing of the concrete in compression, followed by debonding of the steel plate.

3. Test Program

A total of five concrete beams externally reinforced with bonded steel plates were tested under monotonically increasing static loads to determine their strength and behavior.

The variables in the test program were (1) the thickness of the external steel plate, (2) the length of the initial interface crack and its location, and (3) the degree of the surface preparation of the steel plate.

The thickness of epoxy was kept constant as 0.15in., since analytical work conducted by Anandarajah and Hamoush [16,17] showed that thickness of adhesive has no major effect on the behavior of the externally reinforced concrete beam. The summary of the test program is presented in Table 1.

3.1 FABRICATION AND INSTRUMENTATIONS

A total of five plain concrete beams, each 8 feet long, 4x6 inch in cross-section, were cast for the tests. The damage in concrete beams was simulated by inducing flexural cracks during the casting of the beams. The flexural cracks were concentrated in the mid span in the constant moment region. The height of the flexural cracks was equal to 2/3 the height of the beam. The spacing of the flexural cracks was 4 inches. The flexural cracks were induced by the use of thin metallic strips during casting and were later removed after the beam had cured. The beams were cured for 7 days in the curing room. The induced flexural cracks were extended through the epoxy to the steel plate. In each beam the steel plate was bonded to the tension side of the plain concrete beam after the steel surface to be bonded was sand-blasted (see Fig. 1). The concrete surface was prepared by casting the concrete against rubber-mat with a maximum size of indentation of 1 inch. Then the concrete surface was hammered by a needle hammer in order to abrade the aggregate. The interfacial cracks, between either the adhesive and the steel plate of the adhesive and the concrete, were induced by the use of thin plastic strips during casting of the epoxy and were later removed after the adhesive had hardened.

For obtaining the epoxy adhesive, the resin and the hardener were hand mixed with a spatula, as power mixing causes heating due to viscous friction that is associated with epoxy resin system. Immediately after mixing, the resin was applied to the concrete beam. The steel plate was then placed on the beam and weighted down with a weight of approximately 40 lb/ft. Although full curing of the resin takes place within 48 hours under the prevailing laboratory conditions of 20°C and 65 percent relative humidity, the beams were cured for a minimum period of 7 days before the weights were removed. The beams were then instrumented with 1/2 inch long strain gauges at several locations on steel plate, and one gauge on the concrete compression face.

3.2 MATERIALS AND PROPERTIES

The materials used were concrete, steel plate reinforcement and epoxy adhesive.

1. Concrete: The mix design ratio for the concrete beam was chosen such that the compressive strength of 4 ksi was obtained after 7 days. The cement used in the concrete beam was normal type ordinary portland cement. The coarse aggregate was a 0.5 inch (12.5mm) maximum size gravel and the fine aggregate had a maximum size of 0.1 inches (.25mm).
2. Plate reinforcement: Two thicknesses 1/8" and 3/16" of steel plate were used. The width of the steel plates was 4 inches and length was 8 feet. All steel plates were of mild steel type with yielding strength of 60 ksi. The elastic modulus was 29,000 ksi.
3. Epoxy adhesive: The epoxy used was ATLAC-382 type with two parts hardening and resin. The mixture was obtained from using equal parts by volume of the resin and hardener. The flexural strength of the epoxy adhesive was 15,000 psi; the tensile strength was 10,000 psi; the elastic modulus was 0.43×10^6 psi, and Poisson's ratio was 0.45.

3.3 TEST PROCEDURE

All the beams were tested under simply supported conditions over a span of 8 feet with two equal point loads at two feet apart at center as shown in Fig. 2a. The mid-span deflections were measured by dial gauge, and the strains in the steel and concrete were measured by electrical resistance (strain gauges). The test set up is shown in Fig. 2b. The loading was monotonically applied in increments of 500 lbs. until failure was achieved by either flexure or separation of the steel plate.

4. Results

The test results include, initial cracking loads, load-deflection curves up to failure, load-strain relationship in steel plate, load-strain relationship in concrete, the failure load, and mode of failure.

4.1 INTERFACE CRACKING AND LOAD-DEFLECTIONS

In beams 1 and 2, initiation of interface shear crack between the steel plate and the glue was noted at loads of about 5.5 kips. The interface shear crack was located at the base of the vertical flexural crack which was closer to the vertical load being applied to the beam.

In beam 3 the initially simulated interfacial crack of 2 inch length between the steel plate and the epoxy adhesive extended all the way along the interface as the loading progressed.

The interfacial crack in beam 4 remained at the interface of the steel and the epoxy adhesive, but the interfacial crack in beam 5 extended along the interface of concrete and epoxy up to 6 inches from the end of the beam where the crack changed to be between the steel and the epoxy adhesive.

Fig. 3 shows the load mid-span deflection curves. For beam 1, the behavior was similar to that of under-reinforced sections where large deformation was recorded before failure. For beam 2 the load-deformation behavior was similar to that of an over-reinforced section. For beam 3, the deflection was still in the elastic linear region when the beam failed. Failure of the beam occurred before yielding or crushing of the concrete. For beam 4, the load deflection curve was linear till failure. For beam 5, the load deflection curve was linear till a load of 6300 lbs. after which it exhibited non-linearity.

4.2 PLATE STRAINS IN STEEL AND CONCRETE

Fig. 4 shows the variation of the steel plate strains with the increasing loads for beams 1, 2 and 4. In the areas of high strain gradient, the maximum strain at a point may be much higher than the nominal values obtained from the strain gauge.

In beam 1, the yielding of steel plate occurred at 6.120 kips and the maximum value of strain in the steel plate was 0.0083. For beam 2, the maximum strain in the steel plate was 0.00106 which was below the yielding strain of steel indicating an over-reinforced section. For beam 3, the value of plate strain at failure was 0.00082. The behavior of the beam up to failure can be considered as linear elastic behavior. For beam 4, value of strain in steel was lower than the yielding strain. The strain in the steel plate at failure of the beam was 0.00084 as shown in Fig. 4. For beam 5, values of strain in steel were similar to that of beam 4, where the strain in the steel plate did not reach the yielding strain, and a value of 0.00103 was recorded at the failure of the beam.

Fig. 5, shows that plot of the concrete compressive strains versus the loads for beam 1, 2 and 4, respectively. The ultimate concrete strain for beam 1 was 0.0036 in/in and that for beam 2 was 0.0034 in/in. For beam 1 the strain in concrete at the yielding of steel plate was 0.0027 in/in. For beam 3, the ultimate strain in concrete at failure was 0.00154. For beam 4, the ultimate strain in concrete at failure was 0.00176, and the ultimate concrete strain for beam 5 was 0.00197.

4.3 FAILURE LOADS AND MODES OF FAILURE

The failure loads for the beams and modes of failure are given in Table 2. It can be seen from Table 2 that the failure loads varied substantially for the five test beams, because the variables tested have a major effect on the strength of open-sandwich beams.

Beam 1 failed as an under-reinforced section. The failure was initiated by the yielding of the steel plate. Immediately after the yielding of the steel plate, a diagonal crack located under the point load was formed. After further deformation a hair-crack at the interface of the steel plate and the epoxy adhesive occurred followed by sudden separation of the steel plate at its interface with the epoxy adhesive.

Beam 2 failed as an over-reinforced section. The failure occurred by extension of the vertical crack immediately below the crushed zone of the concrete and the spalling of the crushed zone of the concrete began. The sudden crushing of the concrete in the compression zone provided no prior warning. No interface crack at failure was noted.

It can be noted that, for beam 1 and 2, the behavior of the beam changed from that of an under reinforced section to that of an over reinforced section behavior. This change in behavior is due to the change of thickness of the steel plate.

Beam 3 failed by separation of the steel plate before yielding of the steel plate or crushing of the concrete. Interface separation started at the initially simulated interfacial crack location and extended towards the end section of the beam. These initially simulated interfacial cracks of 2 inch length at the epoxy/steel plate interface were symmetrically located with respect to the mid section of the beam (see Fig. 2a). The interface cracks extended slowly as the load increased, then sudden separation occurred causing the failure of the beam. During the test, the vertical cracks extended for about 0.26 inch.

It can be noted that, for beam 3, the surface preparation of the steel plate has a major effect on the strength and the mode of failure. Since the surface preparation was of relatively poor quality, the beam failed at a much lower load by interface debonding.

Beam 4 failed by separation of the steel plate before yielding of the steel plate or crushing of the concrete. The separation started at the initially simulated interfacial crack location between the steel plate and the epoxy adhesive. The interfacial crack propagated rather suddenly to cause complete interface debonding and failure of the beam. The interface remained at the interface of the steel plate and the epoxy adhesive. The vertical cracks extended for about 0.5 inches during the test.

Beam 5 failed by interface separation between the steel and the epoxy adhesive although the initial interfacial crack propagation was between the concrete and epoxy interface. The interface separation started before the yielding of the steel plate or crushing of the concrete. The flexural cracks extended for about 0.5 inches during the test.

Beam 4 failed at a relatively lower load than that for beam 5 indicating the effect of location of the interfacial shear crack on the strength and overall mode of failure of steel plate strengthened concrete beam. Fig. 6 shows the separation surfaces when the crack extends between the steel plate and the epoxy, and Fig. 7 shows the separation surfaces when the crack extends between the concrete and the epoxy.

5. Analytical Investigation

The two modes of failure observed can be classified as (i) flexural failure in which the yielding of the steel plate or the crushing of the concrete takes place before an interfacial separation occurs (ii) debonding failure in which separation of the steel plate occurs (beam No. 4). For the first type of failure mode, the failure loads can be estimated by the flexural theory and for the second type of failure mode the theory of linear fracture mechanics can be used to estimate the failure load.

The fracture mechanics theory is employed to estimate the failure load of beam No. 4 in which the theory is based on the critical strain energy release rate. It was shown by Rice and Smelser [18,19], that for interfacial cracks of two dissimilar materials, the strain energy release rate at failure is constant and it is a material property.

The J_1 contour integral method in bi-material bodies [20, 21] is used to obtain the strain energy release rate which is presented as J_1 integral can be written as

$$J_1 = \int_s (wn_1 - \sigma_{jk} n_k u_{j,1}) ds = 0 \quad (1)$$

where $w = \int \sigma_{ij} \, \partial \varepsilon_{ij}$ is the strain energy density and equals:

$$w = 1/2 \, (\sigma_1 \varepsilon_1 + \sigma_2 \varepsilon_2 + \tau_{12} \gamma_{12}),$$

where σ_{ik} is the stress tensor, $u_{i,j}$ is the strain tensor, and $t_i = \sigma_{ij} n_j$ is the traction components of stress acting on the contour. s denotes any closed contour taken counter-clockwise around the crack tip, u_j is the displacement vectors, and n_k is the unit normal vector to the path s.

Evaluation of J_1 integral is obtained by the second order Gaussian quadrature along a contour passing through Gaussian stations on the element sides. Which suggests that the finite element mesh should be selected at some distance from the crack tip so to avoid the complication of the stress field near the crack tip. Different integration contours paths were used to check the path independently [17].

The failure load for beam No. 4 was obtained by comparing the strain energy release rate obtained from the J_1 integral approach in the finite element model to the critical strain energy release rate obtained from compact tension and shear specimens in [17]. The comparison of the analytically predicted failure load and the experimental result is shown in Fig. 8.

6. Conclusions

The reported work in this paper is of a preliminary nature only and much more data and analysis are required before any general conclusions can be made. The following conclusions are related to the few beams tested.

1. When no debonding occurs, the behaviour of open sandwich beam with bonded steel plate is similar to that of a singly reinforced concrete beam.

2. The case with induced interfacial crack between the steel and the epoxy is more critical than that case when the interfacial crack is between the epoxy and the concrete.
3. The failure load and the mode of failure are dependent on the degree of the surface preparation of the steel plate.
4. When the interface separation occurs before the yielding of the steel plate or crushing of the concrete, the theory of linear fracture mechanics seems well suited to estimate the failure load.

7. Acknowledgements

The authors acknowledge North Carolina State University, Department of Civil Engineering for providing the laboratory supplies.

8. References

1. Berger, R. H., Tallamy, Bvrd, Macdonald, L., Gordon, S. (1978) Federal Highway Administration, "Extension the service life of existing bridges," Transportation Research Record, No. 661-665.
2. Ong, K.C.J., and Mayer, S.G.C., (1982) "Flexural Tests of Steel Concrete Open Sandwiches," Magazine of Concrete Research, Vol. 24, No. 120, September.
3. Fleming, C.J., and King, G.E.M., (1967) "The development of structural adhesives for the original used on South Africa," Bulletin RILEM No. 37, pp. 241-251, December.
4. Jan, J. Dussek, (1980) "Strengthening of bridge beams and similar structures by means of epoxy-resin-bonded external reinforcement," Transportation Research, No. 785.
5. Rybicki, T., Harnaadez, Jr., J., Deiber, J. E., and Knight, R. C., (1987) "Mode I and Mixed Mode Energy Release Rate values for Delamination of Graphite/Epoxy test specimens," Journal of Composite Material Vol. 21, February, pp. 105-123.
6. Mander, R.F., (1974) "Bonded external reinforcement, a method of strengthening structures," Department of the Environmental Report on Quinton interchange for M5 Motorway.
7. Brown, K., (1973) "Japanese style strengthening for M5 bridge", Construction Mews, pp. 21, Oct.
8. Sommerard, T., (1977) "Swanley's steel plate parch up," New Civil Engineer, No. 247, pp. 18-19, 16 June.
9. Parkinson, J., (1978) "Glue solves a sticky problem for Gestetner," New Civil Engineer, pp. 26, 14 Sept.
10. Bresson, J., (1972) "The strengthening bonded reinforcement of the underpass of the CD136 to the Autoroute du Sud," Annales de L'institut Technique de Batiment el des Travaux Publics, Supplemental No. 297, Sept.

11. Macdonald, M.B., (1978) "The flexural behaviour of concrete beams with bonded external reinforcement," TRRL. Report SR 415, Transport and Road Research Laboratory, pp. 1-29.
12. Jones, R., Swamy, R.N. and Ang, J. H., (1982) "Under-and over-reinforced concrete beans with glued steel plates," The international journal of cement composites and light weight concrete, Vol. 4, No. 1., pp. 19-32.
13. Irwin, C.A.K.,(1975) "The strengthening of concrete beams by bonded steel plates," TRRL Supplementary Report 160 UC, Transport and Road Research Laboratory, Department of the Environment, Crowthrone, pp. 8.
14. Solomon, S.K., Smith, M.D.W., Beng, MA, Fice and Crussens, A.R., Bac (Eng), (1976) FICE, FIStructE, FRSE, "Flexural tests of steel-concrete-steel sandwiches," Magazine of Concrete Research, Vol. 28, No. 94, March.
15. Cusen, A.R. and Smith, D.W., (1980) "A study of epoxy resin adhesive joints in shear," The Structural Engineer, Vol. 58, No. 1, January.
16. Anandarajah, A., and Vardy, A.E., (1985) "A theoretical investigation of the failure of open sandwich beam due to interfacial shear fracture," The Structural Engineer, Vol. 63b, No. 3, December.
17. Hamoush, S. (1988) "A facturing model for concrete beams strengthened by externally bonded plates," Ph.D. Thesis, North Carolina State University.
18. Rice, J.R., (1968) "A path independent integral and the approximate analysis of strain concentration by notches a crack," J. Appl. Mech. ASME, Vol. 35, Trans., Vol. 90, Series E, pp. 379-386.
19. Smelser,R.E., Gurtin, M.E., (1977) "On the J-integral for Bi-material bodies", Int. Journ. of fracture. 13, pp. 387-384.

Table 1 Summary of Test Program

Beam No.	b (in)	h (in)	f'_c (Ksi)	Surface preparation	Thickness of steel plate (in)	Thickness of epoxy (in)	Interface crack location	Interface crack length (in)	No. of Flex cracks
1	4	6	4	good*	1/8	0.15	-	-	4
2	4	6	4	good	3/16	0.15	-	-	4
3	4	6	4	poor**	1/8	0.15	Steel/ epoxy	2	4
4	4	6	4	good	3/16	0.15	steel/ epoxy	12	5
5	4	6	4	good	3/16	0.15	concrete/ epoxy	12	5

* prepared by sand blasting

** prepared by hand

Table 2 The Test Results at Failure

Beam no.	Mode of failure	The failure load (Kps)
1	yielding of steel plate	6.22
2	crushing of the concrete	8.92
3	separation between steel and epoxy	3.38
4	separation between steel and epoxy	6.18
5	separation between concrete and epoxy	7.93

Fig. 1 The Prepared Surface for the Steel Plate Magnified 200 Times.

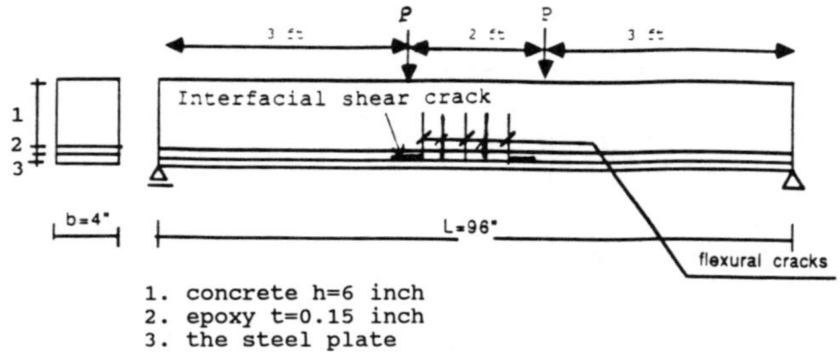

Fig. 2a Description of Test Specimens.

Fig. 2b The Test Set Up.

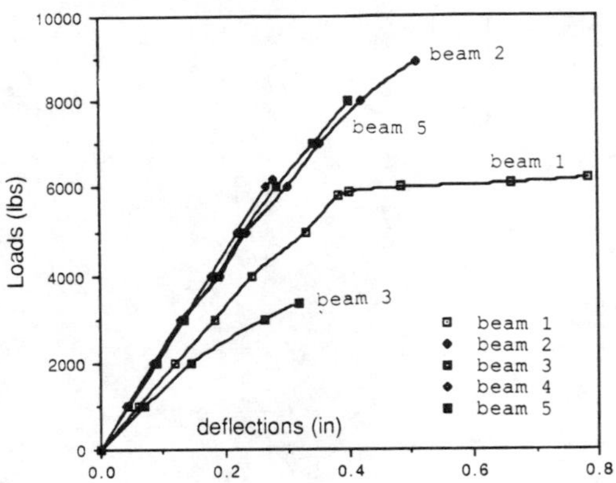

Fig. 3 Load Mid-Span Deflection Curves for Steel Plate Strengthened Concrete Beams.

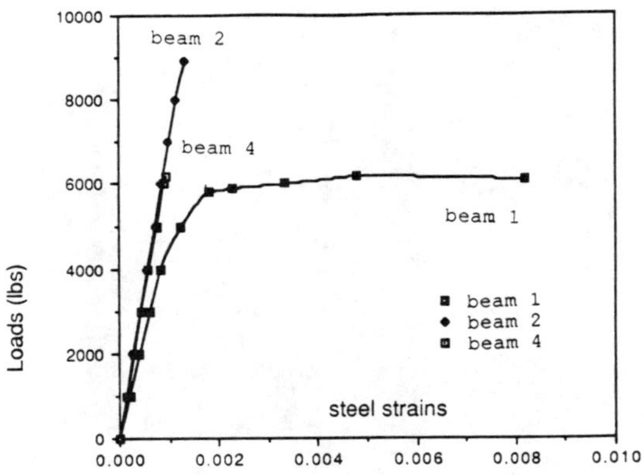

Fig. 4 Load-Plate Strain Curves for Steel Plate Strengthened Concrete Beams.

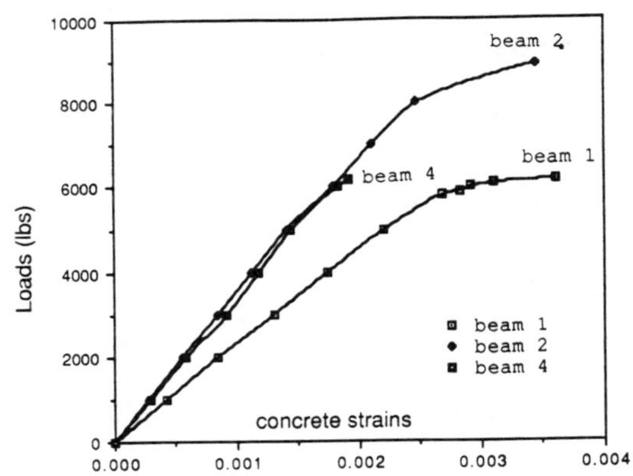

Fig. 5 Load-Concrete Strain Curves for Steel Plate Strengthened Concrete Beams.

Fig. 6 The Steel Epoxy Interface Failure.

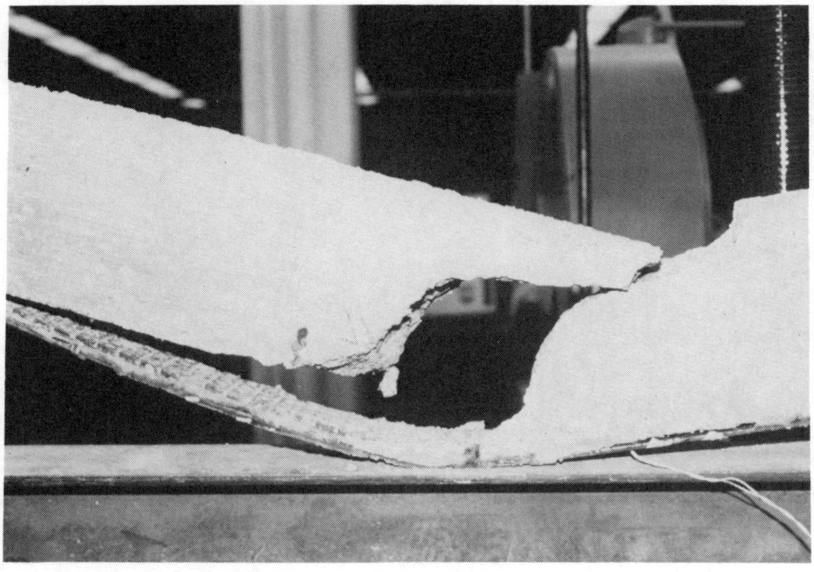

Fig. 7 The Concrete Epoxy Interface Failure.

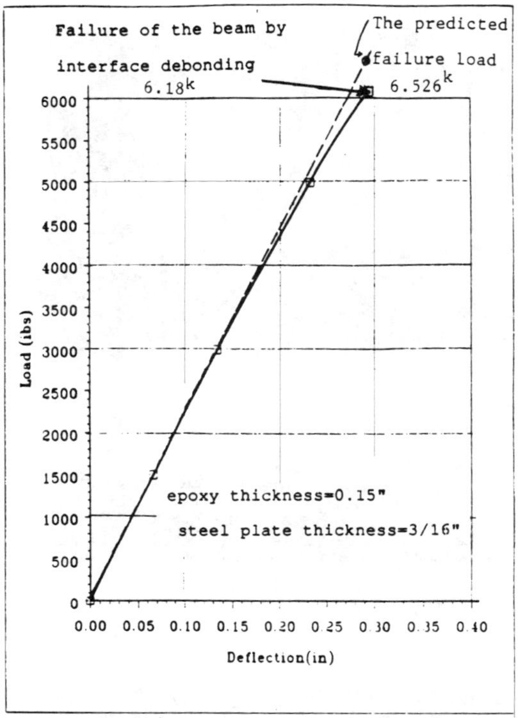

Fig. 8 Comparison of Observed and Predicted Load-Deflection Behavior of Open Sandwich Beam (Beam 4).

V - Repair and Rehabilitation

BRIDGE REPAIR METHODS - U.S./CANADIAN PRACTICE

JOHN W. FISHER and CRAIG MENZEMER
ATLSS Engineering
Research Center
Lehigh University
Herbert R. Imbt Laboratory
Bethlehem, PA 18017 USA

ABSTRACT. Nearly half of the highway bridges in the United States have been classified as structurally deficient or functionally obsolete. The question of functional obsolescence will not be dealt with. Reasons for the structural deficiencies are numerous, as a number of the bridges are over fifty years old and have not been properly maintained. Over a period of time, environmental corrosion has accumulated. In addition, increases in the number of trucks using the highways as well as the rise in allowable truck weights has led to increasing amounts of fatigue cracking. Many structures have welded details that have been identified as susceptible to fatigue cracking only after they were built. Rules used to design some structures often oversimplify member interactions and connection behavior, resulting in distortion-induced fatigue cracking in a large number of different types of bridge structures.

Due to the magnitude and breadth of the bridge deficiency problem, and the prohibitive costs of new construction, a large effort in the repair and rehabilitation of existing bridges is required. Careful examination of the causes of fatigue cracking and corrosion accumulation, coupled with effective technology transfer, will help to prevent the development of similar situations in future construction. This paper briefly reviews the causes of some of the fatigue cracking being experienced in bridge structures. Various repair and retrofit procedures are reviewed, and examples of typical applications of these procedures are examined. It also examines examples of severe corrosion and the inability to detect active corrosion cell activity.

1. Introduction

A number of localized failure have developed in steel bridge components due to fatigue crack propagation which in some instances, has led to brittle fracture [1,2]. A majority of the fatigue cracks can be placed into one of two broad categories. The single largest category is a result of out-of-plane displacements in small, unstiffened segments of web plates. When distortion induced cracking develops in a bridge component, large numbers of cracks can be expected to develop nearly simultaneously in the structure because the cyclic stress is high and the number of cycles needed to produce cracking is relatively small. Displacement induced cracking has

A. S. Nowak (ed.), Bridge Evaluation, Repair and Rehabilization, 495–512.

developed in a wide variety of structures including, suspension, tied arch, two girder floorbeam, multiple beam, and box girder bridges. In general, the cracks form parallel to the primary stress field and are not detrimental to the performance of the structure provided they are discovered and retrofitted before turning perpendicular to the primary stress field.

The second largest category of fatigue damaged members and components comprises large initial defects, cracks, or crack-like conditions. Defects in this category have resulted from poor quality welds that were made before nondestructive test methods were well established. In addition, a number of localized failures in this category have developed because the groove welded component was considered a secondary member or attachment. As a result, weld quality criteria were not established and nondestructive test methods were not employed.

A majority of the remaining failures resulted from the use of low strength details that were not anticipated to have such a low fatigue strength at the time of the original design because of the limited experimental data base upon which specification provisions were based. There are large numbers of structures in this category because of the interstate construction in the 1960 - 70 era.

2. Repair of Fatigue Damaged Components

Any methodology used to repair fatigue damaged details is case specific and is generally dependent on the size and location of the crack(s) at the time of repair [3]. Obviously, this is highly dependent on both inspection procedures and the ability to locate relatively small cracks. Apart from displacement induced fatigue cracks, the majority of remaining fatigue cracking has resulted from low strength details, large initial defects, or detail geometry which simulates sharp crack-like conditions. Figure 1 illustrates the role both initial defect size and stress concentration in the determination of the relative fatigue strength of different detail types [4].

2.1. PEENING

Various methods to retrofit or increase the fatigue resistance of welded details have been employed [5]. For cracks which have propagated from normal in-plane stress fields, peening can be a successful repair as long as the crack depth does not exceed the zone of compressive stress introduced as a result of the peening operation [6]. Laboratory studies have demonstrated that a fatigue crack of up to 3 mm in cover-plated details can be arrested by peening provided the stress range does not exceed 41 MPa. Peened beams with crack depths larger than 3mm usually show no measurable increase in life. Also, peening has been found to be most effective when conducted under dead load and only needs to be effective against live load. In the field, peening may be carried out by mechanical air hammer until the toe of the weld is plastically deformed. Figure 2 illustrates the extent of deformation at a weld toe. Peening intensity can be varied by changing the air pressure and in general, multiple pass peening at lower air pressures has been found to result in a deep compressive residual stress zone.

2.2. HOLE DRILLING

For cracks which have propagated through both the plate thickness as well as away from the local stress concentration, hole drilling has proven an effective retrofit [7]. Holes essentially blunt the tip of the crack, although the size of the hole must satisfy the relationship where ??is the radius of the hole, mm, K is the stress intensity factor assuming the crack size 2a is defined between the edges of the hole, and $_y$ is the yield strength of the plate, MPa.

Otherwise, blunting will probably not be effective and the crack will most likely reinitiate from the edge of the hole. The validity of this retrofit has been studied in the laboratory on full-scale welded beams subjected to variable amplitude loading up to 90 million stress cycles. Sometimes bolts will be inserted in the holes and tightened so as to introduce some local compression around the hole. Other methods attempt to cold work the hole to introduce some beneficial residual stress state around the hole and crack tip [8].

2.3. REPAIR OF DISTORTION DAMAGE

Distortion induced fatigue cracking, is usually the result of primary and secondary member interaction not accounted for in the original design. These effect are usually noticed in small unstiffened segments of web plate where longitudinal and transverse members frame together.

Distortion induced cracking may also be retrofitted by drilling holes, although laboratory studies have shown this to be ineffective when distortion levels are high. For multiple girder bridges, most displacement induced cracking occurs in the web gap region a the end of connection plates in the negative moment region. Lengthening the web gap region effectively increases the flexibility of the connection. However, this has proven effective only if the distortion does not increase as a result of reduced detail stiffness. The most effective, and probably most inexpensive method of retrofit is positive attachment of the connection plate to the flange. By bridging the unstiffened segment of plate, both the distortion and large cyclic stresses are reduced. This bridging may be accomplished by bolting or welding, although bolting is usually preferred over field welding.

3. Example of Increased Gap Length and Peening

Plan and elevation views of the Housatonic River Bridge is shown in Figure 3. Originally, one structure carried bidirectional traffic and was built in the early 1950's. Due to traffic demands of the area, a second geometrically similar structure was built in the late 1970's. Now, each bridge carries traffic in a single direction. Both structures consist of two main longitudinal girders with floorbeams, cantilever brackets, and longitudinal stringers. The earlier bridge is riveted construction while the newer is welded and bolted.

Figure 4 illustrates the main girder -floorbeam -cantilever bracket detail. For the floorbeam to main girder connection, the floorbeam webs are bolted to connection plates which are attached to the girder web with continuous fillet welds. Directly opposite the floorbeam connection, the cantilever bracket web is bolted to a connection plate which is also welded to the longitudinal girder. The main girder web is slotted to allow a tie plate to frame through which attaches the top flanges of the floorbeams and cantilever brackets. Lateral bracing also exists between the main girders and is anchored at the center span of every other floorbeam through a gusset plate welded to the floorbeam bottom flange.

3.1. INSPECTION

After a few years of service, fatigue cracking was reported in the welded structure after routine inspection. No cracking had ever been observed on the riveted structure. Cracks were found in a number of locations around floorbeam and cantilever bracket connection plates, as well as the lateral bracing gusset to floorbeam flange connections. Figure 5 shows typical cracks which propagated along the fillet weld toes of connection plates which attached floorbeam and cantilever bracket webs to the main girders. Figure 6 illustrates cracks which were observed along the inner bolt line and in some instances, extended to the top of the connection plate. Figure 5 illustrates cracking which had occurred along the fillet weld connecting floorbeam web to flange and cantilever bracket web to flange. These cracks were found on the tie plate sides of the beams. These cracks occurred not only along fillet welds of interior floorbeams and brackets, but were largest at pier and abutment locations. Wearing of the main girder web was observed at the slots used for tie plates. Figure 7 shows cracks which formed at the ends of the lateral gusset plate - floorbeam connection where the fillet weld terminates.

3.2. PROBABLE CAUSES OF FAILURE AND RETROFITS

In general, the stiffness of the floor system (deck, stringers, floorbeams, and cantilever brackets) is larger than the main girders. This differential flexibility resulted in relative displacements between the floor and longitudinal girders. Such displacements caused the wearing of the slots, as well as fatigue cracking around the cantilever bracket and floorbeam web connections.

At the floorbeam and bracket girder connections, the relative displacement concentrated in small unstiffened segments of plate referred to as gap regions. A vertical gap was located on the connection plates between the ends of the floorbeam or bracket webs and the fillet weld toe of the connection plates. Differential movement resulted in out-of-plane deformation of the connection plates and double curvature bending of the plate. Large cyclic stresses from the bending caused vertical cracking along both the interior bolt line and fillet weld toe. The deformation is depicted in Figure 8. A horizontal gap also existed on the floorbeam and cantilever brackets between the top of the connection plates and the beams' top flange. Movement between the top flanges and connection plates resulted in crack initiation and propagation along the web-flange fillet weld toes on both front and back faces of the floorbeam and cantilever bracket webs. Out-of-plane distortion of the floorbeam and bracket flanges was equivalent to bending the web plate along the fillet weld toe.

In an attempt to remove and blunt the crack tips in the floorbeam and cantilever bracket web to flange welds, holes were drilled in horizontal gap regions as well as in the vertical gap where fillet welds were used to attach web connection plates to the main girders. In addition, four top bolts of the bracket to main girder connection plate and two bolts from the floorbeam connection plate were removed. These bolts were removed to increase the length of the web gaps to 15.2 cm and accommodate the distortion. Figure 9 shows a repaired floorbeam - main girder connection with the two top bolts removed and holes drilled at the crack tips. The small cracks at the welds connecting the lateral gusset plates to floorbeam flanges were retrofitted by peening. Figure 10 illustrates the procedure.

Subsequent to the initial repair, field measurements were acquired in an attempt to verify the retrofit or indicate a need for further repair. Field measurements for the bracket web - flange connection indicated that the stress range beyond the retrofit holes was not sufficient to reinitiate the crack. In general, measurements on floorbeams, brackets, and connection plates indicated that the web gap flexibility was not adequate removing bolts alone and further change was required. As such, additional bolts were removed from the floorbeam and bracket web connections and 30.5 cm of the connection plate was cut away (Figure 11). Removal of the plate eliminated the fillet weld toe cracks and increased the connection flexibility and capacity to accommodate distortion. Later, the weld toes of floorbeam and bracket connection plates were peened several inches from inches from the flame cut edges.

Lateral gusset plates connected to the bottom flanges of the floorbeams through two fillet welds along the edges of the flange represents an E' detail. Figure 7 shows small cracks found at the fillet weld terminations. The small cracks in the weld terminations were removed by grinding and peened for a second time. Finally, main girder web slots were enlarged at all locations showing wear of the web by the tie plates.

4. Example of Bridging Gap and Retrofit Holes

An elevation and view of the bridge is shown in Figure 12. Built in the early 1960's, each structure carries two lanes of traffic over five continuous spans ranging from 41 m for end spans to 49 m for the interior spans. Each bridge is comprised of two longitudinal plate girders which are haunched at the piers, multiple longitudinal stringers, and transverse floorbeams. Both longitudinal girders and stringers support the reinforced concrete deck and provide partial composite action through embedded flanges. Floorbeams and girders are built-up sections of welded construction.

A typical portion of the framing plan is shown in Figure 13. Floorbeam webs are bolted to transverse connection plates which are attached to the longitudinal girder by continuous fillet welds. These connection plates are not only attached to the girder web, but are fillet weld connected to the compression flanges and cut short of the tension flanges. Lateral bracing frames between main girders and is bolted to horizontal gusset plates. These gusset plates are bolted to the floorbeam flange and to lateral tabs fillet welded to the main girder web. Figure 14 illustrates the connection.

4.1. INSPECTION

Over a three year period, several inspections of the structure were conducted. Initially, cracking was observed in the webs of main girders in negative moment regions. As the transverse floorbeam connection plates were cut short of the main girder tension flanges, an unstiffened segment of girder web was left between the ends of the connection plates and the top flange. Both horizontal and vertical fatigue cracks were found. Horizontal cracks ran along the web flange fillet, and vertical cracks had formed down the sides of the connection plates. Retrofit holes were drilled at the crack tips as shown in Figure 15.

Subsequent inspections revealed additional cracking from some of the retrofit holes. In addition, fatigue cracks were found in the girder web between the lateral web attachments and transverse connection plates. Cracking in this horizontal web gap occurred at the fillet weld toes of the lateral tabs (Figure 16). At nearly every floorbeam - girder connection, vertical cracking was observed on the outside girder web face along the web stiffener weld toe (Figure 17). This vertical cracking corresponded to the inside face horizontal web gap.

4.2. PROBABLE CAUSES OF FAILURE AND REPAIR METHODOLOGY

As the floorbeams in the negative moment regions rotated under traffic loading, the segment of girder web between the girder flange and connection plate was pulled out-of-plane. Such behavior is illustrated in Figure 18. This small portion of web plate was bent in double curvature and large cyclic stresses were introduced. Clip gage readings indicated the deformation to be on the order of 1.5 mm. Strain measurements made after initial attempts to retrofit indicated that while softening of the area by cracking and drilling holes reduced the live load stress range, this reduction was not sufficient to prevent reinitiation. A stronger attachment between the connection plate and top girder flange was required. Positive attachment was provided by bolting angles between the stiffener and flange as shown in Figure 19. Such a retrofit effectively eliminates the distortion and hence, the possibility for further crack growth.

In the positive moment regions, strain readings in the bottom web gap indicated that no further retrofit was required. However, regular inspections were recommended.

Even though the lateral gusset plates were positively attached to the floorbeam bottom flange and girder web, local deformation of the web plate was not prevented. Floorbeam end rotation and out of phase action of the wind bracing resulted in out-of-plane displacements of the girder web at the gusset level. Field measurements indicated that the stress range at the lateral tab termination was less than the design value, but larger than the fatigue limit for a Category E detail. Measured stresses in the horizontal gap between the connection plate and lateral tabs revealed the potential for fatigue cracking at all such locations in the structure. A preliminary retrofit of replacing the lateral tab with a bolted angle increased the fatigue resistance to a Category B detail and eliminated the potential for continued crack growth at the terminations. However, the potential for further crack growth along the weld toe of the transverse connection plate was not eliminated. As such, the gap length, or distance between the bolted angle and connection plate, was increased to accommodate the distortion. To remove any existing vertical cracks, holes were drilled on each side of the stiffener and gusset plates. Figure 20 shows a retrofitted lateral gusset plate.

5. Conclusions

As available funds are often limited, it is usually desirable to repair and retrofit existing structures rather than to design full replacements. A majority of fatigue damaged details can be repaired by weld toe peening, drilling holes, bolting splices over damaged areas, providing positive attachments, or by increasing detail flexibility. In some instances, welding may be used for repair, although care should be taken so that low fatigue resistant details don't result.

Examination of the causes of fatigue cracking coupled with implementation of new design tools, code revisions, and effective technology transfer will help to minimize future structural deficiencies. Simple, realistic models and analysis procedures for member interaction and connection behavior are needed to reduce the incidence of fatigue cracking. Identification of low fatigue resistant details and effective methods for detail classification would ease the burden on designers and limit the occurrence of fatigue cracking.

6. Acknowledgment

This research was partially funded by the NSF Engineering Research Center, Advanced Technology for Large Structural Systems (ATLSS) at Lehigh University.

7. References

1. Fisher, J. W. (1984) Fatigue and Fracture in Steel Bridges, Wiley - Interscience, New York.
2. Fisher, J. W. and Demers, C.(1989) A Survey of Localized Cracking in Steel Bridges 1981 to 1988, Lehigh University, ATLSS Report No. 89-01.
3. Fisher, J. W., Pense, A. W. and Yen, B. T. (1983) Retrofitting Fatigue Cracked Bridge Structures, International Bridge Conference Proceedings, Pittsburgh.
4. Keating, P. (1989) Design of Fatigue Resistant Details, Proceedings - The National Symposium on Steel Bridge Construction, AISC.
5. Fisher, J. W., Hausammann, H., Sullivan, M. D., and Pense, A. W. (1979) Detection and Repair of Fatigue Damage in Welded Highway Bridges, Transportation Research Board, NCHRP #206.
6. Hausammann, H., Fisher, J. W., and Yen, B. (1983) Effect of Peening on Fatigue Life of Welded Details, ASCE Proceedings, W. H. Munse Symposium on Behavior of Metal Structures - Research to Practice.
7. Fisher, J. W., Barthelemy, B. M., Mertz, D. R., and Edinger, J. A. (1980) Fatigue Behavior of Full Scale Welded Bridge Attachments, Transportation Research Board, NCHRP #227.

8. Fisher, J. W. and Mertz, D. R. (1985) Hundreds of Bridges - Thousands of Cracks, ASCE - Civil Engineering, Vol. 5, No. 4.

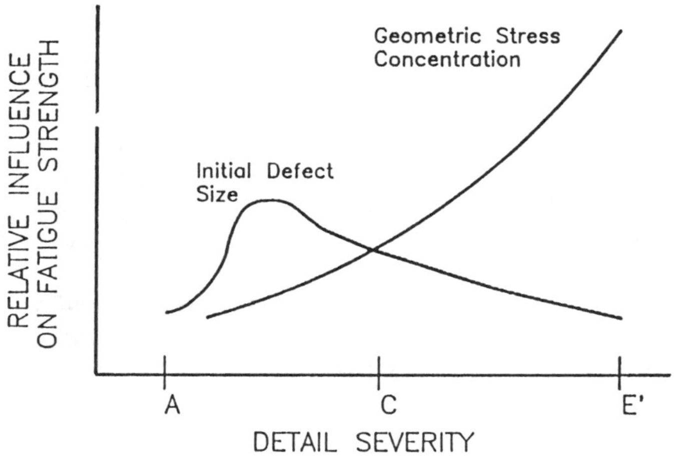

Fig. 1 Relationship Between Fatigue Strength, Detail Type and Defects.

Fig. 2 Cross-Section of a Peened Fillet Weld.

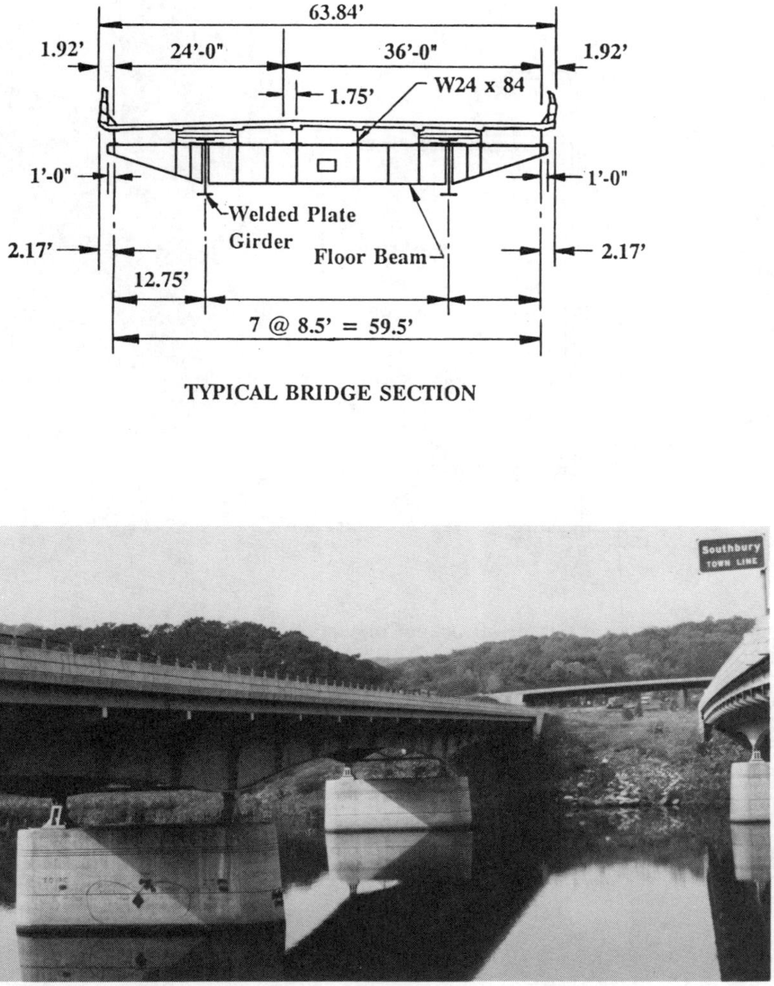

Fig. 3 Elevation and Section Views of the Housatonic River Bridge.

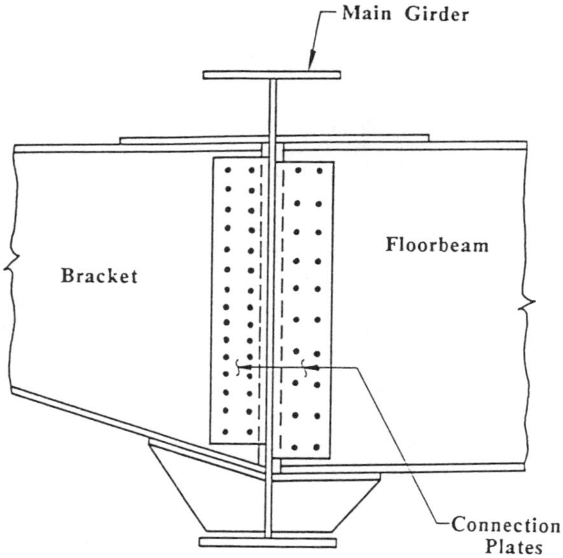

Fig. 4 A Typical Floor Beam - Girder - Bracket Connection.

Fig. 5 Fatigue Cracks Observed Along the Fillet Weld Toes of Connection Plates and Between the Web - Flange Junction of the Floorbeams.

Fig. 6 Cracks Observed Along Bolt Fixity Line on the Floorbeam Connection Plates.

Fig. 7 A Typical Fatigue Crack Found at the Ends of the Lateral Gusset Plate to Floorbeam Connections.

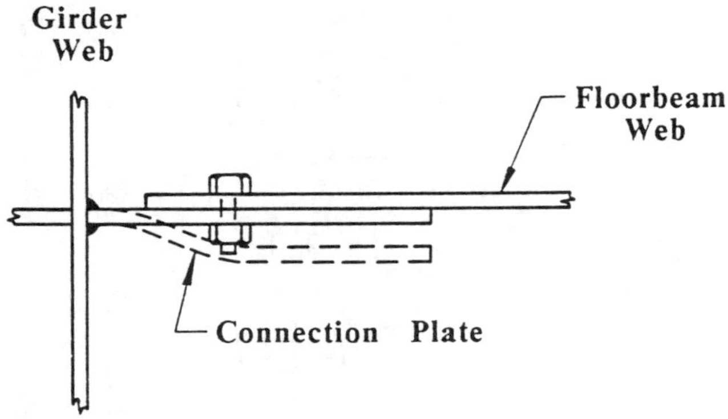

Fig. 8 Out-of-Plane Deformation of a Floorbeam Web and Connection Plate.

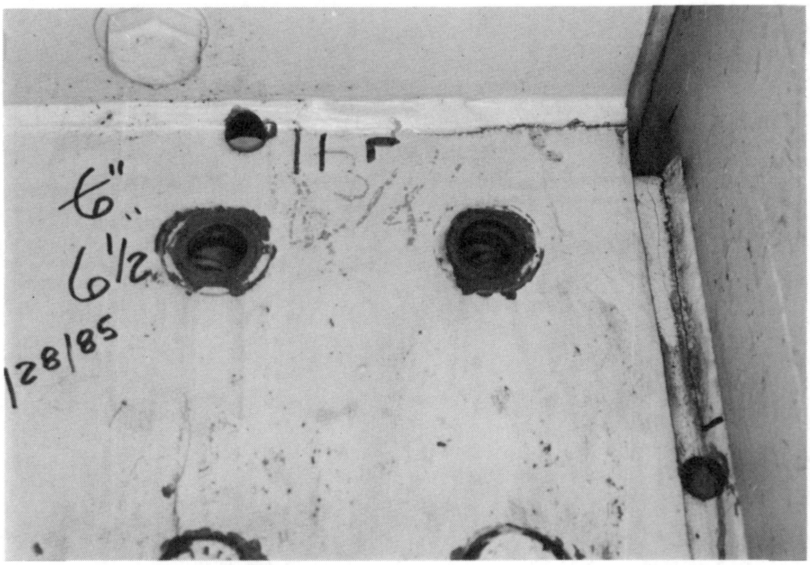

Fig. 9 Preliminary Retrofit of a Floorbeam - Girder Connection, Showing Bolt Removal and Holes Drilled at the Crack Tips.

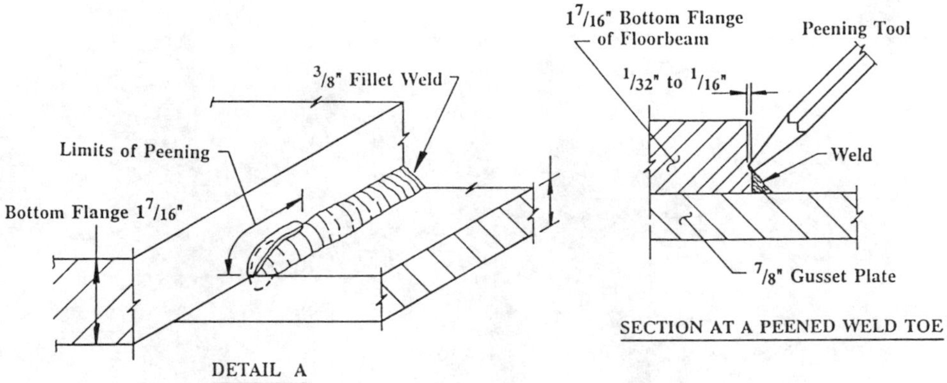

Fig. 10 Peening Procedure for the Lateral Gusset Plate Connections.

Fig. 11 Final Floorbeam - Girder Connection Repair.

Fig. 12 Elevation View of the Structure.

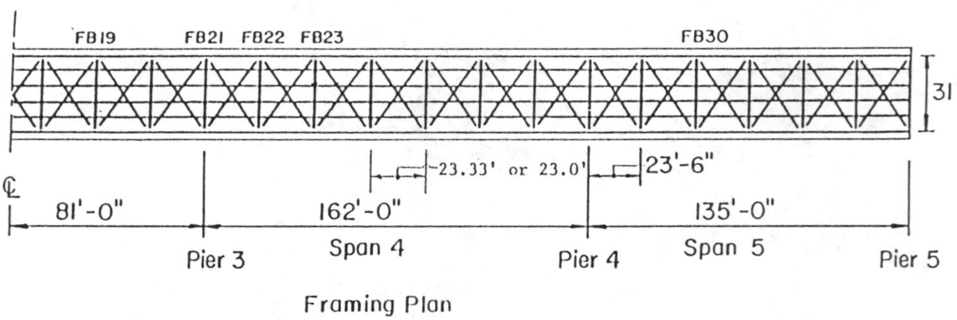

Fig. 13 Typical Portion of the Framing Plan.

Fig. 14 A Typical Lateral Bracing Connection.

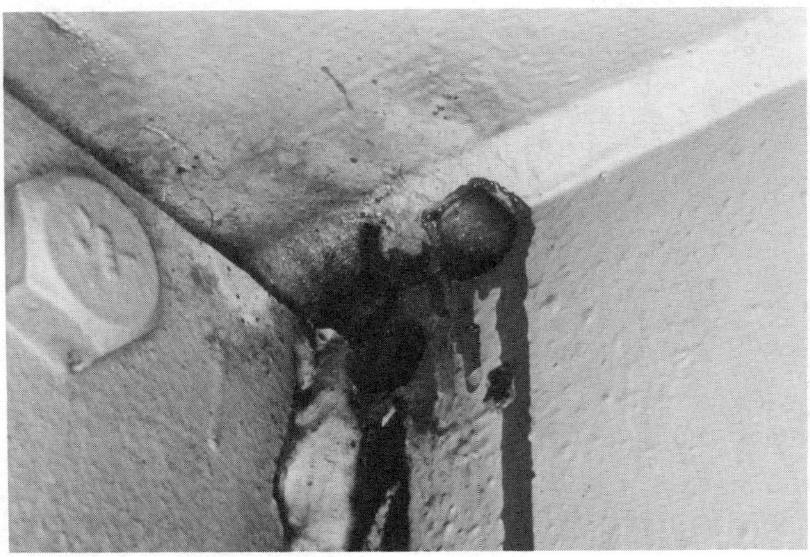

Fig. 15 Retrofit Holes for Fatigue Cracks Found in the Vertical Gap Between the End of Floorbeam Connection Plates and the Girder Flange.

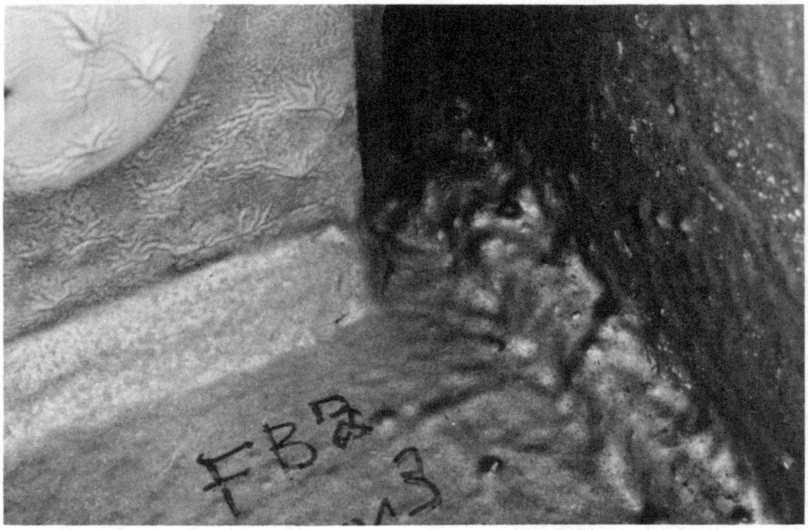

Fig. 16. Cracking in the Horizontal Gap Between the Transverse Connection Plates and the Lateral Tabs.

Fig. 17 Vertical Cracking Observed on the Outside Face of the Girder Web Corresponding to the Inside Horizontal Web Gap.

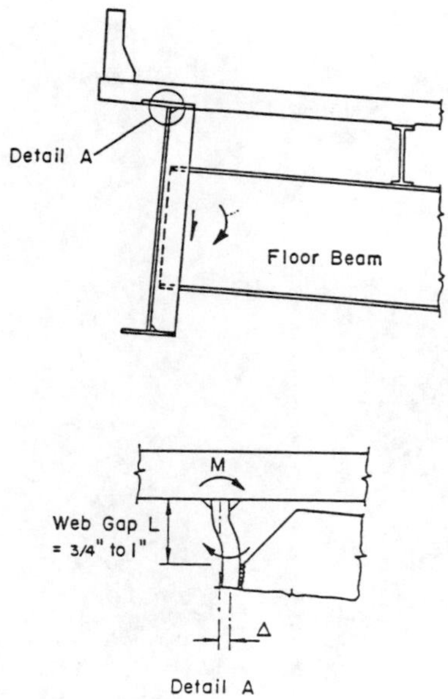

Fig. 18 Example of Web Gap Distortion.

Fig. 19 Positive Attachment Retrofit Between the Floorbeam Connection Plate and Girder Flange.

Fig. 20 Retrofit of a Lateral Gusset Plate.

STRUCTURAL STRENGTHENING BY EXTERNAL PRESTRESSING

J.L. TRINH
C.E.B.T.P
B.P. 37
78470 Saint-Rémy-lès-Chevreuse, FRANCE

ABSTRACT. This paper deals with some examples of activities of the Service d'Etudes des Structures (C.E.B.T.P., France) related to the problem of bridge rehabilitation. This involvement is approached on two levels: a) applied research works carried out in laboratories, and b) investigations performed on some actual bridges, in view of their strengthening by use of external additional prestressing.

1. Studies in Laboratories

Knowledge gained from research works is a source of information of great interest for improving bridge repair techniques. Several test programmes were carried out or are still ongoing in the C.E.B.T.P.'s Centre (Saint-Rémy-lès-Chevreuse, France) on the topic of external prestress, namely:

- fundamental examination of the specific behavior of such a type of structure, or
- check of innovative solutions elaborated by contractors.

As examples of the last case, within R&D projects on innovative types of composite (steel-concrete) structures the application of external prestressing was studied in particular.

- the ability to anchor the prestressing tendons at node junctions of steel truss webs with concrete flanges (Dragages-SGE's project [1]).
- the long-term redistribution of prestressing force inside a composite section, steel plated webs between two concrete flanges (Fougerolle's project [2]).
- the benefit of a longitudinal prestress on serviceability behavior and fatigue resistance of continuous bridges of current composite cross-section (steel caisson with an upper concrete flange).

1.1 RESEARCH PROJECT

More directly related to bridge repair is a part of an extensive experimental programme on externally prestressed structures [3,4]. Single-span beams were initially prestressed with internal cables, then afterwards strengthened with external cables, before being submitted to load testing until failure. The aim was to examine their flexural behavior and available resistance

A. S. Nowak (ed.), Bridge Evaluation, Repair and Rehabilization, 513–523.

(Figure 1). Comparison is achieved with various associated figures as they are defined in Table 1.

The retained parameters are:

- the construction technique of the test elements, monolithically or segmentally cast;
- the reinforcement; prestressing cables are totally inside or partly, or completely outside the concrete; and (bonded) reinforcing steel bars are present or not (which is the important practical case at construction joints of precast segments.

1.2 SOME MAIN RESULTS

- With the increase of the ratio "external cables quantity" over "quantity of internal bonded cables", beyond a certain limit:
  - * the ultimate ductility decreases;
  - * the maximum tension effectively reached in the external reinforcement is lower. Thus the flexural resistance is somehow reduced, this is explained by the second order moment (decrease of the lever arm of external prestress tendons).
- Another important point is the concentration of compressive strain near the edge of construction joints in precast segments. Strains are significantly higher than in the middle part of the segments, hence the concrete crushes at the failure of the models.
- Examples of crack pattern observed for a few cases from Table 1 are shown in Figure 1. a) The segmental beam NM4 is prestressed with external and internal bonded tendons; crack width is thus mainly concentrated in joint sections. b) Beam NM 9 is monolithic but fully externally prestressed and contains no significant reinforcing steel; crack pattern is similar to the former figure. c) In the monolitic beam NM 10, which contains full external prestressing and a certain amount of reinforcing steel, the minimum steel percentage is about 0.55%, which corresponds to the "skin" percentage according to the French Rules BPEL.
- The difference in effect between grouting with cement-based material (bonded cables) and the use of wax (unbonded cables), is that the deformability is greater in the second case. This is due to greater slip of the cable itself relative to the duct at deviators, whereas with cement grouting most of the slip occurs between duct and deviator.

2. History Cases

To illustrate the C.E.B.T.P.'s experiences in this field of investigation of actual construction, two examples are presented below. Special emphasis is given to the particular approaches which were proposed and completed for each situation, and the monitoring techniques used for the assessment of remaining capacity.

2.1 MONITORING TECHNIQUES

Within this scope, special emphasis is given to two particular techniques: the arbalest technique and the stress release technique in concrete.

2.1.1 *Arbalest technique.* It is possible to evaluate the tension force in a tendon by applying the following principle, specially when perfect flexibility is assumed (Figure 2). The axial force P in the cable can be derived, from knowledge of the transversal deflection d caused by an applied force F, by means of the equation:

$$P = \frac{FL}{4d} \tag{4}$$

provided that d is small compared to the length L.

This technique was applied by the CEBTP as a means to estimate the static force in a tensioned member when obviously the access is possible.

In practice a bare steel bar or prestressing cable has in general a relatively low lateral rigidity, then equation (1) is still applicable. The problem becomes even more complex if the tendon is placed within a grouted duct. However, it is thought that the parameters d and F, are still relevant to characterize the axial force P. It is thus necessary to replace equation (1) by an empirical relationship obtained from a calibration test done in the laboratory (Figure 3). Good results and accuracy were generally achieved in past experience with this technique.

2.1.2 *Stress Release Technique.* The C.E.B.T.P. has gained some experience lately on application of this technique to evaluation of local permanent compression in concrete. Its principle consists in making a plane cut in concrete at the considered area by sawing perpendicularly to the compressive stress direction in order to unload material. A small proper flat hydraulic jack is inserted into the slot, to establish, at best, the initial strain field, the pressure needed is one which corresponds to the former compressive stress (Figure 4). However, the heat induced at sawing can highly alter the result accuracy if this aspect is not well-handled in practical application.

2.2 1st EXAMPLE: INVESTIGATION OF AN OLD STRUCTURE EXTERNALLY PRESTRESSED

The structure examined is a small continuous 3 span bridge over the Seine River (Figure 5). It was cast in-situ in the late fifties, and originally prestressed by means of longitudinal parallel wire tendons. The latter were placed inside the caisson-section, but along their most current parts the prestressing steel was not embedded in concrete. The anchorages were embedded in the bulky concrete end blocks. Thus the tensioning of those cables was completed afterwards by lateral deviation of the tendons per couples successively. The relevancy of the tension force was checked with the arbalest technique on separated wires.

Along the free length of cables, the protection provided against corrosion was a simple bituminous paint. After more than 30 years of service life, an inspection of the structure had shown some failures of prestressing reinforcement. It was caused by corrosion, and occurred at right sections of the upper deviators where water was infiltrating (Figure 6). The loss was visually estimated at about 15 to 20%. Small flexural cracks resulting from the prestress decrease were observed in the middle of the central span and in sections over the two intermediate supports.

The strengthening solution of the structure was based on the condition of a total recovering of the initial prestress by means of additional external cables. At the both ends of the structure transversal beams were thus built for the anchorage of the new tendons. Galvanized strands were used and were protected by a superficial grease layer. They were placed inside high density polyethylene ducts.

A question generally arisen at tensioning of new tendons concerns the level of additional prestressing force, in theory a right complement to the residual prestressing in old tendons is needed as concrete resistance still sets an upper bound. Moreover it is obviously economical to provide all prestressing steel at once. It is then decided to achieve in a first phase a sufficient prestress force to close all flexural cracks again (Figure 7). The prestress force will be completed later on as the loss of old steel will reach a certain threshold detected by regular surveys. The idea is quite admissible with the current development of demountable cables (for instance the Freyssinet anchorage system).

2.3 2nd EXAMPLE: PRESTRESS EVALUATION IN AN EXISTING STRUCTURE

In this case the owner had to meet a normal regular increase of service loads, then a regular strengthening of the structure was planned. This structure is comprised of a series of transversal two-span prestressed frames juxtaposed in parallel. They are laterally connected by a general upper reinforced concrete slab cast in-situ.

A first strengthening of prestress had been completed fifteen years ago by means of external additional prestressing; two tendons on both sides of each transverse girder. Although no sign of damage was indeed detected at the moment, due to the expectation of load increase the structure has to be strengthened again. Nevertheless there was a bound here also due to the concrete resistance presently available. Therefore the owner decided to carry out an investigation aiming at obtaining a better knowledge of the actual behavior and capacity of in-situ materials. The main points investigated were:

2.3.1 *Degree of Scattering in Concrete Strength.* An extensive investigation by means of the ultrasonic pulse velocity (u.p.v.) method was completed for many parts of the structure. Cores were also taken to the laboratory for loading tests, in order to properly calibrate the measurements done onsite.

2.3.2 *Measurement of Concrete Compression.* An estimate of the permanent concrete compressive stress, in particular that due to the prestressing action effect, was made by means of the stress release technique. Owing to the thermal effect induced by the concrete sawing, this particular application can provide good approximations at the moment, but it still has to be improved in accuracy.

2.3.3 *Loading Test of the Structure.* The structure was monitored at first at some particular sections with strain gauges, linear transducers for the measurement of displacement (deflections) and rotating transducers.

The suggested monitoring programme had produced overall fruitful results for the structural redesign. In particular it was found that the actual residual prestress is higher than expected (less prestress losses in the first additional cables as there is less concrete shrinkage and creep in old concrete). The strengthening project elaborated was similar to the solution adopted in the example formerly presented.

3. Conclusion

Through this review of some CEBTP experiences the author wishes to underline some general ideas on the subject of bridge rehabilitation.

The use of external additional prestressing is indeed appropriate for rehabilitation of concrete bridges, in particular to re-establish the serviceability behavior under the bending moment effect, although some limitations at ultimate states may be expected.

In this connection the ongoing basic research works may lead to fruitful contributions in improvement of this technique since bridge structures are very diverse.

When investigating works with the everyday devices, sometimes for particular purposes special ones are quite relevant, but they may need calibration tests in the laboratory for better accuracy during site measurement.

4. References

1. Lacroix, R. and Trinh, J. (1987) "Structures à âmes traingulées métalliques', Annales de p'ITBTP, Oct., Paris.
2. Attal, A. and Bertho, Ph. (1987) "Structures à âmes métalliques planes, (BEPAC) Expérimentation sur un caisson mixte précontraint', Annales de p'ITBTP, Oct., Paris.
3. Foure, B. and Virlogeux, M. (1986) "Précontrainte extérieure au béton, comportement jusqu'à rupture de poutres à voussoirs préfabriqués, Travaux, Janv., Paris.
4. Martins, P.C. (1989) "Modèlisation jusqu'à la rupture en flexion de poutres en béton à précontrainte extérieure ou mixte. Thesis presented to the Paris University (E.C.P.), Paris.

Table 1 Experimental Programme

	Prestress tendon lay-out				
	fully external			partly internal	fully internal
Duct	unbonded		cement grouted	cement grouted	
	empty	wax			
segmental	NM 1	3	2	4	5
Monolithic with(1)	/	9	6	7	/
Monolithic without(1)	/	11	8 & 10	/	/

(1) reinforcing steel.

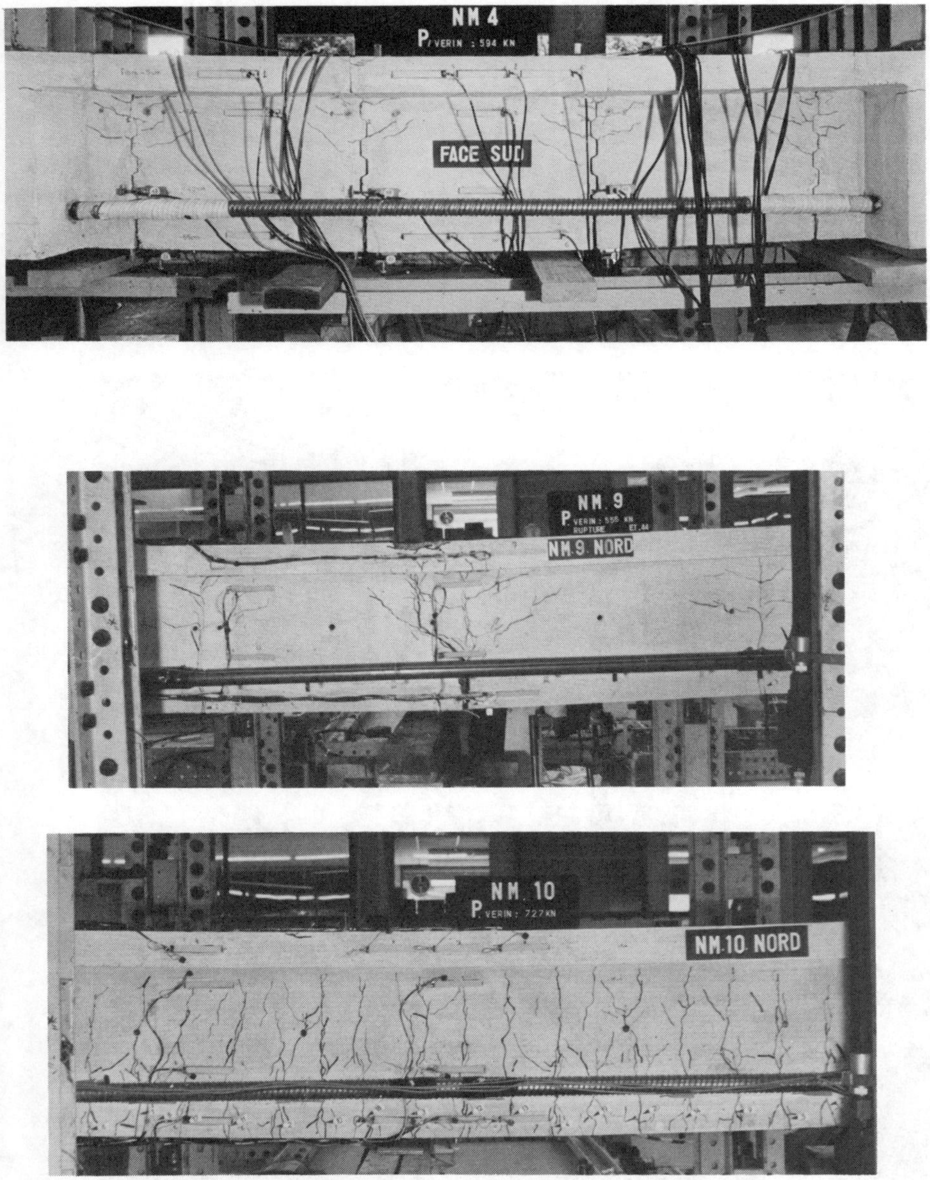

Fig. 1 Views of Flexural Failures of a Segmental or Monolithic Concrete Beam Externally Prestressed According to the Nature of Prestressing Reinforcement and Reinforcing Steel.

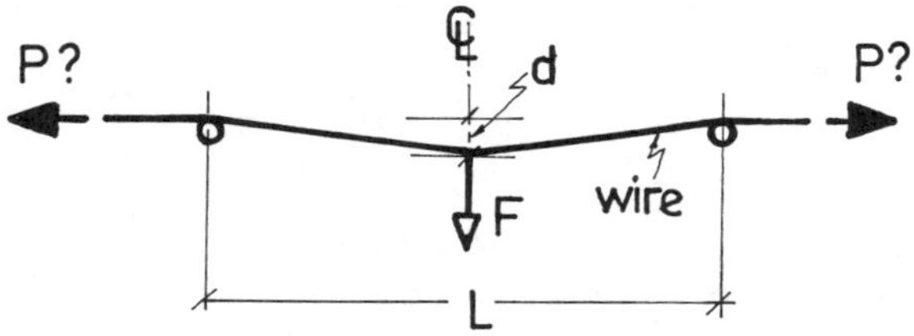

Fig. 2 Principle of the Arbalest Technique.

Fig. 3 Calibrating the Arbalest Technique in the CEBTP's Laboratory (Mr. Durand).

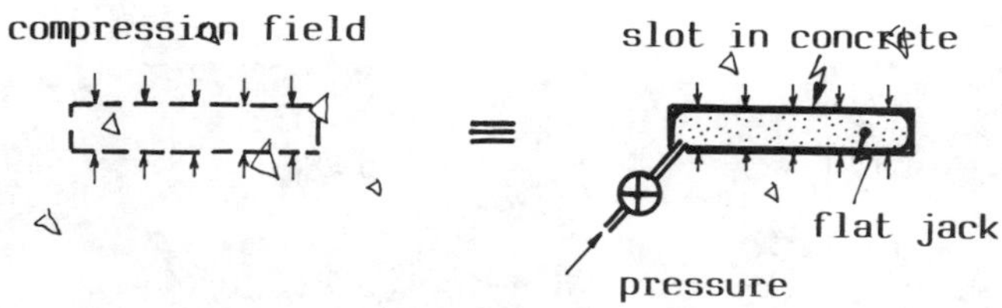

Fig. 4 Stress Release Technique in Compressive Concrete (Calibration in Laboratory, Mr. Chagnaud).

Fig. 5 Bridge Investigated in View of a Strengthening by External Prestress (First Example).

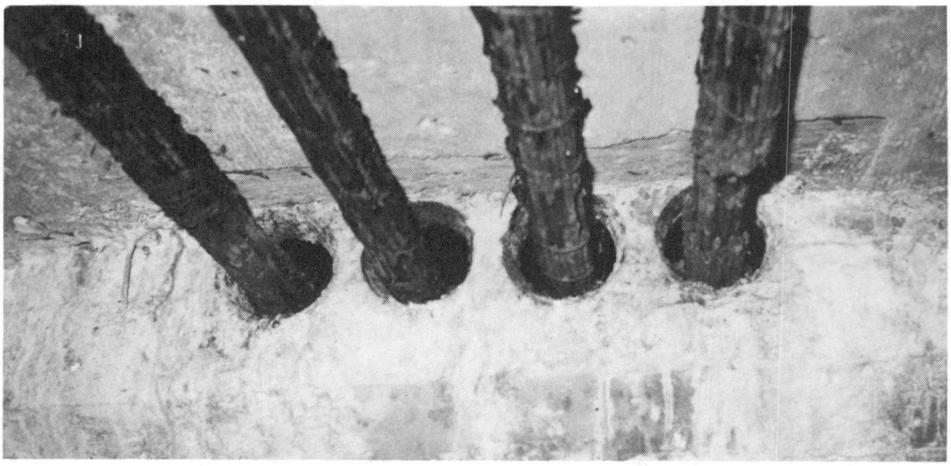

Fig. 6 Examples of Corrosion Damage on External Tendons Protected by a Bituminous Paint, After a 30 Year Service Life.

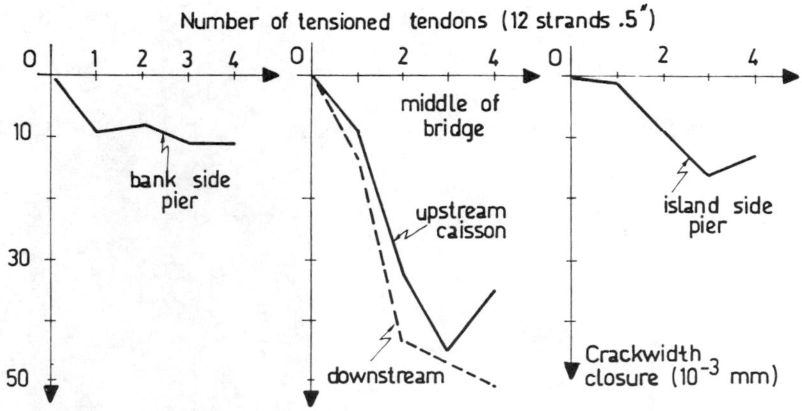

Fig. 7 Checking Crack Closure.

Fig. 8 View of the Additional External Cables.

STRENGTHENING OF A THREE-SPAN CONTINUOUS BRIDGE BY POST-TENSIONING

F. W. KLAIBER and K. F. DUNKER
Dept. of Civil and Constr. Engr.
Iowa State Univ.
Ames, IA 50011 USA

S. M. PLANCK
HDR Engineering
4922 Robert J.
Mathew Parkway
El Dorado, CA 95630 USA

ABSTRACT. The need to upgrade a large number of understrength and obsolete bridges in the United States has been well documented in the literature. Through several Iowa DOT projects, the concept of strengthening simple-span bridges by post-tensioning has been developed. The purpose of the project described in this paper was to investigate the use of post-tensioning for strengthening continuous composite bridges. In a previous, successfully completed investigation, the feasibility of strengthening a continuous, composite bridge by post-tensioning was demonstrated on a laboratory 1/3-scale-model three-span bridge. The project reviewed in this paper can thus be considered the field testing phase. With finite element analysis, a post-tensioning system was developed that required post-tensioning of the positive moment regions of both the interior and exterior beams. During the summer of 1988, the strengthening system was installed along with instrumentation to determine the bridge's response and behavior. In 1989, approximately one year after the initial strengthening, the bridge was retested to identify any changes in its behavior. Although considerably more involved than the post-tension strengthening of single-span bridges, post-tensioning of continuous-span bridges is a feasible, practical strengthening technique.

1. Introduction

According to the Federal Highway Administration, almost 40% of the nation's bridges are classified as deficient and in need of rehabilitation or replacement. Many of these bridges are deficient because their load-carrying capacity is inadequate for today's traffic. Strengthening can often be used as a cost-effective alternative to replacement or posting for restricted loads.

The concept of strengthening single-span, composite steel-beam and concrete-deck bridges by post-tensioning has been developed by the senior authors through several Iowa DOT research projects. The research on strengthening single-span bridges was conducted in three phases:

- Phase I: This feasibility investigation involved the testing of numerous post-tensioning schemes on a 1/2-scale-model bridge [1,2].

A. S. Nowak (ed.), Bridge Evaluation, Repair and Rehabilization, 525–538.

- Phase II: This field study involved the strengthening by post-tensioning and testing of two existing bridges [3,4].
- Phase III: This implementation phase involved the development of a design manual for use by practicing engineers [5,6].

Since the completion of the design manual [5] in 1985, the Iowa DOT has used the allowable stress design methodology for the post-tension strengthening of several single-span bridges. The procedure is now one of the methods in regular use by the Iowa DOT for single-span bridge strengthening.

As a result of the previous success with post-tension strengthening of single-span composite bridges, the senior authors began a research program in 1986 for continuous, composite bridges that was similar to the program for single-span bridges. In Phase I [7], it was verified that continuous bridges can be strengthened by post-tensioning. In most bridges, the desired stress reduction in the positive moment regions as well as in the negative moment regions can be obtained by post-tensioning only the positive moment regions. This was determined theoretically by using a finite element analysis and experimentally by testing various post-tensioning schemes on a 1/3-scale, three-span, continuous bridge.

This paper will review Phase II [8] of this investigation in which a three-span continuous bridge was strengthened by post-tensioning and tested. The primary goals of this study were to design and install post-tensioning on a continuous-span, steel-beam, concrete-deck bridge; instrument the bridge for measurement of deflections and strains; and document the bridge's behavior for a one-year period after the installation of the post-tensioning strengthening system.

2. Description of Bridge

An advisory committee assisted in selecting a three-span continuous bridge in northwest Iowa. The framing plan and a section of the bridge are shown in Fig. 1, while a photograph of the bridge is shown in Fig. 2. The bridge has a total length of 150 ft. (45.72 m), two end spans of 45 ft. 9 in. (13.94 m), and a center span of 58 ft. 6 in. (17.83 m). The four bridge beams are coverplated top and bottom near each of the two piers and are spliced at the center-span nominal dead-load inflection points.

An analysis of the bridge indicated that the bridge required strengthening both to reduce flexural stresses and to improve shear connection for composite action. The post-tensioning scheme designed for the bridge in question involved post-tensioning the positive moment regions of all beams; shown in Fig. 3 are the theoretically required tendon forces. The post-tensioning forces shown in Fig. 3 theoretically will reduce all steel beam stresses below operating load levels, thus removing the need for load posting; in all but a few isolated locations (piers and the immediate vicinity of the brackets), steel beam stresses are actually below inventory stress levels. Note that the forces in Fig. 3 have been modified to account for losses due to tendon relaxation and potential temperature differences.

The layout of the post-tensioning scheme, with brackets and tendons, is given in Fig. 4. For each span, a set of two 1 and 1/4-in.-diameter (37.75 mm) Dywidag threadbar tendons, at less than the manufacturer's rated capacity, provide the required design force. Shown in Fig. 5a is bracket A, which was used in the end spans; this bracket is similar to the bracket used by the authors for post-tensioning single span bridges [3,5]. Bracket B, shown in Fig. 5b, was used in the center span. Because this bracket was bolted to the bottom flange of the beam, stiffeners were added to the beam web at each end of the bracket to prevent an overstress in the beam web.

As previously noted, the bridge did require additional shear capacity. Because the number of load cycles on the bridge was unknown, only strength could be used to determine the amount of additional shear capacity required. The additional capacity was provided by 1-in.-diameter (25.4 mm), high-strength bolt connectors of the type tested and used by the authors in single-span bridges [3]. Added to each interior and exterior beam were 58 and 52, respectively, high-strength bolt shear connectors (220 total on the bridge).

3. Field Testing Program

The bridge in question was initially strengthened and tested during the summer of 1988; approximately one year later the bridge was retested. Except for removing and reapplying the post-tensioning forces the second year, the testing program employed each year was essentially the same. Each year the bridge was subjected to the following four loading conditions to determine its response, strains, and longitudinal and vertical displacements:

1. An overloaded truck at various predetermined locations on the bridge.
2. Various stages of the post-tensioning sequence. Because all 12 beams of the bridge required post-tensioning (2 tendons per beam), and there were only 4 hydraulic cylinders available, it was necessary to post-tension the bridge in 6 stages.
3. The same overloaded truck at the same locations, after post-tensioning of the bridge was completed, to determine the effectiveness of the strengthening system.
4. Two overloaded trucks at various predetermined locations on the bridge, to maximize the moments at various locations.

Although there were some modifications in the instrumentation used in 1988 and 1989, the instrumentation used was essentially the same. Instrumentation for the field tests consisted of electrical-resistance strain gages, direct-current displacement transducers (DCDT's) and crack monitors. Strain gages (two gages per location) were mounted on the lower flanges of all beams near the centerline of each span and near the supports (15 in. from the centerline of bearing at the piers and at the end abutments); thus, there were 14 strain gages per beam. Strain gages were also mounted on the post-tensioning tendons so that the post-tensioning forces could be accurately measured. Deflections of the bridge were determined using DCDT's and crack monitors. The DCDT's were mounted at the center of each beam in each of the 3 spans to determine vertical displacements. Crack monitors (16 total) were positioned to measure the movement (vertical and/or horizontal) of the 4 beams at each support point. The data from the strain gages and DCDT's were recorded by an automatic data acquisition system.

4. Test Results

Because of the page limit, only a portion of the results of this study will be presented in this paper. Details of the results are presented in Ref. 8. As has been previously noted, the bridge was tested in 1988 and 1989; although there were some differences in the weights of the trucks used each year and some differences in the post-tensioning forces applied to the bridge each year, there was still very good agreement between the data obtained in 1988 and 1989. Because of that agreement, this paper presents only 1989 data.

As previously noted, post-tensioning was applied to the bridge in six stages. In Fig. 6, the magnitude and sequence of post-tensioning forces applied are presented. The effect of post-tensioning two beams on the post-tensioning force on beams that have been previously post-tensioned is apparent. This effect is more significant on beams in the same span and on beams in adjacent spans that are "in line" with those being post-tensioned. As may be seen, the greatest loss was 5.9%, while the greatest gain was 2.9%.

Final midspan deflections (south, middle, and north spans) resulting from post-tensioning the bridge are presented in Fig. 7. Also presented in this figure are theoretical deflection curves, which were obtained from finite element analysis, assuming that the ends of the bridge were simply supported or fixed. In the end spans (Fig. 7a and c) the measured deflections were bracketed by the simply supported and fixed-end conditions, indicating the presence of end restraint. In the middle span, measured deflections are relatively close to the theoretical curve, assuming simple support conditions; however, they are less than the theoretical deflections obtained by assuming either end condition. Other than experimental error, the only explanation the authors have for this discrepancy is the contribution of the guardrails.

Shown in Fig. 8 are the bottom-flange strains in an interior beam (Fig. 8a) and an exterior beam (Fig. 8b) resulting from the post-tensioning (all 6 stages completed). Also shown in these figures is the theoretical variation (obtained from a finite-element-analysis) of bottom-flange beam strains assuming simple supports. Except for the midspan, bottom-flange, middle-span strains, there is excellent agreement between the experimental and theoretical strains. Although only the positive moment regions were post-tensioned, strain reduction in the negative moment regions is readily apparent. Also apparent is the presence of considerably more end restraint at the north abutment (pinned end) than at the south abutment (roller support).

Figure 9 illustrates the middle-span east exterior beam and east interior beam tendon force changes as a truck moves across the bridge in two lanes (lane 1 north; lane 4 south). A review of these two figures indicates that the maximum increase in the tendon force in the exterior beam is 10.1%, while the maximum decrease is -2.5%. For the interior beam the maximum percentages of increase and decrease are 6.5% and -1.7%, respectively. The figures verify that the change in tendon force in continuous bridges is a function of both lane and span loading. This is obvious when one considers the deflection patterns that result from the various positions of the truck loading.

Deck crack patterns were observed and mapped before the 1988 post-tensioning, one month after the 1988 post-tensioning, and at the time of the 1989 post-tensioning. Deck cracking increased significantly at each observation; in some instances the final crack length was about three times the initial crack length. The change in crack patterns correlated well with computed stresses in the deck after post-tensioning.

5. Summary and Conclusions

This paper summarizes the research that has been completed in the field testing phase of the investigation of strengthening of continuous, composite bridges by post-tensioning. This procedure has been found (as a result of four years of investigation) to be a viable, economical strengthening technique. Design of the strengthening system for continuous bridges is, however, more complex than for single-span bridges. Note that in some continuous bridges it may be necessary to post-tension the negative moment regions in addition to the positive moment regions to achieve the desired stress reduction. Field testing verified longitudinal distribution of post-tensioning; post-tensioning applied in the positive moment regions is effective in reducing flexural stresses in both the positive and negative moment regions. Deck cracking increased in the strengthened bridge; the possibility of this occurring in a bridge being strengthened by post-tensioning should thus be considered.

6. Acknowledgments

The research reviewed in this paper was conducted by the Bridge Engineering Center, Engineering Research Institute of Iowa State University, and was funded by the Highway Research Advisory Board and the Highway Division, Iowa Department of Transportation, Ames, Iowa. The authors thank the engineers at the Iowa DOT, especially William A. Lundquist, Bridge Engineer, and John P. Harkin, Chief Structural Engineer, for their support, cooperation, and counsel. Special thanks are accorded the numerous students who assisted with the various phases of the project. The opinions, findings, and conclusions expressed herein are those of the authors and not necessarily those of the Iowa Department of Transportation or the Highway Research Advisory Board.

7. References

1. Klaiber, F. W., Dunker, K. F., and Sanders, W. W., Jr. "Feasibility Study of Strengthening Existing Single Span Steel Beam Concrete Deck Bridges." Final Report, ERI Project 1460, ISU-ERI-Ames 81251, Engineering Research Institute, Iowa State University, June 1981, Ames, Iowa, 141 pp.
2. Klaiber, F. W., Dunker, K. F., and Sanders, W. W., Jr. "Strengthening of Single-Span Steel-Beam Bridges." Journal of the Structural Division, ASCE, V. 108, No. 12, Dec. 1982, pp. 2766-2780.

3. Klaiber, F. W., Dedic, D. J., Dunker, K. F., and Sanders, W. W., Jr. "Strengthening of Existing Single Span Steel Beam and Concrete Deck Bridges." Final Report--Part I, ERI Project 1536, ISU-ERI-Ames 83185, Engineering Research Institute, Iowa State University, Feb. 1983, Ames, Iowa, 185 pp.
4. Dunker, K. F., Klaiber, F. W., Beck, B. L., and Sanders, W. W., Jr. "Strengthening of Existing Single-Span Steel-Beam and Concrete Deck Bridges." Final Report--Part II, ERI Project 1536, ISU-ERI-Ames 85231, Engineering Research Institute, Iowa State University, March 1985, Ames, Iowa, 146 pp.
5. Dunker, K. F., Klaiber, F. W., and Sanders, W. W., Jr. "Design Manual for Strengthening Single-Span Composite Bridges by Post-Tensioning." Final Report--Part III, ERI Project 1536, ISU-ERI-Ames 85229, Engineering Research Institute, Iowa State University, March, 1985, Ames, Iowa, 102 pp.
6. Dunker, K. F., Klaiber, F. W., and Sanders, W. W., Jr. "Post-Tensioning Distribution in Composite Bridges." Journal of Structural Engineering, ASCE, V. 112, No. 11, Nov. 1986, pp. 2540-2553.
7. Dunker, K. F., Klaiber, F. W., Daoud, F. K., Wiley, W. E., and Sanders, W. W., Jr. "Strengthening of Existing Continuous Composite Bridges." Final Report, ERI Project 1846, ISU-ERI-Ames 88007, Engineering Research Institute, Iowa State University, July, 1987, Ames, Iowa, 220 pp.
8. Klaiber, F. W., Dunker, K. F., Planck, S. M., and Sanders, W. W., Jr. "Strengthening of an Existing Continuous-Span, Steel-Beam, Concrete-Deck Bridge by Post-Tensioning." Final Report, ERI Project 3030, ISU-ERI-Ames 90210, Engineering Research Institute, Iowa State University, Feb. 1990, Ames, Iowa, 165 pp.

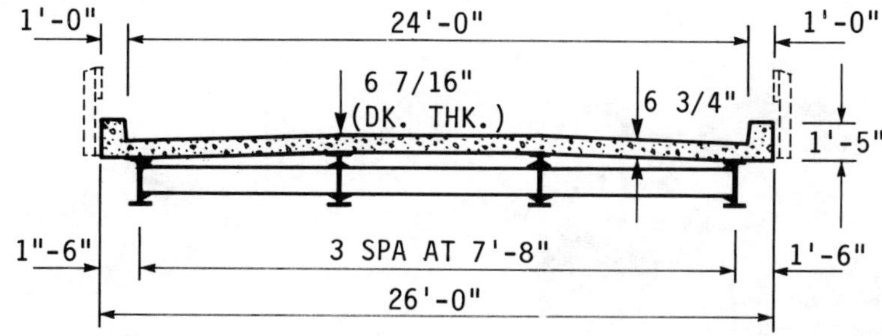

a. BRIDGE SECTION

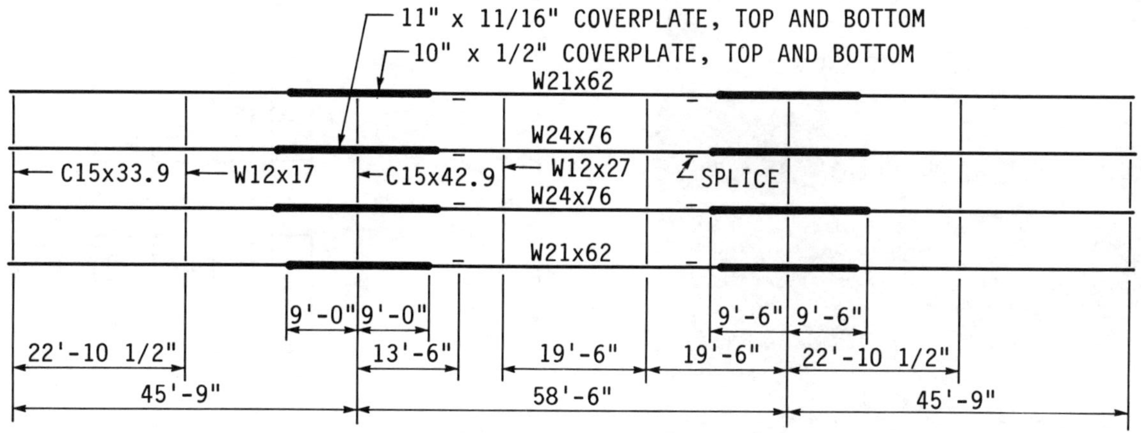

b. BRIDGE FRAMING PLAN

Fig. 1. Framing plan of bridge (1 ft = 0.305 m).

Fig. 2. Photograph of bridge.

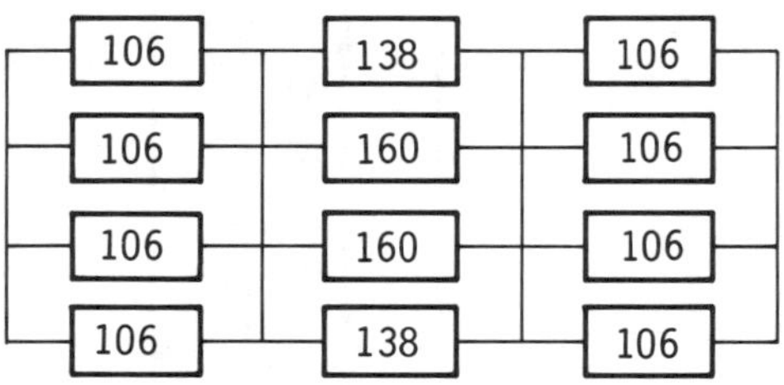

Fig. 3. Post-tensioning force theoretically required per beam (1 kip = 4.45 kN).

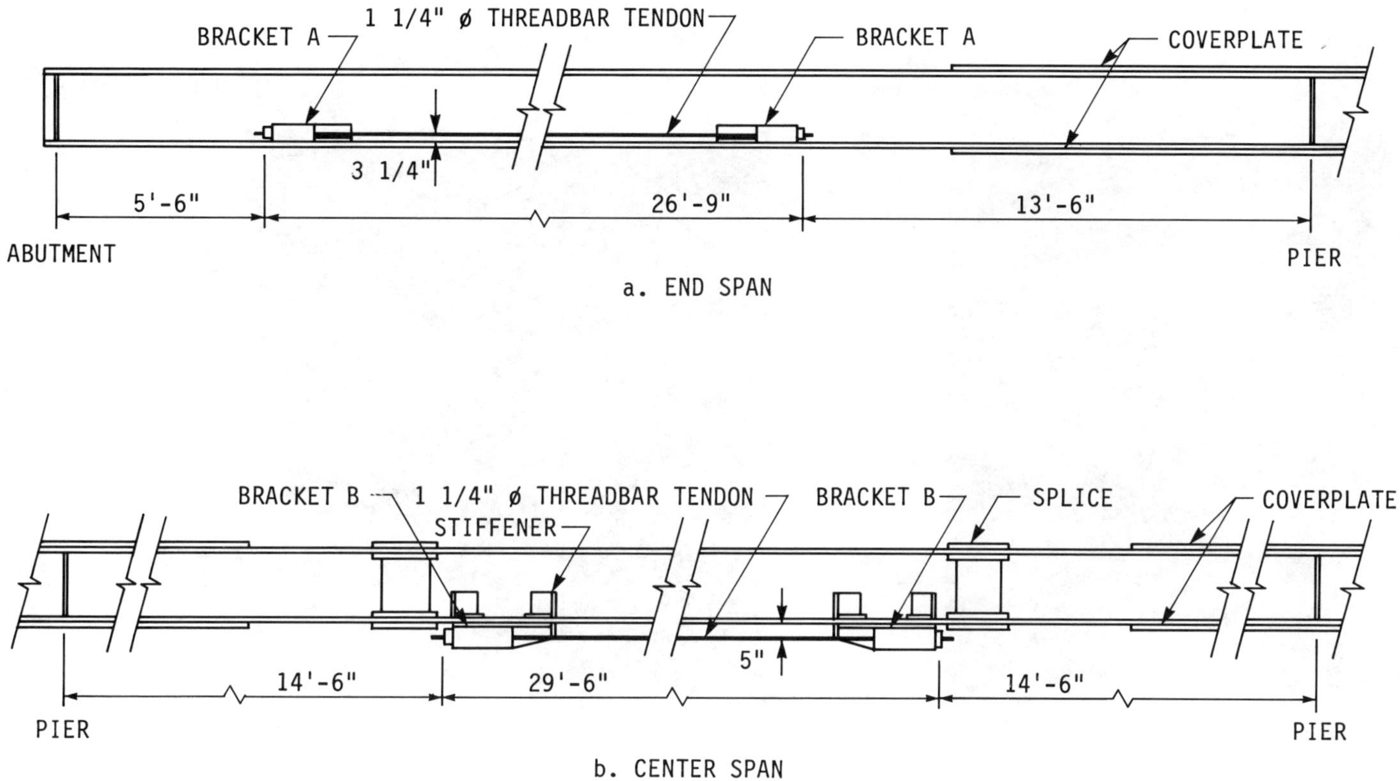

Fig. 4. Post-tensioning layout (1 ft = 0.305 m).

a. BRACKET A IN END SPAN

b. BRACKET B IN CENTER SPAN

Fig. 5. Post-tensioning system in place.

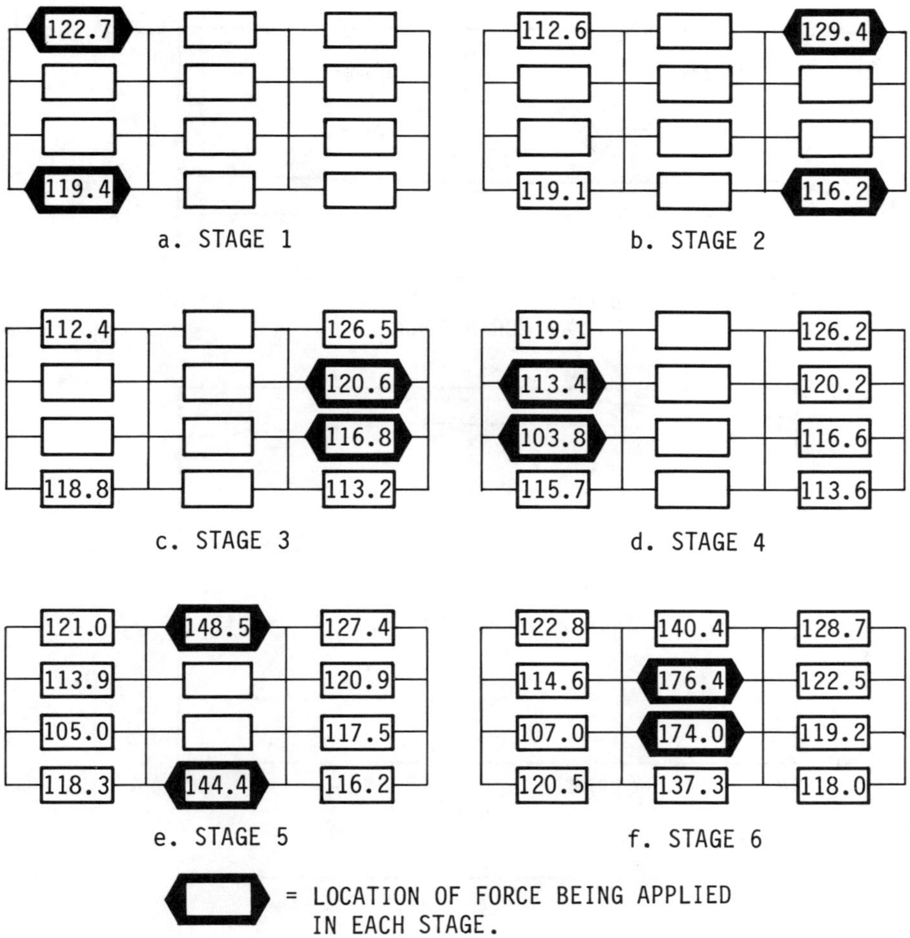

Fig. 6. Post-tensioning forces applied to each beam: force in kips (1 kip = 4.45 kN).

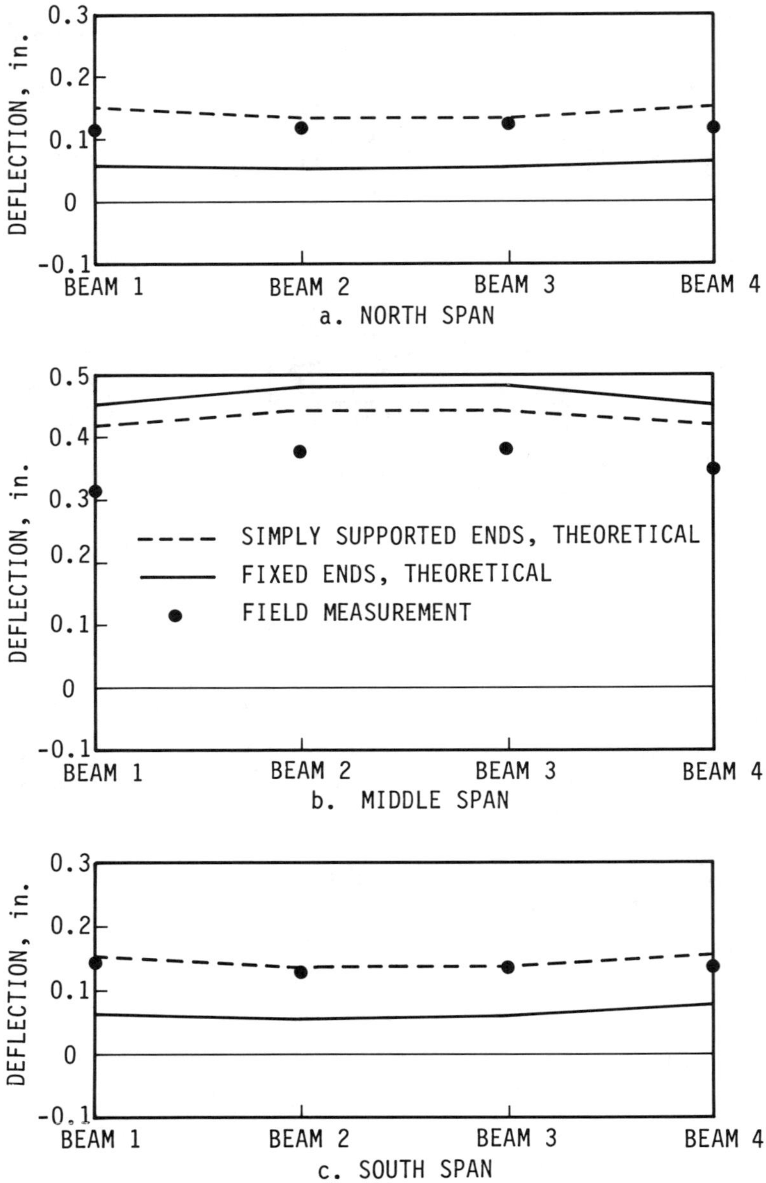

Fig. 7. Comparison of experimental and theoretical post-tensioning deflection data (1 in. = 2.54 cm).

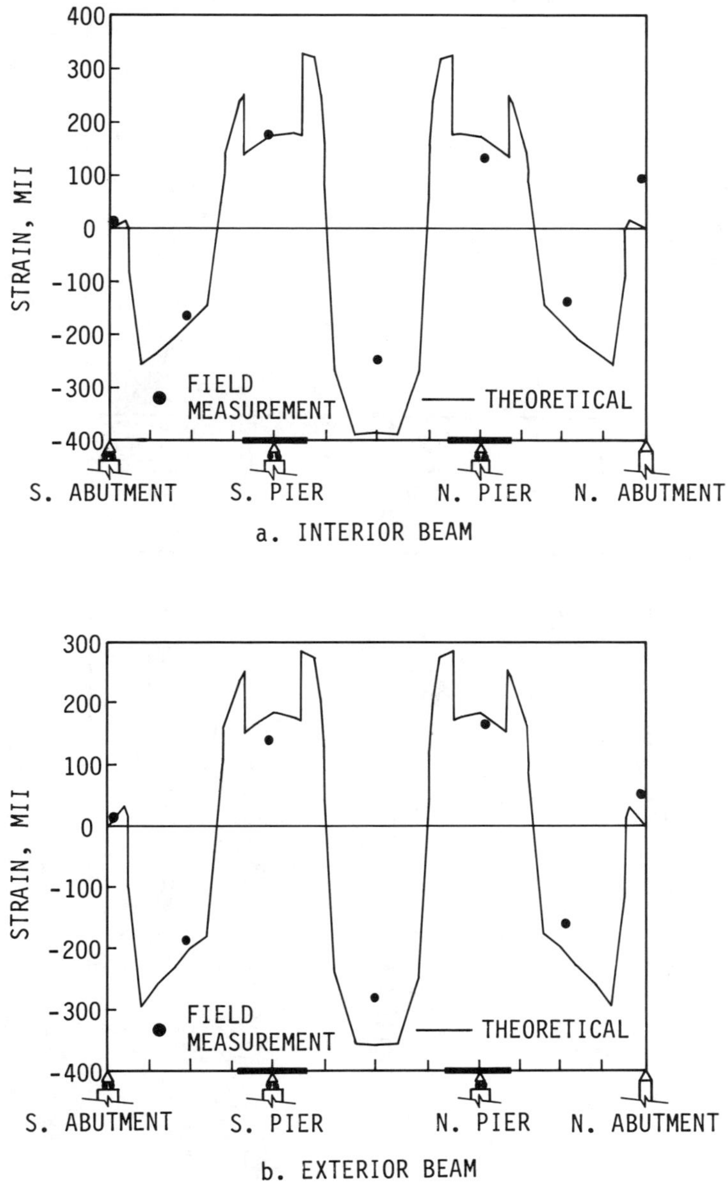

Fig. 8. Post-tensioning bottom flange beam strains.

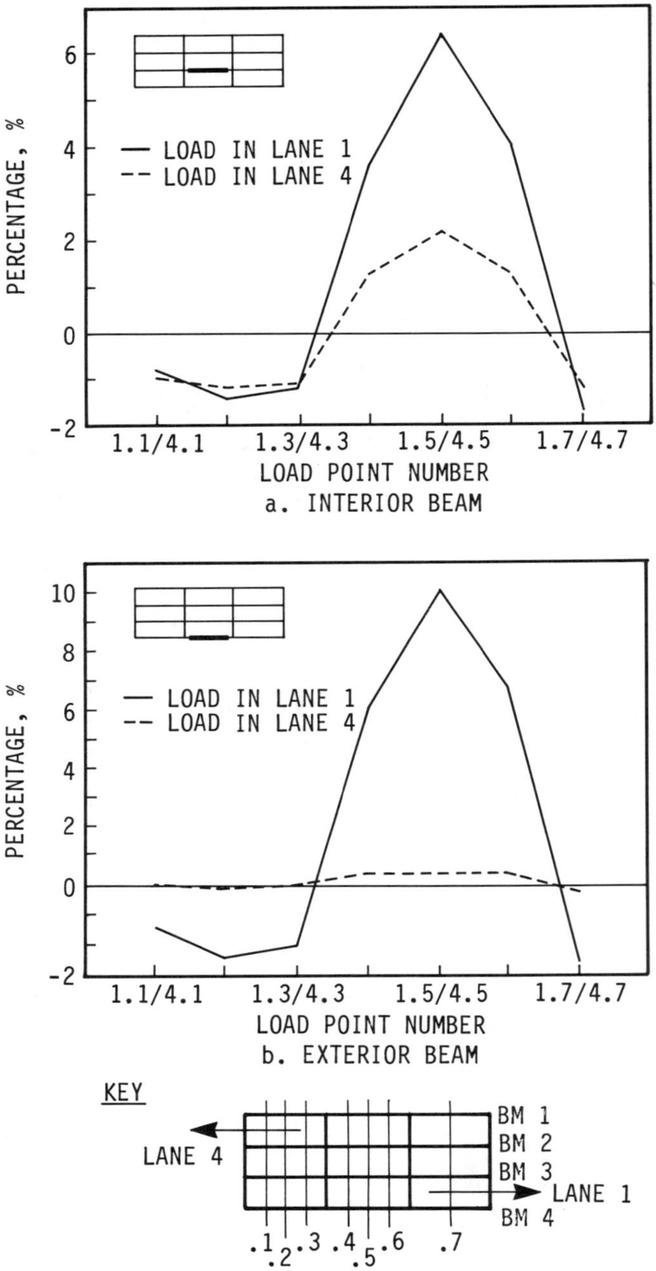

Fig. 9. Change in middle span tendon force due to vertical loading.

CATHODIC PROTECTION OF REINFORCED CONCRETE STRUCTURES - EXPERIMENTAL SYSTEM INSTALLED ON A VIADUCT IN SERVICE ON THE IRI-ITALSTAT NETWORK

M. ROMAGNOLO and S. FERRARI
Societa Autostrade S.p.A.
Roma 00159, ITALY

MAURIZIO GRANDI
Strada Anulare
Milano 20090, ITALY

ABSTRACT. Cathodic protection is a very effective system for keeping structure electric potentials at such levels as in prevent development of corrosion phenomena. The AUTOSTRADE Company was the first in Italy to install an experimental system of this nature, adopting the most advanced implementation types and technologies (computer-controlled monitoring, photovoltaic feed and special coating grouts). Significant advantages of the cathodic protection system include partial purging of de-icing salts from concrete, protection from possible stray currents, and elimination of the need to waterproof surfaces exposed to aggressive agents, as well as easy and timely quality control of structure protection.

1. Types of Deterioration

Deterioration of reinforced concrete structures due to external causes depends essentially on environmental-type factors and actions induced by traffic. Those most closely related to the topic under examination here are mainly of the former type. It is thus useful to recall for a moment the following specific phenomena:

- freeze/thaw cycles: these cause deterioration of concrete by increasing the volume (9%) due to the moisture present;
- action of sulphates: present in the water and frequently in the aggregates, these react with the cement mixtures, forming crystals of greater volume, resulting in a breakdown of the concrete;
- action of chlorides: abundantly present in de-icing salts, and occasionally also in the aggregates, these penetrate into the cement and, reaching the reinforcing, cause depassivation and hence the commencement of corrosion;
- action of CO_2: in the presence of water, carbon dioxide transforms calcium oxide contained in the cement into calcium carbonate, thus lowering the pH;
- leaching: dissolves and carries away soluble salts present in the cement and thus increases porosity;

A. S. Nowak (ed.), Bridge Evaluation, Repair and Rehabilization, 539–554.

- alkali-silica reaction: this occurs between the alkalies (Na, K) in the cement and certain types of aggregates containing amorphous or imperfectly crystallized silica (opal, flint), giving rise to an expansive gel.

In reality these actions do not occur in isolation, and their effects are amplified by those due to the dynamic actions of traffic, (cracking, fatigue). Moreover, the complex mechanism of deterioration evolves in increasingly rapid fashion in real time, such as to necessitate timely implementation of repair measures.

2. Electro-Chemical Mechanism of Corrosion

At the time of construction the metal reinforcing, embedded in a concrete free of chlorides, benefits from alkaline environment (pH = 12.5 - 13.5) which permits the formation of a film of passivating oxides which act to preserve it from corrosion. Over time, however, the evolution of the degradation phenomena described above results in lowering the pH below the protection threshold (pH = 11.5), favouring the onset of corrosive phenomena. The electrochemical mechanism which develops is the result of three processes.

- anodic, with oxidation of the steel and liberation of electrons ($2\ Fe = 2\ Fe^{++} + 4e$);
- cathodic, which consumes said electrons for the reduction of the oxygen dissolved in the water ($O_2 + 2H_2O + 4e = 4OH^-$);
- passage of current from the anode to the cathode in the concrete and in the opposite direction in the reinforcing.

The resulting overall corrosion reaction is

$$2\ Fe + O_2 + 2H_2O = 2\ Fe^{++} + 4OH^- = 2Fe(OH)_2$$

which leads to the formation of iron hydroxide, and this latter is subsequently transformed, with progressive increases in volume, as illustrated in Figure 1.

3. Mapping of Potential

To check the extent to which corrosion has spread through the reinforcing in reinforced concrete structures, one can apply a nondestructive technique better known as potential mapping. This method can predict whether the reinforcing rods are in ideal condition (i.e. passive) or if they are suffering corrosion attack; on the other hand, it does not provide any quantitative information regarding the speed of the corrosion process or on the remaining thickness of the reinforcing rods. Using a high-impedance voltmeter, this method permits measurement of the potential present locally in the reinforcing with respect to a half-cell potential at the nodes of a conventionally-defined ideal grid of 0.5 m per side. Prior to installing the system it is necessary to verify, for each span, that there is electrical continuity among all pieces of reinforcing present, whether old or new. The measurements obtained on the data logger are then fed into specially-designed PC programs which convert the data, for each structural element and part thereof, in the form of a contour map showing the potential and the distribution of the number of measurements by interval of potential (see Figure 2).

The interpretation of the measurements is strictly correlated to the electro-chemical mechanism of the corrosion phenomena, and the key for reading the potentials in relation to corrosion is as follows:

- up to -200 mV (with respect to the Ag/AgCl) passive reinforcing not suffering corrosion;
- from -200 mV to - 350 mV (with respect to the Ag/Agcl) situation of doubtful corrosion (50% probability);
- over - 350 mV (with respect to the Ag/AgCl) 95% probability of corrosion.

This key is based on the indications of ASTM standard C876-80, to which the reader is referred for a complete description of the method.

4. Constituent Elements of the Cathodic Protection System

A cathodic protection system requires the following basic components:

- insoluble disperser element to act as anode;
- power supply system;
- reference electrodes to measure potential;
- system for processing and recording measurements;
- electrical connections.

Currently preference is given to wire mesh systems consisting of conductive polymers or titanium activated with ruthenium oxides, anchored by a layer of suitable cement mix (see Figure 3).

Among the advantages of this system are such things as more uniform distribution of current, easier installation, and the possibility of application on vertical surfaces. The more commonly used power supply systems differ according to the type of power source utilized, i.e. electrical energy or photovoltaic generators.

In the first instance, the system consists of a transformer, rectifier, voltage or current regulator and measuring instruments (amperometer and voltmeter) according to the diagram in Figure 4.

In the second instance the system consists of photovoltaic panels, charge regulators, batteries, a current or constant potential regulator, and measuring instrumentation, as shown in Figure 5.

So as to control over time the potential assumed by the reinforcing, it is necessary to ensure that reference electrodes be installed at a suitable number of measuring points within the structure. In general, a good reference electrode should have the following characteristics:

- be stable and non-polarizable;
- not contaminate the environment in which it is installed;
- be sturdy and durable in view of the conditions in which it is installed.

The types of electrodes most commonly used are those of silver/silver chloride, copper/copper sulphate and graphite.

It is preferable to utilize several different technologies on the same structure for the reason that the Ag/AgCl electrodes have a longer working life, but over time provide increasingly less accurate measures of potential, whereas those of $Cu/CuSO_4$ ensure more reliable measurements but offer only limited stability.

It is furthermore possible to equip the structure with suitable galvanic probes or macrocouples consisting of additional elements of reinforcing steel covered with a cement mix containing a high concentration of chlorides (NaC1 = 1 to 3%).

This galvanic probe connected to the reinforcing gives rise to concentration macrocouple in which under natural conditions a corrosion current flows.

With the startup of the cathodic protection system, an electric field is formed which tends progressively to reduce the corrosion current until it becomes inverted, at which point the structure is protected.

The monitoring system records periodically the measurements of the potential and drainage current.

The date acquired permit the scheduling of control tests providing indications of performance and any need to correct current/supply voltage levels, and in cases where the pre-established thresholds are exceeded, provide for shutting down the power supply. These functions are performed by a special automatic microprocessor installed near the structure which can also be remotely operated by means of a mode linked to the telephone network or by radio.

Installation of the cathodic protection system naturally also includes a series of electrical connections between the various elements using suitably protected shielded cables.

5. Limits of Applicability

In prestressed reinforced concrete structures cathodic protection is adopted only with particular caution for the reason that, if it is not calibrated correctly, it can cause breaks in the reinforcing as the result of the mechanism of hydrogen embrittlement. This mechanism develops according to the reaction $2H^+ + 2e^- \rightarrow 2H \rightarrow H_2$, which allows the protons to penetrate into the lattice of the metal, weakening the atomic bonds and permitting the propagation of cracks. Other factors which can favour the onset of this mechanism are:

- presence of chlorides or sulphates in the concrete;
- high mechanical strength, especially in tempered and drawn steels;
- the magnitude of strains applied, especially if slowly increasing.

Nevertheless, experience shows that cathodic protection has no effect on this phenomenon as long as the potential in the reinforcing is not less than -0.9 V (with respect to $Cu/CuSO_4$), even if this limit poses significant design and management problems.

Other constraints affecting the installation of cathodic protection systems are the need to ensure the electrical continuity of the reinforcing, the limited reliability over time of the reference electrodes, and finally the time required to install the system. This latter activity interferes with both traffic and ordinary maintenance programs, and this to greater or lesser degree depending on the season of the year and the different methods required.

6. Description of the Structure

The structure chosen for the experimental installation of the cathodic protection system is located on a section of the motorway running across the Apennine range between Florence and Bologna, specifically at km no. 216. The section in question carries some of the most intense traffic volumes of the entire network, and in winter is characterized by frequent and heavy snowfalls, a high number of days with constant sub-zero temperatures, rapid weather variations and abrupt changes in humidity levels. All these conditions require maximum use of chemical de-icers to ensure adequate user safety levels. For reasons cited elsewhere, these products are highly aggressive to the structure itself. Moreover, the second bay passes over the Florence-Bologna express railway line which runs along an artificial embankment.

The structure, in operation for around 30 years, has undergone a series of interventions over the years both to repair and to upgrade it to support evolving service conditions and safety requirements. Besides this, the structure was also equipped with a cathodic protection system to prevent future deterioration and possible damage due to stray currents to the slab and related kerbs as described in detail in the following sections.

7. Methods of Execution

The areas selected for installing the electrodes were:

- the heads of each single span where reinforcing density was greatest;
- the midpoint of the second span;
- the longitudinal prestressed reinforced concrete beams, (for control of possible overprotection).

Each electrode or group of electrodes had its own connection to the reinforcing, so as to ensure faithful measurement of the potential, and avoid excessive ohmic drops. All the signal cables (carefully shielded) were exited from the slab's lower surface and were strung through special conduits to the control room located at the base of the bridge. Each span thus had its own isolated and independent cathodic meshwork.

In regard to the current distribution straps, so as to ensure similar power to both the slab and the kerbs, two transversal power supply points (titanium straps) were provided for each span, located 7 m from the heads. Thus no point on the meshwork lay more than 7 m from the strap, and this interval provided sufficient ohmic drop along the anode. The techniques used to install the anodic mesh on the SOUTH and NORTH carriageways were different, because of the different seasons of the year when installation took place.

7.1 WET INSTALLATION

The installation of the mesh in the winter period entailed the following stages:

- pouring the concrete to a level sufficient to suitably cover the reinforcing steel (thickness above rods of 2 to 2.5 cm);

- checking to ensure that all reinforcing steel is suitable covered, and remedying the situation if such is not the case;
- rolling out and installing mesh and titanium straps (suitably fastened to the reinforcing of the heads with insulating material); this operation is complicated by the need to perform it while the concrete is still wet, thus requiring the use of planks and supports, for example, so as not to damage the poured concrete;
- finishing by pouring a concrete cover over the anodic mesh for a thickness of 2 to 2.5 cm.

7.2 DRY INSTALLATION

This method provides for installation of the mesh in the summer period on an already hardened and cured concrete slab. The stages are as follows:

- checking for possible emergence of reinforcing steel from the concrete and remedying of such situations;
- roughening of the surface so as to ensure perfect bounding of the cement material to cover the mesh;
- washing of the surface at medium pressure (100 to 150 BAR);
- laying of mesh and power supply straps, fastening these to the slab with plastic nails; this operation is rather delicate insofar as good adhesion of the mesh to the slab is required so as to keep the thickness of the concrete cover layer to a minimum;
- washing of the slab at medium pressure;
- spray application of a specially modified cement mixture in an initial coat of some 8 mm in thickness, providing a finish roughened manually is a longitudinal direction. This product should guarantee a perfect adhesion to the underlying layer, good electrical conductivity and particular waterproofing properties;
- spray application (at an interval of no less than 1 hour) of a second coat of about 4 mm, with a smooth manual transversal finish in the direction of maximum slope, so as to avoid possible ponding of water between the product and the subsequent pavement layer.

7.3 PECULIARITIES OF THE TWO METHODS

The particular characteristics of the two methods can be summarized as follows:

Wet installation - Advantages:

- fewer work stages for installation, thus greater rapidity;
- few risks regarding adhesion between covering layer and substrate;
- lower cost due to lack of need for costly special covering layer.

Wet Installation - Disadvantages:

- greater pouring time, with increased risk related to changing weather conditions and detrimental effects on quality of casting;
- complexity of organization between repair works to structure and electrical works to implement cathodic protection;

- with high temperatures (summer), risks of too-rapid drying of surface of concrete prior to final finishing, and thus risk of breaks and detachment of covering layer;
- application is limited to horizontal surface.

Dry Installation - Advantages:

- no interference with construction and/or repair operations;
- cover layer provides greater conductivity compared to concrete, for improved electric field distribution though the structure;
- can be applied to any surface (horizontal or vertical);
- may permit dispensing with application of pavement.

Dry Installation - Disadvantages:

- greater costs due to use of special cement mixes;
- more work stages and hence less rapid;
- greater risks of poor adhesion between cover layer and substrate.

8. Power Supply and Monitoring

The cathodic protection system of the viaduct was powered as follows:

- 1 solar cell feeder for the first span (NORTH and SOUTH carriageway) which charges a buffer battery of 6 V, which in turn supplies current via the regulators;
- 3 switching type network feeders for the 2nd, 3rd and 4th spans (NORTH and SOUTH carriageway), current stabilized, from 0 to 10 A and from 0 to 15V, which supply the six spans via the corresponding regulators.

The feeders receive 220 VAC single-phase current network via a general switch, protective fuses and filters which eliminate disturbances by radio signals. Each charge regulator for the spans has on its front panel a digital instrument which functions as an amperometer or voltmeter depending on the position of its control switch. The charge current is regulated by means of a set point trimmer. Each regulator possesses a logical gate to automatically turn it on or off, as well as a manual ON/OFF switch. The feeder-regulator rack is of modular type manufacture to facilitate maintenance and replacement.

The monitoring system is controlled by a "Data Logger" which performs the following functions:

- gathers all the measurements of potential associated with the reference electrodes;
- measures the protection current density values by means of the galvanic probes and interface shunts (10 ohms);
- periodically compares the absolute values of the potentials read from the Ag/AgCl electrodes, located on the beams, with the threshold values;
- memorizes possible alarms and controls the shut off of the feeder of the span possibly affected by over protection;
- controls the system feeders to annul the protection current in the four hour survey of the potential drift according to the "NACE 100 mV method";
- prints out the significant data of the various measurements.

All the functions described are performed automatically according to the various programs memorized on magnetic tape at the time the system is brought into service. The Data Logger has a backup battery to safeguard the programs and recorded data during power failures. The data maintenance capacity of the battery is around 60 days. An internal power on reset capability reactivates the system after possible power failures of the network. The Data Logger was installed on the monitoring and control panel to which all the cables from the electrodes and the power cables are connected. The printouts can be collected at regular intervals by staff in charge of periodic inspections.

9. Startup

The cathodic protection system was brought on stream in 3 stages:

- measurement of the natural potentials of the reinforcing (cathodic protection system deactivated and monitoring system activated);
- powering up of cathodic protection system on each span, with construction of the potential/current curve (E-LogI) for an initial determination of the protection current;
- startup of the automatic monitoring system with measurement of the depolarization in 4 hours.

It was noted that after the repair of the slab, bridge reinforcing measured in the passivity range and thus was not under corrosion.

Initially the cathodic protection system was designed to operate on constant current until the electrical behaviour induced in the structure by external influences was determined, at which time an assessment will be made to determine the suitability of regulating the system by constant tension so as to maintain a correct state of protection

Figure 6 shows an example of so-called Tafel curves obtained by reporting Log I (density of current per m^2 of slab) as a function of the potentials of the reference electrodes installed in the structure.

From these curves one can initially estimate the protection current density, taking as a starting point the straight line section of the curves. Values obtained range from 6 to 8 mA/m^2, corresponding to an initial protection current of 3A per span.

The cathodic protection system was therefore regulated on this variable-tension constant-current value, and at the same time the monitoring system was started up for automatic verification of the state of protection using the experimental criterion of 100 mV recommended by NACE.

The automatic sequence of the 100 mV test proceeds as follows. Every five days in which the cathodic protection system is in operation, the potentials are measured, and always at the same intervals. The Data Logger interrupts the operation of the cathodic protection feeders for a period of four hours. At the beginning of this period the Data Logger acquires all the data on the potentials of the structure, measured a few seconds before and after the interruption of the current, and prints out the "delta-potential" values. At the end of the four hour period the Data

Logger calculates the difference between the value of the potential of each point just a few second before interruption of the protection current and as measured again after four hours.

If for each measuring point the final difference obtained exceeds 100 mV, it can be affirmed that the structure is adequately protected. On the other hand, if in any point of the structure the difference amounts to less than 100 mV, this indicates that locally there is a lack of protection against corrosion. The results of these operations are illustrated in the end-of-test printout (see Table 1). Following these checks, the Data Logger switches the feeders back on for another five days until the next check stage.

10. Monitoring

The two carriageways (North and South) were restored subsequently, for obvious reasons of service, and hence the operating data for the South carriageway were collected over a period of about 2 years, whereas the data for the North carriageway are over about 1.5 years. Figure 7 shows the typical depolarization values for a span.

The increased depolarization values obtained with the same output current (3A per span) shows that cathode polarization has occurred. The protection current was thus regulated at levels of around 2A per span, corresponding to 25 to 30 mA/m^2 of anode surface area, 2 mA/m^2 of cathode surface area (reinforcing rods) and 5 mA/m^2 of bridge slab surface area.

The supply voltage was around 1.6 to 2 volts for the South carriageway, and rose to around 2.5 to 3 volts for the North one.

This phenomenon was attributed to the two different installation methods used on the two carriageways, which resulted, in the case of the North one, in a greater distance between the anode (titanium mesh) and the cathode (reinforcing), and a consequently greater ohmic drop.

The half-cell potential showed good performance over time, including the graphite ones, which can frequently prove unstable (see Figure 8).

The peaks in potential (positive values) correspond to the depolarizations of the systems over the 4 hours (Delta NACE).

The chart also shows the effect of the electrical field also on the electrode located in the longitudinal beams of prestressed reinforced concrete at a distance of about 1.5 m from the anodic mesh.

The only abnormalities noted at the electrodes were those recorded at two Ag/AgCl sensors which proved reluctant to depolarize; this phenomenon was attributed to the close proximity of the electrode to a macrocell which exercised a constant protective effect in its vicinity.

The numerous half-cell potentials installed also made it possible to show the electric field attenuation field over the span (see Figure 9).

This phenomenon, due to the fact that technical problems limited current feed to only one side, diminished considerably with the drop in the protection current following polarization of the structure, which serves to confirm the general principles of cathodic protection. The photovoltaic power supply system performed adequately and reliably over time, confirming its validity for use in systems of this kind.

11. Conclusions

As shown, the application of cathodic protection requires very high technology materials for the anodes, and continuous, computer-guided monitoring which will permit the automatic operations necessary to optimize protection.

From the management standpoint it should be noted that in our country the costs of this treatment are quite high:

- due to the cost of the anodes currently not produced in Italy;
- due to the cost of the control systems, if used to monitor works of limited dimensions.

Nonetheless, having verified the validity of the system in protecting reinforcing steel form corrosion (the primary factor reducing the working life of structures, and further aggravated by the use of deicing salts), and given the availability of tested techniques, we hope to utilize this method in the future:

- to protect a series of structures connected to one another, using a single computerized control system so as to reduce the costs per square meter;
- to produce anode materials so as to cut down supply costs.

In view of the above, the system will thus be employed mainly on large structures at the design or construction stage.

12. Bibliography

1. --------, (1980) ASTM C876-80, Standard Test Method for Half-Cell Potentials of Reinforcing Steel in Concrete, American Society for Testing and Materials, Philadelphia.
2. --------, (1987) Standard Recommended Practice, Design Considerations for Corrosion Control of Reinforcing Steel in Concrete, N.A.C.E., Houston.
3. Cigada, A. (1988) Corrosione, fatica e infragilimento da idrogeno e protezione catodica, Milano.
4. DiTomaso, A. (1988) Prove sperimentali, Istituto di Scienza delle Costruzione, Bologna.
5. Ingvarsson, H. and Westerberg, B. (1986) Operation and Maintenance of Bridges and Other Bearing Structures, LinKoping, Sweden.
6. Pedeferri, P. (1986) La Corrosione delle armature nel c.a. e c.a.p., C.I.S.M., Udine.
7. Pedeferri, P. (1987) La protezione catodica delle armature nel calcestruzzo, Milano.
8. Pedeferri, P. and Pastore, T. (1984) Corrosione delle armature nel cemento armato o precompresso e protezione catodica, Milano.
9. Siniscalchi, A. (1984) Impianti di protezione catodica e criteri per il loro esercizio e manutenzione, Milano.

Table 1 Delta-Potential values (NACE Methods)

C	01	DELTA NACE	97	MV	C	18	DELTA NACE*	12	MV
C	02	DELTA NACE	108	MV	C	19	DELTA NACE*	17	MV
C	03	DELTA NACE*	33	MV	C	20	DELTA NACE*	10	MV
C	05	DELTA NACE	74	MV	C	22	DELTA NACE	147	MV
C	06	DELTA NACE	92	MV	C	23	DELTA NACE	145	MV
C	07	DELTA NACE	186	MV	C	25	DELTA NACE	199	MV
C	09	DELTA NACE	152	MV	C	26	DELTA NACE	201	MV
C	10	DELTA NACE	208	MV	C	27	DELTA NACE*	17	MV
C	12	DELTA NACE	169	MV	C	29	DELTA NACE	156	MV
C	13	DELTA NACE	183	MV	C	30	DELTA NACE	162	MV
C	14	DELTA NACE*	10	MV	C	31	DELTA NACE	440	MV
C	16	DELTA NACE	192	MV	C	32	DELTA NACE*	32	MV
C	17	DELTA NACE	196	MV					

* = Electrodes installed on the reinforced concrete beams

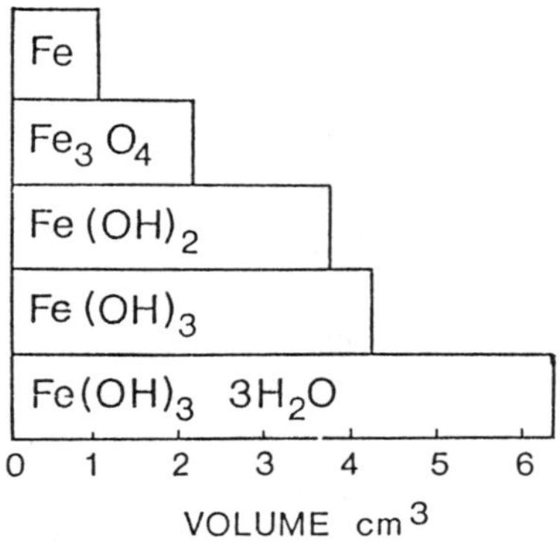

Fig. 1 Volume Relative to the Various Corrosion Products with Formation of Ferric Hydroxide.

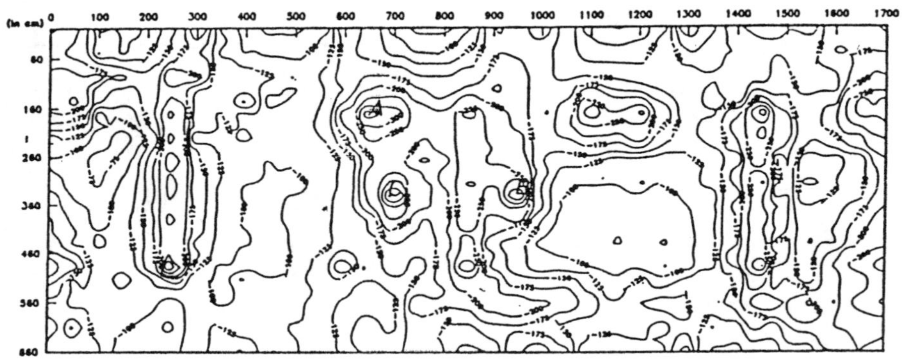

Fig. 2 Mapping of the Potential of a Slab.

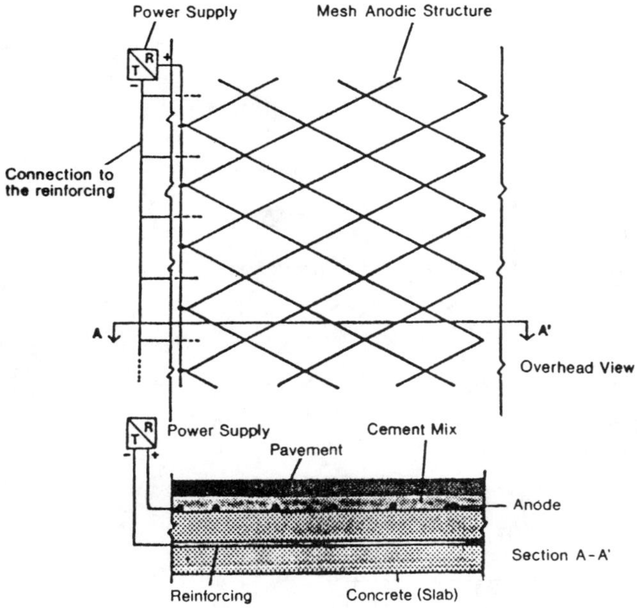

Fig. 3 Cathodic Protection System with Wire Mesh Anode.

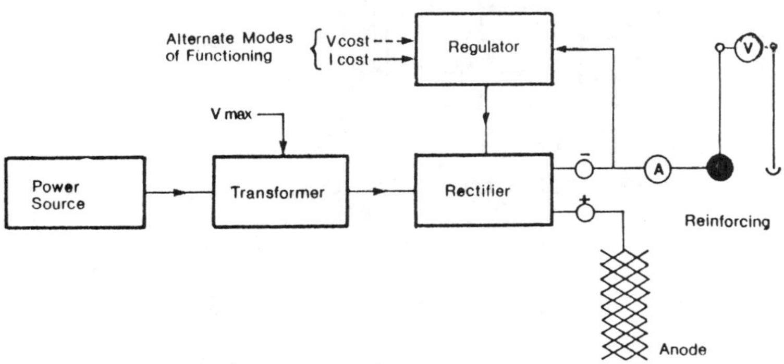

Fig. 4 Power Supply System from Mains.

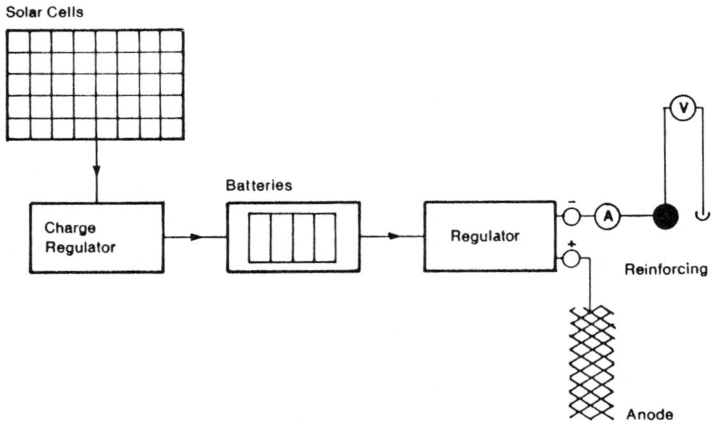

Fig. 5 Supply System Based on Photovoltaic Generators.

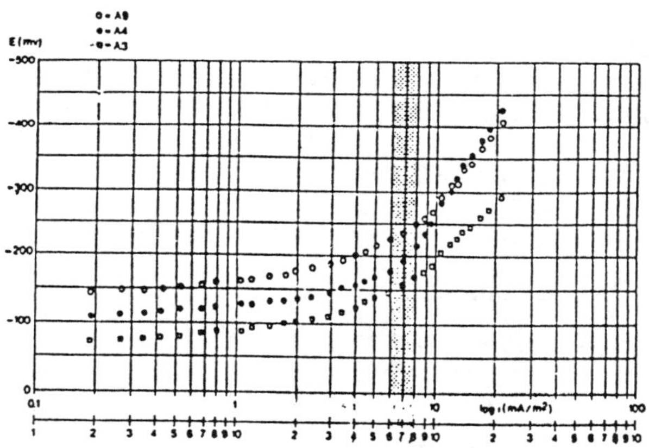

Fig. 6 Tafel Curves Obtained with the Different Reference Electrodes.

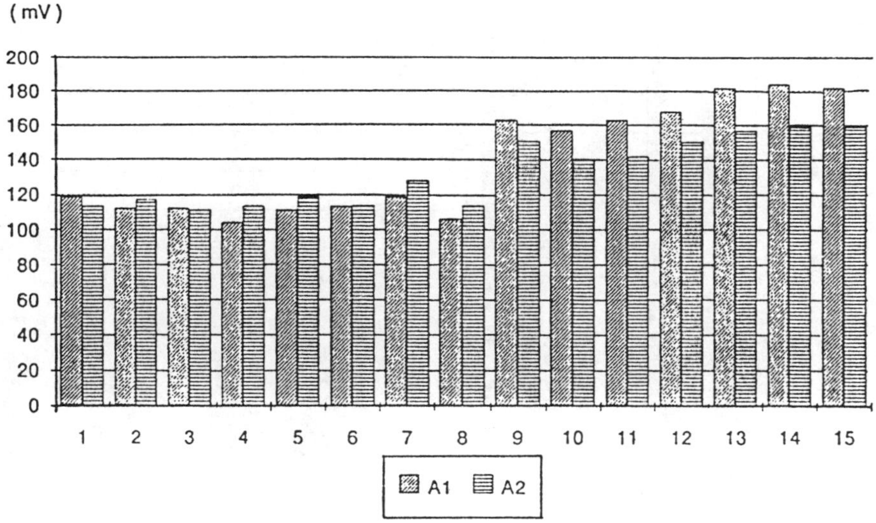

Fig. 7 DELTA NACE - Span 1 South.

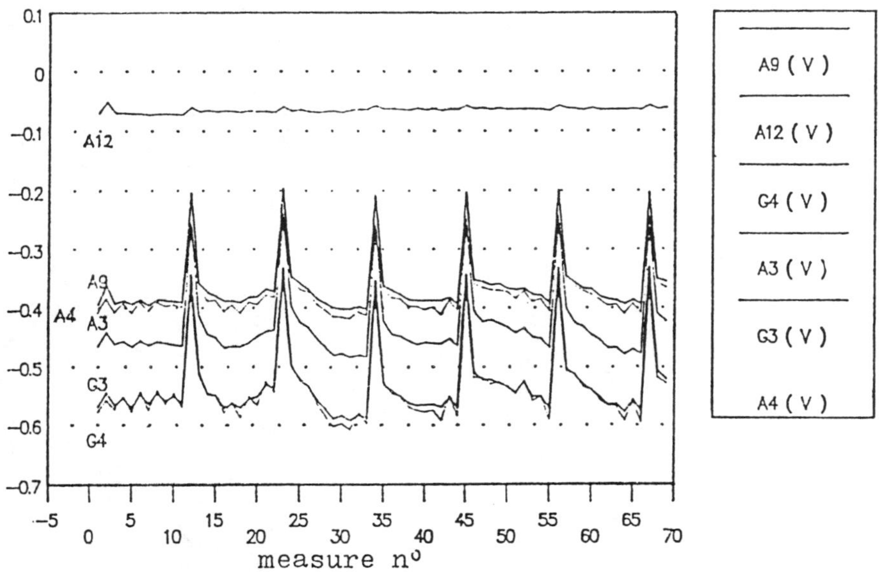

Fig. 8 Monitoring Span 2 South - Readings 19/5/89 to 19/6/89.

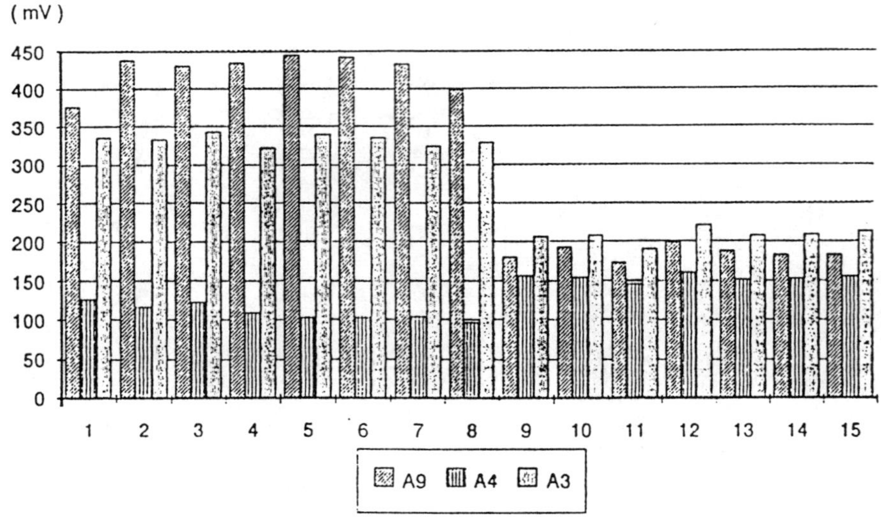

Fig. 9 DELTA NACE Span 2 South.

REHABILITATION OF STEEL TRUSS BRIDGES IN ONTARIO

B. FARAGO
Ministry of Transportation of Ontario
1201 Wilson Avenue, Atrium Tower
Downsview, Ontario, M3M 1J8 CANADA

ABSTRACT. The paper deals with the rehabilitation of truss bridges, that often involves strengthening as well. The key question, whether to rehabilitate or replace a bridge is complex.

The engineer must have a good knowledge of the condition of the bridge, the actual load capacity and reliable cost estimates for rehabilitation. Increasingly present value based on life-cycle costing is used in the decision making. Alternative methods of strengthening and the decision-process followed in Ontario is illustrated through case-histories.

1. Introduction

From the early stages of road-construction until the early fifties truss bridges were used extensively for medium and long spans. However, since the development of prestressed concrete and welded steel plate and box girders, truss bridges have only been built in exceptional circumstances.

Over the past years many of the shorter truss bridges have been replaced with girder type structures, rather than rehabilitated, particularly if the bridge had insufficient width and/or vertical clearance.

Most of the truss bridges still in use today, are major structures, whose maintenance and upgrading to present-day standards is of prime importance, because of the difficulties and high costs involved with their replacement.

There are 3251 structures on the provincial road system in Ontario, 2455 are concrete and 796 are steel. Of these only 106 are truss bridges. Their ages are given in Table 1.

Most of the trusses are of the through or half-through (pony-truss) type, and only a few are of the deck type. Most of these structures are simply-supported; over larger openings cantilever and suspended span structures are used. Only a few of the trusses are of continuous construction or other forms, such as truss-arches.

A. S. Nowak (ed.), Bridge Evaluation, Repair and Rehabilization, 555–566.

2. Deficiencies of Truss Bridges

All the bridges on the provincial highway system are inspected at two-year intervals. Their condition and deficiencies are recorded and maintained in a computerized file system.

Deficiencies may fall into one or more of the following categories:

1. Inadequate width or vertical clearance
2. Inadequate load capacity, based on original design
3. Loss of strength due to local deterioration or damage

The determination of deficiencies involves inspection and evaluation.

2.1 INSPECTION OF TRUSS BRIDGES

Because of their size and location, the inspection of trusses presents a formidable challenge. Visual inspection, which in the past was often done from river shores or from a boat, using binoculars, is no longer acceptable and scaffoldings, swing stages or inspection vehicles are used for detailed inspection.

Inspection involves determination of the extent of loss of coating and, where corrosion occurs, the loss of section. Rivets are checked for looseness by tapping and bolts by torquing. Detection of corrosion or fracture in pins is done by ultrasonic testing. Expansion bearings are often of the nest of roller-types. Where small diameter rollers were used, bearings soon became inoperative. Inspection involves checking if movement is taking place.

2.2 EVALUATION OF THE LOAD CAPACITY OF TRUSS BRIDGES

Analysis of trusses frequently indicates over-stress even when there is no visible sign of failure. To assess the real behaviour of older bridges the Ministry regularly resorts to load tests. In most cases tests indicate substantially lower stresses in many members, but in particular, where interaction of chord members with the floor system takes place. On some occasions, structural behaviour changes significantly, such as when simply supported members become fixed because of frozen bearings.

The knowledge gained from load testing is often used to reduce the amount of strengthening, but on occasions, it also helps to avoid major failures, as illustrated in the following two examples:

2.2.1 *Keewatin Channel Bridge.* The Keewatin Channel Bridge is located on Highway 17 near Kenora. The bridge is 275m long and consists of a centre three-pin steel truss arch, flanked by multiple concrete arch approach spans (Figure 1). The steel arch has a span of 104.1 m and a rise of 13.2 m and carries a 9.14 m wide two-lane roadway with a sidewalk on one side. The reinforced concrete deck is supported on three stringers and on the top chords of the trusses. Analytical evaluation of the structure indicated overstressing in the top chord members, due to the combined action of bending and axial forces, and in some of the diagonals.

Visual inspection of the bridge did not reveal any weakness, therefore it was decided to carry out a load-test in line with current Ministry policies. A description of the test program and the findings were published by Tharmabala (1986). According to this report the interaction of the floor system with the top chord resulted in a significant reduction of the stresses in top chord members. Stress levels in all diagonal members were below calculated values. It was resolved not to strengthen any of the members, but only a few member-connections.

The cost of the load test was only a fraction of the saving that was achieved through this testing program.

2.2.2 *The Finney Bridge.* Test of the Finney Bridge was described by Bakht, (1986). This is a pin-connected single-span steel truss bridge built about 90 years ago. The span is 51.2 m and height 6.7 m. The top chord is a continuous built up section, but the tension chord and diagonals are composed of a pair of unconnected components, each with a square or rectangular cross-section.

To confirm its load-capacity, the bridge was load-tested in 1985. The load test revealed that some of the twin tension members, because of a lack of fit, did not share the load equally. For this reason these members will carry less than the calculated load.

3. Rehabilitate or replace? - The decision process.

Following inspection and determination of the condition of a bridge, the load capacity is determined analytically and, if necessary, by a load-test. If a structure is found deficient, a decision has to be made on how to upgrade the structure.

If the structure is too narrow or has insufficient vertical clearance, and widening it is not feasible, the only option available is to maintain the structure as is, or replace the bridge.

If load capacity is inadequate, because of design, deterioration of structural components, or both, there are many solutions available to increase the capacity.

In every case a preliminary cost-estimate is prepared of the proposed strengthening and rehabilitation. Often it is necessary to carry design to an advanced stage so that reliable estimates can be made. Reel and Muruganandan (1989) developed a Financial Analysis Manual incorporating a computer program to allow comparison to be made between the cost of rehabilitation and replacement. The program calculates present day values of different alternatives on the basis of life-cycle costing.

In spite of all the available tools, it is still a difficult task to decide whether to rehabilitate or replace a bridge. In addition to the purely financial aspects, other factors such as historic significance, inconvenience to the users and interruption of usual business patterns by detouring, affect the decision.

4. Rehabilitation involving strengthening

The Ministry has used the following methods to increase the load-capacity of existing bridges:

1. Adding plates to existing members or replacing weak members. This is an expensive method, because it requires field-drilling and is used where only a few members require strengthening.
2. Using external prestress where the majority of members requires strengthening. External prestressing can significantly increase the load carrying capacity.
3. Increasing capacity by reducing dead load by removing the concrete deck and stringers and replacing them with a longitudinally laminated and prestressed timber deck supported by the floorbeams; significant increase in live load capacity can be gained. This method has been used successfully on secondary highways in the northern part of the province.
4. Erecting additional trusses to carry the increased load. Inserting an additional truss between two existing trusses helped to increase live load capacity of one bridge significantly.

In addition to strengthening, there are a number of specific problems common to most truss bridges.

1. Replacement of bearings - Most major bridges require replacement of non-functioning bearings. In most instances no provisions were made for jacking and the floor beams are not strong enough to be used as jacking beams. In such circumstances it is necessary to develop a jacking bracket by attaching a frame to the existing trusses.
2. Protection of the structure from impact - Most through and half-through type bridges contain only nominal railings. Truss diagonals are often damaged by over-size loading or in vehicle accidents. As these bridges are single load-path structures, the Ministry is replacing handrails with barriers, where feasible.
3. Elimination of open expansion joints - On longer structures the use of open finger-plate joints was common. These joints cause extensive corrosion to steel members and are being replaced with sealed joints, often modular expansion joints.
4. Coatings - In order to prevent corrosion from occurring and subsequent loss of section, it is important to maintain coatings in good condition. Rehabilitation of trusses is normally followed by a separate coating contract in the following year.

The coating system in use currently for existing structures is a three-coat epoxy-zinc, vinyl system.

Experience shows that good quality, durable re-coating is difficult to achieve. The life of a new coating depends on the competence and thoroughness of the applicator and presence of the inspector. Environmental protection requirements often necessitate the construction of complete enclosures that increase the cost of coating by 20 - 30%. The high cost of providing a durable coating system and the need for more frequent re-coating makes coating a major component in the life-cycle costing of truss bridges.

5. Case Histories

5.1 BURLINGTON SKYWAY BRIDGE

The Burlington Skyway is a high level crossing for the Queen Elizabeth Way, the main link between Toronto and the Niagara Peninsula, over the shipping channel to Hamilton Harbour. The three span continuous steel truss of variable depth has a central span of 150.9 m and side spans of 83.8 m each. The original roadway section consisted of two 7.3 m roadways and a 1.8 m centre strip, and was supported on steel floor beam trusses, spanning 19.1 metres between the main trusses (Figure 2).

In 1983 a second structure was built and it was decided to convert the existing structure to 5 lanes of one-directional traffic. An evaluation of the existing structure revealed no weakness in the main trusses, but due to the heavier design live loads of the Ontario Highway Bridge Design Code (1983) and increased number of lanes, some of the floor beam truss members were over-stressed by as much as 40%.

According to Holowka (1988) and Sadler (1990) several methods of increasing the strength of the floorbeam trusses were investigated. These included:

a) Individual replacement of deteriorated or under-strength components,
b) Addition of new structural steel to increase capacity of existing sections,
c) Replacement of entire floorbeam truss panels, and
d) The use of external prestressing.

Overstressing was eliminated by the application of external prestress. Only those vertical and diagonal members affected by the highly localized prestress forces needed strengthening. Different prestressing configurations were considered. The alternative selected consisted of a draped prestressing profile with the anchorages embedded in the concrete deck. This had the greatest drape and, therefore, provided the largest vertical load component. The vertical and diagonal truss members framing into the location of the lower prestressing anchorages required strengthening. The external loads from prestressing were applied at the first lower interior node of each floorbeam truss.

Details of the arrangement are given in Figure 3 and 4. A steel plate assembly was designed to allow for the transfer of load from the prestressing bars to the floorbeam trusses. Dywidag 36 mm diameter high strength threadbars were used. The effective force in each bar was 330 kN or just over 60 percent of its tensile strength. All six ties were stressed simultaneously.

5.2 SOUTH MUSKOKA RIVER BRIDGE

This two-span deck type steel truss bridge was built in 1952. The spans are 20.5 and 30.8 metres with 8.6 m approach spans on either end. The two trusses support a 12.0 m wide deck, that consists of a reinforced concrete slab supported on stringers and floorbeams. The bridge carries the two northbound lanes of Highway 11 (Figure 5).

A condition survey of the bridge revealed the deterioration of the deck to the point that replacement was required. The structural steel was found to be in fair condition but required re-coating. Structural evaluation of the trusses, however, indicated overstressing in most truss members and in all floorbeams.

Alternatives considered included replacement of the bridge or rehabilitation by strengthening of members, prestressing the bridge and adding a third truss located between the two existing trusses. Based on preliminary cost-estimates, this third alternative was selected. It had the advantage that by halving their spans, floorbeams did not require any strengthening and strengthening of existing trusses was limited to four members per truss. This solution was possible because it was decided to detour traffic around the bridge.

The work involved removing all bracing members; erecting the new truss piece-meal and installing new bracing members (Figure 6).

Cost of the rehabilitation (based on the average of the three lowest bids), in 1987 was as follows:

removal of existing structure	$CAN121,914
roadwork (including detour and paving of bridge deck)	94,133
rehabilitation of structure	593,051
Total	$CAN809,098

Unit cost of the new truss (installed) was $6,000 per tonne, while new plate girders at the time cost only $3,000.

Unit cost of strengthening existing members was $24,000 per tonne, however total steel involved amounted to 3.8 tonne only. On similar projects the unit cost of strengthening varied between $16,000 and $20,000 per tonne.

The cost of coating the bridge in 1988 was $182,197

The cost of rehabilitation (excluding removal and roadwork) increased to $775,248 or $940 per m2.

For comparison, the average cost of a comparable bridge superstructure at the time was $800 to $850 per m2.

In hindsight, perhaps the bridge should have been replaced, rather than rehabilitated once the decision was made that traffic could be detoured during construction. Building a new superstructure would have taken less time than to rehabilitate the existing trusses. No value was placed at the time on the additional construction time and the hardship that a detour creates.

The Ministry is now looking at ways to place a financial value on saving in construction time, particularly on high-volume roads, where reduction of traffic lanes and/or detouring can increase travel-time significantly.

5.3 BUSKEGAU RIVER BRIDGE

Rehabilitation of this three-span deck type truss bridge was planned in 1988. The spans are 23.4 + 33.0 + 23.4 m and the deck is 9.75 m wide, carrying two lanes of traffic (Figure 7).

According to the condition survey, the deck had to be replaced and some of the badly rusted members required strengthening. It was proposed that rehabilitation be carried out immediately, to prevent further deterioration. Calculated load-capacity of the bridge was estimated to be 60 % to 70% of that required for this road.

Following investigation of a number of alternatives, it was decided to maintain single lane operation during construction and to strengthen the bridge by pre-stressing. This methods still required strengthening of 25% of the members. The rehabilitation was planned as a two-year contract.

Following completion of the design, cost estimates were prepared and it was found that strengthening would cost more than originally anticipated. A two-year construction period not only increased costs but meant two years of hardship to the road-users.

After preparing life-cycle costing and taking the high cost of coating into account, it was concluded that very little could be gained by rehabilitating the existing bridge. Replacement on the present alignment will require a Bailey-bridge detour, but the project could be completed in one year.

The original decision was reversed at this stage and it was decided to delay reconstruction up to five years. Some further deterioration could be tolerated without endangering safety of the bridge because the bridge will be scrapped.

Delaying construction for five years and replacing the bridge at that time provided the most economical solution, as indicated below:

Cost of rehabilitation (2 year contract)	$CAN1,300,000
Coating (in following year)	210,000
Cost of new structure (width increased to 12m from 9.75m)	1,900,000
Detour (single lane)	200,000

For a life-cycle cost comparison, four options were considered:

1. Rehabilitation in year 1, re-coat in year 3, replace bridge in year 30 and minor rehabilitations every 15 years.
2. Replace bridge in year 1, minor rehabilitation in years 15, 30 and 45.
3. As option 2, but delay replacement to year 3.
4. As option 2, but delay replacement to year 5, and spend $100,000 on critical repairs now.

Table 2 shows the calculation of net present values based on a discount rate of 6%. The lowest net present value was obtained for option 4.

5.4 CONFEDERATION DRIVE BRIDGE

This is a two-span steel pony-truss bridge, with spans of 23.4 m each. The bridge was built in 1904 and it is located in a historic park along the Rideau Canal in the Town of Smiths Falls. The bridge is only 5.3 m wide, carrying a single lane of traffic Figure 8).

A condition survey revealed extremely severe corrosion in stringers, floor-beams and some bottom chord members. The bridge, which was posted for a 5 tonne load limit, was closed down immediately.

While the superstructure could have been replaced economically with steel beams, the Town wanted to keep the trusses for historic reasons. The rehabilitation of the bridge involved replacing badly deteriorated members and construction of a longitudinally laminated and prestressed timber deck bearing on the floorbeams. The lightweight deck allowed increasing the capacity of the bridge.

Because of its size, the contractor dismantled the bridge after the removal of the deck, and moved the trusses into his shop where the trusses were rebuilt and recoated over the winter and re-erected in the spring.

The cost of rehabilitating the steelwork, including new sidewalk brackets and re-coating, was $CAN 112,428 and the timber deck was $CAN 73,000 (1987 prices) or $CAN600 per m2.

As reported by Taylor (1989) the Ministry has used the same technique on other steel truss bridges to replace concrete with timber deck and gain additional load capacity without strengthening.

6. Conclusions

The most important decision is whether to rehabilitate or to replace a bridge. When extensive strengthening is involved, it is difficult to arrive at a reliable cost estimate without completing a preliminary design.

Past experience indicates that where extensive strengthening is involved and it is possible to detour traffic at a reasonable cost, it is more economical to replace than to rehabilitate.

It is often the high cost of coating that tips the balance between rehabilitation and replacement. The duration of time involved in detouring traffic is also an important factor in the decision.

On the other hand, heritage bridges and major structures that require little or no strengthening and where it is important to maintain traffic at all times, will be kept in use for a long time.

7. References

Holowka, M. (1988), Rehabilitation of the Existing Burlington Skyway - Strengthening of the Structural Steel. Report SO-88-4. Ontario Ministry of Transportation.

Tharmabala, T. (1986) Performance Studies of Keewatin Channel Bridge. Report SRR-85-05 Ontario Ministry of Transportation.

Bakht, B. (1986) Testing of Two Pin-Connected Truss Bridges. Report SRR-86-06. Ontario Ministry of Transportation.

Taylor, R. (1988) "Field Applications of Prestressed Laminated Wood Bridge Decks," Can. J. Civ. Eng. Vol 15.

Ontario Highway Bridge Design Code (1983) Ontario Ministry of Transportation.

Reel, R.S. and Muruganandan, M.C. (1989) Structural Financial Analysis Manual. Ontario Ministry of Transportation.

Sadler, C. (1990) "Post-Tensioning a Steel Bridge: Strengthening of the Burlington Bay Skyway," Proceedings, Third International Conference of Small and Medium Span Bridges.

Table 1 Year of construction of truss bridges

Year	No. of structures
before 1940	45
1940 - 1949	15
1950 - 1959	35
1960 - 1969	9
1970 - present	2

Table 2 BUSKEGAU River Bridge, Present Value of Different Options in $CAN1,000.

Yr.	OPTION 1		OPTION 2		OPTION 3		OPTION 4	
	Cost	Pres. Value	Cost	Pres. Value	Cost	Pres. Value	Cost	Pres. Value
0	1,300	1,300	2,100	2,100	0	0	100	100
3	210	176	0	0	2,100	1,763	0	0
5	0	0	0	0	0	0	2,100	1,569
15	200	83	100	41	0	0	0	0
18	0	0	0	0	100	35	0	0
20	0	0	0	0	0	0	100	31
30	2,100	365	200	34	0	0	0	0
33	0	0	0	0	200	29	0	0
35	0	0	0	0	0	0	200	26
45	200	14	100	7	0	0	0	0
48	0	0	0	0	100	6	0	0
50	0	0	0	0	0	0	0	0

Fig. 1 Keewatin Channel Bridge.

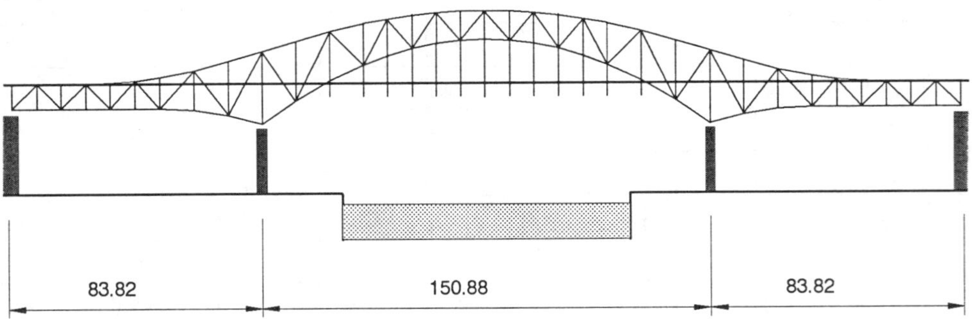

Fig. 2 Burlington Skyway Bridge - Elevation.

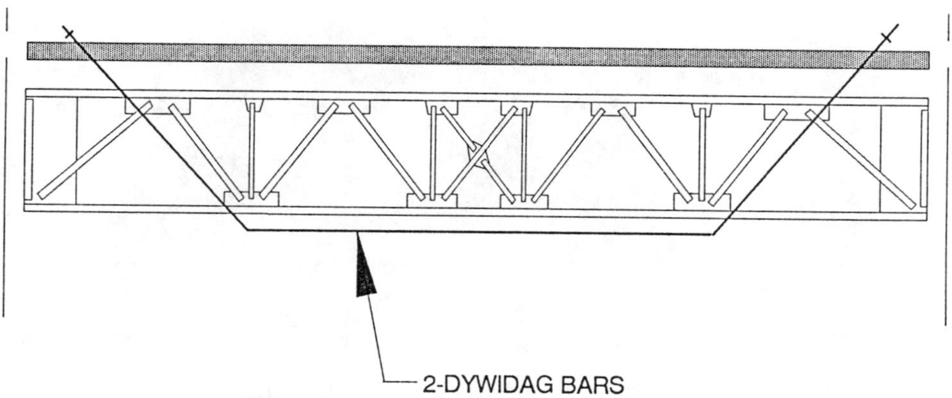

Fig. 3 Burlington Skyway Bridge - Prestressing.

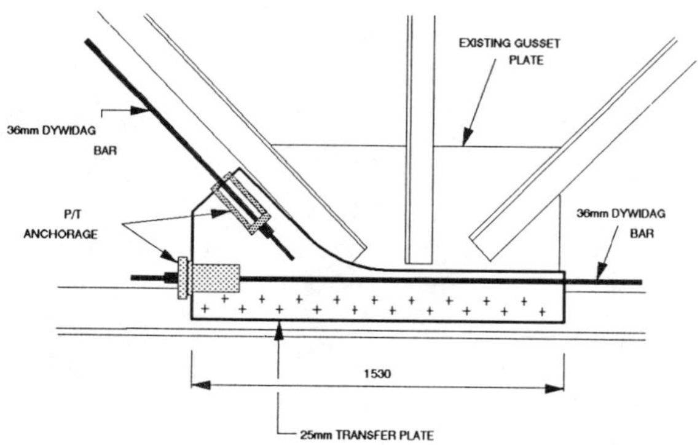

Fig. 4 Prestressing Floor Beams of the Burlington Skyway Bridge.

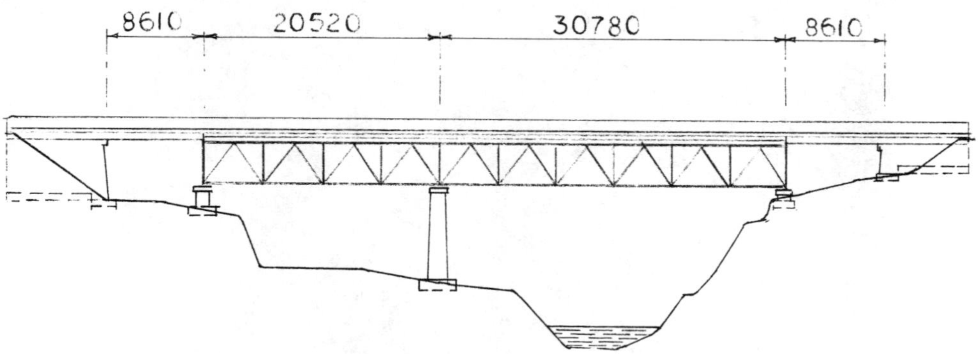

Fig. 5 South Muskoka River Bridge.

Fig. 6 South Muskoka River Bridge - Installing New Truss at Centre.

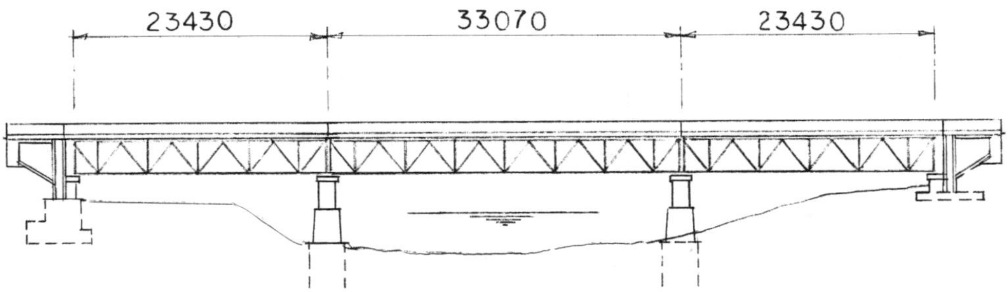

Fig. 7 Buskegau River Bridge.

Fig. 8 Confederation Drive Bridge.

REINFORCEMENT OF N9 MOTORWAY BRIDGES OVER THE LUTRIVE VALLEY NEAR LAUSANNE- SWITZERLAND

JEAN-PHILIPPE FUZIER
FREYSSINET
52-54, rue de la Belle Feuille
92100 Boulogne-Billancourt, FRANCE

ABSTRACT. Constructed between 1971 and 1973, two bridges carry the N9 motorway across the Lutrive Valley to the North-East of Lausanne. There are two parallel bridges about 400 m long, consisting of three portals resting on twin-column piers.

Inspection of the centre of the portals showed that the camber, introduced during construction, had long since been reabsorbed and that the tendency to deflect was continuing. In 1986 the authorities commissioned a consulting engineer to analyze the behavior of the structure and to propose a remedy.

This remedy came in the form of 18 T15 tendons running the whole length of the structure and anchored in the abutment zones. The works included:

- drilling holes for passage of cables and cars,
- suppling and installating steel deviators,
- casting the anchorage blocks and fixing them to the structure by means of Macalloy-Freyssinet bars,
- suppling and installating the prestressing.

They were carried out during the second half of 1988 without interruption in traffic. The tendons were successfully tensioned in December 1988 and the behaviour of the structure was monitored for a period of six months under the effect of the additional prestressing. The tendons were then grouted with cement-grout.

1. Introduction

The Lutrive Valley bridges were built between 1971 and 1973 and opened to the N9 motorway traffic in 1974. These two parallel bridges have an approximate length of 400 m; the horizontal alignment is curved with a 1000 m radius, the vertical profile is also curved with a 12, 500 m radius.

Each of these two bridges consist of three portal frames resting on twin-column piers spaced at 12 m. The cantilever lengths are 55, 66 and 58 m, thus providing two main spans 117.5 m and 131.5 m long (Figure 1). Linkage between cantilevers is provided by means of a concrete hinge. The cross section consists of two prestressed box girders 13.40 m wide with a depth

A. S. Nowak (ed.), Bridge Evaluation, Repair and Rehabilization, 567–577.

varying from 8.50 m at piers 2 and 3, to 2.50 m at the mid-span and at the abutments. The pier heights vary from 3.0 m for pier 1 to 36 m for pier 3.

2. Deflection analysis

A 110 mm camber was provided during the cantilever construction of the structure.

However, the deflection survey which was done from 1977 to 1987, shows a deflection equal to 165 mm in the 131 m span (Figure 2).

The client, Canton de Vaud Motorways Department, commissioned a consulting engineer to analyse the behaviour of the structure and to propose a remedy.

2.1 CAUSES OF EXCESSIVE DEFLECTION

Various causes created these excessive deflections; they are classified according to their relative importance.

- The real creep factor was higher than the SIA code specified creep factor at the time of construction.
- The creep evolution was longer than the one specified.
- Prestressing losses (elastic or time dependent) were higher than the ones used in the calculations.
- This type of structure, cantilever with hinge at mid-span, is very sensitive to the factors listed above. This is due to the fact that the vertical deflection is equal to the difference between two large values; deflection due to dead load and superimposed loads and deflection due to prestressing. So, a small variation in one of these figures may create an important deflection variation. Extra thickness of the wearing surface has also been noticed. - It is usual to neglect the deflections due to shear force in this type of structure. However, because of the small thickness of the webs, calculations have shown that deflection due to shear at the mid-span is on the order of 15% of the total deflection due to dead load, superimposed loads, and prestressing.

  Moreover, the small cracks, which appeared in the 25 to 40 cm thick webs, increase the deflection because of the decrease in the moment of inertia. The compressive concrete struts are highly sloped and act in the same manner, taking also into account the creep effect.
- Creep of piers and foundations, difference between the two columns of the same pier and differential shrinkage between upper and lower face of the deck, have little effect.

2.2 PROPOSED REMEDY

Several solutions have been considered, each of them proposing the use of additional longitudinal tendons.

2.2.1 *Hinge Closure.* This solution would have required a prestressing force higher than 1,500 tons in the area of the positive moments. It would have been difficult to anchor these forces in concrete sections of small thickness. Geometry problems due to the curvature in plan would have also to be solved. The positive results in terms of upward deflection was not sufficient to justify this solution.

2.2.2 *Additional prestressing in the main spans only.* Another solution would have been to keep the structure with its hinges, but provide additional longitudinal tendons in the main spans only where the deflections were most important. This would have required anchoring the tendons in piers 1 and 3 as close as possible to the upper slab, 8.0 m above the soffit (difficult construction conditions).

Moreover, this solution would not have improved the side spans which showed light bending cracking.

2.2.3 *Additional prestressing anchored in the abutments.* It was finally decided to anchor the tendons in the abutments. The prestressing layout permits this configuration to take advantage of the vertical component of tendon forces.

This layout was optimized to take into account relative displacements and rotation of deviator sections under live and thermal gradient loads. So the tendons were kept as high as possible, close to the upper slab in the main spans and taken down in the side spans in order to balance the positive moment.

3. Remedial Works

3.1 PRESTRESSED FORCE

The additional longitudinal tendons consisted of 4 Freyssinet tendons, 18 strands of T15, which lead to a total initial prestressing force equal to 13,240 kN.

This prestressing force is sufficient to provide upward deformation without creating excessive compressive stresses in the concrete hinge (compressive stress was brought from 3 N/mm^2 to 10 N/mm^2.

3.2 ANCHORAGES

In order to transfer these high prestressing forces to the deck, it was necessary to utilize two reinforced concrete blocks cast against the webs (Figure 3). The 3,310 kN (18 x 0.6") are divided in to three units of 1,100 kN each, anchored in the concrete block.

3.3 PRESTRESSING LAYOUT

In the main spans, the prestressing tendons are held in place at a constant distance from the top slab by means of steel deviators. These deviators are attached to the top and bottom slab in the thickened areas close to the webs, but no forces are introduced in the webs. In the side spans, the tendons are lowered down toward the abutments. Prestressing forces are introduced in this way below the neutral axis. Due to hyperstatic bending moments, a negative bending moment appeared in the vicinity of the abutments thus reducing the tensile stresses in the bottom fibers.

3.4 DEVIATORS

Because of the horizontal and vertical curvature of the bridge, the deviators are placed according to the following principles.

- Sufficient number of deviators to keep the tendons close to the webs.
- Avoidance of possible conflicts with a main water pipe.
- Constant spacing between the deviators in order to have almost equal deviation forces and consequently the same deviators. The spacing was chosen equal to about 48 m. There are in this manner 9 deviators per tendon:
  - 1 deviator in the side span
  - 2 deviators in the main spans
  - 1 deviator at each pier.

  The maximum deviation forces applied on a deviator are:

  Horizontal force: 165 kN

  Vertical force: 13 kN

  The friction coefficient was taken as 0.2, although tests have shown a value of 0.15.

Each deviator consists of a bent steel pipe used to deviate the tendon and a horizontal or vertical steel beam attached either to the slabs or the webs. The steel pipe connection to the beam permits a slight rotation; it is possible also to detension the tendons after grouting, allowing their replacement if required. A test was carried out on a scale of 1 to 1 to demonstrate the ability of the system.

3.5 PRESTRESSING ACTION

Additional prestressing has several favorable effects on the structure. First of all, it will permit a vertical elastic displacement equal to 50 mm at mid-span of spans 2 and 3. Moreover it is possible to expect some positive creep effect. The creep coefficient may be assumed to be equal to 1.0 to 1.2.

Another favorable effect is the compression of the tensile sections and the reduction of shear forces which may close the existing cracks. However, this is not true for the section close to the hinges. This is why two additional prestressing tendons made of 7 strands of T13 have been placed in the lower corners of the boxes shown in Figure 4.

4. Conclusion

A check was made of the structure with the additional prestressing. Most of the sections are in a total compressive state; passive reinforcement is sufficient to carry the small tensile stresses of the other ones. It has not been possible however to recover the complete initial deflection.

The cost of rehabilitation represents about $130/m^2 of deck.

5. Acknowledgement

* Authority: Canton de Vaud Motorway Dept.
* Consulting Engineer: Realini + Bader et Associés (M.R. Beyloune).
* General contractor: Freyssinet S.A.
* Principal Sub-Contractors: Bertholet & Mathis SA, Zwahlen & Mayr SA, Stump S.A.

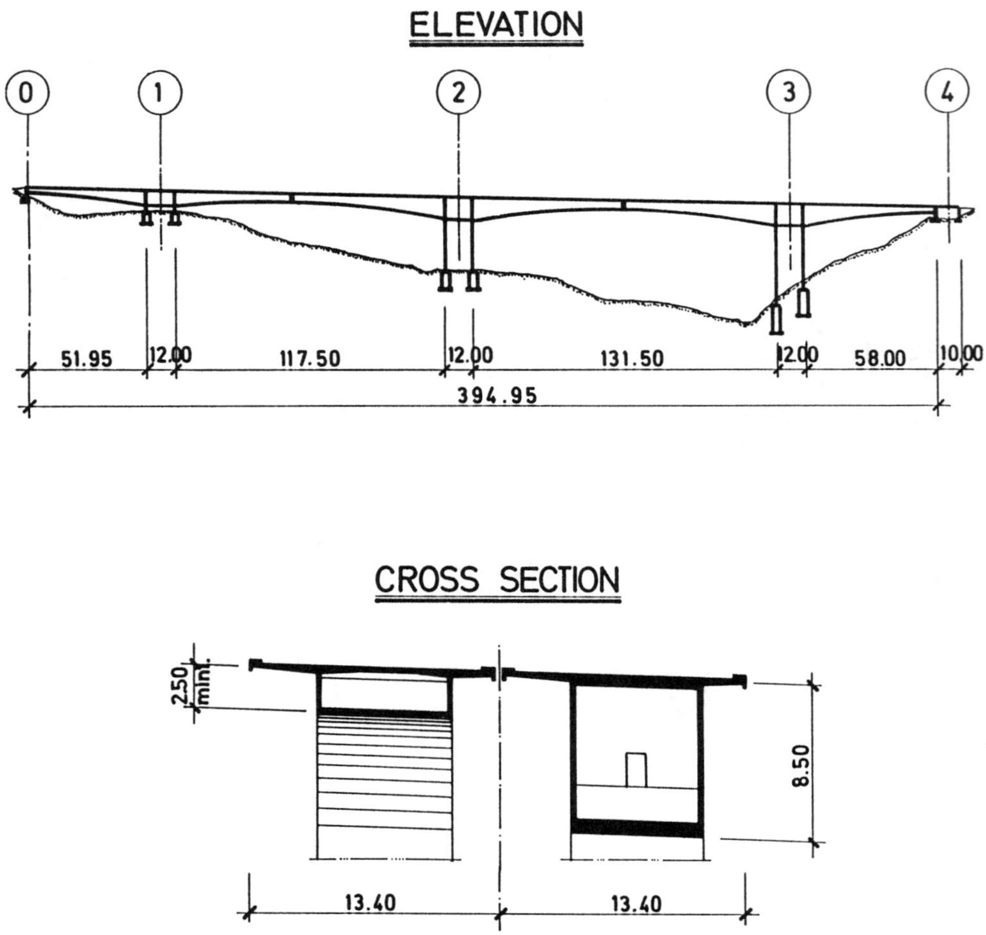
ELEVATION
0
1
2
3
4
51.95
12.00
117.50
12.00
131.50
12.00
58.00
10.00
394.95
CROSS SECTION
2.50 min.
8.50
13.40
13.40

Fig. 1

Fig. 2 Steel Deviators.

Fig. 3 Preparation of the Anchorage Block in the Abutment.

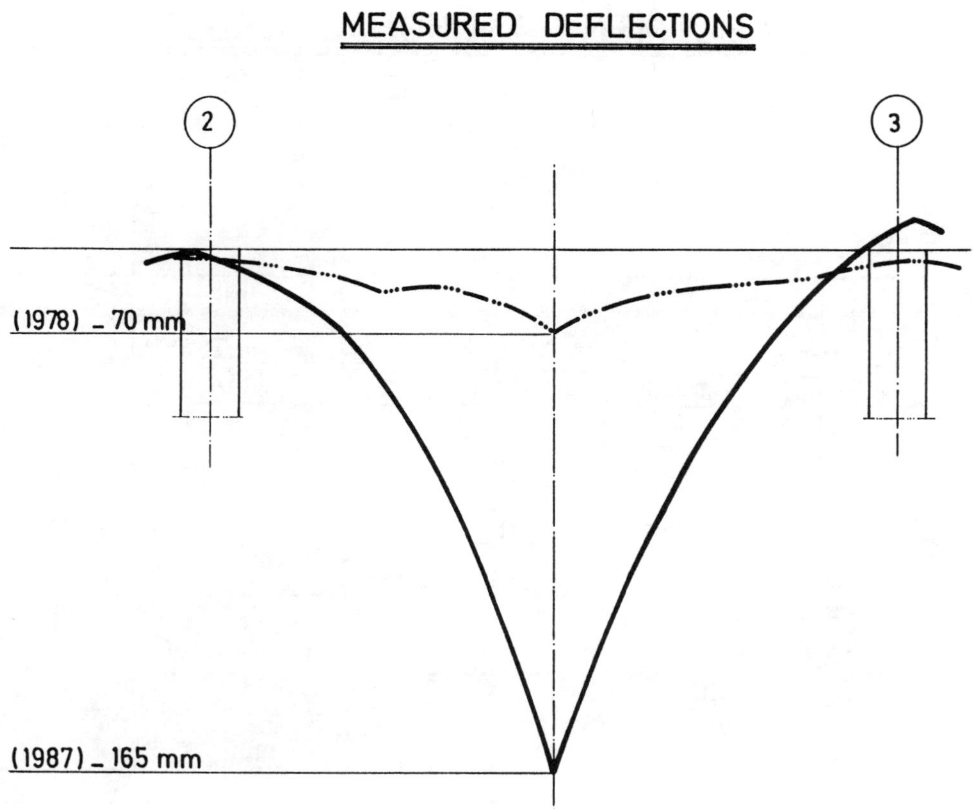

Fig. 4

ABUTMENT ANCHORAGE BLOCK

PLAN VIEW

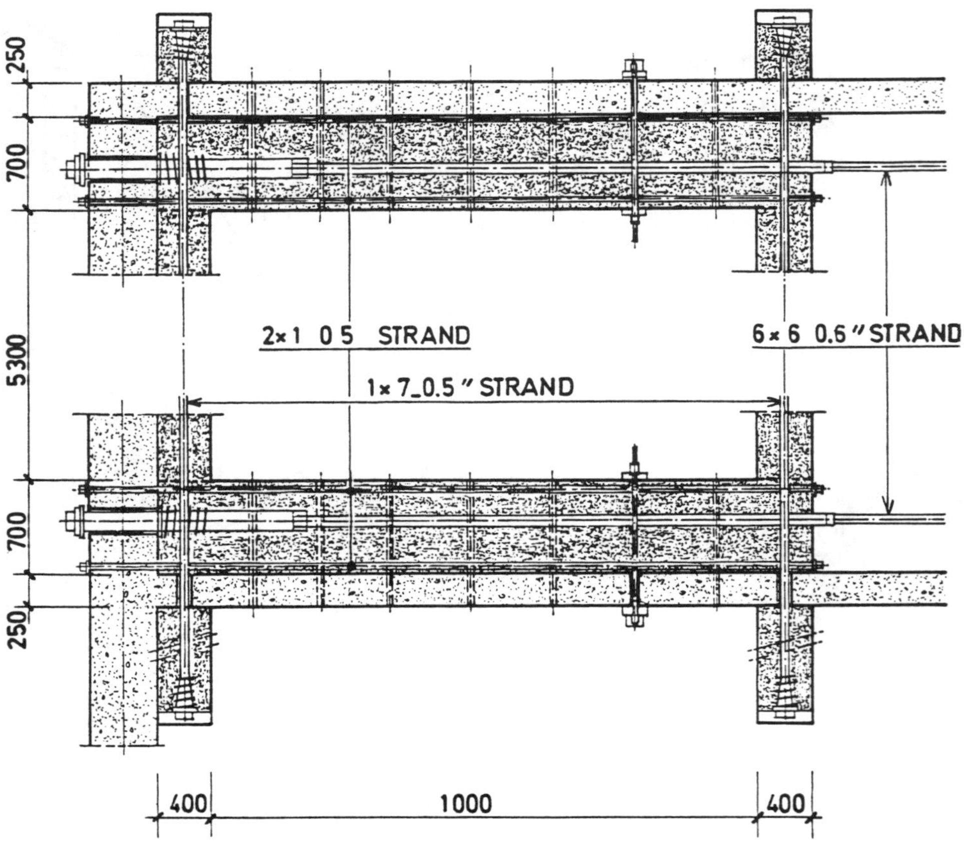

Fig. 5

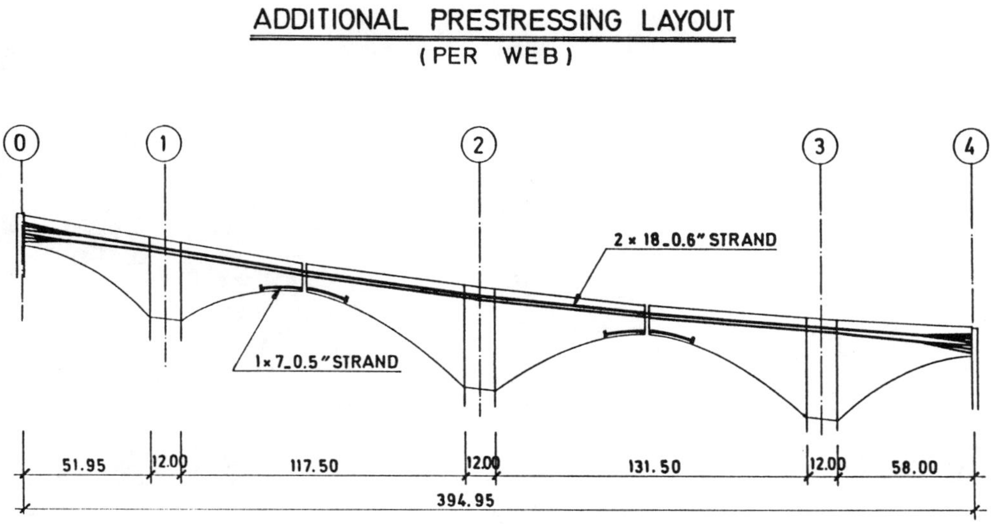
ADDITIONAL PRESTRESSING LAYOUT
(PER WEB)
0
1
2
3
4
2 × 18_0.6" STRAND
1× 7_0.5" STRAND
51.95
12.00
117.50
12.00
131.50
12.00
58.00
394.95

Fig. 6

FIBER COMPOSITES FOR BRIDGE REHABILITATION

MOHAMMAD R. EHSANI and HAMID SAADATMANESH
Department of Civil Engineering
University of Arizona
Tucson, AZ 85721 USA

ABSTRACT. Results from tests of one rectangular and one T-beam are reported. The beams were strengthened by epoxy-bonding a 1/4 in. (6 mm) thick Glass Fiber Reinforced Plastic plate along the tension face. It is shown that the nominal load carrying capacity of the beams was significantly increased. The parametric study presented indicates that the method is most effective when applied to beams with low flexural steel reinforcement ratios.

1. Introduction

About one-half of the 575,607 inventoried highway bridges in the United States are more than 30 years old and are in need of replacement or rehabilitation (FHWA (1986)). Many of these bridges were originally designed for lighter loads and lower traffic volumes than are common today. In order to adequately serve the present and future traffic needs, these bridges must be either replaced or strengthened to comply with current loads. In most cases, repair and retrofitting is more economical than replacing the structure with a new one.

Among the methods used to strengthen girders in existing structures are external post-tensioning and the addition of epoxy-bonded steel plates to the tension flange. External post-tensioning by means of high-strength cables is an effective method for increasing the flexural strength of girders and has been reported for application in steel bridges (Saadatmanesh et al. (1989a, b), Mancarti (1984), Berridge and Donovan (1956)). The method is, however, complicated because it requires external anchorage and protection against corrosion of the cables. Maintaining the lateral stability of the girder during the post-tensioning process adds more difficulty in the field.

A different approach for strengthening of girders is plating. In this technique, a steel plate is externally attached to the tension face of the beam by means of epoxy, bolts, or a combination of the two. The flexural capacity of the member is increased by a moment couple consisting of the tension force in the steel plate and an equal compression force in the flange. Several studies have reported on this technique (Macdonald and Calder (1982), Raithby (1980), Swamy et al. (1987), Maeda et al. (1980)). Many problems associated with the first method, such as anchorage of cables and concerns regarding lateral stability are eliminated here. The only drawback

A. S. Nowak (ed.), Bridge Evaluation, Repair and Rehabilization, 579–588.

to this repair technique is the maintenance required to prevent the corrosion of the steel plate. Such corrosion could adversely affect the bond between the plate and the beam and cause the failure of the beam.

To overcome the above shortcoming, corrosion-proof Glass Fiber Reinforced Plastic (GFRP) plates can be substituted for the steel plate. The light weight of GFRP plates (about 1/4 that of steel) will further simplify the construction. Additional benefits can be realized if the beam is cambered prior to attachment of the plate. This, in effect, results in external prestressing of the beam and introduces initial stresses in the beam which oppose those caused by service loads. This paper describes the results of a recent experimental and analytical study of this method conducted by the authors.

2. Fiber-Reinforced Plastics

Recent advancements in the fields of plastics and composites have resulted in the development of high-strength fiber-reinforced plastics (FRP) that surpass the strength and fatigue properties of steel. These materials have been successfully used in a variety of industries such as aerospace, automotive, and shipbuilding for almost two decades. However, their use in civil engineering application has been virtually nonexistent.

Fiber composites offer unique advantages for solving many civil engineering problems in areas where conventional materials fail to provide satisfactory service life. Unlike steel, FRP's are unaffected by electrochemical deterioration and can resist the corrosive effects of acids, alkalies, salts, and similar aggressive materials under a wide range of temperatures. Fiber composites are made of small fibers bonded together with a resin matrix. The fibers provide the composite with its unique structural properties; the resin serves as the bonding agent. The mechanical properties of composites vary with the amount and orientation of fibers in different directions.

Glass fibers are commonly used to reinforce FRP's. The modulus of elasticity of glass fiber-reinforced plastic (GFRP) plates is relatively low (about one-fifth that of steel). Therefore, structures made of GFRP will most likely deflect beyond acceptable limits. However, GFRP's can perform very well for certain structural applications when combined with conventional materials as presented here.

Other fibers such as kevlar or graphite can be used in lieu of glass to achieve significantly stiffer and stronger plates. But the cost of such composites at present is substantially higher than that of GFRP which costs approximately $2.00 per pound. Considering the low density of GFRP (about one-fourth that of steel), this material is a viable substitute for steel for the proposed application.

3. Experimental Study

Results from tests of two beams are presented here. The beams were simply supported with a clear span of 15 ft (4.6 m). One of the beams had a rectangular cross section 8-in. wide and 18-in. high (200x460 mm). The other beam was a T-section with the same web thickness and height as the rectangular beam. The flange was 24-in. wide and 3-in. thick (610x75 mm). Shear reinforcement for each beam consisted of No. 4 rectangular ties spaced at 6 in. (150 mm). A pair of equal concentrated loads 2-ft (600 mm) apart and symmetrically placed about the midspan were applied to each beam.

3.1. MATERIAL PROPERTIES

The average compressive strength of concrete in the specimens at the time of testing was 5040 psi (35 MPa). The average yield strength for the steel was 66.2 ksi (456 MPa). Samples of the GFRP plate were tested in uniaxial tension. The stress-strain response was linear with a modulus of elasticity of 5400 ksi (37200 MPa) and a breaking strength of 58 ksi (400 MPa).

In order to select the appropriate epoxy for the attachment of the GFRP plate to the beam, several small-scale beams were tested (Saadatmanesh and Ehsani 1990). It was concluded that tough epoxies are most suitable for this application and an appropriate epoxy was identified to be used for the two beams reported here. Each beam was strengthened with a GFRP plate 6 in. (150 mm) wide, 1/4 in. (6 mm) thick, and 14.5 ft (4.4 m) long. The average thickness of the epoxy was about 1/16 in. (1.5 mm).

3.2. TEST RESULTS

The rectangular beam was cambered prior to bonding of the GFRP plate. To simplify the cambering procedure, the beam was loaded upside down while in loose contact with an epoxy-coated GFRP plate and was held in that position until the epoxy cured completely. The beam was then rotated right-side up and placed in the test frame.

Flexural reinforcement for this beam consisted of 2 No. 4 (13 mm) bars for tension and compression. As a result, only a small camber could be induced in the beam. In field applications, due to the large self weight of structures, a larger cambering moment can be applied to offset all or parts of the dead load stresses. Plot of load vs. deflection at mid-span of the beam is shown in Fig. 1. The initial negative deflection indicates the camber in the beam after the beam was turned right-side up. The beam was loaded to 24 kips (107 kN); it was unloaded and reloaded to the failure load of 42 kips (187 kN). Failure of the beam was accompanied by shear failure of a layer of concrete above the epoxy bond line (Fig. 2). Similar behavior has been observed in some beams reinforced with steel plates (Oehlers 1988).

A computer program was developed to predict the load-deformation response of the beams. The program assumes linear variation of strains along the height of the beam and does not account for the possible slip that may take place at the epoxy line. The predicted response for this beam is also shown in Fig. 1, along with the theoretical behavior of the beam if the plate were not added. There is excellent agreement between the computer model and the measured data. The beam responded almost linearly up to a load of 22.5 kips (100 kN), correspond -ing to the yield of the tension steel. The post-yield response was also linear but with a reduced stiffness. It is noted that the yield load was increased by nearly two and a half times from 40 to 100 kN. In addition, while there is little difference between the yield and nominal strength of the unrepaired beam, for the strengthened beam, there is a significant reserved strength beyond the yield load.

Figure 3 shows the variation of strain in the GFRP plate at the mid-span. Up to the yield point, the measured data are in good agreement with the values predicted by the analytical model. However, for higher loads, and due to possible slip at the epoxy bond line, the measured strains in the plate are lower than the expected values. The plot does indeed verify the contribution of the GFRP plate to the load carrying capacity of the repaired beam.

Flexural reinforcement for the T-beam consisted of 2 No. 8 (25 mm) bars for tension and 3 No. 4 (13 mm) bars in the compression flange. The composite plate was bonded to the tension flange of this beam with no camber. The load vs. mid-span displacement for this beam is shown in Fig. 4. There is good agreement between the analytical and measured results. Plot of the load vs. strain in the GFRP plate, shown in Fig. 5, indicates a significant change of slope beyond the yield load of 47 kips (210 kN). This indicates that the contribution of the plate increases greatly after the steel yields. Due to possible inadequate beam surface preparation prior to bonding, the GFRP plate separated from the beam at a load approximately 90% of the theoretical nominal strength. Nevertheless, plating almost doubled the nominal strength of the beam.

It is noted that the study reported here has examined the static strength of retrofitted beams only. Additional investigations are required to evaluate the performance of such beams subjected to fatigue loading.

4. Parametric Study

In order to examine the effect of different design variables such as steel reinforcement ratio, concrete compressive strength, and plate area and strength on the strength of the retrofitted beams, a comprehensive parametric study was conducted. Due to space limitations, only the results for a T-beam constructed with 3000 psi (21 MPa) concrete (i.e., T3 series) are presented here. The beam section was 18 in. (460 mm) high, with a 32 in. (810 mm) by 4 in. (100 mm) flange and an 8 in. (200 mm) thick web. The section was analyzed using the computer program mentioned previously and moment vs. curvature diagrams were plotted for each section. Failure was assumed when either the concrete reached a strain of 0.003 or when the GFRP plate reached its ultimate strength.

Each beam in the T3 series is identified with three additional characters. The first letter (L or H) corresponds to the beam flexural reinforcement ratio being low (0.5%) or high (1.5%). The next numeral (5 or 10) represents the modulus of elasticity and tensile strength of the GFRP plate; 5 is used for plates with a modulus of 5,000 ksi (34450 MPa) and tensile strength of 60 ksi (413 MPa); 10 is used when both of these values have been doubled to indicate a stiffer and stronger plate. The last letter (L, M, or H) indicates the area of the GFRP plate as a percentage of the beam cross sectional area with the ratio being 0.25% for L, 0.5% for M, and 1.5% for H.

Figure 6 shows the results for a beam with low steel ratio, and various plate areas. For comparison, the plot for a beam without any plate (T3L) is also included. It is clear that the addition of the plate increases the yield and nominal moments of the section with the latter increasing as much as 247% when a large plate area is used. As shown in Fig. 7, for a section with high steel ratio, the nominal moment could be increased by a maximum of 83%. This indicates that the method is most effective for sections with low steel reinforcement ratios.

The effect of plate strength and stiffness can be observed by comparing Figs. 6 and 8. For the best case shown in Fig. 8 (T3L10H), the nominal moment capacity was increased by some 500%. This clearly indicates the advantages of utilizing stiffer and stronger plates for such applications.

5. On-going Studies

Composite materials offer unique properties for many structural engineering applications. One such study currently in progress at the University of Arizona is strengthening of columns. In conjunction with strengthening of girders in a bridge, the piers may require strengthening too. In addition, many older bridge piers located in seismic zones, may require retrofitting to supplement the insufficient ties used in the original design. The 1979 Loma Prieta Earthquake in California is the most recent case in which insufficient lateral ties were a major contributing factor to the collapse of the Cypress Street Viaduct.

An effective and economical method to increase the ductility and shear resistance of concrete columns in existing bridges is to externally confine and/or laterally prestress the columns with composite straps. The straps are constructed from high-strength and flexible fibers weaved to form a fabric-like material of desired width and length. Materials such as kevlar or graphite are excellent candidates for this application. Kevlar fibers, for example, are extremely strong in tension and can reach an ultimate strength of well above 400 ksi. The reinforcement is performed by wrapping straps of desired widths around the columns. The straps can be wrapped in a continuous spiral and/or discontinuous rings. The fabrics can be made very thin, i.e. less than 50 mils thick. This results in sufficient flexibility for the straps to be wrapped around circular as well as rectangular columns.

Additional advantages can be gained at the time of strengthening by laterally prestressing the columns. The prestressing can be accomplished by wrapping the straps around the column while under tension. The lateral confining pressure produced in such a manner will increase the compressive load carrying capacity of the concrete considerably. In the event of severe seismic loading, for example, such confinement will prevent the shell region of the column from spalling. This will prevent buckling of the longitudinal bars and will make the entire concrete section effective in resisting the loads. A pilot study on this application is currently underway where the gain in strength and ductility of concrete cylinders is being investigated.

6. Summary and Conclusions

Test results of a rectangular and a T-beam are reported. Each beam was strengthened by epoxy bonding a 1/4-in. thick Glass Fiber Reinforced Plastic plate to its tension flange. In the case of the rectangular beam, the beam was cambered prior to the attachment of the plate. The flexural strength of the repaired beams was significantly increased in each case.

A parametric study was also reported on the effect of the flexural strength ratio, stiffness and strength of the plate, and the area of the plate as a percentage of the beam cross sectional area. It was shown that most gain in nominal strength can be obtained for beams with low flexural reinforcement ratios.

7. References

Berridge, P.S.A., and Donovan, L.H. (1956) "Prestressing Restores Weakened Truss Bridge," Civil Engineering, ASCE, 26(9), pp. 48-49.

FHWA (1986) "Highway Bridge Replacement and Rehabilitation Pro-gram," 8th Annual Report of the Secretary of Transportation to the Congress of the United States, Bridge Division, Office of Engineering, Federal Highway Administration, Washington, D.C.

Macdonald, M.D., and Calder, A.J.J. (1982) "Bonded Steel Plating for Strengthening Concrete Structures," International Journal of Adhesion and Adhesives (Guilford), 2(2), pp. 119-127.

Maeda, Y., Matsui, S., Shimada, I., and Kato, H. (1980) "Deterioration and Repairing of Reinforced Concrete Slabs of Highway Bridges in Japan," Technology Reports (Osaka University), 30(1599), pp. 135-144.

Mancarti, G.D. (1984) "Strengthening of California Steel Bridges by Prestressing," Transportation Research Record No. 950, Transportation Research Board, pp. 183-187.

Oehlers, D.J. (1988) "Reinforced Concrete Beams with Steel Plates Glued to their Soffits: Prevention of Plate Separation Induced by Flexural Peeling," Report No. R80, Department of Civil Engineering, the University of Adelaide, Adelaide, Australia.

Raithby, K.D. (1980) "External Strengthening of Concrete Bridges with Bonded Steel Plates," Supplementary Report No. 612, Trans-port and Road Research Laboratory, Crothorne, 17 pp.

Saadatmanesh, H., Albrecht, P., and Ayyub, B.M. (1989a) "Experimental Study of Prestressed Beams," Journal of Structural Engineering, ASCE, 115(9), pp. 2344-2364.

Saadatmanesh, H., Albrecht, P., and Ayyub, B.M. (1989b) "Analyti-cal Study of Prestressed Beams," Journal of Structural Engineering, ASCE, 115(9), pp. 2365-2382.

Saadatmanesh, H., and Ehsani, M.R. (1990) "Fiber composite plates can strengthen concrete beams," ACI, Concrete International, 12(3), pp. 65-71.

Swamy, R.N., Jones, R., and Bloxham, J.W. (1987) "Structural Behavior of Reinforced Concrete Beams Strengthened by Epoxy-Bonded Steel Plates," The Structural Engineer (London), 65A(2), pp. 59-68.

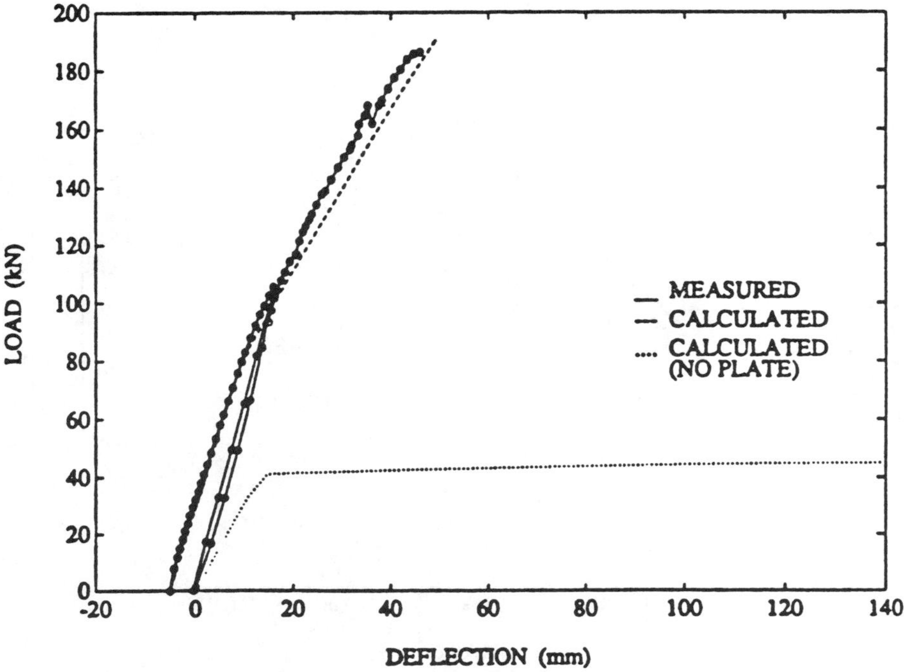

Fig. 1 Load vs. Deflection at Midspan for the Rectangular Beam.

Fig. 2 Rectangular Beam at Failure.

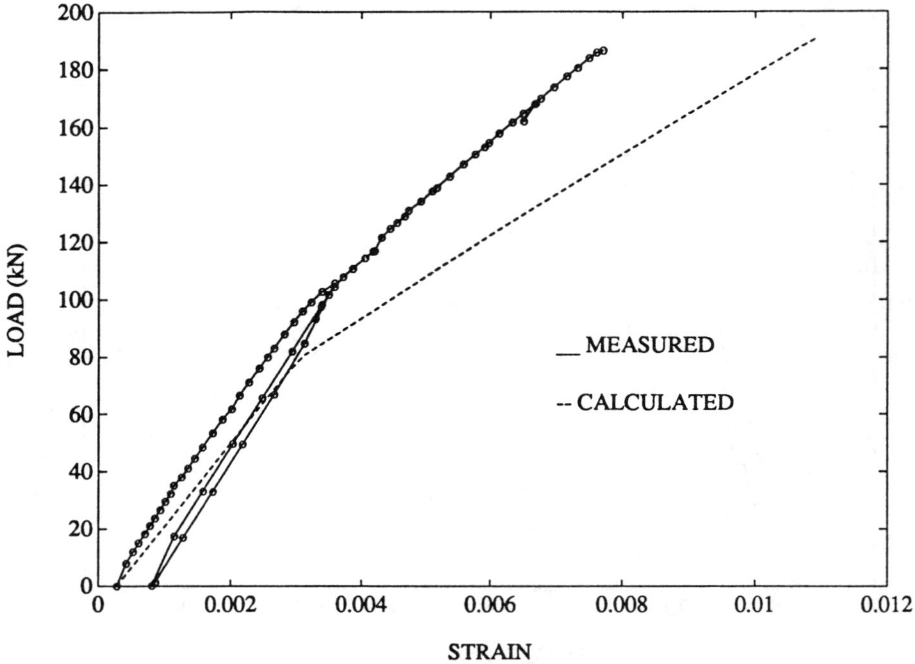

Fig. 3 Load vs. Strain in GFRP Plate at Midspan for the Rectangular Beam.

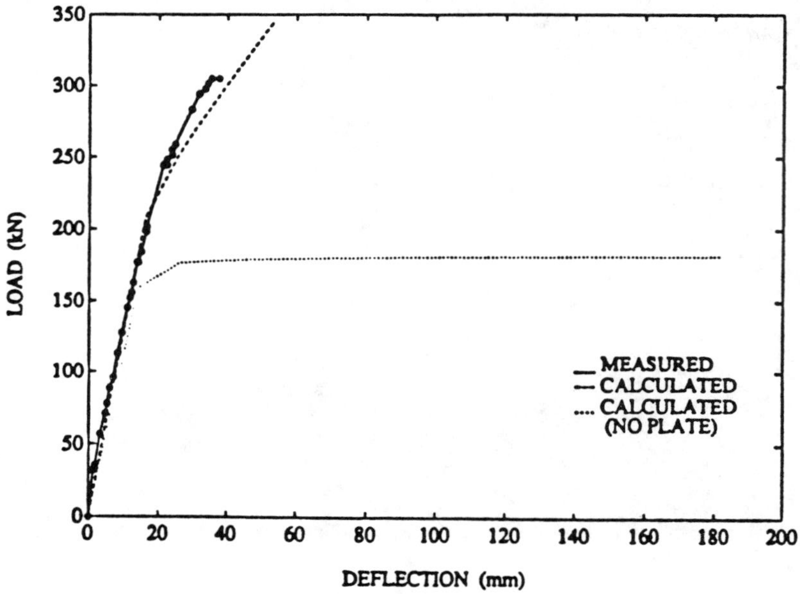

Fig. 4 Load vs. Deflection at Midspan for the T-Beam.

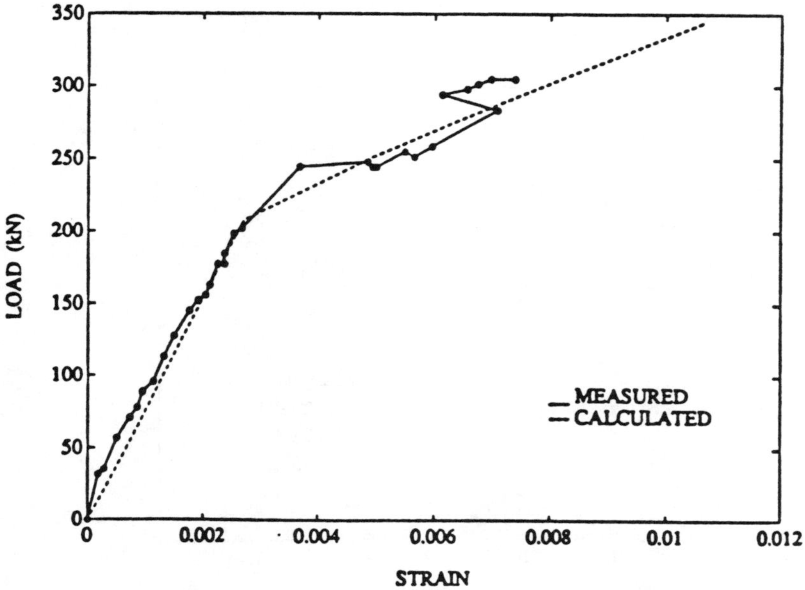

Fig. 5 Load vs. Strain in GFRP Plate at Midspan for the T-Beam.

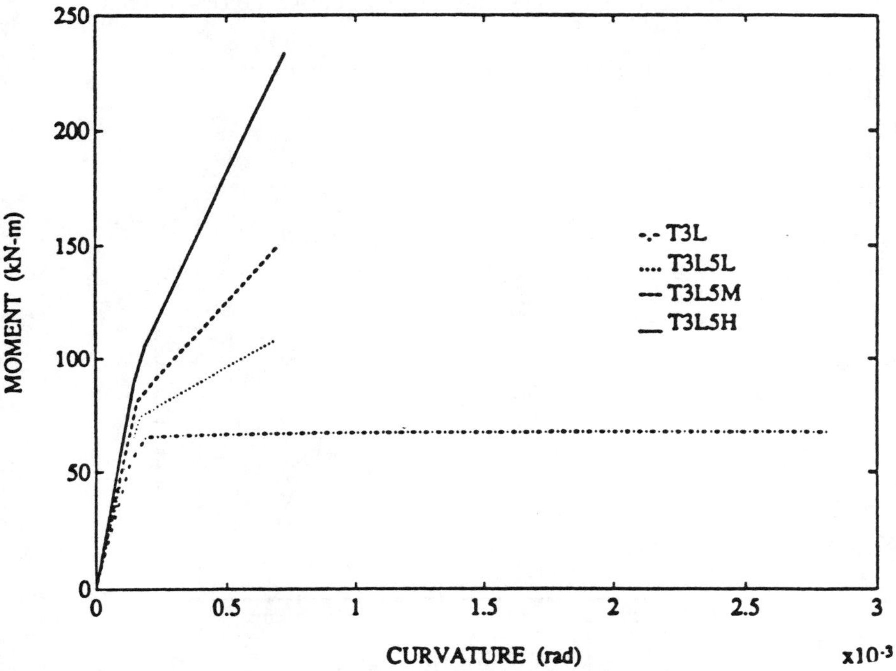

Fig. 6 Moment vs. Curvature for Specimens with Low Flexural Reinforcement Ratio.

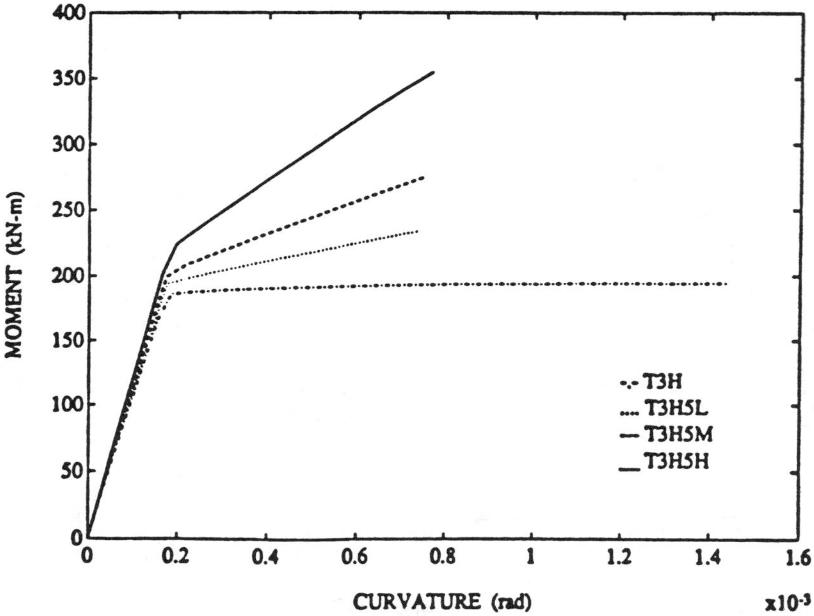

Fig. 7 Moment vs. Curvature for Specimens with High Flexural Reinforcement Ratio.

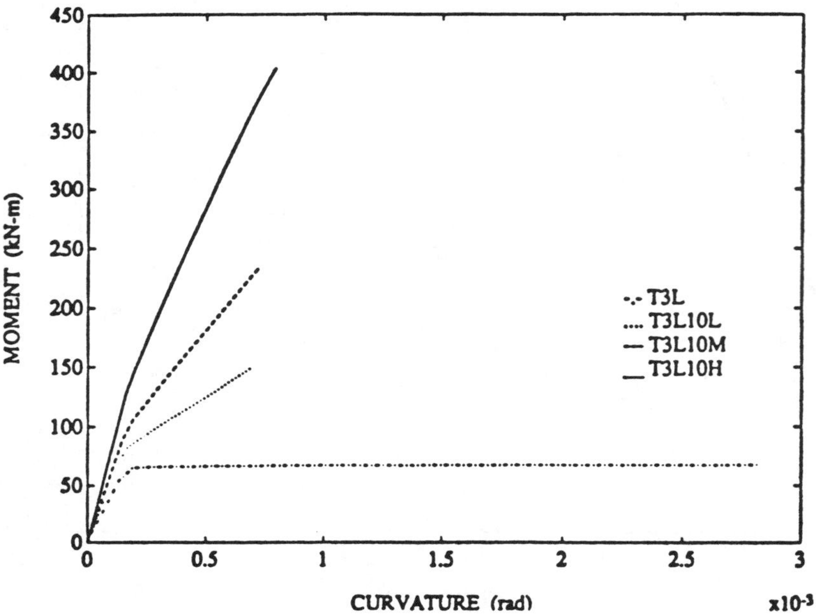

Fig. 8 Moment vs. Curvature for Specimens with Stiffer and Stronger Plates.

REHABILITATION OF THE HARRY W. NICE MEMORIAL BRIDGE

JAMES G. LUTZ
Greiner, Inc.
2219 York Road
Timonium, MD 21093 USA

ABSTRACT. After 43 years of service, the deck of the two-lane Harry W. Nice Memorial Bridge was in need of rehabilitation. The Maryland Transportation Authority also desired to strengthen the bridge, widen the roadway and perform other necessary repairs. Since the bridge is revenue bond project, it was essential to minimize loss of tolls caused by the construction. Preliminary concepts for traffic control plans and deck rehabilitation indicated disruptions to traffic over an estimated 45 months. During design, innovative alternative concepts were proposed by Greiner, Inc. After review and an exchange of information with the public, the Authority accepted these proposals, which significantly reduced construction time, construction costs, and loss of tolls. The history of the project, development of these innovative concepts, other rehabilitation techniques, and the construction operations are summarized.

1. History of the Project

1.1 DESCRIPTION OF THE BRIDGE

The Harry W. Nice Memorial Bridge carries U.S. 301 across the Potomac River in Charles county, Maryland approximately midway between Baltimore, Maryland and Richmond, Virginia.

It was constructed in 1940 as a revenue bond project and is a part of Maryland's system of toll facilities currently administered by the Maryland Transportation Authority.

The bridge is 10,050 feet long. The superstructure consists of 60 simple beam spans totalling 3,689 feet, 12 Wichert girder spans totalling 1,765 feet, 10 Wichert deck truss spans totalling 3,057 feet, and a three-span through cantilever truss unit 1,539 feet long over the river's channel (Figure 1).

The original roadway was 24 feet wide with steel safety walks and railings. In the beam spans the original deck was reinforced concrete, while in all other spans the deck was a concrete filled steel grid with a 3/4 inch overfill.

A. S. Nowak (ed.), Bridge Evaluation, Repair and Rehabilization, 589–602.

1.2 BRIDGE TRAFFIC

When the bridge was opened to traffic, the Virginia approach was a gravel road. As traffic steadily increased, the approach roadways were improved, ultimately to four-lane dual highways.

In 1964, prior to the opening of I-95, a parallel route to the west of U.S. 301, the annual traffic had reached 3,300,000 vehicles. With the opening of I-95 there was a sharp decline in traffic to a low of 1,9000,000 vehicles in 1967. After that, recovery was steady to a high of 3,600,000 vehicles in 1985, with a peak daily traffic of approximately 19,000 vehicles.

1.3 PLANNING FOR REHABILITATION

By 1982, after more than 40 years of service, the deck was in need of rehabilitation. In conjunction with the deck work, the Authority desired to increase the structure's live load capacity, to widen the roadway as much as practical, to replace the original steel safety walks and railings, and to perform any other needed repairs.

Since the traffic volumes had only recently recovered from the diversion of traffic to I-95, the Authority was particularly concerned to minimize the adverse effects of the rehabilitation on users of the bridge.

Greiner was retained to perform a detailed inspection of the substructure and superstructure of the bridge, excluding the deck, to perform rating analyses, to identify elements requiring repairs or strengthening, and to determine the feasibility of widening the roadway.

The Authority also contracted with another consultant to perform a deck rehabilitation study including concepts techniques for deck rehabilitation. The Authority's criteria for traffic control included a requirement to maintain at least one lane of traffic at all times and a limitation to no more than one zone of restricted traffic at any one time.

Subsequently, Greiner was retained by the Authority to provide designs and contract documents for the rehabilitation contract based on the investigative engineering contracts.

2. Design of the project

2.1 DECK REHABILITATION STUDY

In conformity with the Authority's criteria, the consultant for the deck rehabilitation study recommended the traffic control plan shown in Figure 2 for all deck rehabilitation operations. It required the closing of one lane for a single construction zone of 250 feet and the controlling of alternating directional traffic in the other lane with traffic signals. Temporary concrete safety varies and traffic drums separate the traffic from the work and protect against drop-off. With the completion of all work in a zone, the entire 1,225 foot system would be advanced 250 feet.

During the first stage, all work was to be completed in one lane for the full length of the bridge while alternating traffic was maintained in the available 10'-2" traffic lane at a maximum speed of 40 miles per hour. In the second stage the roles of the two lanes were to be reversed with a slightly wider lane available for traffic.

Two deck rehabilitation procedures were recommended. In the girder and truss spans the concrete overfill in the steel grid was to be scarified, deteriorated grid fill or grid were to be replaced, and a two-inch latex modified concrete wearing surface placed. In the original beam spans (Figure 3), it was recommended that the cast-in-place reinforced concrete deck be replaced with longitudinally prestressed full span length (61'-6") precast concrete panels bearing on the original cross beams, with transverse post-tensioning through the panels, and a latex modified concrete wearing surface. Precast concrete safety barriers were to be anchored to the exterior panels (Figure 4).

Replacement of the entire bridge deck by these methods and with the proposed traffic control plan was estimated to require 45 months.

2.2 GENERAL FINDING OF RATING ANALYSES

The structure was originally designed according to AASHO Design Specifications of the era using AASHO H-20 truck train loading or equivalent loading. The rating analyses by Greiner demonstrated that for AASHTO HS-20 loading under current AASHTO Design Specifications, it was not practical to significantly widen the roadway, and even without widening, extensive superstructure strengthening was required.

2.3 ALTERNATIVE PROCEDURES IN BEAM SPANS

As shown in Figure 4 strengthening of the beam spans required two additional stringers. At this point, Greiner proposed an alternative method of construction and an alternative traffic control plan for the beam spans.

If the bridge could be closed to all traffic for limited hours during periods of minimal traffic, entire spans - deck and structural steel - could be removed and immediately replaced by preconstructed spans - deck and new structural steel. Although this violated the Authority's original criteria to maintain traffic in one lane at all times, they deemed the alternative worthy of consideration. Traffic and structural studies were authorized.

Two experienced contractors were asked to review Greiner's preliminary concept for span replacement. As envisioned, a typical work period would involve removal of a temporary span set in the previous work period, setting of a replacement span, removal of an adjacent original span and its replacement with a temporary span. Both contractors confirmed that these operations could be accomplished during an eight-hour nighttime closure. The temporary spans were considered necessary to provide for setting new anchor bolts and replacement of deteriorated bearing pads.

Recognizing that the preconstructed span work could proceed in the same calendar period with the two-stage procedure in the other spans. the estimated total period for restrictions to traffic could be reduced by the 17 months estimated for deck replacement in the beam spans. Detailed analysis of the available toll records, and an origin and destination survey showed that the combination of 60 nightly closures for the setting of the spans and traffic restrictions for only 28 months decreased the estimated loss of tolls be $1,380,000 from that estimated for 45 months of restricted traffic during the two-stage plan for the full length of the bridge.

Preliminary designs for the preconstructed spans included compositely designed welded steel girders 42" deep, and lightweight reinforced concrete deck and safety barriers, Figure 5. Composite design, the deeper girders customized to the loading and the lightweight concrete all minimized the new structural steel for this alternative. In fact, the weight of the new structural steel for the three girders is approximately the same as that of the two additional beams required to strengthen the original system.

Comparative estimates for the two proposed beam span rehabilitation systems indicted the preconstructed span alternative offered a construction cost saving of $765,000 in approximately $5.5 M.

After contacts with local governments and civic organizations, and public meetings attended by private citizens, businesses and interested associations, the preconstructed span alternative was adopted by the Authority with the proviso that there would be no more than one full bridge closure per week from midnight Saturday to 8:00 a.m. Sunday.

2.4 FINAL PLANS AND CONTRACT DOCUMENTS

Final plans and contract documents for the rehabilitation of the Harry W. Nice Memorial Bridge included:

- Traffic control plans for two-stage construction in the girder and truss spans.
- Traffic control plans for full bridge closures during beam span replacement on Saturday nights.
- Signing in the vicinity of the bridge to advise motorists of traffic restrictions and closures, plus remote signing at principal decision points in the vicinity of Baltimore, Washington, Richmond, and Fredericksburg for highways leading to U.S. 301.
- Beam span replacement with preconstructed spans including stringers, framing, roadway joints, and lightweight reinforced concrete deck. Alternatives for lightweight concrete safety barriers to be precast and anchored, of cast on the new deck before installation of the spans.
- Details for a temporary beam span with bearing clear of the bearing pads on the bents for the original three stringers.
- Removal of the concrete overfill in the concrete filled steel grid deck of the girder and truss spans, repairs to the fill and grid, placement of a 2" latex modified concrete wearing surface and modifications to roadway joints.
- Removal of the original steel safety walks and rails in the girder and truss spans and replacement with precast lightweight concrete safety barriers attached to the existing deck crossbeams.
- Strengthening of the existing floorbeams.
- Strengthening of Wichert girders be the installation of vertical struts in the rhomboids to make the spans continuous and the addition of cover plates as necessary.

- Strengthening of Wichert deck truss spans by the installation of vertical struts in the rhomboids and the addition of material to certain truss chords.
- Strengthening of the through truss members as necessary the retrofitting of the cantilever span hangers reduce the unit bearing stress at the pins.
- Miscellaneous repairs to corroded secondary members and replacement of corroded or failed rivets.
- Replacement of concrete bearing pads under all stringers in the beam spans.
- Reinforced concrete encasement of cartain cracked bent caps under beam spans and miscellaneous repairs to spalls on the bents.
- Reinforced concrete encasement, placed in the dry, of all bent struts at the waterline.
- Replacement of the existing navigational lighting system.
- Painting of all original and new structural steel.
- Miscellaneous related work.
- There were two incentive and disincentive clauses. One for the span replacements limited to 32 times an established amount per week. The other for all other work requiring restriction to bridge traffic limited to 108 times an established daily amount.

3. Rehabilitation Construction Contract

3.1 BIDDING

Three contractors bid this rehabilitation project in November 1984. The low bid of $17,541,255 by CIANBRO of Pittsfield, Maine was 99% of the engineer's estimate.

3.2 PREPARATORY WORK

Early in 1985 the contractor developed a staging area adjacent to the Maryland abutment. The area provided for offices, material storage, transfer of materials from land to water, two casting beds for the preconstructed spans and an area for demolition of removed spans. Completed spans were stored on barges.

3.3 SPAN REPLACEMENT

The contractor averaged the removal (Figure 6) and replacement (Figure 7) of two spans per week. This rate of progress was due in part to the advance drilling for anchor bolts and approval of a contractor's request to substitute methyl methacrylate polymer concrete for the portland cement concrete specified for bearing pads on the bents. These measures eliminated the need for the setting and removal of temporary spans. The contractor also elected to cast the safety barriers on the new deck before installing the span.

Figures 8 and 9, respectively, provide comparisons of the original and rehabilitated deck sections, and an elevation view of a replacement girder span adjacent to an original beam span.

3.4 SUBSTRUCTURE REPAIRS

Rehabilitation of the bents supporting the replacement girder spans was planned to include encasement of all bent struts at the waterline and encasement of only certain caps where significant cracks had occurred at span anchor bolts. During the course of the work the Authority negotiated an extra work order with the contractor to encase all bent caps with reinforced concrete and to coat the strut encasements with a combination curing compound and protective coating. Figure 10 provides a view of the rehabilitation bents.

3.5 REHABILITATION IN GIRDER AND TRUSS SPANS

Concurrent with the span replacement, rehabilitation was in progress in the girder and truss spans while traffic was maintained in accordance with the two-stage traffic control plan. Figure 11 shows the deck work in progress in the through truss span emphasizing the narrow spaces available to traffic and for the rehabilitation operations. Figure 12 shows the approach to a work zone in the deck truss area where the traffic and work do not appear to be as confined. In the right foreground the latex modified concrete wearing surface has been placed and the precast safety barrier has been installed.

Two factors assisted the contractor to expedite the work in the girder and truss spans. After observing traffic operations at the early work zones, the Authority agreed to extend the work zone from 250 to 350 feet. The contractor also took advantage of the weekly full closures for span replacement to supply and perform some deck rehabilitation operations.

Strengthening of the girder and truss spans from the deck generally kept pace with the deck rehabilitation in those spans. However some operations could be performed from outside of the roadway, such as where the Maryland approach was over land. An example of this type of work is provided in Figures 13 and 14, respectively, showing the original Wichert girder rhomboid and the same rhomboid with a vertical strut installed to create continuity throughout the Wichert girder unit.

3.6 PROJECT PROGRESS

The first limitation to traffic were in April 1985. The last replacement span was set on December 1, 1985 and the last restriction to traffic caused by the two-stage construction procedure was July 18, 1986. Traffic disruption only extended over a 15 1/2 month period rather than the originally estimated 45 months. The contractor earned the full amounts for incentive established in the incentive-disincentive clauses of the contract. All other original contract work was completed within the allotted contract time and authorized extra work was completed during January 1987.

3.7 PROJECT COST

The final value of the contract was $19,801,853 including the incentive payments (945,000) and encasement of the bent caps ($637,964). Variances in quantities and miscellaneous extra work amounted to $677,635 or less than 4% of the bid.

4. Summary

The rehabilitation of the Harry W. Nice Memorial Bridge providing increased structural capacity, extended service life and improved safety was accomplished with a minimum of disruption to traffic and loss of toll revenues. These goals were achieved because the Maryland Transportation Authority was willing to accept innovative ideas from its engineer and the contractor, and because the public, given an opportunity to understand the problem and proposed alternative solutions, was able to accept short-term inconveniences for long-term improvements.

Fig. 1 The Harry W. Nice Memorial Bridge, U.S. 301 over the Potomac River, Charles County, Maryland.

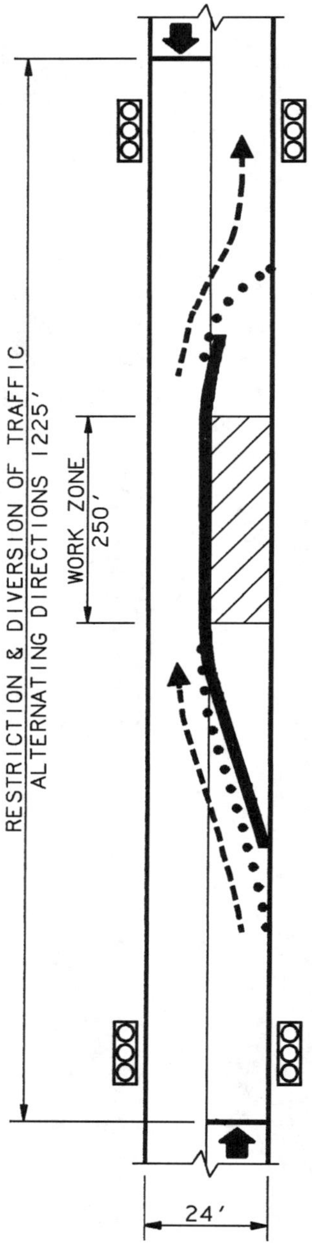

Fig. 2 Traffic Control Plan for Two-Stage Construction.

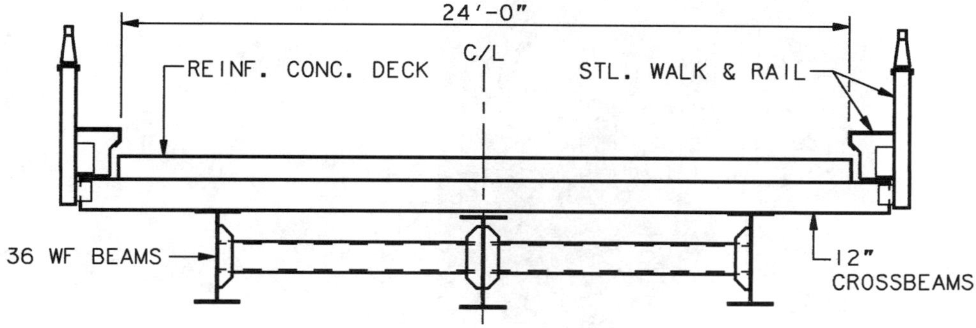

Fig. 3 Original Beam Span Cross Section - 60 Spans at 61'-6".

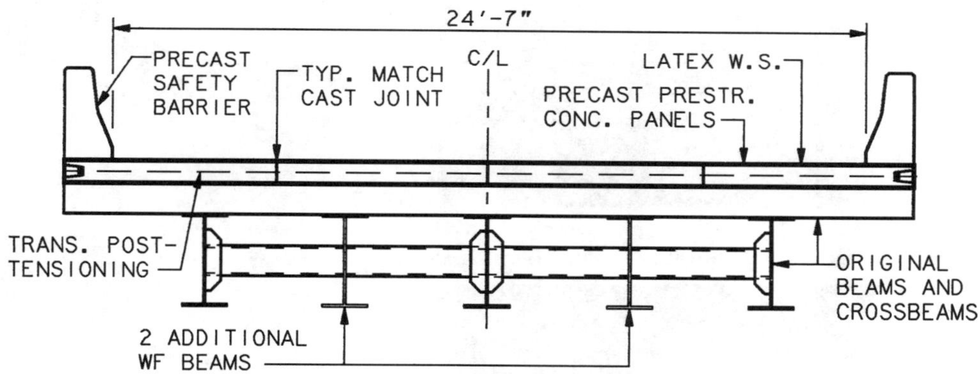

Fig. 4 Preliminary Concept for Strengthening and Deck Replacement in Beam Spans with Two-Stage Construction.

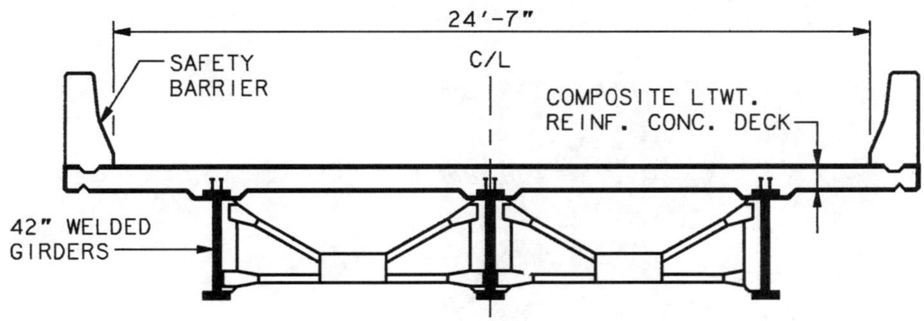

Fig. 5 Final Concept for Strengthening and Deck Replacement in Beam Spans Using Preconstructed Girder Spans.

Fig. 6 Night time Removal of Original Beam Span.

Fig. 7 Same Night Replacement of New Preconstructed Girder Span.

Fig. 8 Comparison of Original and Replacement Deck Sections.

Fig. 9 Comparison of Original and Replacement Span Elevations.

Fig. 10 Overview of Replaccment Spans also showing Encasement of Bent Caps and Struts.

Fig. 11 Two-Stage Traffic Control Plan in Operation in Through Truss Spans.

Fig. 12 Two-Stage Traffic Control Plan in Operation in Deck Truss Spans.

Fig. 13 Original Rhomboid of Wichert Girder Span.

Fig. 14 Rhomboid after Installation of Strut to Achieve Continuity in Wichert Girder Span Unit.

DURABILITY OF CONCRETE BRIDGE DECKS WITH REDUCED REINFORCEMENT

H. DAGHER, W. HAMILTON, and S. KULENDRAN
Civil Engineering Department
Boardman Hall
University of Maine
Orono, ME 04469 USA

ABSTRACT. The premature deterioration of concrete bridge decks in regions where deicing chemicals are used has created a multi-billion dollar problem in the United States. In this paper, a simple method to extend the life of new concrete bridge decks is explored: the reduction of deck reinforcement. When a reinforcing bar corrodes, its volume will increase, and eventually, the surrounding concrete will crack. With less steel, there will be less material corroding and expanding, and a longer deck life may result. Rather than attempt a precise proof for this thesis, a number of supporting arguments are made. An approximate procedure to estimate the possible gains in deck life resulting from reduced reinforcement is described. The procedure is based on a smeared-cracking finite element model.

1. Introduction

Concrete bridge deck durability is of particular concern in regions where deicing chemicals are used. In these regions, the heavy use of deicing salts accelerates the corrosion of the reinforcement, and leads to a premature deterioration of the deck. Corroding bars expand and cause the surrounding concrete to crack. When cracks originating from two or more adjacent bars meet, the concrete spalls.

In this paper, the potential of extending the life of new concrete bridge decks by reducing the reinforcement below the current AASHTO requirements [9] is assessed. With less steel, there will be less material corroding and expanding inside the deck, and a longer deck life may result. While the idea is simple, it is extremely difficult to prove, short of building a number of decks with reduced reinforcement, and monitoring their performance over an extended period of time.

Rather than attempt a precise proof for the above thesis, a number of supporting arguments are made. In addition, it is shown how a smeared-cracking finite element model can be used to study the relationship between deck life and the amount and distribution of deck reinforcement.

A. S. Nowak (ed.), Bridge Evaluation, Repair and Rehabilization, 603–615.

2. Amount of Steel Reinforcement in Concrete Decks

Throughout this paper, "less reinforcement", or "reduced reinforcement" refers to a reduction in deck reinforcement below the current AASHTO requirements, but not below .3 percent. Also a "deck" refers to an exposed reinforced concrete bridge deck.

The amount of reinforcing steel specified for concrete bridge decks varies considerably between the AASHTO and the Ontario bridge codes [9,10]. The Ontario method, referred to as isotropic reinforcement, calls for placing equal amounts of steel in both directions in quantities necessary to limit shrinkage and temperature cracking [10,18,20]. The amount of steel that can be saved by this method is up to 60% that of the current AASHTO requirements [2,4]. The savings over current AASHTO designs can be realized because the deck load-carrying capacity is controlled not by flexure, but by punching shear and deck arching [7,8,12,14,18,20]. Figure (1) shows a typical isotropic bar placement.

The use of isotropic reinforcement for concrete bridge decks has saved the Ontario Ministry of Transportation and Communication more than $1 million annually in steel reinforcing costs [4]. It has been estimated that the method would save Florida $7 million per year [4]. Despite the savings in steel costs that isotropic reinforcement can realize, the current AASHTO bridge specifications [9] do not cover its use. As a result, only a handful of states have experimented with isotropic reinforcement.

While the Ontario bridge code has adopted a reduction in deck reinforcement [10], the tendency in AASHTO throughout the years [13] has been to increase the distribution steel in order to control deck cracking (Figure 2). The major question slowing down the acceptance of reduced deck reinforcement in the AASHTO bridge specifications is not a strength issue. It is very well established that decks with isotropic reinforcement have adequate static strength [7,8,12,14,18,20]. The pivotal concern is the effect of reduced reinforcement on long-term deck durability.

3. Arguments Supporting the Use of Reduced Steel

3.1 EFFECT OF REDUCED REINFORCEMENT ON DECK DURABILITY

The prospect of using less steel in bridge decks is of particular significance in regions where deicing chemicals are used. With less steel, less material will be subject to corrosion and expansion, and a longer deck life may result. While a longer deck life is possible, some arguments are made to the contrary. Of particular concern are: (a) the possibility that less steel will result in more cracking of the deck, thereby allowing deicing chemicals easier access to the steel and (b) the possibility that less steel will reduce the fatigue life of the deck. The research reported in this paper did not include a direct study of the fatigue problem, however, both concerns are addressed below.

3.1.1 *Relationship Between Reduced Steel and Deck Cracking, Excluding Fatigue.* While reduced steel may result in more deck cracking, it is unclear whether reduced steel will increase deck cracks of the kind that can accelerate corrosion. Of all cracks, those that run parallel and directly over the reinforcing are considered most capable of accelerating corrosion.

On one hand, less steel may result in more flexural, temperature and shrinkage cracking. On the other hand, less steel causes less subsidence cracking [23]. Flexural, temperature and shrinkage cracks do not necessarily run parallel and directly over the reinforcement; however, subsidence cracks run parallel to and directly over the reinforcement, and are more likely to initiate corrosion than flexural, temperature and shrinkage cracks [23].

3.1.2 *Relationship Between Reduced Steel and Deck Cracking, Including Fatigue.* Deck spalling has often been reported to be concentrated in heavily travelled areas of the deck, e.g. [1]. Hence, the deck fatigue loading is possibly an important factor contributing to concrete cracking and spalling. Of concern here is the potential increased detrimental effect of this fatigue loading on a deck with reduced reinforcement.

The available experimental evidence relating the deck fatigue life to the amount of reinforcement is very limited. Existing experimental research on concrete bridge deck fatigue life has not accurately modeled actual traffic wheel loads. Fatigue tests have used a single pulsating concentrated load, and in one case, a stepwise-moving pulsating load. Recently, limited fatigue tests on reinforced concrete bridge decks using a more realistic simulated single moving wheel load have been conducted [3,5]. It was concluded that one passage of a single moving wheel load is equivalent to between 34 and 1800 fixed pulsating load cycles. The study however used a polyurethane-coated steel wheel which may have caused excessive damage to the deck.

Contrary to what one might expect, the moving-wheel-load study showed that less steel may not be detrimental to the fatigue/cracking life of concrete decks. Results from the small-scale models indicated that the fatigue life of decks reinforced using the Ontario Code is about 20 times the fatigue life of decks reinforced according to the AASHTO requirements [3,5]. These results are based on 1/6 scale models and have been questioned by many. However, even if these results are inaccurate by as much as a factor of 20, the fatigue/cracking life of the "Ontario deck" will still be at least comparable to that of the "AASHTO deck".

3.2 EFFECT OF CRACK WIDTH ON CORROSION RATE

Even if less deck reinforcement results in larger crack width, it is still not clear how crack width affects long-term corrosion rates. After reviewing evidence from exposure tests on reinforced concrete members relating to the influence of cracking on corrosion, Beeby concluded that there is "no reason to believe that any relationship exists between crack width and corrosion" [15]. Crack width will affect the time it takes for corrosion to start, i.e., the time from construction to initiation of corrosion. However, this depassivation period is usually small compared to the life of the deck (about 5 to 10 percent of the life of the unprotected deck). Once corrosion starts, it is assumed to proceed at a relatively constant rate [15,21] which depends on the environment, the cover, the bar size, the bar arrangement, and the concrete quality.

Beeby gives the following hypothetical example [15]. Consider two cracks A=.15 mm wide and B=.30 mm wide, present near the steel. Crack A being the smallest, it may take two years for the steel near the crack to become depassivated, whereas near crack B, it may take only one year for corrosion to start (these are typical numbers for exposed concrete). If the reinforcing near both cracks is examined three years after placement, bars in the vicinity of crack B will have more corrosion than those near crack A. Therefore, at an early stage, the crack width appears to be an important factor contributing to the amount of corrosion. However, if the same reinforcement is examined 20 years after placement, it will be difficult to see a difference in the amount of corrosion near cracks A and B.

4. Increase in Deck Life Resulting from Reduction in Reinforcement: Analytical Model.

Determining the possible increase in deck life that can be attributed to a reduction in deck reinforcement is not easy. Short of building a number of decks and monitoring their performance for a long period of time, the exact answer may probably never be known. In this section, an approximate analytical approach that has been developed for this purpose is described.

The design life T of the deck is expressed as follows [15]:

$$T = t_0 + t_1 \tag{1}$$

in which:

t_0 = time from construction to initiation of corrosion
t_1 = time from initiation of corrosion to the occurrence of unacceptable corrosion (deck spalling)

A reduction in reinforcement will trigger a number of mechanisms which will affect the deck's life. One such mechanism will cause significant gains in deck life during the period t_1: With less material corroding and expanding, it will take longer for expansive forces to cause deck spalling. Moreover, for the many reasons discussed before, other mechanisms triggered by less reinforcement can only cause negligible reductions in deck life: (a) Larger crack width will only cause a reduction of the time t_0, which in itself is only a small portion of the unprotected deck's life [15] and (b) increased flexural, temperature, and shrinkage cracking will be offset by reduced subsidence cracking [23].

Perhaps the above arguments are best supported by preliminary results of a survey of Ontario- and AASHTO-type decks built within the past five years in Ontario. This survey, which is being conducted by Mr. Agarwal of the Ontario Ministry of Transportation and Communication, shows no "significant difference in cracking or corrosion" between the Ontario deck and the AASHTO deck. This information was obtained through phone conversations between the P.I. and Mr. Agarwal.

The relative gain "G" in deck life resulting from a reduction in reinforcement below current AASHTO requirements is defined as follows:

$$G = \frac{\text{Expected deck life } T_R \text{ of a deck section with reduced reinforcement}}{\text{Expected deck life } T_A \text{ of the same deck section designed with AASHTO reinforcement}} \quad (2)$$

in which the subscripts "R" and "A" indicate Reduced or AASHTO reinforcements, respectively. Using Equation (1), the relative gain in deck life becomes:

$$G = \frac{(t_{0R}+t_{1R}) \text{ with reduced reinforcement}}{(t_{0A}+t_{1A}) \text{ with AASHTO reinforcement}} \quad (3)$$

Following Beeby's arguments [15], t_0 is much smaller than t_1 for unprotected decks with either AASHTO or reduced reinforcement. Therefore, Equation (3) may be simplified to:

$$G = \frac{t_{1R} \text{ with reduced reinforcement}}{t_{1A} \text{ with AASHTO reinforcement}} \quad (4)$$

The gain G can be computed for various reinforcement ratios and various arrangements of the reinforcement. In estimating G, a constant corrosion rate is assumed after depassivation, irrespective of the amount of reinforcement [15,21]. The constant corrosion rate is modelled as a constant radial expansion rate of the reinforcing bars from which cracking of the deck is determined. A smeared-cracking two-dimensional finite element model has been developed for the calculations [6,11,16,19,22,24,25,26]. To avoid numerical difficulties, a reduced shear modulus term is being retained in the constitutive matrix of the cracked element. The smeared cracking finite element model is described in more detail in the next section.

A major advantage of the formulation proposed in Equation (4) is that the actual corrosion rate need not be known, and the actual time t_1 need not be computed. In fact Equation (4) can be expressed as:

$$G = \frac{N_R}{N_A} = \frac{\text{Number of radial expansion increments until failure for a deck section with reduced reinforcement}}{\text{Number of radial expansion increments until failure for the corresponding deck reinforced according to AASHTO}} \quad (5)$$

in which N_R and N_A for any deck section are obtained as follows:

(a) Select an arbitrary small increment for the radius of all reinforcing bars. This increment approximates an arbitrary amount of bar expansion due to corrosion buildup, and is the same for all deck sections and for all bars within a deck section.

(b) Impose the increment of radial expansion at the location of all bars in the finite element model.
(c) Compute the resulting principal stresses in each element and set the stiffness of all cracked elements equal to zero in the direction perpendicular to the crack.
(d) Repeat steps (b) and (c) until cracks from adjacent reinforcing bars meet, or until cracks originating in the vicinity of a single bar reach the top surface of the deck.
(e) The number of radial increments "N" that were needed for the deck to spall is proportional to the time t_1 in Equation (4).

5. Smeared-Cracking Finite Element Model

In this section a two-dimensional smeared-cracking finite element model that can be used to estimate the gain in deck life G as expressed in Equation (5) is described. First the assumptions are stated, then the model is described, and some results are shown.

5.1 PHILOSOPHY OF THE APPROACH AND ASSUMPTIONS

When a reinforcing bar corrodes, its volume will increase, and eventually the surrounding concrete will crack. With less steel, there will be less material corroding and expanding, and a longer deck life will result. The objective of this model is not to determine exact corrosion rates nor accurate crack distributions. Rather, it is to compare the overall effect of different amounts and distributions of reinforcing steel on corrosion-induced cracking.

Since the objective is to determine general trends rather than exact crack sizes and locations, accuracy is traded for simplicity and practicality. The following assumptions are made:

(a) Even though the real problem is three-dimensional, a two-dimensional plain strain smeared-cracking finite element model is used.
(b) Stresses resulting from bar expansion, dead and live loads, shrinkage and temperature may be included in the model; however, fatigue effects caused by live loading on the bridge and freeze-thaw effects are not considered.
(c) Although the model can accept any shape for the corrosion around a bar, it is deemed sufficient to consider a uniform corrosion buildup. Corrosion is generally more pronounced near the top of the bars, however, this assumption will have a reduced impact since only relative gains in deck life are computed.
(d) For the range of reinforcement ratios considered, fatigue-induced cracking will be assumed insensitive to the amount of reinforcement. Recent testing has shown that reduced reinforcement may not impair the deck's fatigue/cracking life [3,5].
(e) The long-term effects of deicing chemicals on concrete quality and strength are not included at this time. These effects can be incorporated by using time-dependent concrete strength properties. However, since the proposed model will estimate relative gains rather than absolute gains in deck life, this assumption is justified.

5.2 DESCRIPTION OF THE MODEL

5.2.1 *Purpose.* The model determines the cracking induced in a concrete deck by the corrosion and subsequent expansion of the reinforcing bars. It is used to determine the gain in deck life G as described in Section 4.

5.2.2 *Hardware/software.* The Program is written using FORTRAN 77 and currently runs on an IBM 3090-180E-VS mainframe.

5.2.3 *Input.* The input to the model are the physical geometry and mechanical properties of the concrete deck section, an increment of radial expansion for the reinforcing bars, and the number of increments. The increments of radial expansion simulate the expansion of a reinforcing bar due to corrosion.

5.2.4 *Output.* For each increment, the program outputs the stresses at the center of each element, the crack location and direction, and the nodal displacements .

5.2.5 *Post-processing.* A post-processing module plots the cracks on the finite element mesh for each expansion increment. Figure (4) contains sample crack propagation plots from the post-processing module.

5.2.6 *Analysis Method.* A conceptual flowchart of the program is shown in Figure (3). The program conducts a 2-dimensional nonlinear finite element analysis using a smeared-cracking approach. Plane strain isotropic elements are used for the uncracked concrete whereas plane strain orthotropic elements are used for the cracked concrete. The program conducts a stepwise linear analysis between expansion increments. The global stiffness matrix is updated at the end of each step to include the effect of the cracked elements. A two-dimensional failure surface in principal stress space is used to identify a cracked element. The maximum principal strain is used to identify the direction of the smeared cracking. A deck is considered spalled when cracks from adjacent bars meet, or when cracks originating near a single bar reach the top of the deck.

5.3 SAMPLE RESULTS

Figure (4) contains sample crack propagation plots from the post-processing module. The plots show a concrete section 4.375 inches deep with one #7 bar at the center. The #7 bar is given radial expansion increments of .0001 inch and the corresponding cracks are plotted at the center of each element. In each figure, only the new cracking that has formed during the corresponding .0001 inch bar expansion increment is shown. By comparing the figures, one can see how the cracks propagate from the vicinity of the bar all the way to the surface of the deck as the bar expands due to corrosion.

6. Conclusions

In regions where deicing chemicals are used, concrete decks with reduced reinforcement may last significantly longer. This paper presented a number of arguments supporting the thesis that a reduction in deck reinforcement may increase a deck's life. It also showed an approximate method to estimate the possible gains in deck life resulting from reduced reinforcement. The method is based on a smeared-cracking finite element model. If deck life can be extended by reducing reinforcement, there is a potential for very significant savings in deck repair and replacement costs, as well as initial steel costs.

7. Acknowledgments

This paper presented results to date of a project that is developing durability-based design criteria for reinforced concrete bridge decks for regions where deicing chemicals are used. The criteria will specify amounts and distributions of reinforcing steel that will result in maximum deck life. The project is sponsored by the Federal Department of Transportation under Grant DTOS 88-G-0001. The assistance of Mr. Nathan Collins and Mr. Michael Seals, which were sponsored by the National Science Foundation through the REU Grant NSF ENG-8712705, is gratefully acknowledged.

8. References

1. Borgard, B., Warren, C., Somayaji, S. and Heidersbarch, R., "Correlation Between Corrosion of Reinforcing Steel and Voids and Cracks in Concrete Structures," Transportation Research Record, No. 1211, 1989.
2. Csagoly, Paul F., "Future Bridge Deck Slab Construction," Concrete Construction, January 1989.
3. Perdikaris, Phillip and Beim, Sergio, "RC Bridge Decks Under Pulsating and Moving Loads", Journal of structural Engineering, ASCE, Vol. 113, No. 3, March, 1988.
4. "Small Bars Lead to Large Savings", TR News, September-October 1987, p 18-19.
5. Perdikaris, P.C., and Beim, S., "Fatigue Strength of Reinforced Concrete Bridge Decks," Presented at the ACI Fall Convention, Physical Models in Engineering Education and Practice, Baltimore, Md., Nov., 1986
6. "Finite Element Analysis of Reinforced Concrete Structures," Proceedings of a Seminar Sponsored by the American Society of Civil Engineers and the National Science Foundation, Meher and Okamura Editors, 1986, 696 pp.
7. Fang, I.K., "Behavior of Ontario-Type Bridge Decks on Steel Girders," Thesis Presented to the Department of Civil Engineering, the University of Texas at Austin, in partial Fulfillment of the requirements for the Degree of Doctor of Philosophy, 1986.

8. Beal, David B., "Reinforcement for Concrete Bridge Decks," NY State Department of Transportation, Engineering Research Division Bureau, Report No.:RR-105 FHWA/NY/RR-83/105, July 83, 32 p.
9. AASHTO, Standard Specifications for Highway Bridges, 13th Ed., American Association of State Highway and Transportation Officials, Washington, D.C., 1983
10. Ontario Highway Bridge Design Code., 2nd Ed., Ministry of Transportation and Communication, Highway Engineering Division, Ontario, Canada, 1983.
11. "Finite Element Analysis of Reinforced Concrete," State-of-the-Art Report, Task Committee on Finite Element Analysis of Reinforced Concrete Structures, Arthur Nilson Chairman, American Society of Civil Engineers, 1982.
12. Beal, David B., "Load Capacity of Concrete Bridge Decks," Journal of the Structural Division, American Society of Civil Engineers, Vol. 108, No. ST4, April 1982, p 814-832.
13. Abbott, S.P., "Concrete Deck Study," Report, Bridge Division, Maine Department of Transportation, October 1981.
14. Beal, David B., "Strength of Concrete Bridge Decks," Research Report, NY State Department of Transportation, Engineering Research Division Bureau, July 1981.
15. Beeby, A.W., "Corrosion of Reinforcing Steel in Concrete and its Relation to Cracking," Structural Engineer, March, 1978.
16. Bashur, F.K., and Darwin, D., "Nonlinear Model for Reinforced Concrete Slabs," Journal of the Structural Division, ASCE, Vol. 104, No. ST1, January, 1978.
17. Tuuti, Kyosti, "Cracks and Corrosion," Swedish Cement and Concrete Research Institute at the Institute of Technology, Stockholm, Sweden, 1977.
18. Batchelor, Barrington deV. and Hewitt, Brian E., "Tests of Model Composite Bridge Decks," Journal of the American Concrete Institute, Vol. 73, No. 6, June 1976.
19. Kabir, A. F., "Nonlinear Analysis of Reinforced Concrete Panels, Slabs, and Shells for Time Dependent Effects," UC-SESM Report No. 76-6, University of California at Berkeley, December, 1976.
20. Hewitt, Brian E. and Batchelor, Barrington deV., "Punching Shear Strength of Restrained Slabs," Journal of the Structural Division, American Society of Civil Engineers, No. ST9, September 1975, p 1837-1853.
21. Shiessl, P., "Admissible Crack Width in Reinforced Concrete Structures," Contribution II 3-17, Inter-Association Colloquium on the Behavior in Service of Concrete Structures, Preliminary Reports, II, Liege, 1975.
22. Lin, C. S., and Scordelis, A., "Nonlinear Analysis of Reinforced Concrete Shells of General Form," Journal of the Structural Division, ASCE, No. ST3, March 1975.
23. Dakhil, F. H., Cady, P., and Carrier, R. E., "Cracking of Fresh Concrete as Related to Reinforcement," ACI Journal, August, 1975.
24. Hand, F.R., Pecknold, D.A., and Schnobrich, W.C., "Nonlinear Layered Analysis of Reinforced Concrete Plates and Shells," Journal of the Structural Division, ASCE, Vol. 99, No. ST7, July 1973.

25. Vallipan, S., and Doolan, T.F., "Nonlinear Stress Analysis of Reinforced Concrete," Journal of the Structural Division, ASCE, Vol. 98, No. ST 4, Proc. Paper 8845, April 1972, pp. 885-898.
26. Cervenka, V., and Gerstle, K. H., "Inelastic Analysis of Reinforced Concrete Panels, Part I: Theory," International Association of Bridge and Structural Engineering Publications, Vol. 31-11, 1971, pp. 31-45.
27. Rashid, Y. R., "Analysis of Prestressed Concrete Pressure Vessels," Nuclear Engineering and Design, Vol. 7, No. 4, April 1968, pp. 334-344.

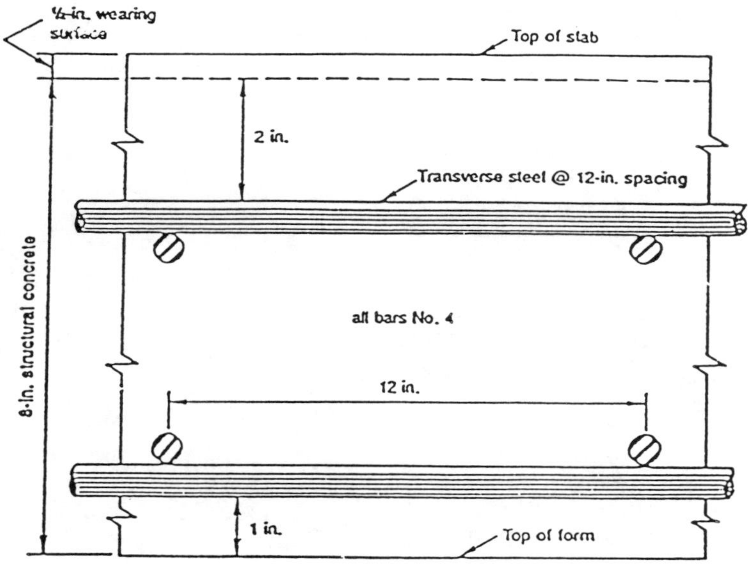

Fig. 1 Typical Isotropic Bar Placement (Ref. [4]).

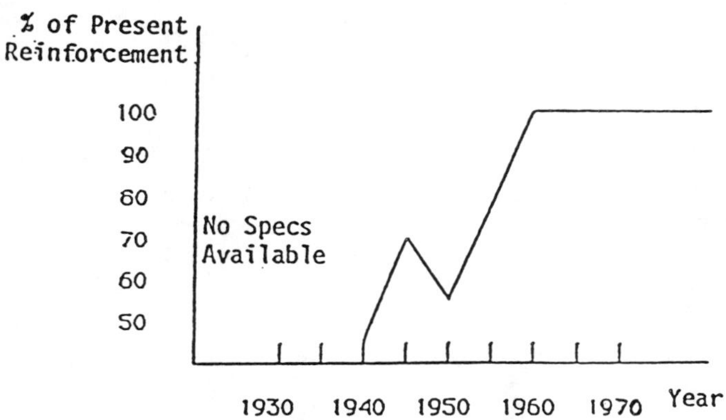

Fig. 2 History of AASHTO Distribution Reinforcement.

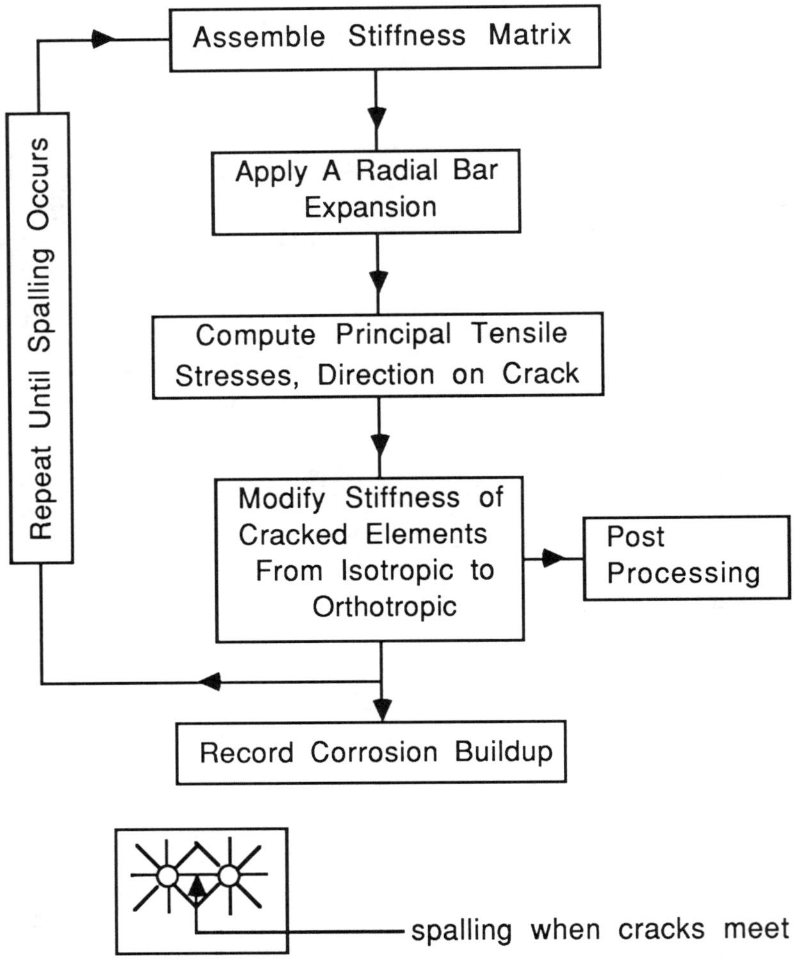

Fig. 3 Smeared Cracking FE Model.

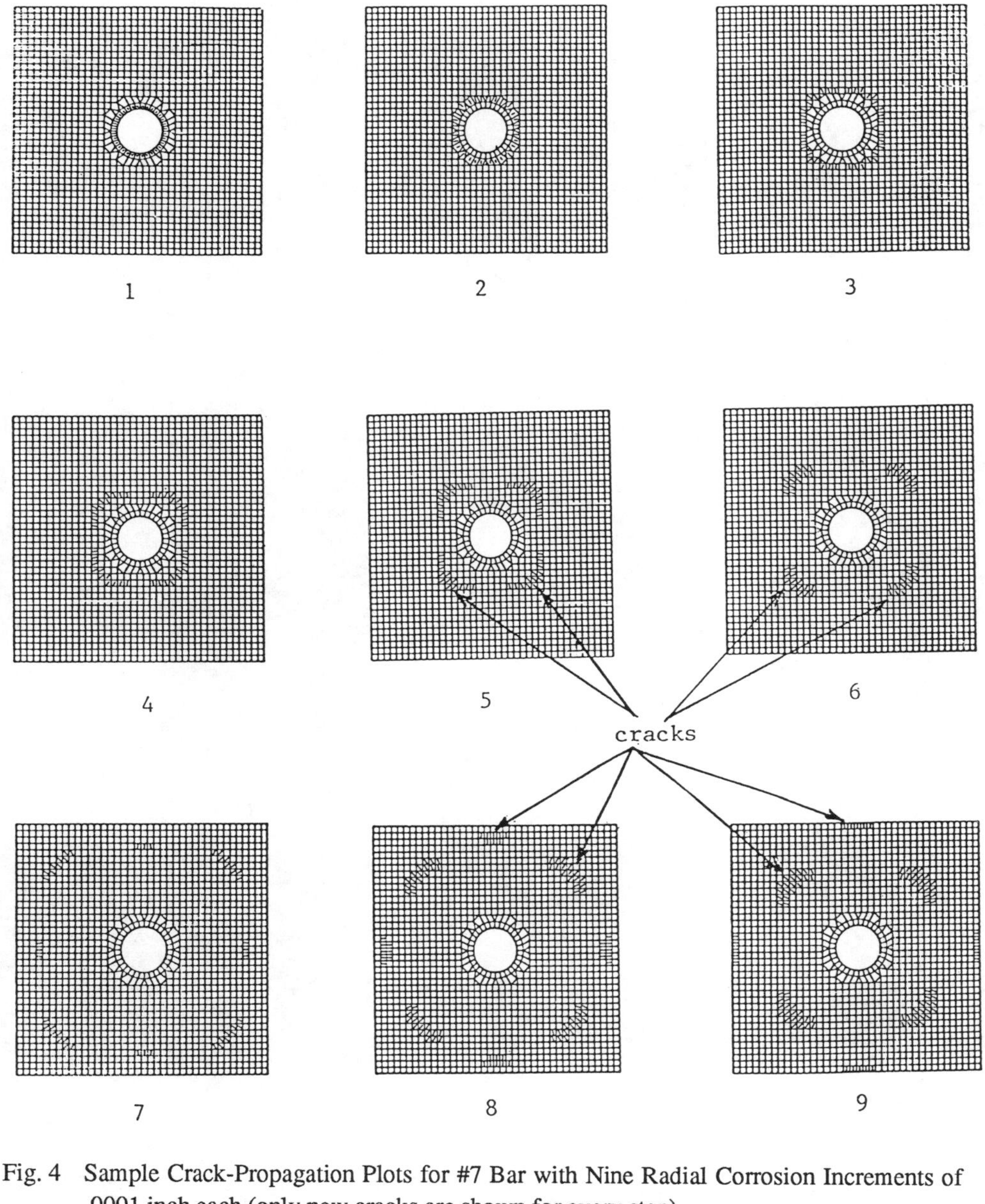

Fig. 4 Sample Crack-Propagation Plots for #7 Bar with Nine Radial Corrosion Increments of .0001 inch each (only new cracks are shown for every step).

VI - Working Group Reports

I. BRIDGE MANAGEMENT

Working Group Report

Co-chaired by N.H. Andersen and C. Loveall

Management becomes an important part of bridge engineering. Bridge management system (BMS) provides support for decisions regarding evaluations, maintenance, repairs, rehabilitation and replacements. Computerized BMS's are being developed and implemented in the United States, Canada and Europe. Topics identified for further development include:

1. Need for identification of necessary data to drive a bridge management system. Data necessary for assessment is not necessarily the same as for bridge management.
2. Need for development of deterioration models.
3. Need for better tools for field assessment of deterioration, i.e. better description of component states.
4. Need for a good competitive bridge management system that is easy to input, easy to operate, and is understandable.

A. S. Nowak (ed.), Bridge Evaluation, Repair and Rehabilization, 619.

II. BRIDGE DIAGNOSTICS

Working Group Report

Co-chaired by A. S. Nowak and R. Favre

Diagnostics and monitoring are important procedures providing information about the actual strength, loads and degree of deterioration.

Monitoring of the structural performance will reduce the probability of an unexpected failure. The critical moment can be detected prior to disaster, so that the structure can be closed for traffic, repaired or replaced. Discussion focused on problems requiring solution and available techniques.

1. Fatigue damage accumulation. A serious problem in steel bridges, in prestressed concrete does not seem to be as important. The growing number of heavy trucks may cause critical load levels.
2. Corrosion of prestressing tendons. Often ducts are not grouted properly, there is a need to check if tendons are affected by corrosion (sensors).
3. Proof-loading to check load carrying capacity.
4. Load monitoring; Weigh-in-motion (WIM) techniques. There is a need for reliable site-specific data.
5. Monitoring of long-term deformations. There is a need for warning when deformations exceed critical levels (e.g. settlements).
6. Measurement of temperature effects. The problems include distribution of temperature in the structure, behavior in low temperatures (expansion and contraction coefficients are not the same).
7. Actual properties of materials. There is a need for accurate data (yield strength, modulus of elasticity, dimensions).
8. Monitoring and checking water related problems (waterproofing). Effort should be directed to waterproofing the bridge (macro-approach) rather than steel (micro-approach).
9. Bridge decks require a special consideration (economic life, covers problem preventive techniques and insulation).

A. S. Nowak (ed.), Bridge Evaluation, Repair and Rehabilization, 621.

III. BRIDGE EVALUATION

Working Group Report

Co-chaired by B. Bakht and G. P. Tilly

1. Need for Evaluation

While in the USA and Canada, bridges are routinely evaluated, it seems that in several European countries, evaluation is undertaken often after inspection has brought to light some apparent distress in the structure. Some emphasis is placed in France on "material inspection".

2. Quantification of Bridge Strength

In the USA and Canada, the bridge strength is quantified by a dimensionless parameter Rating Factor (RF). Similar parameter is also used in the U.K., although it is called the Reduction Factor.

Evaluation calculations in the U.K. are carried with the help of a code referred to as the Assessment Document.

3. Redistribution of Loads at the Ultimate Limit States (ULS)

France:	No redistribution permitted.
Spain:	10% redistribution permitted in the longitudinal direction only if there is sufficient ductility.
UK:	Same as Spain, but the code is not specific.
USA:	Longitudinal redistribution of moments permitted through the Autostress Method.
Canada:	8% redistribution permitted in the transverse direction for slab-on-girder bridges with concrete decks and compact section steel girders.
Italy:	Permits analysis by any method which can analyze the failure of the structure.

There was a considerable discussion on the blanket approval of non-linear methods. Many members of the discussion group agreed that the partial factors used in evaluation should be consistent with experience and confidence in the analysis methods.

A. S. Nowak (ed.), Bridge Evaluation, Repair and Rehabilization, 623–624.

There were different definitions of the ULS. Some recommended that the onset of yield in one component or an "acceptable" degree of damage in a localized portion of the bridge, should constitute the ULS. There were others who favored the formation of a mechanism to correspond to the ULS.

There was a general agreement that the consequences of uncertainty should be reflected in the various safety factors.

4. Recommendations for Evaluation and Rehabilitation Design

It was recommended that the next workshop should deal with the following specific topics:

a. Definition of the limit states for evaluation.
b. Non-linear methods of analysis.
c. Data base of in-situ tests.
d. Uncertainty in the behavior of bridges.
e. Soil-structure interaction dealing with spandrel-filled brick and masonry arches of Europe and corrugated metal plate buried structures of North America.
f. Segmental concrete bridges, in particular evaluation of the condition of prestressing steel.

5. Research Needs

a. Develop means of inspecting the "inside" of a bridge.
b. Develop methods of determining the strength of concrete components with Alkali reactive aggregate.
c. Develop non-linear methods of analysis.
d. Refine reliability models.
e. Develop methods of dealing with systems reliability.
f. Develop methods of determining strengths of cable with localized corrosion.
g. Develop means of evaluating foundation.

IV. BRIDGE REPAIR

Working Group Report

Co-chaired by J. W. Fisher and G. Konig

Bridge repair methods were discussed. Decision about repair, rehabilitation or replacement has clear economic implications, and it is a part of the bridge management system. Closely related are bridge diagnostics (intelligent gauges, remote sensing) and evaluation (non-destructive testing, mathematical models, damage accumulation, inspection, proof loading).

Some important topics requiring a cooperative research effort were identified as follows.

1. External prestressing can be used to repair a variety of structural types. Special problems include corrosion and fatigue limit.
2. Treatment of concrete decks.
3. Repair of substructure.
4. Repair of bearings and joints.
5. Evaluation of stress during repair, damage accumulation, and actual strength.
6. Cracking or distress in orthotropic decks.
7. Minor repairs on small bridges, repair due to impact damage.
8. New materials (nonmetalic tendons).
9. Seismic repair, retrofit and rehabilitation.
10. Prediction of remaining life.
11. Repair of corroded bars, tendons and sections.
12. Cathodic protection in concrete decks and piers in tidal zones.

A. S. Nowak (ed.), Bridge Evaluation, Repair and Rehabilization, 625.

INDEX OF SUBJECTS

INDEX OF AUTHORS